Wetlands: Ecology, Conservation and Management

Volume 12

Kerrylee Rogers, School of Earth & Environmental Sci
Univ of Wollongong Wollongong NSW, Australia
Rebecca Woodward, 35 Percent Chalford, UK

The recognition that wetlands provide many values for people and are important foci for conservation worldwide has led to an increasing amount of research and management activity. This has resulted in an increased demand for high quality publications that outline both the value of wetlands and the many management steps necessary to ensure that they are maintained and even restored. Recent research and management activities in support of conservation and sustainable development provide a strong basis for the book series. The series presents current analyses of the many problems afflicting wetlands as well as assessments of their conservation status. Current research is described by leading academics and scientists from the biological and social sciences. Leading practitioners and managers provide analyses based on their vast experience.

The series provides an avenue for describing and explaining the functioning and processes that support the many wonderful and valuable wetland habitats, such as swamps, lagoons and marshes, and their species, such as waterbirds, plants and fish, as well as the most recent research directions. Proposals cover current research, conservation and management issues from around the world and provide the reader with new and relevant perspectives on wetland issues.

Babafemi Raphael Babaniyi
Sesan Abiodun Aransiola
Ebunoluwa Elizabeth Babaniyi
Naga Raju Maddela
Editors

Wetland Ecosystems: Conservation Strategies, Policy Management and Applications

Springer

Editors
Babafemi Raphael Babaniyi
Bioresources Development Centre
National Biotechnology Development
Agency (NABDA)
Abuja, Nigeria

Sesan Abiodun Aransiola
Department of Microbiology
Faculty of Science
University of Abuja
Abuja, Nigeria

Ebunoluwa Elizabeth Babaniyi
Department of Biology Education
Adeyemi College of Education Ondo
(Obafemi Awolo University Ile Ife)
Ondo, Nigeria

Naga Raju Maddela
Departamento de Ciencias Biológicas
Facultad de Ciencias de la Salud
Universidad Técnica de Manabí
Portoviejo, Ecuador

ISSN 1875-1261 ISSN 1875-127X (electronic)
Wetlands: Ecology, Conservation and Management
ISBN 978-3-031-91981-7 ISBN 978-3-031-91982-4 (eBook)
https://doi.org/10.1007/978-3-031-91982-4

This Springer imprint is published by the registered company Springer Nature Switzerland AG
The registered company address is: Gewerbestrasse 11, 6330 Cham, Switzerland

Foreword

Wetlands, commonly known as the "renal system of the planet," are crucial ecological systems that sustain a diverse range of organisms and offer numerous benefits to our world. This book, "Wetland Ecosystems: Conservation Strategies, Policy Management and Applications," endeavours to investigate the multifaceted essence of wetlands, illuminating their ecological significance, the obstacles they encounter, and the measures essential for their preservation. Wetlands are brimming with life, from the luxuriant flora flourishing in their aquatic environments to the varied assortment of fauna that inhabit these ecosystems. The abundant biodiversity present in wetlands serves as evidence of their ecological worth, acting as a vital sanctuary for numerous species. This book digs into the diverse network of life within wetlands, analysing the connections among plants, animals, and microorganisms. Grasping these interactions is crucial for recognizing the function of wetlands in enhancing environmental well-being.

The ecological roles of wetlands are irreplaceable. They function as natural absorbers, soaking up surplus water in times of flooding and discharging it during dry spells, thereby managing water circulation and averting floods. Wetlands also operate as natural purifiers of water, straining out impurities and enhancing water quality. Moreover, they play a substantial part in carbon storage, amassing carbon and alleviating the impacts of climate change. The significance of these roles cannot be emphasized enough, and this book strives to examine their critical essence. Despite their significance, wetlands confront numerous challenges. Anthropogenic activities, such as urban expansion, agriculture, and industrial activities, have resulted in the deterioration and decline of wetlands globally. Pollution from various sources further compounds these challenges, deteriorating water and soil quality and disrupting the fragile equilibrium of these ecosystems. Climate change presents another notable hurdle, modifying weather patterns and influencing the well-being and operations of wetlands. This publication thoroughly examines these challenges, emphasizing the pressing necessity for measures to safeguard and rehabilitate wetlands.

The preservation and rehabilitation of wetlands necessitate a multifaceted strategy. This publication deliberates on diverse approaches for conserving these vital ecosystems, ranging from legal frameworks and policies to community-driven conservation initiatives. It stresses the significance of comprehending the composition and generation of wetlands and the role of wetland soils in sustaining life and sustainable agriculture. By spotlighting successful conservation and restoration endeavours, this publication strives to motivate and direct forthcoming initiatives. Wetland fauna, encompassing their habitats, migration patterns, and pivotal species, are scrutinized to comprehend their ecological significance. The historical and cultural importance of human engagements with wetlands is also deliberated, unveiling the profound correlation between human civilizations and these natural habitats. From an economic standpoint, wetlands deliver substantial ecosystem services that enrich human welfare, underscoring the imperative for their safeguarding.

Technological progressions in wetland surveillance, such as remote sensing and GIS, are indispensable for evaluating wetland well-being and steering conservation endeavours. The book examines the repercussions of climate change on wetlands, concentrating on susceptibilities, adaptation, and mitigation approaches. Engineered wetlands for sewage treatment are accentuated as a sustainable remedy for improving water quality and controlling pollution. Pollution and remedial tactics within wetlands are pivotal for upholding their well-being, while the wetland microflora plays a notable part in water purification. The incorporation of wetlands into urban scenery presents inventive resolutions for regulating stormwater overflow and urban pollutants, displaying the significance of wetland-cantered remedies for urban water administration. The publication also debates the socioeconomic gains of wetland-focused water treatment, accentuating the economic, social, and environmental benefits. Prospective dilemmas and prospects for wetland preservation are examined, alongside the notion of phytoremediation in wetlands, involving the utilization of plants to absorb and detoxify pollutants. Ultimately, the book contemplates the multifaceted utilities that wetlands offer, encompassing agriculture, food safety, industrial, and medicinal applications. By comprehensively addressing these subjects, this book endeavours to provide an exhaustive comprehension of the myriad crucial functions of wetlands and the urgency of their preservation for the benefit of both nature and humanity.

The editors have done a great job by inviting 76 experienced experts from 12 different countries to contribute to this volume. Therefore, I am confident that the wealth of knowledge of all the contributors of chapters in this book are of valuable importance and will have a wide readership. Besides, it is a good source of information to students, scientists, researchers, agriculturists, professionals, and environmentalists for many years to come.

Executive Director, M. Dinesh Kumar
Institute for Resource Analysis and Policy (IRAP) ,
Hyderabad, India
August 1, 2024

Preface

Wetlands represent one of the most crucial and diverse ecosystems globally, playing a fundamental role in preserving environmental well-being and supporting a wide array of flora, fauna, and microorganisms. This book explores extensive biodiversity found in wetlands, investigating the intricate connections and interplays among different species and the microorganisms pivotal for sustaining the health of these ecosystems. The significance of wetlands is paramount, given their indispensable functions like flood regulation, water purification, and carbon sequestration, all of which are essential for mitigating the effects of climate change and upholding environmental equilibrium. Nevertheless, these invaluable ecosystems confront numerous challenges. Human activities, pollution, and climate change stand out as primary threats jeopardizing wetlands. This book extensively discusses these challenges, emphasizing the pressing necessity for conservation and restoration efforts to safeguard these ecosystems. By evaluating the structure and genesis of wetlands, alongside the chemistry of wetland soils, we can comprehend their role in promoting sustainable agriculture and their capacity to sustain a diverse range of organisms. Additionally, the book scrutinizes wetland fauna, their habitats, migration behaviors, and key species to grasp their significance within the ecosystem. The text deliberates on the historical and cultural significance of human engagements with wetlands, illustrating the profound link between human societies and these natural environments. Economically, wetlands offer substantial advantages that augment human welfare, necessitating legal frameworks and policies to ensure their protection.

Innovations in wetland surveillance, such as remote sensing and Geographic Information Systems (GIS), are pivotal for evaluating water treatment and overseeing wetland preservation. The text addresses the repercussions of climate change on wetlands, focusing on vulnerabilities, adaptation strategies, and mitigation measures. Constructed wetlands designated for wastewater treatment exploring the role of these ecosystems in enhancing water quality and filtration. Strategies for pollution prevention and remediation within wetlands are critical for preserving their vitality, while the wetland microbiome significantly contributes to water treatment. Integrating wetlands into urban settings provides solutions for managing stormwater runoff and urban contaminants, examining the significance of wetland-oriented approaches for urban water treatment. Future challenges and prospects for wetland preservation are discussed, alongside the concept of phytoremediation in wetlands, which involves utilizing plants to absorb and detoxify pollutants.

Lastly, the book contemplates the diverse array of applications that wetlands offer, encompassing agriculture, food safety, and industrial purposes. By thoroughly addressing these subjects, this book aims to furnish a comprehensive comprehension of the myriad crucial roles of wetlands and the immediate necessity to conserve them for the betterment of both the nature and humanity.

Ogbomoso, Nigeria — Babafemi Raphael Babaniyi
Abuja, Nigeria — Sesan Abiodun Aransiola

Ile-Ife, Nigeria — Ebunoluwa Elizabeth Babaniyi
Ondo, Nigeria
Portoviejo, Ecuador — Naga Raju Maddela

Acknowledgments

The Editorial team extends warm gratitude to all the contributors of this book project; their collective cooperation and prompt responses through the project eased the path to the successful completion of the book. Their expertise in the subject matter enhanced the quality of the chapters and their dedication to meeting deadlines is highly commendable. We would also like to thank the reviewers for their valuable input and efforts in enhancing the overall quality of the chapters. We are grateful for the support and fulfilling experience gained from working with the Springer Editorial and Production unit. Finally, we would like to relay our deep appreciation to the Management of our respective institutions, National Biotechnology Development Agency (Nigeria), University of Abuja, Abuja (Nigeria), Obafemi Awolowo University (Nigeria), Adeyemi College of Education (Nigeria), and Universidad Técnica de Manabí (Ecuador) for their firm support throughout all the stages of the book preparation process, from the proposal to the completion of the project.

Babafemi Raphael Babaniyi
National Biotechnology Development
Agency (NABDA)
Ogbomoso, Nigeria

Sesan Abiodun Aransiola
Department of Microbiology
University of Abuja
Abuja, Nigeria

Ebunoluwa Elizabeth Babaniyi
Department of Biology
Obafemi Awolowo University
Ile-Ife, Nigeria
Adeyemi College of Education
Ondo, Nigeria

Naga Raju Maddela
Departamento de Ciencias Biológicas,
Facultad de Ciencias de la Salud
Universidad Técnica de Manabí
Portoviejo, Ecuador

Contents

Editors and Contributors

About the Editors

Babafemi Raphael Babaniyi is a graduate of Industrial Chemistry (BSc) at Caritas University Amorji-Nike, Emene, Enugu State, Nigeria, in 2008. He later received his master's degree (MTech) in Environmental Chemistry at Federal University of Technology, Akure, Ondo State, Nigeria, in 2021. Currently, he is a PhD candidate in the Department of Chemistry (Environmental Chemistry) Federal University of Technology, Akure, Ondo State, Nigeria. He had over 32 publications including book chapters and articles of international repute. Additionally, he is an assistant chief scientific officer at Bioresources Development Centre, National Biotechnology Development Agency, Nigeria. His areas of interest include Environmental Chemistry with research area in Green Chemistry, Phytoremediation, Chemical Biology, Environmental Science, Bioremediation of contaminated soil, plant protection and animal health, Environmental impact assessment, and Biodegradable plastics.

Sesan Abiodun Aransiola is a PhD holder in Environmental Microbiology from the Department of Microbiology, Federal University of Technology, Minna, Nigeria. He obtained his first degree (BTech) and master's degree (MTech) from the same department in 2009 and 2014, respectively. He has demonstrated his research expert in the production of vermicasts from vermicoposting (Vermitechnology) of organic wastes to assist plants in the remediation of polluted soil with heavy metals. Currently, he is a Lecturer at the Department of Microbiology, University of Abuja, Nigeria. His areas of interest include Environmental Microbiology with research area in phytoremediation, vermicomposting, biosorption, and bioremediation of soil contaminated environment. He is an award-winning researcher and has over 80 publications including book chapters and research and review articles of good international repute with high impact factor. He has co-edited scientific books of global interest which include *Microbial Biotechnology for Bioenergy* (Elsevier); *Ecological Interplays in Microbial Enzymology* and *Prospects for Soil Regeneration and Its Impact on Environmental Protection* (Springer); *Microbial Biofilms—Applications and Control, Marine Greens: Environmental, Agricultural, Industrial and Biomedical Applications, Emerging Contaminants in Food and Food Products, Phytoremediation in Food Safety: Risks and Prospects* (CRC Press) with many more in the process. Also, he has worked to the rank of an assistant chief scientific officer at Bioresources Development Centre, National Biotechnology Research and Development Agency, Nigeria, where he was involved in the uses of vinasse (a by-product of ethanol) as biofertilizer to reclaim polluted soil for agricultural purposes. Abiodun is a member of Nigerian Society for Microbiology and American Society for Microbiology, among others.

Ebunoluwa Elizabeth Babaniyi holds a BSc Ed in Biology from Adeyemi College of Education, Ondo, an affiliate of Obafemi Awolowo University, Ile Ife, Nigeria (2014), and an NCE in Biology/Geography from the College of Education, Ikere-Ekiti (2009). She has been professionally registered with the Teachers Registration Council of Nigeria (TRCN) since 2019. With significant experience in education, Ebunoluwa has made notable contributions to the field, as evidenced by her published research manuscripts.

Naga Raju Maddela received his MSc (1996–1998) and PhD (2012) in Microbiology from Sri Krishnadevaraya University, Anantapuramu, India. During his doctoral program in the area of Environmental Microbiology, he investigated the effects of industrial effluents/insecticides on soil microorganisms and their biological activities. He is working as Faculty in Microbiology since 1998, teaching undergraduate and postgraduate students. He worked as Prometeo Investigator (fellowship received from SENESCYT) at Universidad Estatal Amazónica, Ecuador during 2013–15, and has received "Postdoctoral Fellowship" (2016–2018) from Sun Yat-sen University, China. He also received external funding from "China Postdoctoral Science Foundation" in 2017 and internal funding from "Universidad Técnica de Manabí" in 2020. He participated in national/international conferences and presented research data in China, Cuba, Ecuador, India, and Singapore. Currently, he is working as a Full Professor at the Facultad de Ciencias de la Salud, Universidad Técnica de Manabí, Portoviejo, Ecuador. He has been actively publishing scientific articles, books (authored and edited), and chapters since 2007. As of now, he has published 70 articles, 12 books, and 40 book chapters.

Contributors

Orimisan Emmanuel Abata Department of Chemistry, Federal University of Technology, Akure, Nigeria

Abdulhakeem Idris Abdulhakeem Department of Fisheries and Aquaculture, Federal University, Dutsin-Ma, Nigeria

Mustapha Abdulsalam Department of Microbiology, School of Sciences and Information Technology, Skyline University, Kano, Nigeria

Alaaya Abdulrasheed Abdulwaheed Department of Surveying and Geoinformatics, Faculty of Environmental Technology, Federal University of Technology, Minna, Niger State, Nigeria

Joshua Ibukun Adebomi Department of Biological and Chemical Engineering, University of Saskatchewan, Saskatoon, SK, Canada

A. A. Adelekun Department of Chemical Sciences, Achievers University, Owo, Nigeria

Adedayo Olufemi Adesola Department of Mathematics and Statistics, University of Massachusetts, Amherst, MA, USA

Paul Ojo Adewoye Bioresources Development Centre, National Biotechnology Research and Development Agency (NBRDA), Ogbomoso, Nigeria

A. Adewumi Federal College of Freshwater Fisheries Technology, New Bussa, Nigeria

A. E. Adeyemi Texas A&M University, College Station, TX, USA

Gado Mustapha Ali National Biotechnology Development Agency, Abuja, Nigeria

Sikirulai Abolaji Akande Department of Mathematic, Federal University of Technology, Minna, Niger State, Nigeria

Joshua Oluwatobi Akinola Department of Agriculture and Natural Sciences, University of Maryland, College Park, MD, USA

Ulelu Jessica Akor Department of Animal Production, University of Ilorin, Ilorin, Nigeria

Gema Alexandra Anchundia-Vélez Maestría en Química mención Química Ambiental, Facultad de Posgrado, Universidad Técnica de Manabí, Portoviejo, Ecuador

Chinedu Christian Anyene Department of Agricultural and Bio-environmental Engineering, Federal Polytechnic, Oko, Anambra State, Nigeria

Adejoke Blessing Aransiola Department of Surveying and Geoinformatics, Faculty of Environmental Science, University of Abuja, Abuja, Nigeria

Sesan Abiodun Aransiola Department of Microbiology, Faculty of Science, University of Abuja, Abuja, Nigeria

Helen Shnada Auta Department of Microbiology, Federal University of Technology, Minna, Niger State, Nigeria

Mamud Baba Department of Surveying and Geoinformatics, Faculty of Environmental Technology, Federal University of Technology, Minna, Niger State, Nigeria

Babafemi Raphael Babaniyi Bioresources Development Center, National Biotechnology Development Agency, Abuja, Nigeria

Ebunoluwa Elizabeth Babaniyi Department of Biology Education, Adeyemi College of Education, Ondo, Nigeria

Obafemi Awolo University, Ile Ife, Nigeria

Gabriel Gbenga Babaniyi Department of Agricultural Development and Management, Agricultural and Rural Management Training Institute (Armti), Kwara, Nigeria

Sunday Zeal Bala Department of Education, University of Maryland Eastern Shore, Princess Anne, MD, USA

F. V. Bekun Istanbul Gelisim University, Istanbul, Turkey

Lebanese American University, Beirut, Lebanon

Anderson Javier Pazmiño Castro Departamento de Procesos Agroindustriales, Facultad de AgroCiencias, Universidad Técnica de Manabí, Universidad Técnica de Manabí, Chone, Ecuador

BioGea Research Group, Chone, Ecuador

Juan Carlos Vélez Chica Departamento de Recursos Naturales Renovables, Facultad de Acuicultura y Ciencias del Mar, Universidad Técnica de Manabí, Sucre, Ecuador

BioGea Research Group, Sucre, Ecuador

Eduardo Antonio Caicedo Coello Universidad Laica Eloy Alfaro de Manabí, extensión Sucre, Sucre, Ecuador

BioGea Research Group, Sucre, Ecuador

Danny L. Cuenca-Nevárez GenteOil Ecuador, Quito, Ecuador

BioGea Research Group, Quito, Ecuador

Gerardo J. Cuenca-Nevárez Departamento de Recursos Naturales Renovables, Facultad de Acuicultura y Ciencias del Mar, Universidad Técnica de Manabí, Sucre, Ecuador

BioGea Research Group, Sucre, Ecuador

Ayodeji Adams Dahunsi Department of Microbiology, Faculty of Science, University of Abuja, Abuja, Nigeria

Oluwatosin Emmanuel Daramola Department of Chemistry and Biochemistry, Texas Tech University, Lubbock, TX, USA

Oluwatosin Olaoluwa Daramola Bioresources Development Centre, National Biotechnology Development Agency (NABDA), Akure, Nigeria

Shehu-Alimi Elelu Department of Chemistry, Howard University, Washington, DC, USA

O. I. Eresanya Bioresources Development Centre, National Biotechnology Development Agency (NABDA), Akure, Nigeria

Abdulmujeeb Bolaji Hamzat Bioresources Development Centre, National Biotechnology Development Agency (NABDA), Akure, Nigeria

Nigerian Institute for Oceanography and Marine Research, Victoria Island, Lagos, Nigeria

Madinat Hassan Foresight Institute of Research and Translation (FIRAT), Kigali, Rwanda

Ganiyat Omotayo Ibrahim Department of Chemistry, Nottingham Trent University, Nottingham, UK

Mohammed Oludare Idrees Department of Surveying and Geoinformatics, Faculty of Environmental Science, University of Abuja, Abuja, Nigeria

Chukwuka O. Igwegbe University of Hull, Hull, UK

Musa Ojeba Innocent Department of Microbiology, School of Sciences and Information Technology, Skyline University, Kano, Nigeria

Gbolahan Olamide Isaac Bioresources Development Centre, National Biotechnology Development Agency (NABDA), Akure, Nigeria

Femi Joseph Department of Microbiology, Federal University of Technology, Minna, Niger State, Nigeria

Kemi Catherine Adesorioye Department of Chemistry, Federal University of Technology, Akure, Nigeria

Salami Olaitan Lateefat Department of Chemistry and Molecular Biology, University of Gothenburg, Gothenburg, Sweden

Miracle Uwa Livinus Department of Biochemistry, School of Sciences and Information Technology, Skyline University, Kano, Nigeria

Naga Raju Maddela Departamento de Ciencias Biológicas, Facultad de Ciencias de la Salud, Universidad Técnica de Manabí, Portoviejo, Ecuador

Instituto de investigación, Universidad Técnica de Manabí, Portoviejo, Ecuador

Grupo de investigación en Biodiversidad y Ecología de Ecosistemas Acuáticos, Departamento de Acuicultura, Pesca y Recursos Naturales Renovables, Facultad de Ciencias Veterinarias, Universidad Técnica de Manabí, Bahía de Caráquez, Ecuador

Faith Ann Miracle Department of Microbiology, Faculty of Science, University of Abuja, Abuja, Nigeria

Carlos Augusto Morales-Paredes Laboratorio de Análisis Químicos y Biotecnológicos, Instituto de Investigación, Universidad Técnica de Manabí, Portoviejo, Ecuador

Auwal Sagir Muhammad School of Informatics, Xiamen University, Xiamen, China

Adamu Mustapha Department of Microbiology, Federal University of Technology, Minna, Niger State, Nigeria

Madaki Enyojo Nathan Department of Microbiology, Faculty of Science, University of Abuja, Abuja, Nigeria

Víctor Hugo Nevárez-Barberán Universidad Laica Eloy Alfaro de Manabí, extensión Sucre, Sucre, Ecuador

BioGea Research Group, Sucre, Ecuador

D. T. Ogundele Department of Chemistry and Industrial Chemistry, Kwara State University, Malete, Nigeria

O. D. Ogundele Department of Chemical Sciences, Achievers University, Owo, Nigeria

Oluwafemi Temitope Ojo Bioresources Development Center, National Biotechnology Development Agency, Abuja, Nigeria

Olaniran Victor Olagoke Department of Agricultural Development and Management, Agricultural and Rural Management Training Institute (Armti), Kwara, Nigeria

V. A. Olagunju Department of Chemistry and Biochemistry, Miami University, Oxford, OH, USA

Kehinde Oluwasiji Olorunfemi Bioresources Development Centre, National Biotechnology Development Agency (NABDA), Akure, Nigeria

O. J. Olorunleke Federal College of Freshwater Fisheries Technology, New Bussa, Nigeria

Temitope Oyadokun African University of Science and Technology, Abuja, Nigeria

O. J. Popoola Department of Geological Sciences, Achievers University, Owo, Nigeria

T. J. Popoola Department of Environmental Management, School of Health and Life Sciences, Teesside University, Middlesborough, UK

Joan Manuel Rodríguez-Díaz Laboratorio de Análisis Químicos y Biotecnológicos, Instituto de Investigación, Universidad Técnica de Manabí, Portoviejo, Ecuador

Departamento de Procesos Químicos, Biotecnología y Alimentos, Facultad de Ingeniería y Ciencias Aplicadas, Universidad Técnica de Manabí, Portoviejo, Ecuador

Job Oloruntoba Samuel Department of Microbiology, Federal University of Technology, Minna, Niger State, Nigeria

Mariana Andreina Valle Suárez Gobierno Autónomo Descentralizado Provincial de Pastaza, Puyo, Pastaza, Ecuador

Abdulrahman Itopa Suleiman Department of Science Laboratory Technology, Kogi State Polytechnic, Lokoja, Nigeria

Maritza Viviana Talledo-Solórzano Departamento de Recursos Naturales Renovables, Facultad de Acuicultura y Ciencias del Mar, Universidad Técnica de Manabí, Sucre, Ecuador

BioGea Research Group, Sucre, Ecuador

Abd'Gafar Tunde Tiamiyu Department of Mathematic, The Chinese University of Hong Kong, Ma Liu Shui, Hong Kong

Bata Timothy Department of Geology, Abubakar Tafawa Balewa University, Bauchi, Nigeria

Daniel Oraeloka Udorah Department of Civil and Environmental Engineering, Federal University of Technology, Akure, Ondo State, Nigeria

Department of Agricultural and Bio-environmental Engineering, Federal Polytechnic, Oko, Anambra State, Nigeria

Imeime Innocent Uyo Institute of Photogrammetry and Geoinformation, Leibniz University, Hannover, Germany

Rodolfo Patricio Panta Velez Departamento de Recursos Naturales Renovables, Facultad de Acuicultura y Ciencias del Mar, Universidad Técnica de Manabí, Sucre, Ecuador

BioGea Research Group, Sucre, Ecuador

Munachimso Odenakachi Victor-Ekwebelem Department of Microbiology, Alex Ekwueme, Federal University, Ndufu-Alike, Ikwo, Ebonyi State, Nigeria

Katimu Yusuf Department of Microbiology, Federal University, Gashua, Yobe State, Nigeria

Juan José Bernal Zambrano Departamento de Recursos Naturales Renovables, Facultad de Acuicultura y Ciencias del Mar, Universidad Técnica de Manabí, Sucre, Ecuador

BioGea Research Group, Sucre, Ecuador

Part I
Ecological Functions of Wetlands

Chapter 1
Ecosystem Structure and Formation of Wetland

Musa Ojeba Innocent, Job Oloruntoba Samuel, Mustapha Abdulsalam, Miracle Uwa Livinus, Sikirulai Abolaji Akande, Adamu Mustapha, Helen Shnada Auta, Bata Timothy, Abd'Gafar Tunde Tiamiyu, and Adedayo Olufemi Adesola

Abstract Wetlands are complex natural systems with distinct flora and animals growing on their soggy soils. They may be found in places where land and water meet. This work investigates the intricate processes that give rise to wetlands and how these vital ecosystems are arranged. The interplay of geological, hydrological, and biological factors results in the creation of wetlands. Geological activities include river dynamics, tectonic movements, and glacial activity influence the base of the terrain. The water conditions required for wetlands are determined by hydrological factors, which include the flow of surface water, groundwater discharge, and rainfall distribution. The environment is also changed by biological processes, particularly the development of plants and the activities of microbes. Wetland ecosystems consist of a complex web of both living and non-living (abiotic and biotic) components. These include elements like the chemical composition of the water, the soil's chemistry, and the vast array of organisms that have evolved to survive in wet

M. O. Innocent (✉) · M. Abdulsalam
Department of Microbiology, Skyline University Nigeria, Kano State, Nigeria

J. O. Samuel · A. Mustapha · H. S. Auta
Department of Microbiology, Federal University of Technology, Minna, Niger State, Nigeria

M. U. Livinus
Department of Biochemistry, Skyline University Nigeria, Kano State, Nigeria

S. A. Akande
Department of Mathematic, Federal University of Technology, Minna, Niger State, Nigeria

B. Timothy
Department of Geology, Abubakar Tafawa Balewa University, Bauchi, Nigeria

A. T. Tiamiyu
Department of Mathematic, The Chinese University of Hong Kong, Ma Liu Shui, Sha Tin District, Hong Kong

A. O. Adesola
Department of Mathematics and Statistics, University of Massachusetts, Amherst, MA, USA

B. R. Babaniyi et al. (eds.), *Wetland Ecosystems: Conservation Strategies, Policy Management and Applications*, Wetlands: Ecology, Conservation and Management 12, https://doi.org/10.1007/978-3-031-91982-4_1

environments. The many types of wetlands, their biological functions, and the significant ecosystem services they provide such as carbon storage, flood control, and water purification are all examined in this review. Despite their biological importance, wetlands face a number of difficulties, including pollution, habitat loss, and climate change. Acquiring a thorough understanding of the structure and functioning of wetlands is crucial for effectively maintaining, managing, and revitalising these distinct ecosystems, so ensuring their continued existence for posterity.

Keywords Ecosystem · Pollution · Wetland · Flora

1.1 Introduction

Wetlands are distributed around the globe, being present in every climate zone and continent, according to Balwan and Kour (2021). The presence of water at or near the surface for a long amount of time characterises these unique ecosystems, creating a climate that supports the establishment of a diverse array of specialist species of organisms. Wetlands remain critical to the preservation of biodiversity, carbon sequestration, and the global cycle of water. Thus, in order to deal with the increasing environmental challenges, we face, it is crucial to research and protect wetlands (Xu et al. 2019; Gimba et al. 2024).

According to Moomaw et al. (2018), the idea of wetlands has evolved throughout time in tandem with our growing understanding of these complex systems. Wetlands are specifically defined as "areas of marsh, fen, peatland, or water, whether they occur naturally or are man-made, and whether they are permanent or temporary" by the Ramsar Convention, an international agreement for the conservation of wetlands. These places may contain fresh, brackish, or saltwater, and the water may be stagnant or flowing. Areas of seawater which are no higher than 6 m at low tide are also included in this definition. According to Gell et al. (2023), the term is broad and encompasses a wide range of habitats, from inner bogs to coastal mangroves, each with unique characteristics and biological functions. Geology, hydrological, and biological variables interact in a multifaceted way throughout the technique of creating wetlands. Wetlands are created by the alteration of the ground brought about by geological events such river erosion, tectonic movements, and glacier passage. The water regime that defines the conditions of wetlands is determined by hydrological variables, including precipitation patterns, groundwater discharge, and surface water flow. The construction of distinct soil types in wetlands and habitats is influenced by biological processes, particularly the microbial community and the emergence of plants (Chakraborty et al. 2023).

The complex interactions between biotic (living) and abiotic (non-living) elements form the overall framework of wetland ecosystems. The abiotic variables that determine the chemical and structural foundation of the wetland ecosystem include the properties of soil, water chemical science, and weather patterns. The diverse range of organisms, from bacteria to plants and animals that have evolved to live inside such water-saturated habitats is referred to as the "biotic components." According to Yang et al. (2019b), the interplay between these constituents gives rise to distinct ecological processes, such as primary production, decomposition, and nutrient cycling that are distinct from those documented in fully aquatic or terrestrial settings. Due in large part to the variety of settings in which they develop and persist, wetlands exhibit an astounding array of shapes and characteristics. Marshes, swamps, bogs, and fens are the primary categories; they are distinguished by their unique hydrological patterns, soil properties, and biological populations. Wetlands are widely distributed around the globe, as shown by their presence in every climate zone and on every continent save Antarctica. From the vast peatlands of the north to the tropics mangrove forests, wetlands demonstrate remarkable resilience to a variety of climatic conditions (Hopkinson et al. 2019). Numerous biological activities and services that wetlands provide are very useful to human societies as well as natural systems. These ecosystems serve as natural water purifiers, ridding bodies of water of impurities and excess nutrients. For a wide range of species, including many rare or endangered species, they serve as essential habitats. Because they collect large volumes of water and release it gradually, wetlands are essential for controlling flooding.

Coastal regions provide as a buffer against the effects of storm surges and sea level rise. Furthermore, wetlands play a critical role in the carbon cycle by efficiently storing large amounts of carbon in their plants and soils. They so significantly aid in the mitigation of climate change (Alikhani et al. 2021). Despite their biological importance, wetlands face a number of threats worldwide. Wetlands have significantly decreased in number and quality as a consequence of human activities including resource exploitation, urbanisation, and draining of agricultural land. By altering hydrological patterns and perhaps shifting the geographic distribution of different wetland groups, climate change poses additional challenges. These fragile ecosystems are made even more vulnerable by pollution and exotic species. Recent years have seen an increased emphasis on conservation as a reaction to these threats. In order to protect and restore wetland ecosystems, this has meant putting international agreements, national laws, and local efforts into action (Xu et al. 2019). To effectively execute management and conservation efforts, it is essential to get a thorough knowledge of the complex processes involved in the formation of wetlands and the complicated organisation of their natural systems. This information is used to guide and advise restoration projects, predict environmental changes in ecosystems with accuracy, and make well-informed policy choices. Our awareness of these dynamic ecosystems' importance in maintaining the health of our planet and the many advantages they provide to human civilization grows as we continue to explore their mysteries (Ballut-Dajud et al. 2022). This research seeks to carry out a thorough examination of the composition and development of wetland ecosystems. Our goal is to improve our knowledge of these essential ecosystems by examining the geological, hydrological, and biological processes that lead to the development of wetlands as well as the complex web of abiotic and biotic elements that make up their ecosystem structure. The aim of this study is to add to the increasing amount of information that backs up the protection of wetlands and the use of sustainable management techniques. This will guarantee that these unique ecosystems are preserved for next generations.

1.2 Formation of Wetlands

Wetlands are created by the delicate interaction of hydrological, biological, and geological elements. This is a dynamic and complex process. According to Hopkinson et al. (2019), these components cooperate across different periods to produce the special circumstances that define wetland ecosystems.

1.3 Geological Processes

Geology affects critical processes including water flow, land form, and the cycling of chemicals in the environment, all of which are essential to the development and functioning of wetlands. Wetlands' location, kind, and characteristics are directly influenced by the substrate's geological and lithological features (Reddy et al. 2022). Different geophysical and geochemical processes, including heat, pressure, melting, crystallisation, weathering, erosion, and deposition, are responsible for the creation of the three main classes of rocks: igneous, sedimentary, and metamorphic. These processes also supply the landscape characteristics required for the formation of wetlands, hence shaping the Earth's surface (Manuel 2018; Khalef et al. 2022). In particular, sedimentary rocks have a major role in the formation of wetlands. They often make their home in low-lying areas, leaving behind depressions and basins that may hold water deposits. According to Majumdar et al. (2020), sedimentary rock stacking may lead to the creation of impermeable barriers called aquitards, which force groundwater to rise to the surface and produce wetlands. Wetlands are mostly the result of glacial processes like sedimentation and erosion, which wear down the surrounding terrain. Wetlands may develop as a result of the till and outwash that glaciers deposit in hollows, which can impede water movement. Wetlands' development and evolution are influenced by post-glacial events such isostatic rebound and sea level changes (Job and Sieben 2022). The kind and properties of wetlands are greatly influenced by the age and pace of geological processes. Older wetlands still have developing soils and plants, but newly formed wetlands have not yet reached that stage. For example, the age and stage of development of the wetland are directly related to the pace at which peat accumulates in bogs and fens (Job and Sieben 2022).

According to Lisenby et al. (2019), geological processes and events are crucial in forming the physical basis for the creation of wetlands and in modifying the landscape. These long-term processes provide the groundwork for the creation of wetlands.

Glacial activity The creation of wetlands has been significantly influenced by the glaciers' retreat at the end of ice eras, especially in northern latitudes. Uneven topography resulted from the huge ice sheets' retreat, which left behind glacial till and hollowed out depressions in the terrain. These "kettle holes," which are depressions that collect water from precipitation and melting ice, are the foundation of many wetlands. Furthermore, valleys may be dammed by glacial moraines, which would allow for the upstream formation of wetlands (Byun et al. 2021).

Tectonic movements Changes in the crust of the Earth may produce basins or modify current drainage patterns, which can result in the development of wetlands. Subsidence may lower land surfaces and provide ideal circumstances for the creation of wetlands. It is often brought about by the compaction of sediments or the dissolving of underlying rock. On the other hand, new floodplains and altered river channels might result from tectonic uplift. Sometimes, abrupt topographic changes brought on by earthquakes might result in depressions that could eventually turn into wetlands (Job and Sieben 2022).

Processes that flow Wetland development is greatly aided by the constant erosion and deposition of rivers and streams on the land. When meandering rivers veer off course, they may leave behind curved volumes of water that often turn into wetlands, known as oxbow lakes. Rivers build up their floodplains with sediments that they deposit during flood occurrences, which fosters the growth of wetland plants. Large-scale wetland complexes may be created in delta areas by rivers continuously depositing sediments (Job and Sieben 2022).

Dynamics of the coast Wetland development is facilitated by the interaction of sediment deposition, erosion, and sea level fluctuations along coasts. Sand builds up parallel to the shore to create barrier islands, which often result in lagoons behind them that eventually turn into wetlands. Estuaries are dynamic places where freshwater and saltwater combine, which makes them ideal for the development of wetland ecosystems. Estuaries are formed when rivers meet the sea. Mangrove swamps arise in intertidal zones in tropical and subtropical climates due to specialised plants that adjust to the salinity levels (Job and Sieben 2022).

Karst topography The dissolving of rock may produce sinkholes and depressions that may fill with water and form wetlands in areas with soluble bedrock, such as limestone. According to Gutiérrez et al. (2014), these karst wetlands often have intricate hydrology and are connected to subterranean water networks.

1.4 Hydrological Elements

The process of creating wetlands is greatly influenced by hydrology, which determines the duration, depth, and timing of water saturation or inundation. The water sources and flows into, through, and out of a wetland are essential to its development and functioning (Lane et al. 2018). Wetlands primarily get their water from precipitation, surface water (from lakes, rivers, and tides), and groundwater. Depending on the kind of wetland and where it is in relation to other landscape features, each source has a different relevance (Berkowitz et al. 2020). Bogs and fens are wetlands that get their water mostly from snowmelt and rainfall, hence precipitation is essential to their formation. Ombrotrophic wetlands are what these wetlands are called. Conversely, water from the surface or below, carrying dissolved nutrients and minerals, is usually used to irrigate mineral wetlands, such as marshes and swamps (Lane et al. 2018). Topography and soil permeability are two landscape elements that affect the flow and distribution of water within wetlands. There is a greater likelihood that wetlands may emerge in concave depressions and areas with impermeable soils or bedrock. The flow of water is also influenced by the topography and gradient of the ground; areas where water converges, such foot slopes and toe slopes, are more likely to become wetlands (Berkowitz et al. 2020).

An important factor is hydrodynamics, which describes the movement of water into and out of a wetland. The variety of plants and animals that may thrive in a wetland, the distribution of nutrients, and the deposition of silt can all be influenced by the speed and direction of water flow. Wetlands with slowly flowing or stagnant water often accumulate organic matter and develop into wetlands that produce peat. Conversely, wetlands that experience rapid water flow are more likely to have mineral soils and support a wider variety of plant life (Möller and Christie 2019; Yang et al. 2019a). The particular kind of wetland that forms is greatly influenced by the length and timing of water saturation or inundation. Wetlands that are intermittently wet or have fluctuating water levels will not have the same features as wetlands that are always flooded or saturated.

The species composition and ecological success of wetland colonisation may also be influenced by the temporal availability of water, such as the start of spring melt or the monsoon season (Hondula et al. 2021). Wetlands are characterised by water, and hydrological elements play a critical role in both their creation and upkeep: Patterns of precipitation: Wetland formation may be supported by regions with heavy rainfall or distinct wet seasons, particularly when paired with inadequate drainage. The hydroperiod of wetlands is influenced by the timing, amount, and length of precipitation episodes, which is critical to their biological features (Pereira et al. 2021).

Groundwater discharge Springs and seeps may sustain wetland conditions all year long in areas where the water table meets the surface of the ground. When opposed to wetlands that are largely supplied by surface water, these groundwater-fed wetlands often have more consistent water levels and chemistry. Wetlands and groundwater may interact in complicated ways; certain wetlands operate as regions where groundwater recharge occurs at particular times of the year (Pereira et al. 2021).

Flow of surface water Wetland hydrology is influenced by rivers, streams, and runoff, which often results in either seasonal or permanent flooding. Soil properties and plant communities in wetlands may be influenced by the frequency and length of flooding occurrences. According to Hayashi et al. (2016) and Papa and Frappart (2021), there are instances in which man-made structures like dams and levees may change natural flow patterns, resulting in the creation of new wetlands or the alteration of existing ones.

Retention of water Bedrock or impermeable soil layers may obstruct drainage, causing water to build up and the creation of wetlands. Even in places where the regional water table is much lower, wetlands may be supported by perched water tables created by clay-rich soils or hardpans. According to Papa and Frappart (2021) the existence of organic matter in soils may also enhance their ability to retain water, hence promoting the conditions of wetlands.

Evapotranspiration The preservation of wetland hydrology depends on the equilibrium between water inputs (precipitation, surface flow, groundwater) and outputs (evaporation, transpiration). Wetlands may develop in places where evaporation rates are low enough to permit water storage in certain climates, such as desert regions (Hayashi et al. 2016; Wu et al. 2016).

1.5 Biological Influences

According to Job and Sieben (2022), geology and hydrology have the most effects on the evolution of wetlands, but biological activities and interactions also have a big impact on how these ecosystems shape over time. Among the most important biological consequences, vegetation is essential to the establishment of wetlands. Plants that are acclimated to wetland conditions—emergent macrophytes, floating-leaved plants, and submerged aquatics, for example—can grow and thrive in areas with plenty of water. These plants contribute to the soil's organic matter accumulation as they rot and develop, which may lead to the formation of peat (Lesiv et al. 2020; Rodrigo 2021). In certain cases, the growth and decomposition of marsh vegetation may actively alter the ecosystem's water dynamics, creating a positive cycle that fortifies the wetland habitat. For example, the accumulation of peat in fens and bogs may raise the water table and provide more hospitable conditions for wetland plants. Similarly, an abundance of aquatic plants may obstruct water movement, which can lead to silt deposition and the creation of less dynamic and shallower conditions (Rodrigo 2021; Temmink et al. 2021).

Wetlands are created and maintained in large part by animals, which is an important biological component. The ability of beavers to build dams that hold water and create wetlands is well known (Law et al. 2019; Bashinskiy 2020). By their mating and feeding habits, other animals such as muskrats and ducks may similarly modify wetland ecosystems (McNicholl 2018; Feierabend 2020). Wetlands' development and functioning depend on microorganisms as well. An essential step in the production of peat is the breakdown of organic waste by anaerobic bacteria in the absence of oxygen. Other microbes are involved in the cycling of nutrients, particularly nitrogen and phosphorus, which are essential for the growth of the flora and fauna of wetlands (Yarwood 2018).

The development of wetland species and the acquisition of characteristics appropriate to wetland habitats during long epochs have also influenced the origins and characteristics of these ecosystems. For example, aquatic vegetation's ability to create buoyant and projecting growth structures has allowed it to successfully establish and thrive in wetland habitats. In order to live in the oxygen-deficient environments that are often found in wetland soils, wetland plants have evolved adaptations including aerial roots and gas exchange tissues (Vahsen et al. 2023; Hovick et al. 2023). Living things are important for the creation and development of wetlands because they often alter their surroundings in ways that support the conditions found in wetlands:

Establishment of vegetation Pioneer plant species settle in damp regions, stabilising sediments and fostering the emergence of further plants. Through transpiration and water intake, their root systems affect the local hydrology and aid in the trapping of organic materials and sediments. Plant decomposition and death aid in the creation of organic-rich wetland soils and the generation of new soil. The numerous vegetation patterns seen in wetlands are a result of different plant species' varying adaptations to wetland environments, such as the use of aerenchyma tissue for oxygen transfer to roots (Popoff et al. 2021).

Peat accumulation Over time, the gradual breakdown of plant material under damp, anaerobic conditions in certain kinds of wetland, especially bogs and fens, results in the accumulation of peat. Raised bogs that are hydrologically separated from nearby groundwater may be produced by this process, which can drastically change the hydrology and environment. Peat buildup, which may take thousands of years, is an essential step in the sequestration of carbon (Cronan 2023).

Animal activity A few species have a direct impact on the creation and upkeep of wetlands. The most famous example is probably beavers, whose dam-building endeavours may result in ponds and marshes that last for decades or even centuries. These wetlands, formed by beavers, often include unique flora groups and hydrological patterns. Waterbirds are among the other species that might affect the creation of wetland via seed distribution and sediment bioturbation (Cronan 2023).

Microbial processes In wetland soils, microorganisms have an impact on decomposition rates, nutrient cycling, and chemical reactions, all of which are related to soil formation and plant development. The creation of hydric soils, which are typical of wetlands, is caused by the activity of microorganisms, especially in anaerobic environments. Additionally, microbial activities are essential to biogeochemical cycling because they affect the availability of nutrients and the chemistry of water (Cronan 2023).

Biofilm development Microbial biofilm growth may affect nutrient cycling and sedimentation patterns in some kinds of wetland, especially in places with shallow water. By stabilising sediments, these biofilms may provide microhabitats for other species (Yang et al. 2019b).

1.6 Components of the Wetland Ecosystem

Both biotic and abiotic components of wetland ecosystems are essential for preserving biodiversity and the sustainability of the environment. While biotic components include the many plant species and microbial communities present in wetlands, abiotic components include the physical factors like water, soil, and nutrients (Gupta et al. 2020; Boru et al. 2024). Wetlands are dynamic ecosystems that are significantly impacted by natural disasters, human activity, especially floods and droughts. These occurrences result in changes to wetlands' physical characteristics as well as the ecological services they provide to inhabitants (Hammer and Bastian 2020; Sarkar et al. 2022). These ecosystems provide a wide range of ecological, economic, and social benefits that are essential for sustainable development and rural livelihoods. They act as stores of food, shelter, and living space for several biological species (Wen et al. 2024). Understanding and protecting wetland ecosystem components is essential to maintaining their ecological soundness in the face of ongoing environmental stresses and changes in land use.

1.7 Structure Physical

Physical structures have a crucial role in the creation of habitats and the efficiency of ecological processes, making their existence in many ecosystems imperative. Physical elements are important in many domains, from landform classification in evolutionary geomorphology (Corenblit et al. 2010) to enhancing soil fertility in agricultural systems (Zhang et al. 2020). Lakes are home to a variety of plant, animal, and rock formations that enhance the habitat's diversity and support the existence of many micro- and macroorganism groups. Furthermore, accurate understanding and characterization of physical components—specifically, tensors—are essential for accurate physics analysis (Morganti 2018).

1.8 Hydrology

According to Lane et al. (2018), hydrology is the main element influencing the development and functioning of wetlands. The kind, location, and characteristics of wetlands are largely determined by the sources, movements, and timing of water. There are three primary water sources for wetlands: precipitation, surface water from lakes, rivers, and tides, and groundwater. The kind and location of the wetland determine each source's relevance (Lisenby et al. 2019). Topography and soil permeability are two landscape elements that affect the flow and distribution of water within wetlands (Cohen et al. 2016). Regions with impermeable soils or bedrock, as well as concave topographic characteristics, are more likely to store water and develop into wetlands. The speed and direction of water flow, or hydrodynamics, affects nutrient distribution, biodiversity of plants and animals, and sediment deposition (Jackson et al. 2014). The kind of wetland that forms is mostly dependent on the length and timing of water saturation or inundation. Wetlands that are always submerged will grow differently from wetlands that are only sometimes wet or have fluctuating water levels. The kinds of species that may live in and prosper in a wetland may depend on the seasonal availability of water, such as the timing of spring snowmelt or monsoon rains (Lisenby et al. 2019).

1.9 Composition of Soil

Wetland soils, often referred to as hydric soils, are an important abiotic component that influences and reflects the ecosystem's biology and hydrology (Trettin et al. 2020). According to Vereecken et al. (2022), these soils have characteristics including lower oxygen concentrations, higher organic matter contents, and the presence of redoximorphic patterns, which show periodic saturation. The sources and movements of water, together with the growth and decomposition of wetland plants, all have an impact on the mineral and organic content of wetland soils (Walton et al. 2020). The majority of wetland soils are formed by sedimentary rocks and glacial deposits; however, the presence of decomposing plants can lead to the formation of peat through the accumulation of organic matter. The types of plants and microbes that can thrive in a wetland environment are directly influenced by the properties of the soil, including texture, pH, and nutrient availability. For example, lower oxygen levels in hydric soils encourage the growth of anaerobic bacteria, which are essential to the cycling of nutrients and the production of peat (Vereecken et al. 2022).

1.10 The Components of Biological Structure: Biotic

The biological structure of wetlands, or its biotic components, is essential to the ecosystems' ability to operate. The variety of plant and animal species that wetlands sustain increases their productivity and biodiversity (Bhowmik 2019). In wetlands, microorganisms are essential to the ecosystem's general health and nutrient cycling. They support the elimination of impurities and guarantee water purity (Lu et al. 2015; Liang et al. 2017). When cultivated in artificial wetland structures, wetlands plants benefit the environment by acting as a habitat, helping to manage sewage, and improving water quality via their capacity for absorption and degradation (Bhowmik 2019). It is possible to assess and maintain the health of wetland biotic components via the use of monitoring systems based on natural resource asset components, which ensures their preservation and long-term viability (Lu et al. 2015).

1.11 Plants

Hydrophytic plants, also referred to as wetland vegetation, are an essential component of these ecosystems and play a major role in their development and functioning (Gopal 2016). Because of the unique soil and hydrological circumstances present in wetlands—such as sporadic flooding, low oxygen levels, and restricted nutrient availability—wetland plants have evolved to withstand these conditions. According to Guntenspergen et al. (2020), typical wetland plant species include submerged aquatics like pondweeds and watermilfoil, floating-leaved plants like water lilies, and emergent macrophytes like cattails and bulrushes. The development and decomposition of the plants contribute to the accumulation of organic matter in the soil, which may lead to the formation of wetlands that create peat, such as bogs and fens, and the production of peat (Osman and Osman 2018). Wetland plants may affect the hydrology of the system by obstructing water flow and promoting sediment deposition, which can lead to the formation of shallower and less dynamic conditions. Wetland plants have evolved adaptations throughout long epochs, such gas exchange tissues and aerial roots, that enable them to thrive in the oxygen-deficient circumstances often seen in these settings (Hopkinson et al. 2019).

1.12 Animals

Numerous animal species, including fish, amphibians, birds, mammals, reptiles, and invertebrates, find home in wetlands. Many species have adapted to survive in the unique conditions found in wetlands, such as the abundance of plants, stagnant water, and high productivity (Feierabend 2020). Invertebrates, such as insects, crustaceans, and mollusks, play crucial roles in the food chain and nutrient cycling in wetlands. For fish, particularly commercially important species, wetlands are essential habitats because they rely on these places for feeding, spawning, and rearing. The moist and semi-wet habitats present in wetlands are ideal for the survival of reptiles and amphibians, including frogs, turtles, and snakes (Job and Sieben 2022). Wetlands provide vital habitats for breeding, resting, and food finding, which is why they are so important to the migration of birds. Wetlands are vital to many species of wading birds, shorebirds, and waterfowl. Because of the way they dig and eat, mammals such as otters, muskrats, and beavers have a big influence on wetland ecosystems. Since wetlands have a direct influence on the variety and abundance of aquatic animals, their maintenance and restoration are essential to preserving the general health and operation of the ecosystem. This emphasises how important it is to maintain these vibrant, fruitful ecosystems (Job and Sieben 2022).

1.13 Trophic Levels and Flux of Energy

Wetland ecosystems include trophic levels that may be seen, which show how nutrients and energy travel through the food web (Yang and Chen 2013). The base of the pyramid is made up of primary producers, which include phytoplankton, algae, and aquatic plants. They are in charge of using photosynthesis to convert solar energy into chemical energy (Han et al. 2016). Immediately above producers at the trophic level are herbivores, also referred to as main consumers. These animals directly devour the producers; examples of these are mollusks, hippopotamuses, certain fish, and zooplankton. They move nutrients and energy up the food chain when they consume the producers (Han et al. 2016). Carnivores, or secondary consumers, make up the third trophic level. These species, which include bigger fish, reptiles, and certain birds, are the primary consumers' predators. By eating primary consumers, secondary consumers help transport nutrients and energy up the food chain (Han et al. 2016). At the top of the wetland food chain's trophic level are tertiary consumers, which include alligators and big predatory fish. By consuming both primary and secondary consumers, these top-level predators control the populations of creatures at lower trophic levels (Han et al. 2016). Decomposers, including fungi and bacteria, are essential to the recycling process because it allows nutrients to be returned to the environment. In the wetland environment, decomposers break down organic matter that has died, releasing nutrients that producers may use. The cycle of energy and nutrient circulation is efficiently completed by this mechanism (Han et al. 2016). The trophic levels in wetlands are interconnected because nutrients and energy go from producers to consumers and then back to producers via the work of decomposers. Wetland ecosystems depend on the cycling of nutrients and energy for their long-term health (González-Ortegón et al. 2015).

1.14 Ecological Functions and Services of Wetlands

Wetlands provide a multitude of ecological services, making them vital to the environment. In order to cleanse wastewater, reduce nutrient loads, and provide homes for animals, constructed wetlands mimic natural wetlands. According to Agaton and Guila (2023), they provide supplying, regulatory, cultural, and supporting services. Common reed vegetation found in wetlands serves as a habitat for vulnerable bird species. The objective of conservation efforts is to maintain the variety and stability of these wetlands while taking into account the demands of socioeconomic activities and environmental protection (Čížková et al. 2023). Along with other wetlands, the Lashihai Nature Reserve is essential to the provision of ecosystem services, especially those related to climate control. This emphasises how important it is to preserve wetlands for the sake of environmental and economic stability (Cai et al. 2023). Furthermore, wetlands like Jiaxing's Guanjinggang Wetland demonstrate effective water treatment with cutting-edge technology, improving water quality and lowering pollution levels. This demonstrates the many benefits that ecological wetlands provide (Wang et al. 2021). Wetlands have a wide range of purposes, such as protecting biodiversity and purifying water, which highlights how important they are to preserving ecosystems and advancing human wellbeing (Ramachandra et al. 2021). The basic ecological functions and advantages include carbon storage, flood control, water purification, and animal habitat providing.

1.15 Water Purification

Because they filter and remove pollutants, sediments, and excess nutrients from both surface and groundwater sources, wetlands are crucial for improving the quality of water. Wetlands naturally act as cleansers due to their high plant concentration and complex soil composition. Through a combination of physical, chemical, and biological methods, they efficiently collect and degrade contaminants (Cheng et al. 2021). Nutrients like nitrogen and phosphorus may be absorbed and converted by wetland plants and microorganisms, which stops them from flowing into downstream water bodies and lessens the risk of eutrophication (Hammer and Bastian 2020). Because of their capacity to hold sediments, wetlands may reduce water turbidity and keep streams from being obstructed. Furthermore, the slow water flow in wetlands promotes the breakdown of organic materials and the accumulation of suspended particles (Hammer and Bastian 2020). Because they filter water, wetlands are essential for maintaining the health of aquatic ecosystems, protecting the quality of sources of drinking water, and reducing the need for costly water treatment facilities (Cheng et al. 2021). Because wetlands are so effective at removing pesticides, heavy metals, and other toxins from the water, they are especially useful in areas where there is massive industrial or agricultural runoff. One cost-effective way to improve water quality and safeguard water resources for natural and human populations is to restore and conserve wetlands (Cheng et al. 2021).

1.16 Flood Management

Because they act as organic reservoirs, wetlands play a crucial role in lessening the consequences of flooding by retaining excess water and releasing it gradually. Because wetlands can hold onto large volumes of water for a short while after heavy rains or quick snowmelt, they may lessen the risk of erosion and floods later on. Wetlands slow the flow of water and allow it to slowly seep into the ground or evaporate due to the complex soil composition and the presence of vegetation. This process helps to reduce the duration and severity of flooding episodes by spreading the flow of floods across time and reducing flood peaks (Balwan and Kour 2021). Because they are good at absorbing and holding onto water from rivers, lakes, and tidal surges, wetlands located in floodplains and coastal areas are essential for flood control (Hussain et al. 2022). Wetlands provide an essential ecological function by protecting surrounding communities and infrastructure, therefore averting loss of life and property. Extreme weather events are occurring more often and with greater severity as a result of climate change. As a result, wetlands' ability to reduce flooding becomes more and more important. Building resilient communities and lessening the consequences of floods need the preservation and maintenance of these natural flood barriers (Balwan and Kour 2021; Hussain et al. 2022).

1.17 Sequestration of Carbon

Wetlands are a vital component of the global carbon cycle and efforts to slow down climate change because of their remarkable efficiency in absorbing and holding carbon, according to a research done in 2019 by Were et al. Because of their dense plant growth and organic-rich soils, wetlands are essential for photosynthesis—the process by which plants take up and store carbon dioxide from the atmosphere. Peat and other organic materials are used to store this carbon. According to Lolu et al. (2020), wetland plants are highly productive and have the ability to effectively absorb carbon and integrate it into their biomass. Examples of these plants include emergent macrophytes and submerged aquatics. A large amount of carbon is stored in the soil instead of being released into the atmosphere when wetland plants die and decompose. Wetlands retain carbon for a long period, so removing it from the active carbon cycle due to the lack of oxygen and the sluggish breakdown process. Rough estimates suggest that wetlands, which make up about 5–8% of the planet's land area, may store up to 35% of soil carbon worldwide. Peatlands are wetlands that are mostly made of organic matter that has partly decomposed. They are very good at storing carbon. Despite making up less than 3% of the planet's land area, they account for over 30% of the soil carbon reservoir globally (Lolu et al. 2020). We can reduce the effects of climate change by using wetlands' natural capacity to store carbon via preservation and restoration. These important carbon sinks may be efficiently preserved and improved by putting into practice wetland conservation and management strategies, such as rehydrating wetlands that have previously been drained and restricting the conversion of wetlands into other kinds of land (Were et al. 2019).

1.18 Biodiversity's Habitat

Wetlands are very diverse ecosystems that provide as vital homes for a wide range of plant and animal species. Because of their unique combination of aquatic and terrestrial conditions, as well as the abundance of food and water sources, wetlands provide a good habitat for a broad range of species (Balwan and Kour 2021; Aransiola et al. 2024). Aquatic and semi-aquatic plants, such as emergent macrophytes, floating-leaved plants, and submerged aquatics, may find a variety of supports in wetlands. These plants are essential to the general functioning of the wetland ecosystem in addition to providing food and protection to animals. Through photosynthesis, they support the stabilisation of sediment, water filtration, and oxygen generation (Datta et al. 2022).

Wetlands are essential to the life cycles of many different kinds of fish, amphibians, and invertebrates. They serve as feeding grounds, breeding grounds, and juvenile homes for a wide variety of fish species, some of which have substantial commercial value. Wetlands are essential to amphibians like frogs and salamanders for their ability to reproduce, find food, and survive the winter. Because wetlands provide a variety of habitats and an abundance of food supplies, invertebrates like insects, crabs, and mollusks thrive there (Datta et al. 2022). Because they provide habitat for resident species to breed and procreate as well as vital resting spots for migrating shorebirds and ducks, wetlands are essential to the survival of many bird species. A wide variety of bird species, including wading birds, herons, ducks, geese, and songbirds, are drawn to wetlands because of their diverse plant life and abundant food supply (Balwan and Kour 2021). Mammals such as beavers, muskrats, otters, and various deer depend on wetlands for protection, nourishment, and hydration, making them essential to their existence. Due to their burrowing and feeding habits, these organisms have a substantial influence on the composition and functionality of wetland ecosystems (Balwan and Kour 2021). Significant biodiversity found in wetlands is important in and of itself. They also provide a multitude of ecological services beneficial to human societies. Wetlands provide wildlife habitats, promote the expansion of both commercial and recreational fishing, and offer opportunities for ecotourism activities like birdwatching and photography. Maintaining the wide variety of species that relies on these habitats requires the conservation and restoration of wetlands (Datta et al. 2022).

1.19 Conclusion

Examining the composition and evolution of wetland ecosystems reveals the intricate and dynamic features of these vital environments. A complex interplay of hydrological, biological, and geological processes forms wetlands, which give rise to a variety of ecosystems found all over the globe. From interior bogs to coastal mangroves, every form of wetland is distinguished by a unique mix of these important factors, which together contribute to the wide variety of Earth's landscapes and biodiversity. Wetlands are incredibly important from an ecological standpoint. Wetlands are essential for many species' habitats, naturally filtering water, mitigating the effects of floods, storing carbon, and offering a host of other ecosystem services that benefit both human societies and animals. However, these delicate ecosystems face significant threats from the growing effects of human activity and climate change. Developing effective plans for their preservation and management requires a thorough grasp of the mechanisms behind the formation of wetlands and the intricate structure of their ecosystems. This knowledge informs restoration projects, helps us predict how wetlands could respond to changes in the environment, and guides policy decisions meant to protect these vital ecosystems. Wetland ecosystem research must go on, particularly in light of the changing global environment. Through improving our understanding of these complex systems, we can create more effective plans for their preservation and renewal. Not only are wetlands important for the environment, but they also support the ecological balance that keeps life on Earth alive. We can ensure that future generations will continue to benefit from the many services and amazing natural features that wetlands have to offer by using sustainable management practices.

References

Agaton CB, Guila PMC (2023) Ecosystem services valuation of constructed wetland as a nature-based solution to wastewater treatment. Earth 4(1):78–92

Alikhani S, Nummi P, Ojala A (2021) Urban wetlands: a review on ecological and cultural values. Water 13(22):3301

Aransiola SA, Bamisaye A, Abioye OP, Maddela NR (eds) (2024) Marine bioprospecting for sustainable blue-bioeconomy. Springer Nature, Cham. ISBN: 9783031689109. https://link.springer.com/book/9783031689109

Ballut-Dajud GA, Sandoval Herazo LC, Fernández-Lambert G, Marín-Muñiz JL, López Méndez MC, Betanzo-Torres EA (2022) Factors affecting wetland loss: a review. Land 11(3):434

Balwan WK, Kour S (2021) Wetland-an ecological boon for the environment. East African Scholars J Agri Life Sci 4(3):38–48

Bashinskiy IV (2020) Beavers in lakes: a review of their ecosystem impact. Aquat Ecol 54(4):1097–1120

Berkowitz JF, Johnson DR, Price JJ (2020) Forested wetland hydrology in a large Mississippi river tributary system. Wetlands 40(5):1133–1148

Bhowmik S (2019) Ecological and economic importance of wetlands and their vulnerability. In: Current state and future impacts of climate change on biodiversity, p 95. https://doi.org/10.4018/978-1-7998-1226-5.ch006
Boru KA, Ingale LT, Lemt KM (2024) Wetland ecosystem: plant species diversity, services, degradation drivers, and community perception in Sinana District, Oromia region, Southeast Ethiopia. Nat Environ Pollut Technol 23(1). https://doi.org/10.46488/nept.2024.v23i01.004
Byun E, Sato H, Cowling SA, Finkelstein SA (2021) Extensive wetland development in mid-latitude North America during the Bølling–Allerød. Nat Geosci 14(1):30–35
Cai M, Dong Z, Chen Z, Zhao Z (2023) Ecological service function evaluation and sustainable development countermeasures of plateau wetland ecosystem: a case study in Lashihai nature reserve, PRC. E3S Web Conf 393:02007. EDP Sciences
Chakraborty SK, Sanyal P, Ray R (2023) Diversity and classification of wetlands in international and national perspectives. In: Wetlands ecology: eco-biological uniqueness of a Ramsar site (East Kolkata Wetlands, India). Springer International Publishing, Cham, pp 167–226
Cheng K, Xu X, Cui L, Li Y, Zheng J, Wu W et al (2021) The role of soils in regulation of freshwater and coastal water quality. Philos Trans R Soc B 376(1834):20200176
Čížková H, Kučera T, Poulin B, Květ J (2023) Ecological basis of ecosystem services and management of wetlands dominated by common reed (Phragmites australis): European perspective. Diversity 15(5):629
Cohen MJ, Creed IF, Alexander L, Basu NB, Calhoun AJ, Craft C, Walls SC (2016) Do geographically isolated wetlands influence landscape functions? Proc Natl Acad Sci 113(8):1978–1986
Corenblit D, Steiger J, Delmotte S (2010) Abiotic, residual and functional components of landforms. Earth Surf Process Landf 35(14):1744–1750
Cronan CS (2023) Wetland ecosystems. In: Ecology and ecosystems analysis. Springer Nature Switzerland, Cham, pp 161–176
Datta S, Sinha D, Chaudhary V, Kar S, Singh A (2022) Water pollution of wetlands: a global threat to inland, wetland, and aquatic phytodiversity. In: Handbook of research on monitoring and evaluating the ecological health of wetlands. IGI Global, Hershey, pp 27–50
Feierabend JS (2020) Wetlands: the lifeblood of wildlife. In: Constructed wetlands for wastewater treatment. CRC Press, Boca Raton, pp 107–118
Gell PA, Finlayson CM, Davidson NC (2023) An introduction to the Ramsar Convention on Wetlands. In: Ramsar wetlands. Elsevier, Amsterdam, pp 1–36
Gimba UA, Aransiola SA, Oyewole OA, Maddela NR (2024) Marine greens: roles in climate change and global warming mitigations. In: Aransiola SA, Oyewole OA, Maddela NR, Ijah UJJ, Manjunatha B (eds) Marine greens: environmental, agricultural, industrial and biomedical applications. CRC Press. ISBN: 9781032438856. https://www.routledge.com/Marine-Greens-Environmental-Agricultural-Industrial-and-Biomedical-Applications/Aransiola-Oyewole-Maddela-Josiah-Ijah-Manjunatha/p/book/9781032438856
González-Ortegón E, Walton MEM, Moghaddam B, Vilas C, Prieto A, Kennedy HA, Le Vay L (2015) Flow regime in a restored wetland determines trophic links and species composition in the aquatic macroinvertebrate community. Sci Total Environ 503:241–250
Gopal B (2016) Should 'wetlands' cover all aquatic ecosystems and do macrophytes make a difference to their ecosystem services? Folia Geobot 51(3):209–226
Guntenspergen GR, Stearns F, Kadlec JA (2020) Wetland vegetation. In: Constructed wetlands for wastewater treatment. CRC Press, Boca Raton, pp 73–88
Gupta G, Khan J, Upadhyay AK, Singh NK (2020) Wetland as a sustainable reservoir of ecosystem services: prospects of threat and conservation. In: Restoration of wetland ecosystem: a trajectory towards a sustainable environment, pp 31–43. https://doi.org/10.1007/978-981-13-7665-8_3
Gutiérrez F, Parise M, De Waele J, Jourde H (2014) A review on natural and human-induced geohazards and impacts in karst. Earth Sci Rev 138:61–88
Hammer DA, Bastian RK (2020) Wetlands ecosystems: natural water purifiers? In: Constructed wetlands for wastewater treatment. CRC Press, Boca Raton, pp 5–19. https://doi.org/10.1201/9781003069850-3

Han R, Chen Q, Wang L, Tang X (2016) Preliminary investigation on the changes in trophic structure and energy flow in the Yangtze estuary and adjacent coastal ecosystem due to the Three Gorges Reservoir. Eco Inform 36:152–161
Hayashi M, van der Kamp G, Rosenberry DO (2016) Hydrology of prairie wetlands: understanding the integrated surface-water and groundwater processes. Wetlands 36:237–254
Hondula KL, Jones CN, Palmer MA (2021) Effects of seasonal inundation on methane fluxes from forested freshwater wetlands. Environ Res Lett 16(8):084016
Hopkinson CS, Wolanski E, Cahoon DR, Perillo GM, Brinson MM (2019) Coastal wetlands: a synthesis. In: Coastal wetlands. Elsevier, Amsterdam, pp 1–75
Hovick SM, Adams CR, Anderson NO, Kettenring KM (2023) Progress on mechanisms and impacts of wetland plant invasions: a twenty-year retrospective analysis and priorities for the next twenty. Crit Rev Plant Sci 42(4):239–282
Hussain MM, Bhuvaneswari G, Borah RR (2022) Transitioning to Sponge City: a key to water resource management during floods in Assam. ECS Trans 107(1):4999
Jackson CR, Thompson JA, Kolka RK (2014) Wetland soils, hydrology and geomorphology. In: Batzer D, Sharitz R (eds) Ecology of freshwater and estuarine wetlands. University of California Press, Berkeley, pp 23–60. Chapter 2
Job NM, Sieben EJ (2022) Factors controlling wetland formation. In: Fundamentals of tropical freshwater wetlands. Elsevier, Amsterdam, pp 25–41
Khalef RN, Hassan AI, Saleh HM (2022) A review on elemental and isotopic geochemistry. In: Geochemistry and mineral resources. IntechOpen, London. https://doi.org/10.5772/intechopen.105496
Lane CR, Leibowitz SG, Autrey BC, LeDuc SD, Alexander LC (2018) Hydrological, physical, and chemical functions and connectivity of non-floodplain wetlands to downstream waters: a review. J Am Water Resour Assoc 54(2):346–371
Law A, Levanoni O, Foster G, Ecke F, Willby NJ (2019) Are beavers a solution to the freshwater biodiversity crisis? Divers Distrib 25(11):1763–1772
Lesiv MS, Polishchuk AI, Antonyak HL (2020) Aquatic macrophytes: ecological features and functions. Stud Biol 14(2):79–94
Liang Y, Zhu H, Bañuelos G, Yan B, Zhou Q, Yu X, Cheng X (2017) Constructed wetlands for saline wastewater treatment: a review. Ecol Eng 98:275–285
Lisenby PE, Tooth S, Ralph TJ (2019) Product vs. process? The role of geomorphology in wetland characterization. Sci Total Environ 663:980–991
Lolu AJ, Ahluwalia AS, Sidhu MC, Reshi ZA, Mandotra SK (2020) Carbon sequestration and storage by wetlands: implications in the climate change scenario. In: Restoration of wetland ecosystem: a trajectory towards a sustainable environment. Springer, Singapore, pp 45–58
Lu L, Xing D, Ren ZJ (2015) Microbial community structure accompanied with electricity production in a constructed wetland plant microbial fuel cell. Bioresour Technol 195:115–121
Majumdar A, Shrivastava A, Sarkar SR, Sathyavelu S, Barla A, Bose S (2020) Hydrology, sedimentation and mineralisation: a wetland ecology perspective. Clim Change Environ Sustain 8(2):134–151
Manuel BR (2018) Mineral deposits: types and geology. In: Mineral resources: from exploration to sustainability assessment. Springer, Cham, pp 49–119
McNicholl MK (2018) Avian wetland habitat functions affected by water level fluctuations. In: Coastal wetlands. CRC Press, Boca Raton, pp 87–98
Möller I, Christie E (2019) Hydrodynamics and modeling of water flow in coastal wetlands. In: Coastal wetlands. Elsevier, Amsterdam, pp 289–323
Moomaw WR, Chmura GL, Davies GT, Finlayson CM, Middleton BA, Natali SM, Sutton-Grier AE (2018) Wetlands in a changing climate: science, policy and management. Wetlands 38(2):183–205
Morganti M (2018) The structure of physical reality: beyond foundationalism. In: Reality and its structure: essays in fundamentality. Oxford University Press, Oxford, pp 254–272

Osman KT, Osman KT (2018) Peat soils. In: Management of soil problems. Springer, Cham, pp 145–183
Papa F, Frappart F (2021) Surface water storage in rivers and wetlands derived from satellite observations: a review of current advances and future opportunities for hydrological sciences. Remote Sens 13(20):4162
Pereira G, Ramos RDC, Rocha LC, Brunsell NA, Merino ER, Mataveli GAV, Cardozo FDS (2021) Rainfall patterns and geomorphological controls driving inundation frequency in tropical wetlands: how does the Pantanal flood? Prog Phys Geogr: Earth Environ 45(5):669–686
Popoff N, Jaunatre R, Le Bouteiller C, Trinquier M, Paillet Y, Evette A (2021) Ecological succession and fine sediment accretion influence local patch dynamics of a pioneer riparian species (Typha minima Hoppe). Freshw Biol 66(12):2351–2363
Ramachandra TV, Sincy V, Asulabha KS (2021) Accounting of ecosystem services of wetlands in Karnataka state, India. J Resour Energy Dev 18(1–2):1–26
Reddy KR, DeLaune RD, Inglett PW (2022) Biogeochemistry of wetlands: science and applications. CRC Press. https://doi.org/10.1201/9780429155833
Rodrigo MA (2021) Wetland restoration with hydrophytes: a review. Plants 10(6):1035
Sarkar P, Das T, Mandal R, Adhikari D (2022) Change in structural components due to seasonal flooding governs provisioning ecosystem service delivery and livelihood diversification: a case study from a tropical floodplain wetland in Barak Valley Region of India. Anthr Sci 1(4):444–459
Temmink RJ, Dorenbosch M, Lamers LP, Smolders AJ, Rip W, Lengkeek W, van der Heide T (2021) Growth forms and life-history strategies predict the occurrence of aquatic macrophytes in relation to environmental factors in a shallow peat lake complex. Hydrobiologia 848(17):3987–3999
Trettin CC, Kolka RK, Marsh AS, Bansal S, Lilleskov EA, Megonigal P, Gries J (2020) Wetland and hydric soils. In: Forest and rangeland soils of the United States under changing conditions: a comprehensive science synthesis. Springer, Cham, pp 99–126
Vahsen ML, Blum MJ, Megonigal JP, Emrich SJ, Holmquist JR, Stiller B, McLachlan JS (2023) Rapid plant trait evolution can alter coastal wetland resilience to sea level rise. Science 379(6630):393–398
Vereecken H, Amelung W, Bauke SL, Bogena H, Brüggemann N, Montzka C, Zhang Y (2022) Soil hydrology in the Earth system. Nat Rev Earth Environ 3(9):573–587
Walton CR, Zak D, Audet J, Petersen RJ, Lange J, Oehmke C, Hoffmann CC (2020) Wetland buffer zones for nitrogen and phosphorus retention: impacts of soil type, hydrology and vegetation. Sci Total Environ 727:138709
Wang W, Yang T, Guan W, Peng W, Wu P, Zhong B, Yin C (2021) Ecological wetland paradigm drives water source improvement in the stream network of Yangtze River Delta. J Environ Sci 110:55–72
Wen L, Mason T, Powell M, Ling J, Ryan S, Bernich A, Gufu G (2024) Improved wetland mapping of a highly fragmented agricultural landscape using land surface phenological features. Remote Sens 16(10):1786
Were D, Kansiime F, Fetahi T, Cooper A, Jjuuko C (2019) Carbon sequestration by wetlands: a critical review of enhancement measures for climate change mitigation. Earth Syst Environ 3:327–340
Wu CL, Shukla S, Shrestha NK (2016) Evapotranspiration from drained wetlands with different hydrologic regimes: drivers, modeling, and storage functions. J Hydrol 538:416–428
Xu T, Weng B, Yan D, Wang K, Li X, Bi W, Liu Y (2019) Wetlands of international importance: status, threats, and future protection. Int J Environ Res Public Health 16(10):1818
Yang Y, Chen H (2013) Assessing impacts of flow regulation on trophic interactions in a wetland ecosystem. J Environ Inf 21(1):63–71
Yang SL, Fan JQ, Shi BW, Bouma TJ, Xu KH, Yang HF, Shi XF (2019a) Remote impacts of typhoons on the hydrodynamics, sediment transport and bed stability of an intertidal wetland in the Yangtze Delta. J Hydrol 575:755–766

Yang Y, Gao Y, Chen Y, Li S, Zhan A (2019b) Interactome-based abiotic and biotic impacts on biodiversity of plankton communities in disturbed wetlands. Divers Distrib 25(9):1416–1428
Yarwood SA (2018) The role of wetland microorganisms in plant-litter decomposition and soil organic matter formation: a critical review. FEMS Microbiol Ecol 94(11):fiy175
Zhang X, Yang W, Xin X, Zhu A, Ding S (2020) Poor physical structural components restrict soil fertility and crop productivity for wheat–maize cropping. Nutr Cycl Agroecosyst 117(2):169–184

Chapter 2
Ecological Importance and Functions of Wetlands

Gbolahan Olamide Isaac, Abdulmujeeb Bolaji Hamzat, and Oluwatosin Olaoluwa Daramola

Abstract Wetlands are among the most productive ecosystems on Earth, offering a multitude of ecological, economic, and social benefits. These unique environments serve as critical habitats for a diverse array of flora and fauna, including many endangered and migratory species. Wetlands play a vital role in water purification, acting as natural filters that trap sediments, nutrients, and pollutants, thereby improving water quality. They are essential in the global carbon cycle, sequestering significant amounts of carbon and mitigating climate change impacts. Furthermore, wetlands contribute to flood control by absorbing excess rainfall and reducing the severity of floods. Their rich biodiversity and complex food webs support recreational activities, such as birdwatching and fishing, which are valuable for local economies. Despite their importance, wetlands face numerous threats from human activities, including drainage, pollution, and climate change. Conservation and restoration efforts are crucial to maintaining the ecological functions and services that wetlands provide. This paper highlights the ecological significance of wetlands and underscores the need for sustainable management practices to protect these vital ecosystems.

Keywords Biodiversity · Flood control · Vegetation · Conservation

G. O. Isaac · O. O. Daramola
Bioresources Development Centre, National Biotechnology Development Agency (NABDA), Abuja, Nigeria

A. B. Hamzat (✉)
Bioresources Development Centre, National Biotechnology Development Agency (NABDA), Abuja, Nigeria

Nigerian Institute for Oceanography and Marine Research, Victoria Island, Lagos, Nigeria
e-mail: hamzata@futa.edu.ng

B. R. Babaniyi et al. (eds.), *Wetland Ecosystems: Conservation Strategies, Policy Management and Applications*, Wetlands: Ecology, Conservation and Management 12, https://doi.org/10.1007/978-3-031-91982-4_2

2.1 Introduction

Ecology, as a scientific discipline, involves the study of interactions between organisms and their environment, examining relationships among living organisms, including humans, and their surroundings. Within ecology, various branches focus on different aspects of the natural world. One key aspect is environmental monitoring, which involves detecting and analyzing pollutants in the environment (Zou et al. 2021; Gimba et al. 2024). This is crucial for understanding the impact of human activities on ecosystems and developing strategies to mitigate environmental damage. Another important area is the study of ecosystems and the relationships between different species within them, including food webs, energy flow, and nutrient cycling. This research helps scientists comprehend the complex interactions that sustain life on Earth and how disturbances can have far-reaching consequences. Ecosystems are intricate systems consisting of various organisms interacting with each other and their physical environment. The diversity of ecosystems is essential for maintaining ecological balance and providing vital services to both nature and humans. Recent research highlights the importance of understanding different ecosystem types to effectively manage and conserve biodiversity. Advances in ecosystem mapping have facilitated monitoring trends across various ecosystem types, such as aquatic ecosystems like wetlands, enabling a more comprehensive assessment of ecosystem health and changes over time (Nicholson et al. 2021). Studying these ecosystems requires acknowledging the impact of human activities and societal structures on ecological systems. Integrating decolonial and anti-oppressive practices in ecological scholarship is crucial for promoting ethical ecological research and conservation efforts (Trisos et al. 2021).

Wetlands play a crucial role in ecosystems, providing a wide array of environmental benefits. They support diverse plant and animal species, regulate water flow, improve water quality, and sequester carbon. Wetlands are known for their ability to trap sediments and pollutants, enhancing water quality (Martínez-Espinosa et al. 2021). They serve as essential habitats for various species, including migratory birds, fish, and amphibians, contributing to overall biodiversity (Aguirre et al. 2021). A key aspect of wetlands is their role in carbon sequestration and storage. Wetlands are highly effective at capturing and storing carbon, thereby helping mitigate climate change by reducing greenhouse gases in the atmosphere. The organic matter in wetlands undergoes anaerobic decomposition, leading to carbon accumulation in the soil (Martínez-Espinosa et al. 2021). This process aids in climate regulation and contributes to the health of the wetland ecosystem by influencing nutrient cycling and soil fertility. Wetlands also play a vital role in water management and flood control. They act as natural sponges, absorbing excess water during high rainfall and releasing it slowly during dry spells, thus preventing downstream flooding. Wetlands along coastlines help protect against storm surges and erosion by dissipating wave energy and reducing coastal damage (Choi et al. 2021). Wetlands support a wide range of plant and animal species, many of which are specially adapted to the unique wetland environment. They provide critical habitats for numerous species, including endangered and migratory birds, amphibians, and aquatic plants (Aguirre et al. 2021). Preserving and restoring wetlands is essential for maintaining biodiversity and ensuring the survival of these species in the face of habitat loss and degradation. Wetlands also play a significant role in nutrient cycling and water purification by acting as natural filters, removing excess nutrients and pollutants from water through processes such as denitrification and sedimentation (Martínez-Espinosa et al. 2021). This purification function is essential for maintaining the health of aquatic ecosystems and ensuring clean water for wildlife and humans. In the context of wetland restoration, using hydrophytes has been identified as a valuable approach to enhancing wetland ecosystems. Hydrophytes, or aquatic plants, play a crucial role in revegetating wetlands and improving their ecological functions. Reintroducing native hydrophyte species to degraded wetlands aims to enhance biodiversity, stabilize soil, and promote the natural functioning of wetland ecosystems (Rodrigo 2021). This approach not only helps restore the ecological integrity of wetlands but also contributes to the overall health of the surrounding environment by improving water quality and supporting diverse wildlife populations. Protecting and restoring wetlands is crucial for ensuring the continued provision of these ecosystem services and safeguarding the well-being of wildlife and human communities. This paper reviews the ecological importance and functions of wetlands, emphasizing their critical roles in biodiversity support, water quality improvement, carbon sequestration, flood regulation, the need for sustainable management practices to ensure their protection and continued provision of essential ecosystem services.

2.2 Functions of Wetlands

2.2.1 Biodiversity and Habitat Provision

Wetlands are known to provide essential ecosystem services that contribute significantly to biodiversity conservation. They play a crucial role in supporting biodiversity due to their unique characteristics and functions. Wetland ecosystems support a wide variety of plant and animal species, creating diverse habitats that are essential for the survival of numerous organisms (Rodrigo 2021). These habitats serve as breeding grounds, feeding areas, and shelter for a plethora of species, contributing to the overall biodiversity of an area. They act as corridors for wildlife movement, facilitating gene flow and maintaining genetic diversity within populations (Boland and Burwell 2022). The biodiversity supported by wetlands is vital for ecosystem stability and resilience. Biodiversity ensures that ecosystems can buffer the effects of environmental variations, maintaining stability and retaining essential functions such as productivity, carbon sequestration, and pollination. The presence of diverse species within wetlands enhances ecosystem functions and services, contributing to the overall health and sustainability of the environment. Studies have shown that wetlands with higher biodiversity tend to have increased ecosystem functioning, highlighting the importance of preserving and restoring wetland biodiversity (Mao et al. 2023).

According to (Rodrigo 2021) wetlands offer unique environmental conditions that support specialized plant and animal communities, including hydrophytes that are adapted to waterlogged conditions. These hydrophytes play a crucial role in wetland ecosystems by providing food and shelter for various organisms, including insects, birds, amphibians, and fish. The presence of diverse plant species in wetlands contributes to the overall biodiversity of the area, creating a rich and dynamic ecosystem that supports a variety of life forms. Moreover, wetlands act as important sites for nutrient cycling and energy flow within ecosystems, further enhancing biodiversity. The diverse plant and microbial communities in wetlands play a significant role in nutrient recycling, decomposition, and primary production processes (Dervash et al. 2024). These processes support the growth and development of various organisms, creating complex food webs and interactions that contribute to the overall biodiversity of wetland ecosystems. The interconnected nature of wetland habitats allows for the transfer of energy and nutrients between different trophic levels, supporting a diverse array of species. In addition to supporting biodiversity directly, wetlands also play a crucial role in regulating water quality and quantity, which indirectly contributes to biodiversity conservation. Wetlands act as natural filters, trapping sediments, nutrients, and pollutants from water bodies, thereby improving water quality downstream (Malyan et al. 2021). By maintaining water quality, wetlands create suitable habitats for aquatic species and support the overall health of freshwater ecosystems. Wetlands contribute to climate change mitigation and adaptation, which are essential aspects of biodiversity conservation. The preservation and restoration of wetlands can play a vital role in climate change adaptation by providing natural buffers against extreme weather events and sea-level rise. These climate-resilient features of wetlands support biodiversity by ensuring the continuity of habitats and ecosystems in the face of changing environmental conditions. They serve as critical ecosystems that support biodiversity through a variety of functions and services. By providing essential habitats, regulating water quality, supporting nutrient cycling, and contributing to climate change mitigation and adaptation, wetlands play a fundamental role in maintaining biodiversity and ecosystem health. Preserving and restoring wetlands is essential for ensuring the continued support of biodiversity and the sustainability of ecosystems in the face of ongoing environmental challenges.

2.2.2 Flood Regulation and Water Management

Flood control is a crucial function of wetlands due to their ability to mitigate the impacts of flooding through various mechanisms. Wetlands act as natural sponges, absorbing excess water during heavy rainfall and releasing it slowly, thus reducing the risk of flooding in downstream areas (Rodrigo 2021). A study by (Kumar et al. 2021) evaluating the economic and biophysical significance of coastal wetlands in the northern USA revealed that wetlands prevented nearly US$625 million in flood damages during Hurricane Sandy, demonstrating their effectiveness in reducing flood losses. Artificial or constructed wetlands have been employed for flood control by storing and filtering excess water to protect rivers from different types of runoff, underscoring their importance in flood mitigation efforts.

Wetlands has help to regulate hydrological flows, particularly in flood-prone areas, by serving as natural buffers that can absorb and slow down the movement of water, thus reducing the risk of flooding in adjacent areas. Constructed wetlands in suburban areas have been implemented to manage stormwater and decrease the impact of floods, showcasing the practical application of wetlands in urban flood control strategies (Lin et al. 2021). The ability of wetlands to regulate flooding has been acknowledged as a key aspect of nature-based solutions for disaster risk reduction, emphasizing their importance in managing flood events. They contribute to denitrification processes, which are vital for water quality management by reducing nitrogen levels in aquatic ecosystems, the organic-rich soil conditions in wetlands create an ideal environment for denitrification to occur, highlighting the diverse benefits of wetlands beyond flood control. Additionally, wetlands have played a significant role in water phytoremediation efforts, providing a biotechnological approach to treating contaminated water and enhancing water quality by removing heavy metals (Martínez-Espinosa et al. 2021).

In areas where flooding can lead to the resurgence of diseases like schistosomiasis, wetlands play a crucial role in influencing disease transmission dynamics. Flooding can create conducive environments for the proliferation of disease vectors like snails, contributing to the persistence of diseases in regions with extensive wetlands (Igwaran et al. 2024). Therefore, the management and preservation of wetlands are essential not only for flood control but also for preventing the reemergence of waterborne diseases in vulnerable communities. Overall, wetlands serve as natural infrastructure that offers a variety of ecosystem services, with flood control being a significant function that helps mitigate the impacts of flooding on both natural and human systems. By preserving and restoring wetlands, societies can enhance their resilience to floods, improve water quality, and safeguard biodiversity, highlighting the importance of integrating wetlands into comprehensive flood management strategies.

2.2.3 Coastal Wetlands and Storm Surge Protection

Coastal wetlands play a crucial role in protecting coastlines from storm surges, flooding, and erosion. These ecosystems act as natural barriers that help reduce the impact of extreme weather events on coastal communities. Study by Kumar et al. (2021) highlights the efficiency of nature-based solutions, including estuarial and tidal wetlands, in mitigating storm surge and flood damages. Wetlands have been shown to dissipate wave energy, offering coastal protection against storm surges, erosion, and rising sea levels, as emphasized by (Ostrowski et al. 2021). Wetlands such as saltmarshes, mangroves, and seagrass ecosystems are collectively known as 'blue carbon ecosystems' due to their ability to sequester carbon and provide coastline protection from erosion and storm surges.

The protection provided by coastal wetlands is essential in the face of climate change-induced challenges. With the warming world leading to habitat loss and decreased coastal protection, the role of wetlands becomes even more critical. Trebilco et al. (2021) explained that the loss of habitat and coastal protection has resulted in declines in fisheries and increased exposure of coastal infrastructure to inundation and storm surges. Furthermore, Filho et al. (2021) emphasized that the protection of ecosystem services, including wetlands, is vital in preventing the deterioration of coastal and marine ecosystems. In the context of storm surge protection, wetlands offer a range of ecosystem services beyond coastal defense. Wetland ecosystems provide provisioning services such as wetland-dependent fish, shellfish, and waterfowl, as well as regulating services like improving water quality and climate regulation, as highlighted by (Martínez-Espinosa et al. 2021). These services contribute significantly to human well-being by enhancing water security, reducing the risk of floods, and protecting coastlines from storms, hurricanes, and tsunamis. While wetlands are effective in reducing the impact of low-to-moderate-energy storms, their capacity to mitigate high-energy storm surges is still being quantified. According to Choi et al. (2021) tidal marshes and wetlands can reduce wave energy from low-to-moderate-energy storms, but their effectiveness against high-energy storm surges is limited and context-dependent. The combination of nature-based solutions like wetlands with engineered features can enhance their effectiveness in storm surge protection. In the face of increasing anthropogenic pressures, such as microfiber pollution from untreated stormwater, the resilience of coastal wetlands is being challenged. Protecting wetlands from pollution and ensuring their conservation is crucial to maintaining their functions in storm surge protection and overall ecosystem health.

2.2.4 Water Quality Improvement

Improvement of water quality is a crucial function of wetlands, with wetlands playing a significant role in enhancing water quality through various mechanisms. Constructed wetlands utilize physical, chemical, and biological processes to retain phosphorus (P) and improve water quality. These wetlands act as natural filters, trapping sediments, nutrients, and pollutants, thereby improving the overall quality of water that passes through them. Additionally, wetlands contribute to denitrification processes, which are essential for reducing nitrogen levels in water bodies, further enhancing water quality (Martínez-Espinosa et al. 2021). The presence of aquatic plants in wetlands aids in the uptake of nutrients like phosphorus and nitrogen, preventing their accumulation in water systems and thereby improving water quality (Rodrigo 2021). Wetlands also support a diverse range of microbial communities that participate in processes like denitrification and the breakdown of organic matter, further enhancing water quality (Retter et al. 2021). They act as natural buffers against pollutants and contaminants. They help in the removal of heavy metals and other pollutants through processes like phytoremediation, where plants uptake and store contaminants, preventing them from entering water bodies. The presence of wetlands also aids in reducing the impact of waterborne diseases by acting as natural filters, trapping pathogens and preventing them from contaminating water sources (Shayo et al. 2024). This function is particularly important in urban areas where access to safe water is essential for public health. Wetlands are invaluable ecosystems that significantly contribute to water quality improvement through various mechanisms such as nutrient uptake, denitrification, phytoremediation, and microbial processes. Their role in filtering pollutants, retaining sediments, and supporting diverse plant and microbial communities underscores their importance in maintaining the health and quality of water bodies.

2.2.5 *Water Purification*

This is a crucial function of wetlands, with various mechanisms contributing to this process. One significant aspect is the positive effect of turbidity on photolysis, where increased diffusion of light through water aids in purification (Richardson and Soloviev 2021). This phenomenon has long been recognized and plays a vital role in the natural cleansing of water within wetland ecosystems. The use of nanocomposites like graphene-TiO_2 and graphene-ZnO has shown promise in water purification processes, further highlighting the diverse applications of advanced materials in enhancing water quality (Thakre et al. 2021). These nanocomposites have been extensively reviewed for their effectiveness in treating wastewater and removing contaminants. With the utilization of Metal–Organic Frameworks (MOFs), which has emerged as a futuristic option for purifying pollutants in aqueous environments due to their high surface area and tailorable properties (Cao et al. 2021). MOFs offer a versatile platform for various purification processes, including water treatment, showcasing the adaptability of novel materials in addressing environmental challenges. In the context of wetlands, the pollutant removal and transformation processes in constructed wetland systems have been studied, emphasizing the importance of these engineered systems in treating wastewater from different pollution sources (Malyan et al. 2021). Constructed wetlands serve as bioengineered solutions that mimic natural wetlands, demonstrating their efficacy in water purification and pollution mitigation.

Biochar has gained attention for its adsorption efficiency in water treatment, with its porous structure resembling that of activated carbon, a widely used sorbent in water treatment processes (Wang et al. 2020). The intrinsic characteristics of biochar make it a valuable tool in water purification, showcasing the potential of natural materials in sustainable environmental practices. Additionally, protein nanofibrils have been explored for next-generation sustainable water purification, emphasizing the ecological approach of water reclamation for ensuring clean drinking water supplies (Peydayesh and Mezzenga 2021). This innovative use of nanofibers highlights the evolving landscape of water purification technologies towards more sustainable and efficient solutions. In the realm of wetland restoration, the use of hydrophytes has been reviewed by (Rodrigo 2021), for enhancing water quality and biodiversity in natural and constructed wetlands. Wetland restoration not only improves water quality and revegetation, but also contributes to ecosystem health and resilience.

Sequestration and recognition of toxic inorganic water pollutants using hydrolytically stable Metal-Organic Frameworks (MOFs) have been investigated, underscoring the importance of tailored porous architectures in addressing water pollution challenges (Dutta et al. 2021). The ability of MOFs to selectively remove pollutants from water highlights their potential in enhancing water purification processes. Also, the development of membrane-based technologies, such as membrane distillation and nanocomposite-modified electrochemical sensors, has shown significant progress in water quality monitoring and treatment (Yeszhanov et al. 2021). These advancements in membrane technology offer efficient solutions for removing contaminants and ensuring water safety. Furthermore, the synthesis and application of anti-fouling membranes using two-dimensional nanomaterials have been explored, showcasing the potential of innovative membrane technologies in water/wastewater treatment systems (Shahzad et al. 2021). These membranes play a crucial role in preventing fouling and maintaining the efficiency of water purification processes. Moreover, the use of magnetic nanoadsorbents for micropollutant removal in real water treatment has been reviewed, highlighting the importance of advanced purification methods in addressing rising pollution levels and the need for sustainable water management practices (Mudhoo et al. 2021).

The application of nanostructured adsorbents demonstrates a promising approach to tackling water pollution challenges effectively. Solar-induced hybrid energy harvesters for advanced oxidation water treatment have also been investigated, showcasing the diverse mechanisms available for water purification, including peroxy-based, light-driven, and electric-driven advanced oxidation processes (Huo et al. 2021). These innovative energy harvesters offer sustainable solutions for water treatment processes. This processes are significant function of wetlands, with various technologies and materials playing crucial roles in enhancing water quality and addressing pollution challenges. From the use of nanocomposites and Metal–Organic Frameworks to membrane-based technologies and advanced oxidation processes, the field of water purification continues to evolve with a focus on sustainability and efficiency. By leveraging innovative materials and technologies, wetlands can serve as natural purification systems, contributing to the conservation of water resources and the protection of ecosystems.

2.2.6 Wetland as Natural Water Storage

Wetlands plays a crucial role in natural water storage, serving as significant reservoirs that help regulate water flow, store excess water during periods of high precipitation, and gradually release it during drier times (Retter et al. 2021). This function is particularly vital in tropical and subtropical regions where wetlands aid in sustaining water availability for various uses, including agriculture and human consumption. The ability of wetlands to store water is essential for maintaining ecological balance, supporting biodiversity, and providing habitats for various plant and animal species (Rodrigo 2021).

Constructed wetlands, which mimic the functions of natural wetlands, are also increasingly being utilized to improve water quality and enhance water storage capacity. These artificial wetlands not only help in purifying water by filtering out pollutants but also aid in water retention, especially in urban areas where natural wetlands may have been lost due to development (Rodrigo 2021). By storing water, wetlands help in reducing the risk of erosion, maintaining soil moisture levels, and supporting plant growth, which further contributes to the overall health of the ecosystem, The storage of water in wetlands also plays a role in mitigating the impacts of climate change by providing resilience against extreme weather events such as droughts and floods (Retter et al. 2021). Wetlands also act as buffers against sea-level rise by storing excess water and reducing the risk of coastal flooding (Retter et al. 2021). The storage function of wetlands is closely linked to their ability to regulate water flow, which is essential for maintaining the health of aquatic ecosystems and supporting the diverse species that depend on wetland habitats (Rodrigo 2021). They play a critical role in nutrient cycling, with their water storage capacity influencing the distribution of nutrients and organic matter within the ecosystem. This nutrient storage function of wetlands contributes to the overall productivity of the ecosystem, supporting plant growth and providing food sources for various organisms (Retter et al. 2021). Protecting and restoring wetlands is essential for ensuring sustainable water management and preserving the numerous benefits that wetlands provide to both the environment and society.

2.2.7 Slow Water Release

Wetlands possess the ability to release water slowly, which aids in water retention, flood control, and maintaining water quality. This slow water release is facilitated by the unique characteristics of wetlands, such as their hydrology, soil composition, and vegetation cover. Wetlands act as sponges, absorbing excess water during periods of high rainfall or flooding and gradually releasing it over time, thus reducing the risk of downstream flooding (Malyan et al. 2021).

The slow release of water in wetlands is essential for various ecological processes, including denitrification. Wetlands are known for their high organic matter content and anoxic soil conditions, which create ideal environments for denitrification to occur. Denitrification is a vital process that helps remove excess nitrogen from water bodies, preventing eutrophication and maintaining water quality. The slow water release in wetlands allows denitrification to proceed efficiently, contributing to the overall health of aquatic ecosystems (Martínez-Espinosa et al. 2021). They act as natural filters, removing pollutants and contaminants from water through various processes. The slow flow of water in wetlands allows for the settling and precipitation of suspended particles, including heavy metals like iron, manganese, and aluminum hydroxides. These pollutants are effectively trapped and immobilized in wetlands, preventing them from entering downstream water bodies and causing harm to aquatic life (Malyan et al. 2021).

As water moves slowly through wetlands, sediments and nutrients are deposited and retained. This process not only improves water quality but also enhances the wetland's capacity to store and release water gradually. Sediment deposition within wetlands creates a natural buffer that further reduces water flow velocity, aiding in flood mitigation (Walton et al. 2020). Wetland vegetation, including reeds, grasses, and trees, plays a crucial role in the slow release of stored water. The root systems of these plants not only absorb water but also create pathways for infiltration. Additionally, the physical structure of the vegetation acts as a drag force, slowing down water movement and allowing for a gradual release. This intricate interaction between water and vegetation is fundamental to the buffering capacity of wetlands (Hong et al. 2021). Aside water quality regulation, the slow water release in wetlands also supports biodiversity and ecosystem resilience.

2.2.8 Vegetation Role

The presence of vegetation in wetlands aids in the uptake and removal of pollutants, thereby improving water quality and supporting overall ecosystem health. In the context of wetlands and natural hazards, estuarine wetlands, including mudflats and channels, have been studied for their role in reducing the impact of storm surges. Research has focused on modeling the effects of wetland vegetation roughness on storm surges using tools like ArcGIS and numerical models like ADCIRC (Martínez-Espinosa et al. 2021). By simulating how wetlands and their vegetation can act as barriers against storm surges, the importance of preserving and restoring wetland vegetation for coastal protection becomes evident. Restoration efforts in wetlands often involve the strategic use of hydrophytes to revitalize these ecosystems. Seed-based approaches have been highlighted as effective methods for rapidly restoring the vegetation structure in wetlands, which is essential for supporting the diverse functions and services that wetlands provide. By focusing on revegetation strategies, wetland restoration projects aim to enhance biodiversity, improve habitat quality, and promote ecosystem resilience. Coastal wetlands, particularly in regions like North America along the Gulf of Mexico and the Atlantic Ocean, host a variety of vegetation types such as tropical mangrove forests and temperate salt marshes.

These wetlands serve as critical habitats that support a wide range of species and provide essential ecosystem services (Deacon and Samways 2021). The presence of vegetation in these coastal wetlands helps stabilize shorelines, mitigate erosion, and regulate water quality, emphasizing the vital role of vegetation in maintaining the integrity of these ecosystems. Certain plant species, like Eleocharis, are specifically adapted to aquatic and wetland habitats, where they play a significant role in environmental monitoring and water quality improvement (Kuuluvainen and Nummi 2023). The presence of Eleocharis species in wetlands contributes to the overall health of these ecosystems by participating in nutrient cycling, providing habitat for various organisms, and enhancing the overall biodiversity of wetland environments. In constructed wetlands designed for wastewater treatment, emergent plants like Phragmites australis and Typha species are commonly used for their ability to remove pollutants and enhance water quality (Retter et al. 2021). These plants facilitate the removal and transformation of contaminants through various biological processes, highlighting the importance of vegetation in engineered wetland systems. By harnessing the pollutant removal capabilities of these plants, constructed wetlands can effectively treat wastewater and mitigate environmental pollution. Denitrification, a critical process in wetlands that involves the conversion of nitrates into nitrogen gas, is facilitated by the presence of vegetation. Wetland vegetation acts as a sink for nitrous compounds, helping to reduce nitrogen levels in water bodies. This denitrification process, supported by wetland vegetation, plays a vital role in maintaining water quality, preventing eutrophication, and supporting the overall health of aquatic ecosystems. The ability of wetland vegetation to actively participate in nutrient cycling processes underscores its significance in ecosystem functioning. Constructed wetlands utilize a combination of physical, chemical, and biological mechanisms to manage agricultural water quality effectively. By incorporating strategies that involve the retention of phosphorus through polyphosphate accumulation, these wetlands can improve water quality and reduce nutrient pollution. The critical review of polyphosphate management in agricultural water systems emphasizes the role of wetland vegetation in nutrient cycling and pollution control, showcasing the multifaceted benefits of vegetation in engineered ecosystems. In urban areas, the preservation of natural wetlands is essential for mitigating the impacts of urbanization and climate change. By protecting inland wetlands, conserving mangroves, and removing invasive vegetation, cities can enhance their resilience to flooding, maintain water quality, and preserve biodiversity. The strategic conservation of wetland vegetation in urban settings not only supports local ecosystems but also provides valuable ecosystem services that benefit human populations. The integration of wetlands into urban planning is crucial for sustainable development and environmental protection. Wetlands, including rivers, lakes, and marshes, serve as natural fire barriers in forested landscapes, reducing the risk and impact of wildfires. The presence of wetlands with diverse vegetation types helps to create fire breaks that can limit the spread of fires and protect surrounding ecosystems. By incorporating wetlands into ecosystem management strategies, forested areas can enhance their resilience to wildfires and promote biodiversity conservation. The role of wetlands as fire barriers underscores their importance in maintaining ecosystem health and stability. Changes in environmental conditions, such as those driven by climate change, can impact groundwater microbial communities and ecosystem functions.

2.3 Nutrient Cycling

Nutrient cycling in wetlands is a vital ecological function that involves the movement and transformation of nutrients within the ecosystem, playing a significant role in sustaining wetland biodiversity and productivity. They are known for their high organic matter content and regular anoxic soil conditions, which create ideal environments for nutrient cycling processes like denitrification to occur (Martínez-Espinosa et al. 2021). The organic matter present in wetlands serves as a rich source of nutrients that undergo various transformations facilitated by microbial communities, plants, and other organisms. Microbes in wetlands play essential roles in nutrient cycling by promoting plant growth, managing toxicity, and cycling nutrients through processes like nitrogen fixation (Nadarajah and Rahman 2021). Bacteria like endophytes living within plant tissues, contributes to nutrient acquisition and root system development, further enhancing nutrient cycling within wetlands (Verma et al. 2021).

Plants in wetlands also play a crucial role in nutrient cycling by taking up nutrients from the soil and water, thereby preventing nutrient accumulation and maintaining nutrient balance within the ecosystem. The plants have high nutrient uptake rates, especially in constructed wetland systems, where they efficiently remove pollutants and help in nutrient transformation processes (Malyan et al. 2021). Moreover, the presence of aquatic plants like Eleocharis dulcis in wetlands contributes to nutrient cycling through their phytochemical properties and interactions with the surrounding environment (Uddin et al. 2021). These plants not only add to the biodiversity of wetlands but also participate in nutrient uptake and cycling processes. Arthropods, such as insects, are highlighted as key regulators of nutrient dynamics in various ecosystems, including wetlands. Their contribution to nutrient transportation and nitrogen fixation underscores their importance in nutrient cycling processes within wetlands, especially in arid ecosystems (Sagi and Hawlena 2021). The integration of arthropods with microbiota in ecosystems emphasizes the interconnectedness of different organisms in driving nutrient cycling and ecosystem conservation. Terrestrial isopods, for example, play a vital role in litter and soil nutrient cycling processes, accelerating the decomposition of plant material and contributing to nutrient availability in wetlands (Schapheer et al. 2021).

The regulation of nutrient cycling in wetlands is influenced by various factors, including lysosome biogenesis, which is essential for cellular nutrient processing and recycling. Lysosomes, along with endosomes, play critical roles in nutrient sensing and cellular homeostasis, responding to intra- and extracellular cues to regulate nutrient availability and utilization (Yang and Wang 2021). In wetlands, the restoration and maintenance of hydrophytes, aquatic plants adapted to wet environments, are essential for preserving nutrient cycling functions. Strategic approaches involving seed-based restoration techniques can help recover and enhance the vegetation structure in wetlands, supporting vital ecosystem functions like nutrient cycling (Rodrigo 2021). They serve as critical sites for nutrient cycling due to their high organic matter content, which supports diverse microbial communities involved in nutrient transformations. The denitrification processes occurring in wetlands contribute to nitrogen cycling, highlighting the importance of wetlands in regulating nutrient availability and water quality in surrounding ecosystems (Martínez-Espinosa et al. 2021). The interactions among plants, microbes, arthropods, and other organisms facilitates the movement, transformation, and recycling of nutrients within wetland ecosystems, sustaining biodiversity and ecosystem functions.

2.4 Carbon Sequestration and Climate Regulation

Wetlands play a crucial role in carbon sequestration and climate regulation due to their ability to store carbon and mitigate the impacts of climate change. They act as significant carbon sinks, trapping and storing carbon dioxide from the atmosphere through a process known as carbon sequestration (Martínez-Espinosa et al. 2021). This process is primarily facilitated by the presence of waterlogged conditions in wetlands, which slow down the decomposition of organic matter, leading to the accumulation of carbon in the soil. Wetland supports growth of vegetation such as hydrophytes, which further contribute to carbon sequestration through photosynthesis and organic matter accumulation (Rodrigo 2021).

They help regulate climate by influencing local and global climate patterns. They play a role in regulating temperature through evapotranspiration, where water is evaporated from wetland surfaces and transpired by plants, cooling the surrounding environment. Wetlands also act as buffers against extreme weather events such as floods, storms, and hurricanes by absorbing excess water and reducing the risk of flooding in adjacent areas (Lin et al. 2021). The presence of wetlands along coastlines can help protect against storm surges and coastal erosion, providing a natural defense mechanism against the impacts of climate change (Kumar et al. 2021; Aransiola et al. 2024).

In terms of greenhouse gas mitigation, wetlands are essential in removing greenhouse gases from the atmosphere. They are known for their ability to remove and store significant amounts of carbon dioxide, methane, and nitrous oxide, which are major contributors to global warming (Asibor et al. 2021; Gimba et al. 2024). Wetlands achieve this through processes like denitrification, where they convert harmful nitrogen compounds into inert nitrogen gas, thus reducing the release of nitrous oxide, a potent greenhouse gas, into the atmosphere (Martínez-Espinosa et al. 2021). The presence of wetlands helps maintain the balance of greenhouse gases by acting as sinks for carbon dioxide and methane, thereby mitigating the overall impact of these gases on the climate. Wetlands also support biodiversity by providing habitats for a diverse range of plant and animal species, contributing to ecosystem stability and resilience in the face of changing climatic conditions (Rodrigo 2021).

Through their capacity, wetlands play a multifaceted role in carbon sequestration and climate regulation to store carbon, regulate greenhouse gases, mitigate extreme weather events, and support biodiversity, wetlands offer nature-based solutions that are essential for maintaining a stable climate and building resilience to environmental challenges.

2.5 Conservation and Management

Conservation and management are essential for the preservation and sustainable use of wetlands, crucial ecosystems that offer various ecosystem services. Wetlands are known for regulating water flow, enhancing water quality, supporting biodiversity, and providing habitats for diverse species. Effective conservation and management strategies are vital to maintain wetlands' functionality and the services they offer to the environment and society. A key aspect of wetland conservation and management involves restoring degraded wetlands. Wetland restoration often includes reintroducing hydrophytes, plants adapted to aquatic environments. Rodrigo (2021) emphasizes the significance of wetland restoration with hydrophytes, stating that prioritizing wetland restoration over creating artificial wetlands is crucial for maintaining wetlands' ecological functions. By restoring wetlands with native hydrophytes, biodiversity can be enhanced, water quality improved, and habitats provided for various species, contributing to the ecosystem's overall health.

In addition to restoration efforts, wetland conservation entails managing anthropogenic activities that may impact these ecosystems. Wallenius et al. (2021) discussed anthropogenic and environmental constraints on the microbial methane cycle in coastal sediments, underscoring concerns about eutrophication from human activities in coastal regions. Eutrophication can adversely affect wetlands by altering nutrient levels, impacting microbial communities and overall ecosystem health. Effective management practices are crucial to mitigate eutrophication's impacts and other human-induced stressors on wetlands.

Moreover, wetlands face external pressures like climate change, which can influence their ecological functions and resilience. Bikomeye et al. (2021) discussed the positive effects of climate change mitigation and adaptation on human health, stressing the importance of safeguarding and enhancing greenhouse gas sinks and reservoirs, including wetlands. They play a vital role in carbon sequestration and climate change mitigation, making their conservation critical for environmental and human health. Sustainable management practices and conservation efforts can enable wetlands to continue providing valuable ecosystem services and aiding in climate change mitigation.

Another crucial aspect of wetland conservation and management is the use of nature-based solutions to enhance wetlands' resilience to natural hazards. Kumar et al. (2021) evaluated the efficiency of nature-based solutions against natural hazards, highlighting wetlands' role as natural buffers against events like storm surges. Wetlands act as natural flood barriers, absorbing excess water and reducing flooding impacts on surrounding areas. By integrating wetlands into hazard mitigation strategies, ecosystems' and communities' resilience to natural disasters can be improved, underscoring wetlands' importance in disaster risk reduction.

Furthermore, managing wetlands involves addressing specific conservation challenges related to wildlife and habitat preservation. Sousa et al. (2021) explored the role of anthropogenic habitats in freshwater mussel conservation, noting how certain habitats can act as ecological traps for populations. Wetlands offer crucial habitats for various species, including freshwater mussels sensitive to water quality changes and habitat degradation. Effective management practices considering these species' needs are vital for ensuring their survival and the overall health of wetland ecosystems. Conservation and management are pivotal functions for wetlands, crucial for preserving these valuable ecosystems and the services they provide. By implementing restoration efforts, addressing anthropogenic impacts, mitigating climate change effects, incorporating nature-based solutions, and managing wildlife and habitats effectively, the long-term sustainability of wetlands and their contributions to the environment and society can be ensured.

2.5.1 Socio-economic Benefit

Wetlands play a crucial role in providing various socio-economic benefits to communities around the world. The socio-economic value of wetlands, particularly mangroves, has been extensively studied in the context of flood risk management (Kumar et al. 2021). These ecosystems act as natural barriers, reducing the impact of storm surges and protecting coastal areas from flooding, thereby saving lives and preventing property damage. The economic value of wetlands in mitigating natural hazards is significant, as they provide cost-effective solutions compared to traditional infrastructure-based approaches. Moreover, wetlands contribute to carbon sequestration and help mitigate greenhouse gas emissions, which is essential for addressing climate change and its associated socio-economic impacts (Tanneberger et al. 2022). Despite the challenges in implementing measures for greenhouse gas emissions mitigation, the co-benefits of such actions, including those related to wetlands, are crucial for sustainable development. Integrating wetlands into carbon markets can provide additional incentives for their conservation and restoration, leading to enhanced socio-economic outcomes. In addition to their environmental benefits, wetlands also have a direct impact on human well-being and socio-economic status. Research has shown that childhood socio-economic status is a strong predictor of lifelong well-being (Tooley et al. 2021). Access to wetland resources can enhance livelihoods, improve food security, and provide opportunities for economic development in local communities. By supporting various economic activities such as fishing, agriculture, and tourism, wetlands contribute to poverty alleviation and sustainable growth. Furthermore, wetlands have implications for public health and healthcare systems, which are integral components of socio-economic development. Effective wetland management can help prevent water-borne diseases, improve sanitation, and enhance overall community health (Mohanty et al. 2024). By preserving wetlands, societies can reduce the burden of healthcare costs associated with treating water-related illnesses, thereby freeing up resources for other essential services. The socio-economic implications of wetlands extend beyond local communities to a global scale, especially concerning industries like crude oil exploration. Understanding the environmental and socio-economic impacts of activities such as oil drilling in wetland areas is crucial for ensuring sustainable practices and minimizing negative consequences on affected populations. Implementing mitigation strategies based on ethical and sustainable principles can help balance economic interests with environmental conservation. Cost-benefit analyses are essential tools for evaluating the socio-economic value of wetlands and guiding decision-making processes related to their conservation and management (García-Herrero et al. 2022). By quantifying the economic benefits derived from wetland ecosystem services, policymakers can make informed choices that prioritize the long-term sustainability of these valuable habitats. Investing in wetland conservation not only preserves biodiversity but also generates tangible economic returns through ecosystem services. In the context of social protection, interventions aimed at enhancing income and employability through programs like livelihood and skills development can significantly impact socio-economic outcomes. Also incorporating wetlands into social protection strategies, communities can diversify their income sources, build resilience against economic shocks, and improve overall well-being. Promoting sustainable wetland use can create opportunities for sustainable livelihoods and poverty reduction. Wetlands serves as vital ecosystems that offer a wide range of socio-economic benefits to society. From flood risk mitigation and carbon sequestration to livelihood support and public health improvement, the value of wetlands extends across various sectors and disciplines. By recognizing and harnessing the socio-economic functions of wetlands, policymakers, researchers, and communities can work together to ensure the sustainable management and conservation of these critical habitats for the well-being of present and future generations.

2.6 Conclusion

Wetlands are invaluable ecosystems that offer a myriad of ecological, economic, and social benefits. Their ecological importance stems from their ability to support a high level of biodiversity, improve water quality, regulate floods, sequester carbon, and facilitate nutrient cycling. As natural habitats, wetlands provide essential resources and breeding grounds for a wide array of plant and animal species, including many that are endangered or migratory. By acting as natural filters, wetlands trap sediments and absorb pollutants, thereby enhancing water quality and protecting downstream ecosystems.

The flood control functions of wetlands cannot be overstated. They act as natural sponges, absorbing excess water during heavy rains and slowly releasing it, thus mitigating flood risks and protecting infrastructure and communities. Additionally, wetlands play a critical role in climate regulation through carbon sequestration, particularly in peatlands where large amounts of carbon are stored. Given their extensive benefits, the sustainable management of wetlands is crucial. Wetlands face numerous threats, including land conversion for agriculture and urban development, pollution, and climate change. Sustainable management practices are essential to mitigate these threats and ensure the continued provision of wetland services. This includes implementing effective conservation policies, restoring degraded wetlands, and promoting practices that reduce pollution and land-use change impacts. The future outlook for wetland conservation is one of cautious optimism. International treaties, national regulations, and local conservation efforts are increasingly recognizing the importance of wetlands and working towards their protection. Advances in ecological research and technology also provide new tools for wetland restoration and management. However, achieving sustainable wetland conservation will require continued commitment from governments, communities, and individuals to balance development needs with environmental stewardship. They are vital for maintaining ecological balance, supporting biodiversity, and providing essential services that benefit both nature and human societies. Sustainable management practices are necessary to protect these valuable ecosystems from ongoing threats. With concerted conservation efforts and a forward-looking approach, we can ensure that wetlands continue to thrive and deliver their invaluable benefits for generations to come.

References

Aguirre W, Alvarez-Mieles G, Anaguano-Yancha F, Burgos-Morán R, Cucalón R, Escobar-Camacho D et al (2021) Conservation threats and future prospects for the freshwater fishes of Ecuador: a hotspot of neotropical fish diversity. J Fish Biol 99(4):1158–1189. https://doi.org/10.1111/jfb.14844

Aransiola SA, Atta HI, Maddela NR (eds) (2024) Soil microbiome in green technology sustainability. Springer. https://link.springer.com/book/9783031718434

Asibor J, Clough P, Nabavi S, Manović V (2021) Assessment of optimal conditions for the performance of greenhouse gas removal methods. J Environ Manag 294:113039. https://doi.org/10.1016/j.jenvman.2021.113039

Bikomeye JC, Rublee CS, Beyer KM (2021) Positive externalities of climate change mitigation and adaptation for human health: a review and conceptual framework for public health research. Int J Environ Res Public Health 18(5):2481

Boland CR, Burwell BO (2022) Modelling habitat quality of high conservation priority vertebrates in an arid ecosystem with limited baseline data. Asian J Conserv Biol 11(1):106–131

Cao Y, Mi X, Li X, Wang B (2021) Defect engineering in metal-organic frameworks as futuristic options for purification of pollutants in an aqueous environment. Front Chem 9:673738

Choi C, Berry P, Smith A (2021) The climate benefits, co-benefits, and trade-offs of green infrastructure: a systematic literature review. J Environ Manag 291:112583. https://doi.org/10.1016/j.jenvman.2021.112583

Deacon C, Samways M (2021) A review of the impacts and opportunities for african urban dragonflies. Insects 12(3):190. https://doi.org/10.3390/insects12030190

Dervash MA, Yousuf A, Bhat MA, Ozturk M (2024) Bioindicators: the eco-sensors for detecting soil pollution. In: Soil organisms: deciphering the life beneath our feet. Springer Nature Switzerland, Cham, pp 47–62

Dutta S, Let S, Sharma S, Mahato D, Ghosh SK (2021) Recognition and sequestration of toxic inorganic water pollutants with hydrolytically stable metal-organic frameworks. Chem Rec 21(7):1666–1680

Filho W, Azeiteiro U, Balogun A, Setti A, Mucova S, Ayal D et al (2021) The influence of ecosystems services depletion to climate change adaptation efforts in Africa. Sci Total Environ 779:146414. https://doi.org/10.1016/j.scitotenv.2021.146414

García-Herrero L, Lavrnić S, Guerrieri V, Toscano A, Milani M, Cirelli GL, Vittuari M (2022) Cost-benefit of green infrastructures for water management: a sustainability assessment of full-scale constructed wetlands in Northern and Southern Italy. Ecol Eng 185:106797

Gimba UA, Aransiola SA, Oyewole OA, Maddela NR (2024) Marine greens: roles in climate change and global warming mitigations. In: Aransiola SA, Oyewole OA, Maddela NR, Ijah UJJ, Manjunatha B (eds) Marine greens: environmental, agricultural, industrial and biomedical applications. CRC Press. ISBN: 9781032438856. https://www.routledge.com/Marine-Greens-Environmental-Agricultural-Industrial-and-Biomedical-Applications/Aransiola-Oyewole-Maddela-Josiah-Ijah-Manjunatha/p/book/9781032438856

Hong Z, Ding S, Zhao Q, Qiu P, Chang J, Peng L et al (2021) Plant trait-environment trends and their conservation implications for riparian wetlands in the Yellow River. Sci Total Environ 767:144867

Huo ZY, Lee DM, Kim YJ, Kim SW (2021) Solar-induced hybrid energy harvesters for advanced oxidation water treatment. IScience 24(7):102808

Igwaran A, Kayode AJ, Moloantoa KM, Khetsha ZP, Unuofin JO (2024) Cyanobacteria harmful algae blooms: causes, impacts, and risk management. Water Air Soil Pollut 235(1):71

Kumar P, Debele S, Sahani J, Rawat N, Martí-Cardona B, Alfieri S et al (2021) Nature-based solutions efficiency evaluation against natural hazards: modelling methods, advantages and limitations. Sci Total Environ 784:147058. https://doi.org/10.1016/j.scitotenv.2021.147058

Kuuluvainen T, Nummi P (2023) Strategies for the ecological restoration of the boreal forest facing climate change. In: Boreal forests in the face of climate change: sustainable management. Springer International Publishing, Cham, pp 443–466

Lin B, Ossola A, Alberti M, Andersson E, Bai X, Dobbs C et al (2021) Integrating solutions to adapt cities for climate change. Lancet Planet Health 5(7):e479–e486. https://doi.org/10.1016/s2542-5196(21)00135-2

Malyan S, Yadav S, Sonkar V, Goyal V, Singh O, Singh R (2021) Mechanistic understanding of the pollutant removal and transformation processes in the constructed wetland system. Water Environ Res 93(10):1882–1909. https://doi.org/10.1002/wer.1599

Mao B, Wang X, Liao Z, Miao Y, Yan S (2023) Spatiotemporal variations and tradeoff-synergy relations of ecosystem services under ecological water replenishment in Baiyangdian Lake, North China. J Environ Manag 343:118229

Martínez-Espinosa C, Sauvage S, Bitar A, Green P, Vörösmarty C, Sánchez-Pérez J (2021) Denitrification in wetlands: a review towards a quantification at global scale. Sci Total Environ 754:142398. https://doi.org/10.1016/j.scitotenv.2020.142398

Mohanty S, Pandey PC, Pandey M, Srivastava PK, Dwivedi CS (2024) Wetlands contribution and linkage to support SDGs, its indicators and targets-a critical review. Sustain Dev. https://doi.org/10.1002/sd.2948

Mudhoo A, Mohan D, Pittman CU Jr, Sharma G, Sillanpää M (2021) Adsorbents for real-scale water remediation: gaps and the road forward. J Environ Chem Eng 9(4):105380

Nadarajah K, Rahman N (2021) Plant–microbe interaction: aboveground to belowground, from the good to the bad. Int J Mol Sci 22(19):10388. https://doi.org/10.3390/ijms221910388

Nicholson E, Watermeyer KE, Rowland JA, Sato CF, Stevenson SL, Andrade A et al (2021) Scientific foundations for an ecosystem goal, milestones and indicators for the post-2020 global biodiversity framework. Nat Ecol Evol 5(10):1338–1349

Ostrowski A, Connolly R, Sievers M (2021) Evaluating multiple stressor research in coastal wetlands: a systematic review. Mar Environ Res 164:105239. https://doi.org/10.1016/j.marenvres.2020.105239

Peydayesh M, Mezzenga R (2021) Protein nanofibrils for next generation sustainable water purification. Nat Commun 12(1):3248

Retter A, Karwautz C, Griebler C (2021) Groundwater microbial communities in times of climate change. Curr Issues Mol Biol:509–538. https://doi.org/10.21775/cimb.041.509

Richardson M, Soloviev M (2021) The Urban River Syndrome: achieving sustainability against a backdrop of accelerating change. Int J Environ Res Public Health 18(12):6406

Rodrigo M (2021) Wetland restoration with hydrophytes: a review. Plants 10(6):1035. https://doi.org/10.3390/plants10061035

Sagi N, Hawlena D (2021) Arthropods as the engine of nutrient cycling in arid ecosystems. Insects 12(8):726. https://doi.org/10.3390/insects12080726

Schapheer C, Pellens R, Scherson R (2021) Arthropod-microbiota integration: its importance for ecosystem conservation. Front Microbiol 12. https://doi.org/10.3389/fmicb.2021.702763

Shahzad A, Oh JM, Azam M, Iqbal J, Hussain S, Miran W, Rasool K (2021) Advances in the synthesis and application of anti-fouling membranes using two-dimensional nanomaterials. Membranes 11(8):605

Shayo GM, Elimbinzi E, Shao GN (2024) Water-based technologies for improving water quality at the point of use: a review. Wiley Interdiscip Rev Nanomed Nanobiotechnol 16(1):e1940

Sousa R, Halabowski D, Labecka AM, Douda K, Aksenova O, Bespalaya Y et al (2021) The role of anthropogenic habitats in freshwater mussel conservation. Glob Chang Biol 27(11):2298–2314

Tanneberger F, Birr F, Couwenberg J, Kaiser M, Luthardt V, Nerger M et al (2022) Saving soil carbon, greenhouse gas emissions, biodiversity and the economy: paludiculture as sustainable land use option in German fen peatlands. Reg Environ Chang 22(2):69

Thakre KG, Barai DP, Bhanvase BA (2021) A review of graphene-TiO2 and graphene-ZnO nanocomposite photocatalysts for wastewater treatment. Water Environ Res 93(11):2414–2460

Tooley UA, Bassett DS, Mackey AP (2021) Environmental influences on the pace of brain development. Nat Rev Neurosci 22(6):372–384

Trebilco R, Fleming A, Hobday A, Melbourne-Thomas J, Meyer A, McDonald J et al (2021) Warming world, changing ocean: mitigation and adaptation to support resilient marine systems. Rev Fish Biol Fish 32(1):39–63. https://doi.org/10.1007/s11160-021-09678-4

Trisos C, Auerbach J, Katti M (2021) Decoloniality and anti-oppressive practices for a more ethical ecology. Nat Ecol Evol 5(9):1205–1212. https://doi.org/10.1038/s41559-021-01460-w

Uddin MM, Zakeel MCM, Zavahir JS, Marikar FM, Jahan I (2021) Heavy metal accumulation in rice and aquatic plants used as human food: a general review. Toxics 9(12):360

Verma S, Sahu P, Kumar K, Pal G, Gond S, Kharwar R et al (2021) Endophyte roles in nutrient acquisition, root system architecture development and oxidative stress tolerance. J Appl Microbiol 131(5):2161–2177. https://doi.org/10.1111/jam.15111

Wallenius AJ, Dalcin Martins P, Slomp CP, Jetten MS (2021) Anthropogenic and environmental constraints on the microbial methane cycle in coastal sediments. Front Microbiol 12:631621

Walton CR, Zak D, Audet J, Petersen RJ, Lange J, Oehmke C et al (2020) Wetland buffer zones for nitrogen and phosphorus retention: impacts of soil type, hydrology and vegetation. Sci Total Environ 727:138709

Wang X, Guo Z, Hu Z, Zhang J (2020) Recent advances in biochar application for water and wastewater treatment: a review. PeerJ 8:e9164

Yang C, Wang X (2021) Lysosome biogenesis: regulation and functions. J Cell Biol 220(6). https://doi.org/10.1083/jcb.202102001

Yeszhanov AB, Korolkov IV, Dosmagambetova SS, Zdorovets MV, Güven O (2021) Recent progress in the membrane distillation and impact of track-etched membranes. Polymers 13(15):2520

Zou X, Ji Y, Li H, Wang Z, Shi L, Zhang S et al (2021) Recent advances of environmental pollutants detection via paper-based sensing strategy. Luminescence 36(8):1818–1836

Chapter 3
Human Interactions with Wetlands: Historical Perspectives and Cultural Significance

Joshua Oluwatobi Akinola, Miracle Uwa Livinus, Musa Ojeba Innocent, Sunday Zeal Bala, Mustapha Abdulsalam, Abdulrahman Itopa Suleiman, Madinat Hassan, and Katimu Yusuf

Abstract This chapter delved into the historical and cultural significance of wetlands, highlighting their crucial role in human development and societal evolution. Wetlands have served as vital resources throughout history, from ancient civilizations like Mesopotamia and Egypt that relied on their floodplains for agriculture, to indigenous cultures that revered these landscapes for their spiritual and practical benefits. The Venice Lagoon, for instance, exemplifies how wetlands have been integral to urban development, defence strategies, and economic prosperity, showcasing the strategic and commercial advantages they offered to the Venetian Republic. Similarly, the Everglades illustrate the evolution of human interactions with wetlands, from the sustainable stewardship of Native American tribes to the modern

J. O. Akinola (✉)
Department of Agriculture and Natural Sciences, University of Maryland, College Park, MD, USA
e-mail: aoluwato@umd.edu

M. U. Livinus
Department of Biochemistry, School of Sciences and Information Technology, Skyline University, Kano, Nigeria

M. O. Innocent · M. Abdulsalam
Department of Microbiology, School of Sciences and Information Technology, Skyline University, Kano, Nigeria

S. Z. Bala
Department of Education, University of Maryland Eastern Shore, Princess Anne, MD, USA

A. I. Suleiman
Department of Science Laboratory Technology, Kogi State Polytechnic, Lokoja, Nigeria

M. Hassan
Foresight Institute of Research and Translation (FIRAT), Kigali, Rwanda

K. Yusuf
Department of Microbiology, Federal University Gashua, Gashua, Yobe State, Nigeria

B. R. Babaniyi et al. (eds.), *Wetland Ecosystems: Conservation Strategies, Policy Management and Applications*, Wetlands: Ecology, Conservation and Management 12, https://doi.org/10.1007/978-3-031-91982-4_3

conservation efforts aimed at addressing the impacts of historical development and environmental change. By examining these case studies, the chapter underscored the diverse ways in which wetlands have been valued and utilized, providing valuable insights into their historical importance and ongoing relevance. The discussion highlighted the need for incorporating historical and cultural perspectives into contemporary conservation strategies, ensuring that the ecological functions and cultural heritage of wetlands are preserved. Recognizing the multifaceted roles wetlands have played throughout history not only enriches our understanding of these ecosystems but also informs future approaches to their management. This comprehensive perspective is essential for developing sustainable practices that balance ecological integrity with cultural respect, ensuring that wetlands continue to provide vital benefits to both human societies and the natural environment for generations to come.

Keywords Wetlands · Historical significance · Venice Lagoon · Everglades · Conservation · Cultural heritage

3.1 Introduction

Maritime regions of which marshes, swamps, bogs, and floodplains are parts have historically shaped human history and civilization. They were also used for securely obtaining necessary resources within worship and have their significance in promoting the development of commerce (Bănăduc et al. 2022). Historically, wetlands started from the usage by early civilizations for agriculture and trade, up to the modern world that is trying to conserve the wetlands as much as possible. This dynamic interaction brings out the significance of wetlands in the formation of societies and their use in the current and past societies for both ecological and cultural aspects (Verhoeven and Setter 2009; Li et al. 2021).

In the past, wetlands have been an essential part of human lives and early societies' existence. Egyptians and Mesopotamians for instance took advantage of the wetland soils which are referred to as alluvial to practice advanced farming. The Nile Delta wetland systems, for instance, have been known to support the ancient Egyptian civilization by providing fertile land required for growing crops such as wheat and barley (Foster 2019). Likewise, Mesopotamian culture also recognized the value of wetlands and both Tigris and Euphrates River flood plains were used for agriculture and urbanization. Across the African continent, wetlands have been central to the survival of so many communities. The Okavango Delta in Botswana is one of the largest inland deltas that has been critical for the BaSarwa people, who rely on its resources for fishing, hunting and agriculture (Mendelsohn et al. 2012). For example, the Sine-Saloum Delta in Senegal is a habitat for diverse species and residents source fish and salt among other materials from this and other wetlands (Boye et al. 2020).

On the other hand, traditional societies across the globe have always held wetlands in high esteem and had quite several spiritual and cultural reverence for them. Particularly, the Cree and the Ojibwe tribes that inhabited the Canadian and the United States territories recognized the utilization of wetlands as a key component of their existence since the water area helped those fish and hunt (Harris 2021). Australian Aboriginal communities have therefore lived in and around wetlands, with an appreciation of a primaeval spiritual connection to these systems (Burgess et al. 2020). These examples describe an impressive level of reverence for wetlands as givers of life and repositories of culture.

European societies started during the medieval period to alter wetlands for land reclamation, and agriculture expansion. An instance is the conversion of wetlands to farmland, which highlighted a change in the utilization of the land intensively (Pereira 2018). However, some of the European regions still saw the need and importance of wetlands for aesthetic and other benefits, and went ahead and included it in gardening and landscape beauty. In Asia, wetlands were fundamental to agriculture, especially rice, in China and Japan where advanced systems of water control turned wetlands into some of the best rice-producing tracts of land (Yoshida 2017). Such practices show how the people valued the wetland resources for the support of an agricultural-based economy as well as food stability. The discovery of the history and culture of wetlands reveals the social relations between man and the environment about wetlands. This chapter will provide more concrete information about these interactions, or how wetlands have been viewed and used both in time and space by different human societies.

3.2 Historical Overview

3.2.1 Early Human Settlements

Based on the analysis of archaeological evidence and anthropological data, wetlands throughout history played the role of resources for early human settlements and contributed to the development of agriculture. The best example of such systems is that of wetlands in Mesopotamia and Egypt, two of the earliest civilizations. Mesopotamia was defined by two rivers—Tigris and Euphrates—whose floods produced rich alluvial soils that defined early agricultural civilizations. These wetlands that were formed from floods contributed in making the soil fertility, a factor that was of essence in the growth of crops such as barley and wheat. The area was inhabited by the Sumerians around 4500 BC and they needed to control the flood waters which came down from the uplands. The wetland people engineered canals and dikes to regulate the movement of water and as a result, they were able to turn the wetlands into favorable ground for farming. This early engineering did much more than promote the agricultural surplus, but also had an impact on the promotion of scale, density and complexity of the populations (Jacobsen 2019).

In the same manner, in the old world of ancient Egypt, wetland habitats common in the Nile Delta basin contributed to agricultural productivity. The flooding of the Nile River was done annually and the depositing of silt on the delta provided the area with nutrients needed for farming. This was the annual flood which was instrumental in Egypt's prosperity and agricultural practice, being controlled by a well-organized series of canals and irrigation systems. It was because of the Nile River that the subsequent agriculture enabled the cultivation of a stable economy which helped in the creation of one of the ancient world's most advanced civilizations (Winston 2020). The above aspects in combination with the wetlands of the Nile Delta and also the fish resources and hunting supported the complexity of the societal systems and the economics of the society.

3.2.2 Indigenous Cultures

For centuries now, people of indigenous societies have had a rich understanding of wetlands and their value to the environment and society. These cultures have not only adapted to the wetlands to depend on them economically, socially, and religiously but have also incorporated the wetlands into the beliefs and myths of their society (Obiero et al. 2023).

Some cultures from North America, for example, the indigenous tribes, have been known to depend on wetlands for their sustenance and other uses. For instance, the Cree and Ojibwe peoples, who belong to the areas of today's Canada and the North of the United States, have long relied on wetlands for food and materials for chic hunting and fishing grounds. These are the wetlands that consist of bogs, swamps and marshes and they act as a source of important necessities like fish, waterfowl and plant products. The Cree and Ojibwe people have acquired certain indigenous knowledge and wisdom about the ways of utilizing and conserving these wetland resources. Fishing practices and migratory cycles for them entail a lot of knowledge of wetland habitats and other factors (Harris 2021). Also, the native moist area is integrated into the cultural and spiritual relations of many Native American tribes. These places are normally related to creation myths and spiritual practices making them seen as spiritual sanctuaries. For example, from the cultural viewpoint, people of some tribes think that wetlands are home to spirits or spirits of their ancestors and different ceremonies are carried out in these places to pray to these spirits (Harris 2021).

The aboriginal cultures, with the example of West Australia's Eglinton wetland, have also acknowledged the significance of wetlands. This expanse is also characterized by wetlands, of which there are billabongs, swamps, and floodplains in Australia. The wetlands that occur in the country are considered sacred by the aboriginal people including the Yolngu and the Noongar because they depend on them for fishing and collection of plant resources. The Dreamtime of the Aboriginal people enshrines that the wetlands are the holy grounds for these people. These stories and practices demonstrate how Aboriginal peoples perceive wetlands as much more than ecological resources that are exploitable for food but also valuable cultural and religious artefacts (Burgess et al. 2020).

Aboriginal people's uses of wetlands are built around an understanding that requires the protection of these places. There are records of people's utilization of wetland resources, based on traditional ecological knowledge of how to sustainably use the wetlands, such as the practice of burning the vegetation to improve habitat conditions for fish and other animals. This knowledge is an indication of an appreciation of wetland ecosystems and the need to preserve them for the sake of future generations (Burgess et al. 2020).

Finally, looking at the chronology of the human interactions with wetlands it is possible to argue that these areas have always been significant for the early civilizations serving as the grounds for various forms of activities connected with agriculture and economy, as well as involving influences on cultural practices. In the great states of Mesopotamia and Egypt, intricate aquaculture systems have been laid down, with native Americans and Australians using wetlands effectively for management. These historical points of view are quite informative in appreciating the usefulness of wetlands in present-day society and more so the need for preserving and maintaining the wetlands.

3.3 Medieval and Renaissance Periods

3.3.1 European Practices

During the medieval and Renaissance periods, wetlands in Europe were subjected to significant changes due to evolving human needs and technological advancements. These changes reflect a broader shift in how societies interacted with and managed wetland environments.

1. ***Draining for Agriculture and Land Reclamation***

   It was common practice, at least starting from the medieval period in Europe, to reclaim wetlands by draining them for farming. This practice was deemed necessary to enhance the extent of accessible land for cultivation, and, therefore, enhance agricultural production. Wetland reclamation required digging dikes, canals, and drainage systems and was a time-consuming process. This brought about changes to land use whereby large areas came to be devoted to agriculture; this was essential in feeding the growing populations and cities that were being formed (Pereira 2018).

   Dikes and polders are among the key examples of wetlands reclamation, and the Netherlands has been one of the countries with the most elaborate systems of the above. The Dutch started gaining the land from water as early as the thirteenth century whereby vast swamps that were uninhabitable were converted into good agricultural land. It enhanced agricultural yields while at the same time showing an advanced level of comprehension of water flow patterns in the landscape (Middelkoop and Kwast 2018). The subsequent development of Dutch water control and management techniques also persisted as the people struggled to conserve and even develop more 'reclaimed' lands as well as combating problems that are related to problems of silting, subsidence and, more recently, rising sea levels.

In England, similar to the continental area, wetlands were also reclaimed in the medieval age. The existing geographical area includes the Fens in eastern England, a huge region of marshland, where some of the land was reclaimed for farming. This process included drainage treatment; the construction of trenches and ways of blockage. The transformation of the Fens into agricultural fields brought changes with repercussions to the geographical and natural geography of the area (Smith 2020a, b).

2. ***Value of Wetlands for Natural Resources and Aesthetics***

   However, several European societies still perceived these wetlands as valuable resources and as important features in the landscape. However, many of the garden or estate designs in medieval and Renaissance Europe embraced the wetlands. Wetland was positively used in landscape design, indicating people's recognition of the aesthetic value of wetlands and the functions of wetland ecosystems. For instance, in the Renaissance, wealthy landlords and nobility included wetlands in their conduct as gardened features. Some of these wetlands were particularly developed for the beautification of gardens together with the purpose of avian habitat of waterfowl. The incorporation of wetlands into estate planning also showed appreciation of the services they offered in what is now often referred to as 'ecosystem services' such as water purification and flood control (Klein 2019).

   It is possible to name such an example as the gardens at Versailles in France; I should remark that the water such as ponds and canals is an inalienable part of the garden designs. Although these features served primarily an aesthetic function, they also helped contribute to the ecological differentiation of the estate. That such landscapes served as both an ornamental and a working aspect reveals that there was indeed a complicated interaction between wetlands and people (Miller 2017).

3.3.2 Asian Influences

In Asia, wetlands played a crucial role in agricultural development, particularly in the cultivation of rice. The use of wetlands for rice paddies in countries like China and Japan highlights a deep understanding of wetland ecosystems and their potential for sustainable agriculture.

Rice Cultivation in China

Rice cultivation in China from wetlands was a practice that was begun in antiquity, although the precise time is unclear. Wetland use for rice farming required the development of extensive systems of canals to regulate water in the fields. These systems were very important in supporting rice cultivation which was a central crop used in supporting the developmental Chinese civilization. Techniques utilized in Social Engineering Terracing are well demonstrated, especially in the Yangtze River Delta where the terraced rice paddies can be found. All these terraces were intended to allow effective exploitation of the water, and at the same time, prevent soil erosion. Wetland use for rice cultivation helped in the emergence of complex agricultural societies and fed a massive population (Yang and Zhang 2018).

Rice Cultivation in Japan
In Japan, there was the utilization of wetlands in agriculture especially in the growing of rice crops. Specifically, they cultivated rice using a system called 'paddy farming' which sought to produce rice by establishing a flooded environment for the purpose. This method of cultivation was suitable for the climatic conditions of the country and the landscape. The development of paddy fields which are artificial wetlands in Japan showed a broad perception of wetland systems. Farmers maintained water, flood and related physical factors and improved the fertility of the soil in the course of using appropriate old style to cultivate paddy. This reveals the fact that the integration of Wetlands into the agricultural systems helped Japan to transform economically and culturally (Saito 2019).

Finally, the Environmental Utilization Assessment during the Medieval and Renaissance epochs provides an understanding of the European and Asian ways of operations and utilization of wetlands. European wetlands were confirmed for agricultural purposes by peat extraction, but, at the same time, the recreational and amenity value of wetlands were recognized. In Asia, the wetlands were used for rice production showing a higher level of knowledge regarding life in wetland regions. These practices enable one to understand instances where the values of wetlands and their utilization have been dynamic with human requirements or societal cultures over time.

3.4 Cultural Significance from Spiritual and Religious Perspectives

Wetlands have long held a sacred place in the spiritual and religious traditions of various cultures. These landscapes are often imbued with symbolic meanings and are revered as sites of divine presence and purification.

3.4.1 Sacred Spaces

In Hinduism, wetlands are often referred to in context with divine beings and are believed to be a site to cleanse oneself. In Hindu tradition, the water sources are regarded as sacred and include the Ganges, the Yamuna, and the Sarasvati most of which are associated with wetlands and are essential elements in religious rites. These water bodies are considered as the direct form of deities or they have the blessings of deities and hence these water bodies have significance for pilgrimage and ritual activities (Narayan 2021).

In particular, the Ganges River is considered as the goddess Ganga. The river and the adjacent wetlands and floodplain are regarded as 'tapovana' or 'forest of the bo-tree' which is considered extremely holy and taking a bath in these waters is supposed to wash away sins and bless the soul. These performances in the form of offerings and ceremonies at the riverbanks maintain the sensitive perception of these wetlands in Hinduism. This conviction traces its roots to Hindu cosmology where water assumes the properties of a purificatory agent; wetlands in this context fall within a divine grace (Jain 2020).

In Islam, there are philosophical connotations associated with the wetlands. It again depicts rivers and very many trees in the context of the Qur'an may also be taken to mean the state of purification and plenty. They [wetlands] have to qualify [richness, Productiveness, Fertility] of water and living beings, which means they give reference to the paradisal quest and also as an illustration of one's purity. Islam has also heavily incorporated wetlands where they depict the heaven and the blessings from the supreme being (Siddiqui 2018).

3.4.2 Folklore and Mythology

Forests and swamps in particular are typical for folklore and mythological narrations of peoples of the world being attributed to spirits, gods and other supernatural creatures. All these stories capture the various cultures' respect for the wetlands as mysterious and sacred areas. In Celtic tradition bogs and marshes have their gods and spirits of the other world although these are now more anticipated to be Pagan. Simply, the bogs of Ireland and Scotland are believed to be such areas where the density of the otherworld is dense as to make the diffusion between human and otherworld realm quite easy. These wetlands have always been associated with spirits and fairies and to this effect, many of the advanced ancient rituals were performed in these landscapes either to appease these spirits or to gain their blessings. The fact that bogs are often rather mysterious, with their special haziness and secret, further depth also plays a big part in presenting them as magical, and sacred (MacCulloch 2019).

Similarly to the Siberian ones the Japanese folklore places spiritual values on wetlands as well. Some of the spirits are believed to inhabit particularly 'yokai' swamp or pond habitats, hence the name 'yokai'. For instance, the "kappa"—a water deity—is believed to dwell in rivers and swamps. These are depicted as beings that are more often than not, either causing all sorts of havoc or are generally bad, this is due to the traditional wariness/worship of wetland areas. These beings' legends focus on wetlands as spaces that are considered culturally significant and liminal between the mundane and the otherworldly (Nakamura 2021).

Myths and legends of the people of Africa also contain references to the wetlands. For example, the Akan people of Ghana have myths and legends that talk of 'Mmoatia,' which are water spirits, inhabiting swampy and riverine areas. These spirits are much regarded as having potencies to control the forces of nature and aspects of human life. Sacrifices are sometimes carried out to appease these spirits in other to maintain a symbiotic relationship between human society and the wetland ecosystem. The respect to these water spirits therefore portrays the culture's recognition of wetlands as contact zones with the divine powers and more so, the natural world (Amoako 2020).

Similarly in South American indigenous cultures, wetlands play an important role in the religious cosmology. Among the tribes inhabiting the Amazon, floodplains and swamps are spaces that are regarded as dwelling places of the powerful spirits and gods. The Yagua people, for instance, consider the wetlands as zones where the otherworld can be found and they engage in prayer and other practices to appease the spirits which dwell in those areas. The general cultural practices of these wetlands centralize their importance as spiritual and social components of such societies (Lara 2019).

Thus, the aspect of the spirituality and culture of wetlands is evident in depicting the wetlands as religious and spiritual terrains and places where people meet with spirits in different civilizations. Hinduism, Islam, Celts, Japanese and African myths and cultures, as well as South American Indigenous peoples respect wetlands as inextricable links to the human story. These cultural views, therefore, bring out the philosophy that wetlands are more than just geographical structures, they are also sacred places of cultural and religious value.

3.5 Economic and Social Roles

3.5.1 Traditional Fishing Practices

Traditional fishing practices involve the use of wetlands, which different cultures have explored and developed ways and means of fishing that depend on the wetlands. In Southeast Asia, mangroves and their other related ecosystems have always formed prominent features of economic and cultural life. Southeast Asian mangroves are very important to traditional fisheries. People in countries such as Thailand, Indonesia and Malaysia depend on such wetlands as a source of food and income. Gill nets, fish traps and lines are some of the modern fishing methods that have adapted into Cameroon's mangrove ecosystem, which includes less affluent fishing techniques. There are, however, measures that are taken by the fishermen that can have fewer effects on the ecosystem; for instance, the fishermen employ some strategies that only operate during certain seasons and these can be in harmony with the reproductive seasons of the local fishes (Wong 2019).

Sustainable management of mangrove wetlands therefore involves the use of populations' oriented and homers approaches as recommended by the indigenous people. For instance, the practice of 'fish ponds' or 'tambaks' in Indonesia entails the construction of enclosed water bodies so that fish and shrimps can be farmed. This practice captures the concern for the ecological conditions that are prerequisites for sustaining productive wetlands that, at the same time, supply food and income.

3.5.2 Agricultural Practices in the Pacific Islands

In the Pacific Islands, wetlands play a crucial role in traditional agriculture. Taro cultivation is a prominent example of how wetlands have been used to sustain local communities. Taro, a root vegetable that thrives in wetland conditions, has been a staple crop in Polynesian cultures for centuries. Traditional agricultural systems, such as taro patches or "lo'i," utilize wetland areas to grow this important crop.

The cultivation of taro in wetlands involves creating terraced fields that are periodically flooded to support the growth of the plant. This practice not only demonstrates an intimate knowledge of wetland ecosystems but also highlights the economic and social importance of these environments. Taro cultivation supports food security and forms a central part of cultural practices and communal activities in many Pacific Island communities (Krauss 2018). In addition to taro, wetlands in the Pacific Islands have been used for growing other crops, such as yams and sweet potatoes. These agricultural practices reflect a harmonious relationship between people and the environment, where traditional knowledge is applied to manage and sustain wetland resources.

3.5.3 Modern Impacts

Transformation into Urban Areas and Agricultural Lands

The modern era has seen significant changes in wetland management, driven by urbanization, agricultural expansion, and industrial development. The transformation of wetlands into urban areas and agricultural lands has had profound social and economic impacts.

The reclamation of wetlands for urban development has been a common practice in many parts of the world. Cities such as Tokyo, New York, and Shanghai have expanded by filling in wetlands and coastal areas to accommodate growing populations and infrastructure needs. While this transformation has provided space for housing, industry, and transportation, it has also led to the loss of critical wetland ecosystems and their associated benefits. The conversion of wetlands into urban areas disrupts natural habitats, reduces biodiversity, and alters local hydrological cycles.

In agricultural contexts, the draining of wetlands for crop production has also had significant impacts. Historically, the expansion of agricultural lands has often involved the reclamation of wetlands, leading to increased food production but also the loss of valuable ecosystem services. Wetlands provide essential functions such as water filtration, flood regulation, and habitat for wildlife. Their transformation into agricultural lands can diminish these services, leading to issues such as increased flooding, water pollution, and loss of biodiversity (Dixon 2019).

Shifting Priorities and Technological Advancements
The management of wetlands has evolved, reflecting shifting priorities and technological advancements. In the past, wetlands were often viewed primarily as obstacles to development or sources of land for agriculture. However, contemporary perspectives recognize the value of wetlands for their ecological services and cultural significance.

Modern technology has enabled more sophisticated approaches to wetland management and conservation. Advances in remote sensing, GIS, and hydrological modelling allow for better monitoring and understanding of wetland ecosystems. These technologies support efforts to balance development with conservation, helping to identify areas where wetland restoration or protection can be integrated into land use planning.

The growing recognition of the value of wetlands has also led to the development of policies and regulations aimed at their protection. International agreements such as the Ramsar Convention on Wetlands highlight the importance of conserving wetland ecosystems and promote sustainable management practices. These efforts reflect a broader shift towards recognizing wetlands as vital components of the global environment and addressing the impacts of past and ongoing transformations (Meyer 2021).

3.6 Case Studies on Human Interactions with Wetlands

3.6.1 The Venice Lagoon

The Venice Lagoon, a complex and historically significant wetland system located in northeastern Italy, provides a rich case study of how wetlands can influence and shape urban development, defence strategies, and economic activities. The lagoon's management by the Venetian Republic is a testament to the strategic importance of wetlands in shaping the city's history and growth.

Historical Management and Urban Development
The Venice Lagoon is a shallow, brackish water body surrounded by a series of islands that form the city of Venice. Its formation dates back to the Roman era, but the city of Venice itself was established in the 5th century AD. The lagoon's unique

geographical setting, with its network of channels, mudflats, and salt marshes, played a crucial role in the city's development and survival.

One of the primary reasons for Venice's establishment in the lagoon was its natural defence capabilities. The lagoon's waters acted as a barrier against potential invaders, providing the city with a strategic advantage. The Venetian Republic, known for its naval prowess and commercial empire, utilized the lagoon's wetlands to fortify its defences. The city's canals were not only used for transportation and trade but also served as defensive moats. The intricate system of waterways allowed for the control of access to the city and provided early warning systems against naval attacks (Horsley 2021).

Trade and Commerce

The Venice Lagoon also played a central role in Venice's economic prosperity. The lagoon's shallow waters and extensive network of canals facilitated trade and commerce, making Venice a major maritime hub in the medieval and Renaissance periods. The city's strategic location allowed it to control key trade routes between the East and West, contributing to its wealth and influence.

Venetian merchants used the lagoon's natural features to their advantage, establishing a vibrant trading network that extended across the Mediterranean and beyond. The lagoon's wetlands provided resources such as fish, salt, and reeds, which were integral to the city's economy. Salt production, in particular, was a significant industry in Venice, and the lagoon's salt pans were crucial to this trade. The economic activities associated with the lagoon supported the city's growth and development, enabling it to become one of the most powerful maritime republics in history (Pritchard 2019).

Water Management

Effective water management was essential for Venice's survival and prosperity. The Venetian Republic invested heavily in engineering projects to control and manage the lagoon's waters. The construction of an elaborate system of canals, locks, and floodgates allowed the city to regulate water levels and protect against flooding. This engineering prowess was critical in maintaining the city's infrastructure and preventing the inundation of the low-lying areas (Luttwak 2018).

In addition to flood control, the Venetians developed innovative methods for dealing with the challenges posed by sedimentation and erosion. The lagoon's dynamic environment required ongoing maintenance and adaptation of water management practices. The Republic implemented measures such as dredging and the construction of protective barriers to address these issues and ensure the continued functionality of the city's canal system (Smith 2020a, b).

Conservation Challenges

Despite its historical significance, the Venice Lagoon faces ongoing challenges related to conservation and environmental degradation. The impacts of climate change, including rising sea levels and increased frequency of extreme weather events, threaten the delicate balance of the lagoon's ecosystem. Additionally, urbanization and industrial activities have contributed to pollution and habitat loss. Efforts to address these challenges include ongoing conservation projects and international collaborations aimed at preserving the lagoon's unique environment. The establishment of the Venice Lagoon's Protected Area and the implementation of sustainable management practices reflect a commitment to maintaining the ecological health of this historic wetland (Casellato 2022).

3.6.2 The Everglades

The Everglades, located in southern Florida, represent a key case study in the historical interactions between humans and wetlands. This expansive and unique wetland system has played a central role in the region's ecology, economy, and culture, with a history that spans from Native American stewardship to modern conservation efforts.

Native American Stewardship
Long before European colonization, the Everglades were inhabited and managed by Native American tribes such as the Seminole and Miccosukee. These tribes developed sophisticated practices for living in harmony with the wetland environment. They utilized the Everglades for various purposes, including fishing, hunting, and agriculture.

The Indigenous peoples of the Everglades understood the ecological significance of the wetlands and practised sustainable resource management. They used controlled burns to manage vegetation and create favourable conditions for hunting and gathering. The traditional knowledge and practices of these communities reflect a deep connection to the land and an understanding of the importance of maintaining ecological balance (Davis 2018).

European Settlement and Development
The arrival of European settlers in the nineteenth century marked a period of significant change for the Everglades. Efforts to drain and reclaim the wetlands for agriculture and urban development led to substantial modifications of the natural landscape. The construction of canals and levees aimed to control the flow of water and facilitate land use for farming and settlement.

One of the most notable projects was the creation of the Central and Southern Florida Flood Control Project, which began in the 1940s. This massive engineering undertaking sought to manage floodwaters and provide water for agricultural and urban areas. However, these interventions had unintended consequences, including disruptions to the natural flow of water and impact on the Everglades' ecological health (Gunderson 2021).

Conservation Efforts
In response to the ecological degradation caused by development, conservation efforts have been initiated to restore and protect the Everglades. The establishment of Everglades National Park in 1947 marked a significant step towards preserving

the wetland environment. The park, which is now a UNESCO World Heritage Site, encompasses a large portion of the Everglades and serves as a critical area for conservation and research (Keddy 2019).

Restoration projects, such as the Comprehensive Everglades Restoration Plan (CERP), aim to address the impacts of historical development and return the natural flow of water to the wetlands. These efforts involve complex collaborations between government agencies, environmental organizations, and local communities. The goal is to restore the ecological health of the Everglades and ensure the continued provision of essential ecosystem services, such as water filtration and habitat for wildlife (Toth 2020).

Cultural and Economic Importance

The Everglades hold cultural and economic significance for the region. The wetlands are home to diverse plant and animal species, including endangered species such as the Florida panther and the West Indian manatee. The Everglades also support recreational activities such as birdwatching, fishing, and eco-tourism, contributing to the local economy and providing opportunities for environmental education.

The cultural heritage of the Everglades is reflected in the traditions and stories of the Native American tribes, as well as in the broader cultural identity of the region. The preservation of the Everglades is not only important for its ecological value but also for its role in maintaining cultural connections and supporting local livelihoods (Kassas 2022).

3.7 Contemporary Issues and Challenges

3.7.1 Environmental Degradation

Environmental degradation is a critical issue affecting wetlands globally, resulting from a combination of human activities and natural processes that diminish their ecological health. Wetlands are among the most productive ecosystems, providing essential services such as water filtration, flood regulation, and habitat for diverse species. However, various forms of environmental degradation are compromising these functions.

Pollution is a major factor in wetland degradation. Agricultural runoff, containing pesticides and fertilizers, often leads to nutrient enrichment in wetlands, causing eutrophication. This process results in excessive algae growth, which depletes oxygen in the water and creates dead zones that are inhospitable to many aquatic species. In the United States, for example, the Mississippi River Basin's runoff has severely impacted wetlands in the Gulf of Mexico, leading to the formation of large hypoxic zones. Industrial discharges and urban wastewater also contribute to contamination, introducing heavy metals and toxic chemicals that can harm both wildlife and human health (Diaz and Rosenberg 2008).

Habitat destruction due to land reclamation and drainage is another significant issue. Many wetlands have been drained to make way for agriculture or urban development, leading to a loss of habitat and a decline in biodiversity. For instance, the draining of the Florida Everglades for agricultural and urban use has dramatically altered its natural hydrology and diminished its ecological value (Gunderson 2021). Similarly, wetlands in Southeast Asia have been converted into rice paddies and urban areas, impacting the ecological balance and reducing the wetlands' ability to provide critical services (Murray et al. 2018).

Invasive species also pose a threat to wetland ecosystems. Non-native plants and animals can outcompete native species for resources, alter habitat structure, and disrupt ecological processes. In the Great Lakes region, for example, invasive species like the zebra mussel have caused significant ecological and economic damage by outcompeting native species and altering nutrient dynamics. These invasive species often spread rapidly and can be challenging to control or eradicate, further exacerbating the degradation of wetland ecosystems (Ricciardi 2001).

3.7.2 Climate Change Impacts

Climate change is profoundly affecting wetlands through alterations in temperature, precipitation patterns, and sea levels. These changes are influencing wetland ecosystems in various ways, often exacerbating existing environmental challenges.

Rising sea levels threaten coastal wetlands worldwide. As global temperatures increase, polar ice melts and the thermal expansion of seawater contributes to rising sea levels. Coastal wetlands, such as mangroves and salt marshes, are particularly vulnerable to this phenomenon. For example, the mangroves of the Sundarbans in India and Bangladesh face increased inundation and salinity due to rising sea levels, which can lead to habitat loss and decreased biodiversity (Das and Tijssen 2020). Coastal erosion and saltwater intrusion further compromise these critical habitats, impacting the flora and fauna that depend on them.

Changes in precipitation patterns also impact wetlands by altering water availability and flow. In some regions, increased rainfall can lead to flooding and waterlogging, while in others, reduced rainfall can cause droughts and decreased water levels. These shifts can disrupt the delicate balance of wetland ecosystems. For instance, the Australian wetlands are experiencing more frequent and intense droughts, which affect waterbird populations and the overall health of wetland communities (Kingsford 2016). Changes in precipitation can also influence the timing and extent of wetland flooding, affecting plant and animal life cycles and ecological processes.

Temperature increases can alter wetland dynamics by affecting species distributions and ecological interactions. Warmer temperatures may shift the ranges of both aquatic and terrestrial species, potentially leading to mismatches between species and their habitats. For example, in the Arctic, rising temperatures are causing shifts in the distribution of plant and animal species, which can disrupt the food web and ecological balance of wetlands (Prowse et al. 2011). Additionally, increased temperatures can enhance the growth of invasive species and exacerbate problems related to nutrient loading and pollution.

3.7.3 Urbanization and Development

Urbanization and development have had profound impacts on wetlands, leading to their loss and degradation. As cities expand and infrastructure projects increase, wetlands are often seen as obstacles to development rather than valuable ecosystems.

Land reclamation and drainage for urban development frequently destroy wetlands. This process involves converting wetland areas into residential, commercial, or industrial spaces, leading to habitat loss and changes in local hydrology. In cities like Jakarta, Indonesia, and Bangkok, Thailand, extensive land reclamation projects have significantly reduced wetland areas and altered natural water flow, increasing flood risks and impacting local biodiversity (Setiadi et al. 2018). The loss of wetlands due to urban expansion also diminishes their ability to provide essential ecosystem services, such as water purification and flood mitigation.

Infrastructure development also affects wetlands through the construction of roads, bridges, and other structures. These developments can fragment wetland habitats, disrupt wildlife movement, and alter natural water flow patterns. The construction of dams and levees, for example, can change the timing and extent of flooding, impacting the ecological functions of wetlands and the species that rely on them. The introduction of impermeable surfaces, such as roads and buildings, can increase runoff and pollution, further compromising wetland health (Wang et al. 2019).

Economic pressures often drive the conversion of wetlands for development purposes. As populations grow and urban areas expand, there is increasing pressure to utilize available land for housing, commerce, and industry. This pressure can lead to the prioritization of short-term economic benefits over long-term ecological sustainability. Efforts to balance development with conservation are crucial to mitigating the impacts of urbanization on wetlands. Integrating wetland conservation into urban planning and development processes can help preserve these vital ecosystems while accommodating growth (Miller et al. 2020).

3.8 Future Directions

As we advance into an era characterized by rapid environmental changes and increasing pressures on natural resources, sustainable management practices for wetlands are paramount. Effective wetland management must prioritize the preservation of ecological functions while accommodating socio-economic needs. Sustainable practices include adopting integrated water resource management

(IWRM) strategies that balance ecological health with human demands. IWRM promotes the efficient use of water resources while protecting wetlands from overexploitation and contamination. For example, the implementation of riparian buffer zones and best management practices (BMPs) in agricultural areas can help mitigate nutrient runoff, reducing the eutrophication of wetlands (Reddy et al. 2021). Additionally, promoting wetland restoration projects that focus on reestablishing natural hydrological patterns and native vegetation is crucial. Restoration efforts, such as those seen in the Mississippi River Delta, aim to rebuild wetland systems that provide essential services, including flood regulation and habitat support (Gosselink et al. 2021). These practices are vital for maintaining wetland health and resilience in the face of climate change and other stressors.

Integrating historical and cultural perspectives into wetland management is equally important for fostering sustainable practices. Recognizing the traditional knowledge and cultural values associated with wetlands can enhance conservation efforts and ensure that management strategies are both effective and respectful of local traditions. Indigenous knowledge, for example, often includes insights into sustainable resource use and ecosystem management that have been refined over generations. In the Amazon Basin, Indigenous communities use traditional ecological knowledge to guide the management of wetland resources, contributing to the conservation of biodiversity and ecosystem functions (McMichael et al. 2023). Incorporating such perspectives into policy frameworks can lead to more holistic and inclusive approaches to wetland conservation. Additionally, engaging local communities in the decision-making process and acknowledging their cultural connections to wetlands can improve the effectiveness of conservation initiatives. Policies that support community-based management and recognize the cultural significance of wetlands can help bridge the gap between conservation goals and local needs (Dixon et al. 2022).

3.9 Conclusion

The historical and cultural significance of wetlands underscores their profound importance in human society, reflecting their role as vital resources that have shaped and been shaped by human activities. From the defensive and economic roles of the Venice Lagoon to the traditional and modern interactions with the Everglades, wetlands have been integral to the development of societies across the globe. These landscapes have provided sustenance, facilitated trade, and influenced cultural practices, illustrating their multifaceted value. Understanding these interactions offers crucial insights into contemporary conservation efforts, highlighting the need for sustainable management practices that honour both the ecological functions and cultural heritage of wetlands. Recognizing the diverse ways in which wetlands have been valued and utilized throughout history not only enriches our appreciation of these ecosystems but also informs current strategies aimed at preserving and restoring them. As we navigate the challenges posed by environmental change and human impact, incorporating historical and cultural perspectives can guide more effective and respectful approaches to wetland conservation, ensuring that these critical ecosystems continue to support both human and ecological well-being for generations to come.

References

Amoako K (2020) Spirits of the swamps: folklore and beliefs in African wetlands. University of Ghana Press

Bănăduc D, Simić V, Cianfaglione K, Barinova S, Afanasyev S, Öktener A, McCall G, Simić S, Curtean-Bănăduc A (2022) Freshwater as a sustainable resource and generator of secondary resources in the 21st century: stressors, threats, risks, management and protection strategies, and conservation approaches. Int J Environ Res Public Health 19(24). https://doi.org/10.3390/ijerph192416570

Boye A, Ndione J, Sene S, Ba B (2020) Wetland resources and livelihoods in West Africa. Springer

Burgess C, Morrison J, White N (2020) Aboriginal ecological knowledge and wetlands. Australian National University Press

Casellato L (2022) Conservation challenges in the Venice Lagoon. Routledge

Das S, Tijssen S (2020) Impacts of rising sea levels on coastal wetlands: a case study of the Sundarbans. Environ Sci Pol 112:40–51

Davis M (2018) Native American Stewardship of the Everglades. University of Florida Press

Diaz RJ, Rosenberg R (2008) Spreading dead zones and consequences for marine ecosystems. Science 321(5891):926–929

Dixon J (2019) Flood control and wetland management in Florida: a historical perspective. University Press of Florida

Dixon P, Smith K, Harrison J (2022) Community-based wetland management and cultural integration. J Environ Manag 312:114734

Foster S (2019) The Nile Delta and ancient Egyptian agriculture. Cambridge University Press

Gosselink JG, Turner RE, Baumann RH (2021) Restoration of wetlands in the Mississippi River Delta. Wetl Ecol Manag 29(2):179–194

Gunderson L (2021) The Everglades: restoration and conservation efforts. Springer

Harris D (2021) Native American perspectives on wetlands. University of Oklahoma Press

Horsley J (2021) Venetian engineering and urban development. Cambridge University Press

Jacobsen T (2019) The Tigris-Euphrates civilization: historical and environmental perspectives. University of Chicago Press

Jain V (2020) Rituals and purity in Hindu water traditions. Oxford University Press

Kassas M (2022) Cultural heritage and economic impact of the Everglades. University of Miami Press

Keddy P (2019) The Everglades National Park: ecology and conservation. Oxford University Press

Kingsford RT (2016) Ecological and conservation issues in wetlands: the case of Australia. Aust J Ecol 41(5):780–790

Klein S (2019) The aesthetics of landscape design in Renaissance Europe. Routledge

Krauss K (2018) The Venice Lagoon: a historical overview. University of Venice Press

Lara M (2019) Sacred waters: indigenous beliefs and practices in the Amazon. University of Arizona Press

Li Z, Ma L, Gou D, Hong Q, Fai L, Xiong B (2021) The impact of urban development on wetland conservation. Sustain For 14(21):13747. https://doi.org/10.3390/su142113747

Luttwak E (2018) Water management in Medieval Venice. University of Chicago Press

MacCulloch J (2019) The Celtic Twilight: myth and folklore of the British Isles. Routledge

McMichael C, Forbes BC, Fernandez-Llamazares A (2023) Indigenous knowledge and wetland conservation in the Amazon Basin. Environ Conserv 50(1):38–47
Mendelsohn J, el Obeid S, MS (2012) The Okavango Delta: a natural history. Oxford University Press
Meyer J (2021) Engineering Venice: the canal systems and urban design. MIT Press
Middelkoop H, Kwast J (2018) Water management and land reclamation in the Netherlands. Springer
Miller M (2017) Gardens and landscapes of the French renaissance. University of Chicago Press
Miller J, Davis M, Peters J (2020) Balancing urban development and wetland conservation: strategies for sustainable planning. Urban Ecol 15(3):215–229
Murray NJ, Bell J, Wilson KA (2018) Wetland loss and its implications for biodiversity in Southeast Asia. Glob Chang Biol 24(12):5356–5370
Nakamura T (2021) Yokai and the sacred wetlands: Japanese folklore and nature. University of Tokyo Press
Narayan K (2021) The Ganges: a sacred river in Hindu tradition. Harvard University Press
Obiero KO, Klemet-N'Guessan S, Migeni AZ, Achieng AO (2023) Bridging Indigenous and non-Indigenous knowledge systems and practices for sustainable management of aquatic resources from East to West Africa. J Great Lakes Res 49:S128–S137. https://doi.org/10.1016/j.jglr.2022.12.001
Pereira J (2018) Wetland draining in medieval Europe. Routledge
Pritchard T (2019) Economic impact of the Venice Lagoon's salt production. Routledge
Prowse TD, Brown R, Bonsal B (2011) Impacts of climate change on arctic wetlands: review and synthesis. Environ Rev 19(3):251–277
Reddy KR, D'Angelo EM, Hargrove WW (2021) Best management practices for wetland nutrient management. Water Res 198:117123
Ricciardi A (2001) Facilitated species and their impact on Great Lakes ecosystems. J Great Lakes Res 27(1):1–14
Saito K (2019) Paddy farming and rice cultivation in Japan. Springer
Setiadi D, Nur M, Sutanto A (2018) Land reclamation and its effects on wetlands in Jakarta and Bangkok. J Urban Plan 36(2):112–128
Siddiqui A (2018) Water and purity in Islamic tradition. Cambridge University Press
Smith J (2020a) The fens: land reclamation and agriculture in England. Oxford University Press
Smith J (2020b) Venetian water management techniques. Cambridge University Press
Toth L (2020) Restoration projects in the Everglades: progress and challenges. Springer
Verhoeven JT, Setter TL (2009) Agricultural use of wetlands: opportunities and limitations. Ann Bot 105(1):155–163. https://doi.org/10.1093/aob/mcp172
Wang X, Li Y, Zhao X (2019) Effects of impermeable surfaces on wetland health: a comprehensive review. Environ Manag 64(2):210–226
Winston R (2020) Agricultural innovations in ancient Egypt. Cambridge University Press
Wong M (2019) Traditional fishing practices in Southeast Asian mangroves. University of Singapore Press
Yang X, Zhang L (2018) Rice cultivation and wetlands in Ancient China. Cambridge University Press
Yoshida S (2017) Rice cultivation in historical Asia. Springer

Chapter 4
Economic Value of Wetlands: Ecosystem Services and Benefits for Human Well-Being

A. Adewumi, A. E. Adeyemi, F. V. Bekun, and J. O. Olorunleke

Abstract This chapter explored the composite economic value of wetlands by examining the ecosystem services they provide. Wetlands are critical components of the Earth's hydrological and ecological systems, and perform various functions, including water purification, flood control, biodiversity conservation, nutrient cycling, and carbon sequestration. Despite their significant contributions to environmental sustainability and human well-being, wetlands are often undervalued economically, leading to widespread degradation and loss. This necessitates the application of various valuation methods to capture their full value. The chapter highlighted the importance of translating the ecological benefits of wetlands into economic terms to inform better conservation and management strategies. Through a detailed analysis of valuation methods, case studies, and challenges in assessing wetland services, the chapter advocates for more robust policy frameworks that recognize the true value of wetlands in supporting sustainable development and ecological resilience.

Keywords Ecosystem services · Economic valuation · Sustainable management · Policy-making

A. Adewumi (✉) · J. O. Olorunleke
Federal College of Freshwater Fisheries Technology, New Bussa, Nigeria

A. E. Adeyemi
Texas A&M University, College Station, TX, USA

F. V. Bekun
Istanbul Gelisim University, Istanbul, Turkey

Lebanese American University, Beirut, Lebanon

B. R. Babaniyi et al. (eds.), *Wetland Ecosystems: Conservation Strategies, Policy Management and Applications*, Wetlands: Ecology, Conservation and Management 12, https://doi.org/10.1007/978-3-031-91982-4_4

4.1 Introduction

The National Research Council (1995) emphasized the need for a universal definition of wetlands independent of any specific agency, policy, or regulation. They defined wetlands as ecosystems subject to regular or periodic shallow flooding or surface saturation. These ecosystems are marked by the consistent presence of water at or near the surface, resulting in distinct physical, chemical, and biological conditions. Typical indicators of wetlands include hydric soils and hydrophytic vegetation unless modified by natural or human activities. This definition highlights three critical attributes of wetlands: water, substrate, and biota, underscoring their multidisciplinary nature and the diverse interests they encompass, which significantly influence the handling of wetland-related issues. Brinson (1993) echoed this interdisciplinary approach by distinguishing between wetlands' "functions and values" throughout the literature. LePage (2011) further emphasized the need for effective communication among all the different fields as an adaptive management strategy for wetland practitioners.

Wetlands, which are critical components of the earth's hydrological and ecological systems, are often underestimated economically (Mitsch and Gosselink 2000). These complex and dynamic ecosystems provide a multitude of services that are essential not only to environmental health but also to human prosperity and resilience (Barbier 2011; Mitsch and Gosselink 2000). The recognition of these services has grown in recent decades (Mitsch et al. 1998), particularly as global challenges such as climate change, biodiversity loss, and water scarcity have underscored the urgent need for sustainable natural resource management (Grygoruk and Rannow 2017; Xu et al. 2019).

Wetlands have traditionally been undervalued and misunderstood, such as the "geographically isolated wetlands (GIWs)" (Mushet et al. 2015), leading to widespread degradation and loss (Ayanlade and Proske 2015; Ballut-Dajud et al. 2022; McCauley et al. 2013). Historically, they were often viewed as wastelands to be drained for agriculture or urban development, ignoring their intrinsic ecological and economic benefits (Panuccio et al. 2017). This perspective has shifted significantly as scientific research has highlighted the indispensable role of wetlands in regulating natural processes, supporting biodiversity, and sustaining human livelihoods (Bond 1992).

The multifaceted value of wetlands arises from their ability to perform various ecosystem functions that directly and indirectly benefit human societies (Metcalfe et al. 2018). These functions include purifying water, protecting against floods, supporting biodiversity, cycling nutrients, pollution abatement, and sequestering carbon, among others (Arya et al. 2020; Das and Shit 2024; Liu et al. 2024). By performing these roles, wetlands contribute to the stability and productivity of other ecosystems and the well-being of human populations (Maltby and Acreman 2011; Nayak and Bhushan 2022; Sutton-Grier and Sandifer 2019). Moreover, the economic valuation of wetlands has become a crucial tool in natural resource management and policymaking (Barbier et al. 1997). By translating the ecological benefits of wetlands into economic terms, stakeholders can make more informed decisions about conservation, restoration, and sustainable use. This approach highlights the cost-effectiveness of preserving wetlands and aligns with broader environmental sustainability and economic development goals.

4.1.1 Importance of Wetlands in the Context of Ecosystem Services

Wetlands, which include a variety of ecosystems such as marshes, swamps, and bogs, are globally recognized for their exceptional ecological productivity and provision of essential ecosystem services. Maltby and Acreman (2011) referred to these ecosystem services as "natural assets," which include clean air, water, food, and materials, all contributing to social and cultural well-being. Blackwell and Pilgrim (2011) emphasized that the benefits of wetlands remain consistent regardless of their size. These services, derived from natural ecosystems, are broadly categorized into provisioning, regulating, supporting, and cultural services (Jessop et al. 2015; Kumar et al. 2011; Margaryan et al. 2022; Meli et al. 2014). Wetlands excel in delivering these services due to their unique hydrological and biological characteristics (McLaughlin and Cohen 2013).

The significance of wetlands lies in their multifunctional roles within the environment, providing indispensable ecological services. They serve as natural water purifiers, filtering pollutants and sediments to enhance water quality (Dordio et al. 2008; Hammer and Bastian 1989; Sundaravadivel and Vigneswaran 2001). Wetlands are also crucial for flood control, absorbing excess rainfall, and mitigating storm surges (Brandolin et al. 2013; Mitsch and Gosselink 2000; Pattison-Williams et al. 2018; Tang et al. 2020). Moreover, wetlands are biodiversity hotspots, providing habitats for a vast array of plant and animal species, many of which are rare or endangered (Berde et al. 2022; Kingsford et al. 2016; Maltby and Barker 2009; Owethu-Pantshwa and Buschke 2019). Additionally, wetlands play a significant role in carbon sequestration, which is crucial for combating climate change by storing carbon in their biomass and soil (Lolu et al. 2020).

The degradation or loss of wetlands disrupts these essential services, posing significant risks to environmental sustainability and human well-being. Therefore, recognizing and quantifying the value of ecosystem services provided by wetlands is imperative for their conservation and sustainable management.

4.1.2 Overview of the Economic Value of Wetlands

Despite their substantial contributions, the economic value of wetlands is often underestimated (Davidson et al. 2019; Grossmann 2012). This value can be assessed through various measures, including direct market values, indirect use values, non-use values, and option values (Chaikumbung et al. 2016; Gimenez and Mas 2020).

Direct market values pertain to resources directly obtained from wetlands, such as fish, timber, and peat (Mahlatini et al. 2020; Tuan et al. 2009). Indirect use values include benefits such as water purification, flood protection, and recreational opportunities while non-use values relate to the intrinsic worth of wetlands, encompassing biodiversity conservation and cultural significance (Rafiq et al. 2014; Setlhogile et al. 2011).

Moreover, Barbier et al. (1997) and Turner et al. (2000) emphasized the importance of economic valuation of wetlands for informed policy-making and sustainable land use practices. Methods such as contingent valuation, cost-benefit analysis, and ecosystem service modelling are employed to estimate the economic value of wetlands. These assessments highlight the financial benefits of wetland conservation and restoration, thereby supporting the allocation of resources for their protection.

4.1.3 Purpose and Structure of the Chapter

This chapter aims to justify the economic value of wetlands through a comprehensive analysis of the ecosystem services they provide. It highlights their crucial role in maintaining ecological balance and supporting human well-being. This chapter underscores the necessity for effective wetland management and conservation strategies by examining the various ecosystem services and their economic implications.

The chapter is structured as follows: Sect. 4.2 explores the specific ecosystem services wetlands provide, including water purification, flood control, biodiversity conservation, nutrient cycling, and carbon sequestration. Section 4.3 delves into the economic valuation of these ecosystem services. Section 4.4 discusses the benefits of wetlands for human well-being. Finally, Sect. 4.5 synthesizes the findings from all sections and discusses their implications for policy and practice.

4.2 Ecosystem Services Provided by Wetlands

As mentioned, wetlands provide a broad spectrum of ecosystem services indispensable to environmental health and human welfare. These services include water purification, flood control, biodiversity conservation, nutrient cycling, and carbon sequestration. The following subsections provide an in-depth analysis of these services.

4.2.1 Water Purification and Filtration

Hammer and Bastian (1989) noted that wetlands exhibit productivity levels surpassing those of the most fertile farmlands, a phenomenon driven by their capacity to receive, retain, and recycle nutrients continuously washed down from upland areas.

In addition to their high productivity, wetlands function as natural filtration systems, effectively capturing pollutants, sediments, and excess nutrients from water bodies (Langergraber and Haberl 2001; Pappalardo et al. 2017). This purification process is facilitated by intricate physical, chemical, and biological interactions within the wetland ecosystem (D'Arcy et al. 2007; Villar et al. 2012). Plants, soil, and microorganisms play crucial roles in breaking down and assimilating contaminants, significantly enhancing water quality.

The economic value of water purification services provided by wetlands is profound. Clean water is indispensable for human consumption, agriculture, and industrial processes. For instance, a study in China by Zhang et al. (2017) revealed that 51,434 hectares of wetlands in Beijing annually provided 944.01 million cubic meters of water, valued at 1.19 billion RMB. Similarly, a study in Uganda by Kakuru et al. (2013) employed market price, productivity, and contingent valuation methods to estimate the value of various wetland resources, finding that the per capita value of domestic water use amounted to US$ 34 million per year. These figures underscore the critical role of wetlands in sustaining water quality and availability, highlighting their economic importance. By naturally filtering pollutants and sediments, wetlands significantly reduce the need for expensive water treatment facilities and mitigate the harmful effects of pollution on public health and ecosystems. The value of these services can be further appreciated when considering the cost savings from avoided water treatment and the substantial health benefits associated with access to clean water.

4.2.2 Flood Control and Storm Protection

Wamsley et al. (2010) emphasized the vital role wetlands play as natural buffers against floods and storm surges. They acknowledged that while the extent to which wetlands reduce storm surges is still debated in the literature, the effectiveness of this attenuation largely depends on the surrounding coastal landscape and the storm's strength and duration. Numerous studies have corroborated that wetlands can absorb and store significant volumes of water, thereby reducing the impact of extreme weather events on human settlements and infrastructure (Barbier et al. 1997; Brandolin et al. 2013; Mitsch and Gosselink 2000; Pattison-Williams et al. 2018; Tang et al. 2020). During periods of heavy rainfall, wetlands act as sponges, absorbing water and releasing it gradually, thus preventing flash floods. Coastal wetlands, including mangroves and salt marshes, also protect shorelines from erosion and storm surges by dissipating wave energy (Leon et al. 2018).

The economic benefits of wetlands in flood control and storm protection are substantial. A study by Vázquez-González et al. (2019) in Mexico revealed that while the economic costs (2007 values) of flooding from Hurricane Karl were $150.85 million, the value of coastal wetlands ranged from $148,277 to $193,674 per hectare. Similarly, research by Barbier et al. (2013) showed that wetland continuity and vegetation effectively reduce storm surge levels. Their study reported that damages in property reduced in southeast Louisiana by 0.1 m increase in wetland continuity, including New Orleans, by $99–$133, while a 0.001 increase in vegetation roughness decreases damages by $24–$43. Additionally, Fairchild et al. (2021) demonstrated the significant role of salt marshes in reducing flooding across estuaries. Their findings indicated that salt marshes reduced flood extents by 35% and damages by 37% ($8.4 million) during 100-year storms. Across all storm scenarios, wetlands provided mean annual damage savings of $2.7 million per estuary, protecting 3–5 properties per storm on average.

By mitigating the impacts of floods and storms, wetlands significantly reduce the costs associated with disaster recovery and infrastructure damage. They also enhance community resilience to climate change. Valuing these services involves estimating the avoided costs of flood damage and the savings in disaster response and recovery expenditures.

4.2.3 Biodiversity Conservation and Habitat Provision

Wetland ecosystems, including riverine, palustrine, lacustrine, estuarine, groundwater-dependent, and human-made types (Kingsford et al. 2016), are among the most biologically diverse environments, supporting a vast array of plant and animal species (Junk et al. 2006; Maltby and Barker 2009). Often referred to as "nature's supermarkets" due to their extensive food chains and rich biodiversity (Mitsch et al. 2015), wetlands provide crucial habitats for migratory birds, fish, amphibians, and invertebrates (Kakuru et al. 2013). The structural complexity and variety of microhabitats within wetlands create ideal conditions for diverse biological communities.

In a study by Zhang et al. (2017), 51,434 hectares of wetlands in Beijing provided 9587 hectares of habitat, valued at 1.08 billion RMB in 2014. Additionally, a study by Semeraro et al. (2015) on constructed treatment wetlands revealed ancillary benefits such as sustaining wildlife habitats and biodiversity, along with providing recreational and educational opportunities. Wetlands also have cultural, recreational, and educational significance, contributing substantially to human well-being and quality of life. The conservation of biodiversity in wetlands holds both intrinsic and economic value, underpinning ecosystem resilience and productivity.

4.2.4 Nutrient Cycling and Carbon Sequestration

Wetlands are crucial for nutrient cycling, particularly for nitrogen and phosphorus, through key processes like denitrification, nitrification, and sedimentation (Almuktar et al. 2018). Wetzel (1992) aptly described wetlands as "metabolic gates," emphasizing their regulatory function in managing nutrient and organic matter flow to downstream waters. These processes are vital for maintaining nutrient balance in aquatic ecosystems, thereby preventing eutrophication and promoting overall ecosystem health (Arias et al. 2005).

Beyond nutrient cycling, wetlands significantly impact the global carbon cycle by sequestering carbon within their biomass and soils (Mitsch et al. 2014; Were et al. 2019). In an extensive review, Were et al. (2019) highlighted that carbon sequestration in wetlands can be enhanced through both active management and natural processes. The objectives are to boost organic matter inputs, channel carbon into more stable pools, and prolong the residence times of these carbon stores. This capacity for effective carbon sequestration underscores the critical role of wetlands in mitigating climate change.

The economic value of nutrient cycling and carbon sequestration services provided by wetlands is substantial. Wetlands in South Korea provide significant economic value through nutrient cycling and carbon sequestration. Yoo et al. (2022) estimate the total soil organic carbon (SOC) stock in inland wetlands at 16.3 TgC, representing 5.3–5.5% of the national SOC stock. The annual carbon sequestration rate is 60.7 gC m^2, valued at $5.5 million. The total SOC stock's economic value is approximately $2.0 billion, emphasizing the crucial role of wetlands in the carbon budget and ecosystem health.

Moreover, nutrient regulation in wetlands plays a critical role in maintaining water quality, which in turn benefits agriculture, fisheries, and human health (Jamion et al. 2023). Carbon sequestration in wetlands significantly mitigates climate change by lowering atmospheric greenhouse gas concentrations (Pindilli 2021). The economic valuation of these services involves calculating the avoided costs from reduced eutrophication and the financial benefits associated with carbon storage.

Iqbal and Shang (2020) highlighted the crucial role of wetlands, particularly in the Hindu-Kush Himalaya region, as vital carbon reservoirs and sinks for atmospheric carbon. However, they also raised concerns about the balance between carbon sequestration and the potential for methane (CH4) emissions. They emphasized the need for further research to understand the relationship between wetlands' roles in greenhouse gas (GHG) emissions and carbon sequestration.

4.3 Economic Valuation of Wetland Ecosystem Services

Understanding the economic value of wetland ecosystem services is crucial for informed decision-making, conservation efforts, and sustainable management of wetland resources. This section explores various methods for assessing the economic value of wetlands, presents case studies that demonstrate the application of these methods, and discusses the challenges and limitations in valuing wetland ecosystem services. The economic value of wetland ecosystem services is significant, yet often underappreciated in traditional market assessments, necessitating the application of non-market valuation methods to more accurately capture their benefits to society and the environment (Mitsch et al. 2015; Lazo 2002). In this context, various approaches such as contingent valuation and cost-benefit analysis are employed to ascertain the economic worth of these ecosystem services, acknowledging that much of their value lies outside conventional market transactions and therefore requires innovative methodologies for accurate assessment (Sherrouse et al. 2011). For instance, while economists typically focus on quantifying the exchange value of these services through monetary assessments, it is imperative to integrate ecological and social dimensions to fully understand their multifaceted contributions to human well-being and environmental sustainability (Costanza et al. 2011; Sarvašová et al. 2014). To this end, frameworks incorporating both monetary and non-monetary indicators can enhance the understanding of wetland services, as they enable a more comprehensive evaluation of the interplay between ecological health and societal benefits, ultimately informing better policy decisions and resource management strategies

4.3.1 Methods for Assessing the Economic Value of Wetlands

Economists have developed various methods for quantifying the economic benefits of wetland ecosystem services. These include revealed preference techniques, such as travel cost and hedonic pricing, which infer values from market transactions, as well as stated preference methods, like contingent valuation and choice experiments, which directly elicit individuals' willingness to pay for wetland services (Costanza et al. 2011). However, the application of these methods is challenged by the public-good nature of many wetland services, which are not traded in traditional markets. (Lazo 2002).

According to Thapa et al. (2020), market-based methods estimate the value of ecosystem services that have direct market prices. For instance, wetlands provide resources such as fish, timber, and medicinal plants, which can be directly valued using market prices. The market price method is straightforward but limited, as it only captures the value of goods traded in markets and often overlooks non-market benefits like water purification and carbon sequestration (Thapa et al. 2020).

The non-market-based methods are essential for valuing ecosystem services that do not have a direct market price. Some of the prominent non-market-based valuation methods include:

1. **Travel Cost Method (TCM)**: Rusciano et al. (2023) argued that the Travel Cost Method (TCM) evaluates the economic value of recreational services provided by wetlands by assessing the expenses incurred by visitors, such as travel costs and time spent. This method effectively estimates the economic value of recreational activities in wetland areas through demand analysis. It operates on the premise that the recreational value of a site is reflected in the costs associated with visiting that site, implying that the economic value is derived from the visitor's willingness to pay. Specifically, the method calculates this value based on all visitor-incurred costs, including travel expenses and entrance fees, which signify the demand to "consume" the recreational experience. This approach pro-

vides a clear framework for estimating participants' willingness to pay for access to the recreational site. However, it is important to consider the method's key assumption: as the costs increase, the frequency of visits to the site is expected to decrease (Chapagain and Poudyal 2020).

Extensive reviews and studies on TCM have emerged in the scientific literature, beginning with the pioneering work of Clawson (1972) and Woodward and Wui (2001), and continuing through more recent evaluations (Scheufele and Pascoe 2023). Today, TCM is a well-established methodology, widely used to assess various recreational activities in natural and rural environments. Specific applications to wetlands include studies such as Fleming and Cook (2008) for Lake McKenzie (Fraser Island) and Wubalem et al. (2023) for Lake Tana in Ethiopia, as well as Tienhaara et al. (2021) for Lake Puruvesi in Finland. These studies generally provide decision-makers with insights into the overall economic benefits of tourism and recreation while emphasizing the importance of preserving natural resources.

There are two primary approaches within the TCM literature: the individual (ITCM) and zonal (ZTCM) approaches. ITCM focuses on the number of trips per individual annually or seasonally, whereas ZTCM examines the number of trips to a site or zone by a group of people. Clawson (1972) suggested that ITCM is more suitable for studying frequently visited sites, while ZTCM is better suited for assessing remote locations that are typically visited by people from a distance. Skarakis et al. (2023) further elaborated on the differences between these two approaches. Ultimately, the advantage of TCM lies in its feasibility, given its relatively low survey costs and straightforward estimation procedures compared to other techniques.

2. **Contingent Valuation Method (CVM)**: This method involves surveying individuals to determine their willingness to pay for specific ecosystem services. It is often used to estimate the value of recreational and cultural services provided by wetlands (Gimenez and Mas 2020). This method is particularly effective for valuing non-market goods such as recreational, cultural, and existence values associated with wetlands. For example, a study on the Woopo Wetland in Korea used CVM to assess the conservation value of the wetland. The study found that respondents were willing to pay an average of \$2.10 to \$3.05 per household annually for wetland conservation, indicating the public's recognition of the wetland's importance (Kwak et al. 2007). Similarly, in a case study of the Paya Indah Wetland in Malaysia, visitors' WTP for wetland conservation was estimated using CVM. The study highlighted that income level and bid price were significant factors influencing visitors' WTP, demonstrating the method's utility in capturing the economic value of ecosystem services provided by wetlands (Siew et al. 2015).

   However, CVM is not without its challenges. Issues such as hypothetical bias, where respondents may overstate or understate their WTP in hypothetical scenarios, can affect the accuracy of the results. For instance, a study in Australia found that inferred valuation methods could reduce hypothetical bias compared to standard CVM, although both methods provided valuable insights (Gregg and Wheeler 2018).

3. **Hedonic Pricing Method**: This method assesses the impact of wetlands on property values by comparing real estate prices in proximity to wetlands with those in non-wetland areas. The difference in prices reflects the value attributed to the wetland's ecosystem services. This method assumes that the value of ecosystem services, such as scenic beauty, flood protection, and air quality improvement, is reflected in property prices. The difference in prices can be attributed to the wetland's ecosystem services by comparing real estate prices in proximity to wetlands with those in non-wetland areas. For example, wetlands that offer aesthetic and recreational benefits can increase the value of nearby properties, as homeowners are willing to pay more for properties with access to these services. On the other hand, wetlands that mitigate flood risks can also enhance property values by reducing potential damage costs, which is particularly relevant in flood-prone areas.

   The strength of HPM lies in its ability to capture the implicit value of ecosystem services in market transactions. It is nonetheless limited by its reliance on property market data, which may not fully capture all ecosystem services provided by wetlands. This method may also not be applicable in regions where real estate markets are underdeveloped or where data on property transactions is scarce.

4. **Benefit Transfer Method (BTM)**: BTM is a cost-effective approach to estimating the economic value of wetlands by transferring economic values estimated in one study to another similar site with similar characteristics, reducing the need for extensive data collection. This method is particularly useful when time or resources are limited (He et al. 2015). For example, in a study on wetlands in Quebec, Canada, BTM was used to transfer economic values derived from previous studies to estimate the value of ecosystem services provided by wetlands. The study emphasized the importance of considering multiple wetland characteristics when conducting a benefit transfer to ensure accurate value estimation (He et al. 2015).

   BTM is particularly advantageous in policy-making contexts where decision-makers require rapid and cost-effective estimates of ecosystem service values. It is worthy to note that the accuracy of BTM depends on the similarity between the original study site and the target site. Differences in ecological, social, and economic conditions can lead to errors in value estimation if not carefully accounted for.

5. **Production Function Method**: The production function approach links ecosystem services to the production of marketed goods allowing for the estimation of the economic value of these services based on their contribution to economic output. This method is particularly relevant for wetlands that provide services such as water regulation, soil fertility, and nutrient cycling, which directly support agricultural productivity and other economic activities. Barbier (1994) highlighted the role of wetlands in contributing to agricultural productivity by providing water regulation and soil fertility services. This method estimates the

value of wetland services in economic terms through the quantification of the contribution of wetlands to agricultural output.

The strength of this method lies in its ability to directly link ecosystem services to economic activities, making it a powerful tool for demonstrating the economic importance of wetlands. Howbeit, this method requires detailed ecological and economic data, which can be challenging to obtain. It may also oversimplify the complex interactions between wetlands and the goods they produce, leading to potential inaccuracies in value estimation.

4.3.2 Case Studies on Economic Valuation of Wetland Services

The application of the aforementioned methods in real-world scenarios provides valuable insights into the economic valuation of wetland services. Several case studies from different regions highlight the diverse benefits of wetlands and the importance of their conservation.

A study conducted on the Begnas Watershed System in Nepal utilized a combination of market and non-market-based valuation methods to assess the economic value of the wetland's ecosystem services. The study found that the wetland generated an economic benefit worth $3.91 million annually, with recreational services accounting for 85% of the total value. The findings emphasize the importance of wetlands for local communities and the need for investment in wetland restoration projects (Thapa et al. 2020).

In West Asia, a regional assessment of coastal and freshwater wetlands revealed significant economic losses associated with wetland degradation. Using scenario analysis, Eppink et al. (2014) estimated that the present value of regional economic loss due to wetland loss by 2050 could range from $2.3 billion to $7.2 billion. The study highlighted the need for indigenous studies to better understand the social and economic impacts of wetland loss.

In a study by Costanza et al. (1989) on the wetlands of coastal Louisiana, both the willingness-to-pay and energy analysis-based methodologies were employed to estimate the economic value of these wetlands. The study estimated the present value of an average acre of natural wetlands to range from $2429 to $17,000 per acre. The findings underscored the significant economic value of wetlands and the need for careful consideration in wetland management decisions. The researchers contended that achieving highly precise estimates of wetland values is not feasible, despite significant effort, a challenge that also applies to various other natural resource categories.

In the Oristanese Reclamation District of Sardinia, Italy, Rusciano et al. (2023) employed the travel cost method to assess the touristic value of wetlands. Their study estimated that recreational activities at the site generate approximately €1.25 million annually in economic benefits. These findings offer crucial insights for decision-makers seeking to balance the economic gains from tourism with the need for wetland conservation.

4.3.3 Challenges and Limitations in Valuing Wetland Ecosystem Services

Despite the advances in economic valuation methods, several challenges and limitations persist in accurately estimating the value of wetland ecosystem services. One of the primary challenges in valuing wetlands according to Shabman and Batie (1980) is the lack of comprehensive data on ecosystem services as many valuation methods require detailed ecological and economic data, which are often unavailable or incomplete. This data scarcity limits the accuracy and reliability of valuation estimates.

Woodward and Wui (2001) opined that different valuation methods have their own set of limitations. For example, contingent valuation methods are prone to bias, as they rely on hypothetical scenarios that may not accurately reflect actual behavior. Similarly, the researchers further argued that production function approach often oversimplifies the complex relationships between wetlands and the goods they produce, leading to potentially inaccurate estimates.

Wetland ecosystems are highly dynamic and can vary significantly in terms of their services and values. Brander et al. (2012) stated that the variability makes it challenging to generalize findings from one wetland to another. Moreover, the long-term impacts of climate change add another layer of uncertainty to wetland valuation, as changing environmental conditions may alter the provision of ecosystem services.

Economic valuation methods often prioritize monetary values, potentially overlooking the intrinsic and cultural values of wetlands. Additionally, there are concerns about equity, as the benefits of wetland conservation may not be evenly distributed across different communities (Batie and Shabman 1982). This raises ethical questions about who should bear the costs of wetland protection and who should benefit from it.

4.4 Benefits of Wetlands for Human Well-Being

Wetlands are among the most productive ecosystems on earth. They provide numerous benefits to human well-being, ranging from economic contributions to cultural significance and health advantages. This section explores the various ways in which wetlands contribute to local livelihoods, public health, recreation, and cultural practices.

4.4.1 Contribution of Wetlands to Local Livelihoods and Economies

Wetlands are vital to the livelihoods of millions of people worldwide, especially in developing countries where subsistence agriculture and fishing are prevalent. Wetlands provide resources such as fish, timber, and medicinal plants, which are essential for local economies and food security.

In many rural areas, wetlands serve as a critical source of natural capital, supporting agriculture, aquaculture, and forestry. For example, the sustainable harvesting of wetland resources, such as fish and aquatic plants, provides income and food security for local communities. Jenyo-Oni and Agbeja (2004) highlight that wetlands serve as essential rearing and breeding grounds for various natural habitats, animals, and aquaculture resources. They also note that wetlands, particularly *fadama* areas, are highly suitable for farming, especially in the cultivation of rice and fish. Additionally, peatlands, a type of wetland, are used for fuel production. Coastal, lake, and floodplain wetlands are recognized as highly productive spawning, nursery, and feeding grounds for fish. Maltby (2013) emphasizes that these wetlands not only provide a rich food base but also protect fish from strong currents and predators. He further notes that two-thirds of the world's fish catches originate from tidal areas.

In Nigeria, floodplains play a crucial role in farming, fishing, and forestry, with wetland resources directly contributing to livelihoods and well-being (Barbier 2011). Fish consumption is significant, accounting for about 40% of the country's protein intake, with an average consumption of 13.3 kg per person annually (Liverpool-Tasie et al. 2021; FAO 2021).

Aquatic plants from wetlands are primarily utilized as fuel for fish smoking and domestic energy needs. Jenyo-Oni and Agbeja (2004) observed that the stems of *Echinochloa* spp. and *Cyperus papyrus* are commonly used for cooking and fish smoking. The untapped potential for converting aquatic plants into biogas for fish smoking and preservation presents an opportunity for economic development.

4.4.2 Health and Recreational Benefits of Wetlands

Wetlands offer significant health and recreational benefits by improving the quality of human surroundings and providing opportunities for physical and mental well-being. Studies have shown that access to natural environments, including wetlands, is associated with improved physical and psychological health. Wetlands serve as spaces for recreation, exercise, and relaxation, contributing to the overall well-being of individuals and communities.

Healthy wetlands provide ecosystem services that directly benefit human health, such as water purification, flood control, and climate regulation. Sutton-Grier and Sandifer (2018) argued that wetlands act as natural water filters, removing pollutants and improving water quality, which is essential for public health. They also mitigate the impacts of natural disasters, such as floods and storms, by absorbing excess water and reducing the severity of these events.

Recreational activities in wetlands, such as birdwatching, fishing, and hiking, provide physical exercise and mental relaxation. These activities promote a connection with nature, which has been shown to reduce stress, anxiety, and depression. Urban wetlands, in particular, offer accessible green spaces for residents to engage in outdoor activities, contributing to their overall health and well-being (Carter 2015). In Nigeria, the splendor of the wetland is captivating with their lush, watery landscapes that hold a wealth of natural beauty and ecological significance. Jenyo-Oni (2024) stated that the wetlands change with seasons; during the rainy seasons, they swell with water becoming expansive lakes, with the symphony of bird calls, the rustling of reeds, grass swaying in the breeze and the gentle ripple of water—a harmonious blend of colourful life, which form a sanctuary for countless species, and in the dry season, they shrink, revealing cracked mud and hidden treasures. Communities living near wetlands have woven their lives around these watery realms, children play along the banks. Songs, dances and folkore celebrate the wetlands; role in the sustaining livelihoods, as they embrace the natural beauty of Mother Nature.

However, there is also a paradox where wetlands can pose health risks. For instance, Finlayson and Horwitz (2015) argued that urban wetlands that support biodiversity may also harbor mosquitoes and other disease vectors, increasing the risk of vector-borne diseases. Balancing the health benefits of wetlands with potential risks is essential for effective wetland management.

4.4.3 Cultural and Spiritual Significance of Wetlands

Wetlands hold deep cultural and spiritual significance for many communities around the world. They are often regarded as sacred spaces, deeply embedded in local traditions, rituals, and belief systems. Wetlands provide a sense of place and identity, connecting people to their cultural heritage and natural environment.

In many indigenous cultures, wetlands are seen as sacred sites where spiritual rituals and ceremonies are performed. Cunningham (2015) posited that these practices often revolve around the natural cycles of water and the life it supports, symbolizing fertility, renewal, and the interconnectedness of all living beings. Wetlands are also central to the cultural identity of many communities, influencing their traditional knowledge, practices, and lifestyles. Wetlands hold immense significance for their intangible, nonmaterial, and cultural services, including their roles in pilgrimage and spiritual fulfillment. For many Indigenous peoples, wetlands are deeply connected to spirituality, as they are often considered to be inhabited by spirits. Mainstream religions also recognize the spiritual value of wetlands, with places of worship frequently being constructed in these areas (Verschuuren et al. 2012). Pilgrimages through sacred rivers and wetlands can have profound impacts, and religious leaders can play a vital role in wetland conservation. International policy incentives support policymakers and decision-makers in preserving these sacred spaces. For example, Tibetan Buddhists regard high mountain lakes as sacred pilgrimage sites, with the lakes believed to be protected by sacred fish that inhabit them (Jacobsen et al. 2013).

The Millennium Ecosystem Assessment (2005) estimates that wetlands contribute approximately 15 trillion US dollars annually in supporting, provisioning, regulating, and cultural services. These include spiritual enrichment, cognitive development, reflection, recreation, and aesthetic experiences. The Ramsar Convention Secretariat (2007) has acknowledged the cultural importance of wetlands, including their spiritual and religious significance. Despite their importance, spiritual services can be challenging to quantify and are often overlooked in decision-making processes. Wetlands are revered for their spiritual, religious, and sacred significance, linking them to both mainstream religions and localized Indigenous spiritualities. Spiritual values, a subset of sociocultural values, embody intangible and sometimes supernatural dimensions associated with wetlands. These values are expressed and experienced by various stakeholders, including local communities, conservationists, tourists, and fishers who have deep connections with wetlands. As spiritual and religious values are increasingly recognized for their contribution to biodiversity, wetland management has gained importance. Notable examples of wetlands with sacred, religious, and spiritual significance include the Nile River in Egypt, Adigbe and Esiribi Lake in Niger, and Lake Titicaca in Bolivia and Peru.

The cultural significance of wetlands extends beyond indigenous communities. Many urban wetlands have been integrated into public spaces, where they provide cultural and recreational benefits to the broader population. For example, wetland parks and reserves are popular destinations for community events, outdoor activities, and environmental education programs, fostering a deeper appreciation for nature and its role in human well-being (Carter 2015). Wetlands often feature in local folklore and stories, passing down ecological knowledge and cultural values from generation to generation. According to data from the Ramsar Convention Secretariat (2007), nearly all Ramsar sites provide cultural ecosystem services, and over half have culture in the management of wetlands can therefore play a strong role in their conservation and wise use. This is assumed to be non-consumptive uses of which the help to imbibe a sensory experience as well as some socio-cultural values such as the amenity service; where the site serves as the aesthetic recreation, artistic and spiritual fulfilments and intellectual development (Moore et al. 2022).

4.5 Conclusion

This chapter has highlighted the crucial economic value of wetlands through their ecosystem services, which are essential for both environmental sustainability and human well-being. Wetlands provide vital services such as water purification, flood control, biodiversity conservation, nutrient cycling, and carbon sequestration, all of which contribute to the stability and productivity of other ecosystems. The economic valuation of these services is essential for informed decision-making in conservation and natural resource management. Although, there are challenges in accurately valuing wetland services due to data scarcity, methodological limitations, and the complex dynamics of wetland ecosystems. Stakeholders can make more effective policies that promote the preservation and sustainable use of wetlands by translating ecological benefits into economic terms.

The implications for wetland conservation and management are profound. Recognizing the economic value of wetlands can lead to better resource allocation for conservation efforts, ensuring that these ecosystems are protected from further degradation. Effective management strategies should integrate both market and non-market valuation methods to capture the full range of wetland services. Importantly, addressing challenges such as data scarcity, methodological limitations, and equity concerns is crucial for improving the accuracy and reliability of wetland valuations.

Future research should focus on enhancing valuation techniques, particularly in the context of climate change and evolving environmental conditions. More comprehensive studies are needed to better understand the long-term impacts of wetland loss and to develop strategies that balance conservation with economic development. Expanding the scope of research to include the cultural and intrinsic values of wetlands will also help create a more holistic approach to wetland conservation. Ultimately, by advancing our understanding of the economic and ecological significance of wetlands, we can contribute to their preservation for future generations.

References

Almuktar SAAAN, Abed SN, Scholz M (2018) Wetlands for wastewater treatment and subsequent recycling of treated effluent: a review. Environ Sci Pollut Res 25(24):23595–23623. https://doi.org/10.1007/s11356-018-2629-3

Arias CA, Brix H, Marti E (2005) Recycling of treated effluents enhances removal of total nitrogen in vertical flow constructed wetlands. J Environ Sci Health A 40(6–7):1431–1443. https://doi.org/10.1081/ESE-200055882

Arya AK, Joshi KK, Bachheti A, Deepti (2020) A review on distribution and importance of wetlands in the perspective of India. J Appl Nat Sci 12(4):Article 4. https://doi.org/10.31018/jans.v12i4.2412

Ayanlade A, Proske U (2015) Assessing wetland degradation and loss of ecosystem services in the Niger Delta, Nigeria. Mar Freshw Res 67(6):828–836. https://doi.org/10.1071/MF15066

Ballut-Dajud GA, Sandoval Herazo LC, Fernández-Lambert G, Marín-Muñiz JL, López Méndez MC, Betanzo-Torres EA (2022) Factors affecting wetland loss: a review. Land 11(3):Article 3. https://doi.org/10.3390/land11030434

Barbier E (1994) Valuing environmental functions: tropical wetlands. Land Econ 70:155–173. https://doi.org/10.2307/3146319

Barbier EB (2011) Wetlands as natural assets. Hydrol Sci J 56(8):1360–1373. https://doi.org/10.1080/02626667.2011.629787

Barbier EB, Acreman M, Knowler D (1997) Economic valuation of wetlands: a guide for policy makers and planners. https://www.ecosystemmarketplace.com/wp-content/uploads/archive/documents/Doc_445.pdf

Barbier EB, Georgiou IY, Enchelmeyer B, Reed DJ (2013) The value of wetlands in protecting southeast Louisiana from hurricane storm surges. PLoS One 8(3):e58715. https://doi.org/10.1371/journal.pone.0058715

Batie SS, Shabman LA (1982) Estimating the economic value of wetlands: principles, methods, and limitations. J Coast Zone Manag 10(3):255–278. https://doi.org/10.1080/08920758209361920

Berde VB, Chari PVB, Berde CV (2022) Wetland and biodiversity hotspot conservation. In: Research anthology on ecosystem conservation and preserving biodiversity. IGI Global, pp 775–784. https://doi.org/10.4018/978-1-6684-5678-1.ch039

Blackwell MSA, Pilgrim ES (2011) Ecosystem services delivered by small-scale wetlands. Hydrol Sci J 56(8):1467–1484. https://doi.org/10.1080/02626667.2011.630317

Bond WK (1992) Wetland evaluation guide: final report of the wetlands are not wastelands project. Sustaining wetlands issues paper No. 1992-1. https://www.osti.gov/etdeweb/biblio/5893588

Brander L, Bräuer I, Gerdes H, Ghermandi A, Kuik O, Markandya A, Navrud S, Nunes P, Schaafsma M, Vos H, Wagtendonk A (2012) Using meta-analysis and GIS for value transfer and scaling up: valuing climate change induced losses of European wetlands. Environ Resour Econ 52:395–413. https://doi.org/10.1007/S10640-011-9535-1

Brandolin PG, Ávalos MA, de Angelo C (2013) The impact of flood control on the loss of wetlands in Argentina. Aquat Conserv Mar Freshwat Ecosyst 23(2):291–300. https://doi.org/10.1002/aqc.2305

Brinson MM (1993) A hydrogeomorphic clasification for wetlands. http://hdl.handle.net/11681/6483

Carter M (2015) Wetlands and health: how do urban wetlands contribute to community well-being? In: Finlayson C, Horwitz P, Weinstein P (eds) Wetlands and human health. Wetlands: ecology, conservation and management. Springer, Dordrecht, p 5. https://doi.org/10.1007/978-94-017-9609-5_8

Chaikumbung M, Doucouliagos H, Scarborough H (2016) The economic value of wetlands in developing countries: a meta-regression analysis. Ecol Econ 124:164–174. https://doi.org/10.1016/j.ecolecon.2016.01.022

Chapagain BP, Poudyal NC (2020) Economic benefit of wildlife reintroduction: a case of elk hunting in Tennessee, USA. J Environ Manag 269:110808. https://doi.org/10.1016/j.jenvman.2020.110808

Clawson M (1972) Methods of measuring the demand for and value of outdoor recreation. Marion Clawson. Resources for the Future, Inc., 1145 Nineteenth Street, N.W., Washington, D.C. 1959. 36p. 50c. Travel Res Bull 10(3):11–11. https://doi.org/10.1177/004728757201000331

Costanza R, Farber S, Maxwell J (1989) Valuation and management of wetland ecosystems. Ecol Econ 1:335–361. https://doi.org/10.1016/0921-8009(89)90014-1

Costanza R, Kubiszewski I, Ervin D, Bluffstone R, Boyd J, Brown D, Chang H, Dujon V, Granek E, Polasky S, Shandas V, Yeakley A (2011) Valuing ecological systems and services. F1000 Biol Rep 3:1–6. https://doi.org/10.3410/b3-14

Cunningham A (2015) Wetlands and people's wellbeing: basic needs, food security and medicinal properties. In: Finlayson C, Horwitz P, Weinstein P (eds) Wetlands and human health. Wetlands: ecology, conservation and management. Springer, Dordrecht, p 5. https://doi.org/10.1007/978-94-017-9609-5_3

D'Arcy BJ, McLean N, Heal KV, Kay D (2007) Riparian wetlands for enhancing the self-purification capacity of streams. Water Sci Technol 56(1):49–57. https://doi.org/10.2166/wst.2007.435

Das S, Shit PK (2024) Assessing ecosystem services in wetlands through importance–performance analysis: evidence from lower Gangetic plain region, India. Socio Ecol Pract Res 6(2):191–205. https://doi.org/10.1007/s42532-024-00189-7

Davidson NC, van Dam AA, Finlayson CM, McInnes RJ (2019) Worth of wetlands: revised global monetary values of coastal and inland wetland ecosystem services. Mar Freshw Res 70(8):1189–1194. https://doi.org/10.1071/MF18391

Dordio A, Palace AJ, Pinto AP (2008) Wetlands: water living filters? [bookPart]. Nova Science Publishers. https://dspace.uevora.pt/rdpc/handle/10174/6485

Eppink F, Brander L, Wagtendonk A (2014) An initial assessment of the economic value of coastal and freshwater wetlands in West Asia. Land 3:557–573. https://doi.org/10.3390/LAND3030557

Fairchild TP, Bennett WG, Smith G, Day B, Skov MW, Möller I, Beaumont N, Karunarathna H, Griffin JN (2021) Coastal wetlands mitigate storm flooding and associated costs in estuaries. Environ Res Lett 16(7):074034. https://doi.org/10.1088/1748-9326/ac0c45

Finlayson C, Horwitz P (2015) Wetlands as settings for human health—the benefits and the paradox. In: Finlayson C, Horwitz P, Weinstein P (eds) Wetlands and human health. Wetlands: ecology, conservation and management. Springer, Dordrecht, p 5. https://doi.org/10.1007/978-94-017-9609-5_1

Fleming CM, Cook A (2008) The recreational value of Lake McKenzie, Fraser Island: an application of the travel cost method. Tour Manag 29(6):1197–1205. https://doi.org/10.1016/j.tourman.2008.02.022

Food and Agriculture Organization (FAO) (2021) The state of World Fisheries and Aquaculture 2021. http://www.fao.org/state-of-fisheries-aquaculture. 18 Feb 2024

Gimenez FV, Mas CR (2020) The valuation of recreational use of wetlands and the impact of the economic crisis. Int J Environ Res Public Health 17(9):3228. https://doi.org/10.3390/ijerph17093228

Gregg D, Wheeler S (2018) How can we value an environmental asset that very few have visited or heard of? Lessons learned from applying contingent and inferred valuation in an Australian wetlands case study. J Environ Manag 220:207–216. https://doi.org/10.1016/j.jenvman.2018.04.116

Grossmann M (2012) Economic valuation of wetland ecosystem services. Case studies from the Elbe River Basin. Technische Universität Berlin, Berlin. https://depositonce.tu-berlin.de/bitstreams/0681c813-5522-408e-a8c0-9e2f90ac8f84/download

Grygoruk M, Rannow S (2017) Mind the gap! Lessons from science-based stakeholder dialogue in climate-adapted management of wetlands. J Environ Manag 186:108–119. https://doi.org/10.1016/j.jenvman.2016.10.066

Hammer DA, Bastian RK (1989) Wetlands ecosystems: natural water purifiers? In: Constructed wetlands for wastewater treatment. CRC Press, Boca Raton

He J, Moffette F, Fournier R, Revéret J, Théau J, Dupras J, Boyer J, Varin M (2015) Meta-analysis for the transfer of economic benefits of ecosystem services provided by wetlands within two watersheds in Quebec, Canada. Wetl Ecol Manag 23:707–725. https://doi.org/10.1007/s11273-015-9414-6

Iqbal A, Shang Z (2020) Wetlands as a carbon sink: insight into the Himalayan Region. In: Shang Z, Degen AA, Rafiq MK, Squires VR (eds) Carbon management for promoting local livelihood in the Hindu Kush Himalayan (HKH) Region. Springer International Publishing, pp 125–144. https://doi.org/10.1007/978-3-030-20591-1_7

Jacobsen D, Laursen SK, Hamerlik L, Hansen KM, Tsering T, Zhu B (2013) Sacred fish: on beliefs, fieldwork, and freshwater food webs in Tibet. Front Ecol Environ 11(1):50–51. https://doi.org/10.1890/1540-9295-11.1.50

Jamion NA, Lee KE, Mokhtar M, Goh TL, Simon N, Goh CT, Bhat IUH (2023) The integration of nature values and services in the nature-based solution assessment framework of constructed wetlands for carbon–water nexus in carbon sequestration and water security. Environ Geochem Health 45(5):1201–1230. https://doi.org/10.1007/s10653-022-01322-9

Jenyo-Oni A (2024) From aquatic splendor to wealth creation: towards sustainable management of Nigeria's wetland fisheries resources. In: An inaugural lecture at the University of Ibadan. Ibadan University Press Publishing House, University of Ibadan, Nigeria. ISBN: 978-978-8550-29-7

Jenyo-Oni A, Agbeja Y (2004) Functional roles of wetlands: a case study of the coastal wetlands of Lagos State, Nigeria. J Environ Ext 4(1):8–12. https://doi.org/10.4314/jext.v4i1.2727

Jessop J, Spyreas G, Pociask GE, Benson TJ, Ward MP, Kent AD, Matthews JW (2015) Tradeoffs among ecosystem services in restored wetlands. Biol Conserv 191:341–348. https://doi.org/10.1016/j.biocon.2015.07.006

Junk WJ, Brown M, Campbell IC, Finlayson M, Gopal B, Ramberg L, Warner BG (2006) The comparative biodiversity of seven globally important wetlands: a synthesis. Aquat Sci 68(3):400–414. https://doi.org/10.1007/s00027-006-0856-z

Kakuru W, Turyahabwe N, Mugisha J (2013) Total economic value of wetlands products and services in Uganda. Sci World J 2013(1):192656. https://doi.org/10.1155/2013/192656

Kingsford RT, Basset A, Jackson L (2016) Wetlands: conservation's poor cousins. Aquat Conserv Mar Freshwat Ecosyst 26(5):892–916. https://doi.org/10.1002/aqc.2709

Kumar R, Horwitz P, Milton GR, Sellamuttu SS, Buckton ST, Davidson NC, Pattnaik AK, Zavagli M, Baker C (2011) Assessing wetland ecosystem services and poverty interlinkages: a general framework and case study. Hydrol Sci J 56(8):1602–1621. https://doi.org/10.1080/02626667.2011.631496

Kwak S, Yoo S, Lee C (2007) Valuation of the Woopo Wetland in Korea: a contingent valuation study. Environ Dev Econ 12:323–328. https://doi.org/10.1017/S1355770X06003524

Langergraber G, Haberl R (2001) Constructed wetlands for water treatment. Minerva Biotecnol 13(2):123

Lazo JK (2002) Economic valuation of ecosystem services: discussion and application. Drug Chem Toxicol 25(4):349–374. https://doi.org/10.1081/dct-120014788

Leon AS, Tang Y, Chen D, Yolcu A, Glennie C, Pennings SC (2018) Dynamic management of water storage for flood control in a wetland system: a case study in Texas. Water 10(3):Article 3. https://doi.org/10.3390/w10030325

LePage BA (2011) Wetlands: a multidisciplinary perspective. In: LePage BA (ed) Wetlands: integrating multidisciplinary concepts. Springer, Netherlands, pp 3–25. https://doi.org/10.1007/978-94-007-0551-7_1

Liu X, Xiong Z, Ouyang L, He G, Liu W, Cai M (2024) Macrohabitat and microhabitat mediate the relationships between wetland multifaceted biodiversity and multifunctionality. Catena 241:108023. https://doi.org/10.1016/j.catena.2024.108023

Liverpool-Tasie LSO, Sanou A, Reardon T, Belton B (2021) Demand for imported versus domestic fish in Nigeria. J Agric Econ 72(3):782–804. https://doi.org/10.1111/1477-9552.12423

Lolu AJ, Ahluwalia AS, Sidhu MC, Reshi ZA, Mandotra SK (2020) Carbon sequestration and storage by wetlands: implications in the climate change scenario. In: Upadhyay AK, Singh R, Singh DP (eds) Restoration of wetland ecosystem: a trajectory towards a sustainable environment. Springer, pp 45–58. https://doi.org/10.1007/978-981-13-7665-8_4

Mahlatini P, Hove A, Maguma LF, Chemura A (2020) Using direct use values for economic valuation of wetland ecosystem services: a case of Songore wetland, Zimbabwe. GeoJournal 85(1):41–51. https://doi.org/10.1007/s10708-018-9947-3

Maltby E (2013) Waterlogged wealth: why waste the world's wet places? In: Routledge eBooks. Informa. https://doi.org/10.4324/9781315066776

Maltby E, Acreman MC (2011) Ecosystem services of wetlands: pathfinder for a new paradigm. Hydrol Sci J. https://www.tandfonline.com/doi/abs/10.1080/02626667.2011.631014

Maltby E, Barker T (2009) The wetlands handbook, 2 Volume Set. John Wiley & Sons, Chichester

Margaryan L, Prince S, Ioannides D, Röslmaier M (2022) Dancing with cranes: a humanist perspective on cultural ecosystem services of wetlands. Tour Geogr 24(4–5):501–522. https://doi.org/10.1080/14616688.2018.1522512

McCauley LA, Jenkins DG, Quintana-Ascencio PF (2013) Isolated wetland loss and degradation over two decades in an increasingly urbanized landscape. Wetlands 33(1):117–127. https://doi.org/10.1007/s13157-012-0357-x

McLaughlin DL, Cohen MJ (2013) Realizing ecosystem services: wetland hydrologic function along a gradient of ecosystem condition. Ecol Appl 23(7):1619–1631. https://doi.org/10.1890/12-1489.1

Meli P, Benayas JMR, Balvanera P, Ramos MM (2014) Restoration enhances wetland biodiversity and ecosystem service supply, but results are context-dependent: a meta-analysis. PLoS One 9(4):e93507. https://doi.org/10.1371/journal.pone.0093507

Metcalfe CD, Nagabhatla N, Fitzgerald SK (2018) Multifunctional wetlands: pollution abatement by natural and constructed wetlands. In: Nagabhatla N, Metcalfe CD (eds) Multifunctional wetlands: pollution abatement and other ecological services from natural and constructed wetlands. Springer International Publishing, pp 1–14. https://doi.org/10.1007/978-3-319-67416-2_1

Millenium Ecosystem Assessment (2005) Ecosystems and human well-being: wetlands and water. World Resources Institute

Mitsch WJ, Gosselink JG (2000) The value of wetlands: importance of scale and landscape setting. Ecol Econ 35(1):25–33. https://doi.org/10.1016/S0921-8009(00)00165-8

Mitsch WJ, Wu X, Nairn RW, Weihe PE, Wang N, Deal R, Boucher CE (1998) Creating and restoring wetlands. Bioscience 48(12):1019–1030. https://doi.org/10.2307/1313458

Mitsch WJ, Zhang L, Waletzko E, Bernal B (2014) Validation of the ecosystem services of created wetlands: two decades of plant succession, nutrient retention, and carbon sequestration in experimental riverine marshes. Ecol Eng 72:11–24. https://doi.org/10.1016/j.ecoleng.2014.09.108

Mitsch WJ, Bernal B, Hernandez ME (2015) Ecosystem services of wetlands. Int J Biodivers Sci Ecosyst Serv Manag 11(1):1–4. https://doi.org/10.1080/21513732.2015.1006250

Moore AC, Hierro L, Mir N, Stewart T (2022) Mangrove cultural services and values: current status and knowledge gaps. People Nat 4:1083–1097. https://doi.org/10.1002/pan3.10375

Mushet DM, Calhoun AJK, Alexander LC, Cohen MJ, DeKeyser ES, Fowler L, Lane CR, Lang MW, Rains MC, Walls SC (2015) Geographically isolated wetlands: rethinking a misnomer. Wetlands 35(3):423–431. https://doi.org/10.1007/s13157-015-0631-9

National Research Council (1995) Wetlands: characteristics and boundaries. National Academies Press. https://books.google.com/books?hl=en&lr=&id=-kwbfLFfamkC&oi=fnd&pg=PR1&dq=Wetlands:+Characteristics+and+Boundaries&ots=hIA5q_wX1t&sig=02krStjDWN7BhbU8CDH1bjDFGsM

Nayak A, Bhushan B (2022) Wetland ecosystems and their relevance to the environment: importance of wetlands. In: Handbook of research on monitoring and evaluating the ecological health of wetlands. IGI Global, pp 1–16. https://doi.org/10.4018/978-1-7998-9498-8.ch001

Owethu-Pantshwa A, Buschke FT (2019) Ecosystem services and ecological degradation of communal wetlands in a South African biodiversity hotspot. R Soc Open Sci 6(6):181770. https://doi.org/10.1098/rsos.181770

Panuccio M, Foschi F, Audinet J-P, Calò CM, Bologna MA (2017) Urban wetlands: wastelands or hotspots for conservation? Two case studies from Rome, Italy. Avocetta 41:13–18

Pappalardo SE, Ibrahim HMS, Cerinato S, Borin M (2017) Assessing the water-purification service in an integrated agricultural wetland within the Venetian Lagoon drainage system. Mar Freshw Res 68(12):2205–2215. https://doi.org/10.1071/MF16083

Pattison-Williams JK, Pomeroy JW, Badiou P, Gabor S (2018) Wetlands, flood control and ecosystem services in the Smith Creek drainage basin: a case study in Saskatchewan, Canada. Ecol Econ 147:36–47. https://doi.org/10.1016/j.ecolecon.2017.12.026

Pindilli EJ (2021) Ecosystem service co-benefits of wetland carbon management. In: Wetland carbon and environmental management, pp 401–409. https://doi.org/10.1002/9781119639305.ch22

Rafiq R, Alam MS, Rahman MM, Islam I (2014) Conserving wetlands: valuation of indirect use benefits of a wetland of Dhaka. Int J Environ Sci Dev 5(1):64

Ramsar Convention Secretariat (ed) (2007) Wise use of wetlands: a conceptual framework for the wise use of wetlands. Ramsar handbooks for the wise use of wetlands, 3rd edn. Ramsar Convention Secretariat, Gland

Rusciano V, Ruberto M, Baralla S, Fasolino NG, Pellegrini E, Zucaro R (2023) Assessing the touristic activities of wetlands through the travel cost method: a case study. Water 15(23):4146. https://doi.org/10.3390/w15234146

Sarvašová Z, Kovalčík M, Dobšinská Z, Šálka J, Jarský V (2014) Ecosystem services – examples of their valuation methods in Czech Republic and Slovakia. Change Adapt Socio Ecol Syst 1(1):74–83. https://doi.org/10.2478/cass-2014-0008

Scheufele G, Pascoe S (2023) Ecosystem accounting: reconciling consumer surplus and exchange values for free-access recreation. Ecol Econ 212:1–11. https://doi.org/10.1016/j.ecolecon.2023.107905

Semeraro T, Giannuzzi C, Beccarisi L, Aretano R, De Marco A, Pasimeni MR, Zurlini G, Petrosillo I (2015) A constructed treatment wetland as an opportunity to enhance biodiversity and ecosystem services. Ecol Eng 82:517–526. https://doi.org/10.1016/j.ecoleng.2015.05.042

Setlhogile T, Arntzen J, Mabiza C, Mano R (2011) Economic valuation of selected direct and indirect use values of the Makgadikgadi wetland system, Botswana. Phys Chem Earth Parts A/B/C 36(14):1071–1077. https://doi.org/10.1016/j.pce.2011.08.008

Shabman LA, Batie SS (1980) Estimating the economic value of coastal wetlands: conceptual issues and research needs. In: Elsevier EBooks, pp 3–15. https://doi.org/10.1016/b978-0-12-404060-1.50008-3

Sherrouse BC, Clement JM, Semmens DJ (2011) A GIS application for assessing, mapping, and quantifying the social values of ecosystem services. Appl Geogr 31(2):748–760. https://doi.org/10.1016/j.apgeog.2010.08.002

Siew M, Yacob M, Radam A, Adamu A, Alias E (2015) Estimating willingness to pay for wetland conservation: a contingent valuation study of Paya Indah Wetland, Selangor Malaysia. Procedia Environ Sci 30:268–272. https://doi.org/10.1016/J.PROENV.2015.10.048

Skarakis N, Skiniti G, Tournaki S, Tsoutsos T (2023) Necessity to assess the sustainability of sensitive ecosystems: a comprehensive review of tourism pressures and the travel cost method. Sustainability 15(15):1–18. https://doi.org/10.3390/su151512064

Sundaravadivel M, Vigneswaran S (2001) Constructed wetlands for wastewater treatment. Crit Rev Environ Sci Technol 31(4):351–409. https://doi.org/10.1080/20016491089253

Sutton-Grier A, Sandifer PA (2018) Society and our coasts: the role of coastal wetlands in reducing disaster impacts and promoting human health. In: AGU Fall Meeting Abstracts. pp OS33E–1951

Sutton-Grier AE, Sandifer PA (2019) Conservation of wetlands and other coastal ecosystems: a commentary on their value to protect biodiversity, reduce disaster impacts, and promote human health and well-being. Wetlands 39(6):1295–1302. https://doi.org/10.1007/s13157-018-1039-0

Tang Y, Leon AS, Kavvas ML (2020) Impact of size and location of wetlands on watershed-scale flood control. Water Resour Manag 34(5):1693–1707. https://doi.org/10.1007/s11269-020-02518-3

Thapa S, Wang L, Koirala A, Shrestha S, Bhattarai S, Aye W (2020) Valuation of ecosystem services from an important wetland of Nepal: a study from Begnas Watershed System. Wetlands:1–13. https://doi.org/10.1007/s13157-020-01303-7

Tienhaara A, Lankia T, Lehtonen O, Pouta E (2021) Heterogeneous preferences towards quality changes in water recreation: latent class model for contingent behavior data. J Outdoor Recreat Tour 35:100386. https://doi.org/10.1016/j.jort.2021.100386

Tuan TH, Van Xuan M, Nam D, Navrud S (2009) Valuing direct use values of wetlands: a case study of Tam Giang–Cau Hai lagoon wetland in Vietnam. Ocean Coast Manag 52(2):102–112. https://doi.org/10.1016/j.ocecoaman.2008.10.011

Turner RK, van den Bergh JCJM, Söderqvist T, Barendregt A, van der Straaten J, Maltby E, van Ierland EC (2000) Ecological-economic analysis of wetlands: scientific integration for management and policy. Ecol Econ 35(1):7–23. https://doi.org/10.1016/S0921-8009(00)00164-6

Vázquez-González C, Moreno-Casasola P, Peralta Peláez LA, Monroy R, Espejel I (2019) The value of coastal wetland flood prevention lost to urbanization on the coastal plain of the Gulf of Mexico: an analysis of flood damage by hurricane impacts. Int J Disaster Risk Reduct 37:101180. https://doi.org/10.1016/j.ijdrr.2019.101180

Verschuuren B, McNeely J, Oviedo G, Wild R (2012) Sacred natural sites: conserving nature and culture, 1st edn. Routledge, 336 pages. https://www.routledge.com/Sacred-Natural-Sites-Conserving-Nature-and-Culture/Verschuuren-Wild-Mcneely-Oviedo/p/book/9781849711678

Villar MMP, Domínguez ER, Tack F, Ruiz JMH, Morales RS, Arteaga LE (2012) Vertical subsurface wetlands for wastewater purification. Proc Eng 42:1960–1968. https://doi.org/10.1016/j.proeng.2012.07.592

Wamsley TV, Cialone MA, Smith JM, Atkinson JH, Rosati JD (2010) The potential of wetlands in reducing storm surge. Ocean Eng 37(1):59–68. https://doi.org/10.1016/j.oceaneng.2009.07.018

Were D, Kansiime F, Fetahi T, Cooper A, Jjuuko C (2019) Carbon sequestration by wetlands: a critical review of enhancement measures for climate change mitigation. Earth Syst Environ 3(2):327–340. https://doi.org/10.1007/s41748-019-00094-0

Wetzel RG (1992) Wetlands as metabolic gates. J Great Lakes Res 18(4):529–532. https://doi.org/10.1016/S0380-1330(92)71320-3

Woodward RT, Wui Y-S (2001) The economic value of wetland services: a meta-analysis. Ecol Econ 37(2):257–270. https://doi.org/10.1016/s0921-8009(00)00276-7

Wubalem A, Woldeamanuel T, Nigussie Z (2023) Economic valuation of Lake Tana: a recreational use value estimation through the travel cost method. Sustainability 15(8):1–20. https://doi.org/10.3390/su15086468

Xu T, Weng B, Yan D, Wang K, Li X, Bi W, Li M, Cheng X, Liu Y (2019) Wetlands of international importance: status, threats, and future protection. Int J Environ Res Public Health 16(10):Article 10. https://doi.org/10.3390/ijerph16101818

Yoo J, Kim J, Kim J, Lim J, Kang H (2022) Soil carbon storage and its economic values of inland wetlands in Korea. Ecol Eng 182:106731. https://doi.org/10.1016/j.ecoleng.2022.106731

Zhang B, Shi Y, Liu J, Xu J, Xie G (2017) Economic values and dominant providers of key ecosystem services of wetlands in Beijing, China. Ecol Indic 77:48–58. https://doi.org/10.1016/j.ecolind.2017.02.005

Chapter 5
Socioeconomic Benefits of Wetland-Based Water Treatment: Discussing the Economic, Social, and Environmental Advantages Associated with Utilizing Wetlands for Water Treatment

A. Adewumi, J. O. Olorunleke, A. E. Adeyemi, and F. V. Bekun

Abstract Wetland-based water treatment systems provide a sustainable alternative to conventional water treatment methods, offering significant socioeconomic benefits across economic, social, and environmental dimensions. Economically, wetlands—both natural and constructed—are cost-effective, reducing infrastructure and operational costs while also providing long-term financial benefits, such as revenue from byproducts like biomass. Wetlands help mitigate carbon emissions, contributing to cost savings associated with climate change mitigation strategies. Socially, wetland-based water treatment systems enhance community well-being by providing essential ecosystem services that improve public health, support recreational opportunities, and foster educational initiatives. These systems also play a crucial role in maintaining cultural values and aesthetics, particularly for communities that are closely tied to wetland ecosystems. Environmentally, wetlands support biodiversity, create habitats for wildlife, and improve water quality by naturally filtering pollutants. Their role in climate regulation, particularly through carbon sequestration, further emphasizes their importance in addressing global environmental challenges. This chapter provides a comprehensive discussion of these socioeconomic benefits, highlighting the multifaceted advantages of utilizing wetlands for water treatment.

A. Adewumi (✉) · J. O. Olorunleke
Federal College of Freshwater Fisheries Technology, New Bussa, Nigeria

A. E. Adeyemi
Texas A&M University, College Station, TX, USA

F. V. Bekun
Istanbul Gelisim University, Istanbul, Turkey

Lebanese American University, Beirut, Lebanon

B. R. Babaniyi et al. (eds.), *Wetland Ecosystems: Conservation Strategies, Policy Management and Applications*, Wetlands: Ecology, Conservation and Management 12, https://doi.org/10.1007/978-3-031-91982-4_5

Keywords Wetland-based water treatment · Socioeconomic benefits · Ecosystem services · Cost-effectiveness · Public health · and Climate regulation

5.1 Introduction

Wetland-based water treatment systems have emerged as an ecologically sustainable solution to address the growing concerns over water quality and environmental degradation. These systems utilize the natural processes occurring in wetlands—both natural and constructed—to treat contaminated water. Wetlands act as natural filters, removing pollutants, nutrients, and pathogens from water, thus playing a crucial role in maintaining water quality and protecting aquatic ecosystems. The integration of wetlands into water treatment strategies not only enhances the efficiency of pollutant removal but also provides additional ecological benefits, such as habitat creation, carbon sequestration, and biodiversity enhancement (Vymazal 2011; Mitsch and Gosselink 2015).

Wetlands, whether natural or constructed, are characterized by their waterlogged conditions, which support a unique combination of soil, water, plants, and microorganisms. These components work together to break down organic matter, absorb nutrients, and trap sediments and heavy metals. The two primary types of wetlands used in water treatment are natural wetlands and constructed wetlands.

5.1.1 Natural Wetlands

Natural wetlands, such as marshes, swamps, and bogs, have been recognized for their inherent ability to improve water quality. These ecosystems have evolved to perform a range of ecological functions, including water purification, flood control, and climate regulation. Natural wetlands remove pollutants through a combination of physical, chemical, and biological processes, such as sedimentation, adsorption, microbial degradation, and plant uptake (Kadlec and Wallace 2008). However, their effectiveness can be influenced by various factors, including hydrology, climate, and human disturbances.

5.1.2 Constructed Wetlands

Constructed wetlands are engineered systems designed to mimic the natural processes of natural wetlands. They are specifically constructed for water treatment purposes and are often used to treat municipal wastewater, agricultural runoff, stormwater, and industrial effluents. Constructed wetlands are categorized into two main types: surface flow wetlands and subsurface flow wetlands. In surface flow wetlands, water flows above the soil surface, whereas in subsurface flow wetlands, water flows through a porous medium, such as gravel, below the soil surface. These systems are designed to optimize conditions for the removal of specific contaminants, such as nitrogen, phosphorus, and suspended solids, through processes like denitrification, adsorption, and phytoremediation (Vymazal 2010).

Brix (1994) stated that the utilization of wetlands for water treatment offers numerous environmental, economic, and social benefits. From an environmental perspective, wetland-based systems reduce the need for chemical treatment processes, thereby minimizing chemical residues and reducing the risk of secondary pollution. Additionally, wetlands provide critical habitats for wildlife, contributing to the conservation of biodiversity. Economically, wetland-based systems are often more cost-effective than conventional treatment plants, particularly in rural or developing areas where infrastructure costs can be prohibitive. Socially, these systems can enhance community resilience by providing flood protection, improving water quality, and creating green spaces for recreation and education.

5.2 Economic Advantages of Wetland-Based Water Treatment

Wetlands, both natural and constructed, have long been recognized for their ability to purify water through natural processes. In recent years, the application of wetlands for water treatment has gained attention as a viable and economically advantageous alternative to conventional treatment methods. Constructed wetlands, in particular, mimic the natural processes of wetlands to remove pollutants from wastewater, stormwater, and agricultural runoff. The economic advantages of wetland-based water treatment systems are diverse. They are not only cost-effective but also offer long-term financial benefits, including savings on operational costs and the generation of valuable byproducts like biomass. This section highlights these advantages.

5.2.1 Cost-Effectiveness Compared to Conventional Treatment Methods

One of the primary economic advantages of wetland-based water treatment systems is their cost-effectiveness compared to conventional treatment methods. Traditional water treatment systems, such as activated sludge processes, typically require substantial capital investment in infrastructure, energy, and chemicals. These conventional systems involve complex machinery, significant energy consumption, and regular chemical inputs to maintain water quality, all of which contribute to high initial and ongoing costs. In contrast, wetland-based systems utilize natural processes like filtration, sedimentation, and biological uptake to remove contaminants from water, significantly reducing the need for expensive infrastructure and operational inputs.

Wetland-based systems are particularly appealing due to their low construction and maintenance costs. Studies such as that of Griffith (1992) highlighted the economic viability of wetland-based systems in Virginia, where several hundred wetland systems were constructed at a fraction of the cost of conventional wastewater treatment facilities. The study demonstrated that wetlands effectively regulated mine outflows and used their natural filtration capabilities to provide a cost-effective solution to water treatment challenges. This highlights the potential for wetlands to be employed in a range of contexts, offering a more financially sustainable alternative to traditional methods.

Another study by Malaviya and Singh (2012) emphasized the growing popularity of wetland-based stormwater treatment systems due to their cost-effectiveness. The study found that these systems were particularly suitable for urban areas, where they provided an economical solution for managing stormwater runoff. Urban environments often face significant challenges with stormwater management due to the large amounts of impervious surfaces, such as roads and buildings, which increase runoff and lead to flooding and water pollution. Wetland-based systems mitigate these issues by naturally filtering and absorbing excess water, reducing the need for costly stormwater infrastructure. This makes them an attractive option for cities and towns looking to manage water resources efficiently and affordably.

Constructed wetlands require lower capital and operational costs, making them an attractive option for municipalities and industries. Donoso (2018) found that constructed wetlands provided a cost-effective solution for treating agricultural wastewater, with lower operational costs and minimal maintenance requirements. Moreover, wetlands can be integrated into existing landscapes, reducing the need for additional land acquisition. This is particularly relevant in urban areas, where land availability is limited and expensive. The cost savings associated with wetland-based systems extend beyond construction and maintenance, as they also provide long-term benefits such as carbon sequestration and biodiversity conservation.

In addition to lower construction costs, wetland-based systems offer considerable savings in terms of operational and maintenance expenses. Conventional treatment systems require continuous energy input to power pumps, aerators, and other machinery, as well as regular chemical dosing to ensure water quality. These factors contribute to high ongoing operating expenses, making traditional systems financially burdensome in the long term. In contrast, wetland systems are largely passive, relying on natural processes such as sedimentation, filtration, and plant uptake to treat water. This passive nature reduces the need for external energy sources and chemical inputs, resulting in significantly lower operational costs over time. Moreover, wetland-based systems have the advantage of being adaptable and scalable, making them suitable for a wide range of applications, from small rural communities to large urban centers. The flexibility of these systems allows for incremental investments, enabling communities and municipalities to implement wetland-based treatment solutions at a pace that aligns with their financial resources.

5.2.2 Long-Term Financial Benefits and Savings

The long-term financial benefits of wetland-based water treatment systems extend far beyond their initial cost-effectiveness. These systems provide sustained financial savings through a combination of reduced operational costs, lower maintenance requirements, and the potential for revenue generation from byproducts. Unlike conventional water treatment facilities, which often require significant energy inputs and chemical usage, wetland-based systems rely on natural processes, thereby minimizing ongoing expenses. Over time, this reduction in operational costs can translate into substantial savings for municipalities, industries, and communities that adopt these sustainable alternatives.

One of the key long-term financial benefits of wetland-based water treatment systems is their ability to mitigate greenhouse gas emissions, which can help offset the costs associated with carbon mitigation. Rhymes et al. (2023) conducted an extensive study on the long-term carbon gains associated with wetland-based water treatment systems. The researchers found that these systems not only reduced greenhouse gas emissions but also provided significant long-term financial benefits by decreasing the need for costly carbon mitigation measures. They also help capture and store carbon dioxide which is a major contributor to climate change. This capability enhances the financial viability of wetland systems, as organizations and governments seek cost-effective solutions to meet their carbon reduction targets.

Another notable advantage of wetland systems is their potential to generate revenue through the sale of biomass. Wetlands naturally accumulate biomass, including plants such as reeds, cattails, and grasses, which can be harvested and utilized for various purposes. Biomass from wetlands can serve as a valuable resource for bioenergy production, contributing to the generation of renewable energy. Snyder (2019) explored the economic value of biomass generated from wetland systems in the neotropics. The study highlighted how the inclusion of ecosystem service valuations, such as biomass production, in bioenergy cost analysis significantly enhanced the economic viability of wetland-based systems. By converting plant biomass into energy or other marketable products, wetland systems offer an additional revenue stream, further improving their financial sustainability.

Moreover, wetland systems can play a critical role in addressing water scarcity, particularly in arid regions. Water reuse initiatives, which involve treating and reusing wastewater for agricultural or industrial purposes, are becoming increasingly important in areas facing water shortages. Gabr et al. (2023) proposed a wetland-based economic approach for wastewater treatment in arid regions, emphasizing its potential as an alternative source of irrigation water. The study found that wetland systems provided a cost-effective solution for water reuse, reducing the need for expensive desalination or groundwater extraction methods. By enhancing water availability in water-scarce regions, wetland-based systems contribute to both environmental sustainability and economic resilience, making them a valuable tool in modern water management strategies.

5.2.3 Economic Value of Byproducts

Wetland-based water treatment systems offer not only substantial cost savings but also generate valuable byproducts that can be monetized, creating additional economic incentives for their adoption. One of the most significant byproducts of these systems is biomass, which can be harvested and utilized for various applications, including bioenergy production, animal feed, and construction materials. The generation of biomass through wetland-based water treatment systems not only provides a sustainable energy source but also contributes to waste reduction and resource efficiency, key principles of the circular economy.

The economic value of biomass derived from wetland systems is increasingly recognized in both research and practice. A study by Turpie and Kleynhans (2010) examined the broader economic benefits of wetland ecosystem services, emphasizing the importance of biomass production. Their research highlighted that the valuation of wetland services, such as sediment retention, water purification, and biomass generation, significantly enhances the overall economic viability of wetland-based water treatment systems. By quantifying these services, Turpie and Kleynhans demonstrated how wetlands could provide direct financial returns, making them a valuable asset for municipalities and industries seeking cost-effective water treatment solutions. Biomass from wetlands can be converted into bioenergy, offering a renewable energy source that reduces reliance on fossil fuels. This is particularly relevant in the context of global efforts to transition to low-carbon energy systems.

Moreover, research has shown that wetland systems can produce other valuable byproducts beyond biomass. These include fish and wildlife habitats, which can be leveraged for recreational and commercial purposes. For example, the East Kolkata Wetlands in India serve as a model for integrating wastewater treatment with fish farming. In a study conducted by Dey and Banerjee (2018), the researchers estimated the opportunity cost of wetland degradation in Kolkata and highlighted the economic gains from sustainable management practices that integrated wastewater treatment with aquaculture. This innovative approach not only improved water quality but also supported local livelihoods through fish farming, demonstrating how wetland systems can provide multiple economic and social benefits simultaneously. The East Kolkata Wetlands showcase how combining wastewater treatment with aquaculture can yield profitable outcomes while preserving the ecosystem's health. By maintaining the wetlands and leveraging their natural processes, communities can create a sustainable source of income and food security. This case study underscores the potential of wetlands to serve as multifunctional landscapes that address both environmental and economic challenges.

The production of biomass from wetland systems aligns with the principles of the circular economy, which emphasizes resource efficiency, waste reduction, and the reuse of materials. The circular economy model seeks to create closed-loop systems where waste is minimized, and resources are continually cycled back into production processes. In the context of wetland-based water treatment, this means that the byproducts generated, such as biomass, can be reused or repurposed rather than discarded. This approach not only reduces the environmental impact of waste but also enhances the financial sustainability of wetland systems. Sharma et al. (2020) explored the circular economy of sanitation in wetland systems and found that the reuse of biomass and other byproducts significantly contributed to the sustainability of these systems. Their research demonstrated that by integrating circular economy principles, wetland-based water treatment systems could achieve greater efficiency and resilience, ultimately leading to long-term economic and environmental benefits.

In addition to biomass and aquaculture, wetland systems can also generate other marketable products, such as reeds and grasses, which can be used in construction materials, crafts, and even as feedstock for bioenergy production. These materials have economic value, particularly in regions where traditional construction materials are expensive or scarce. The ability to harvest and sell these products provides an additional revenue stream for communities and organizations that manage wetland systems. Furthermore, the multifunctional nature of wetlands allows for the integration of various economic activities, creating synergies that enhance the overall economic value of these ecosystems. For example, wetlands that support tourism and recreation can generate revenue through ecotourism, while also providing ecosystem services such as water purification and flood control. The aesthetic and cultural value of wetlands can attract visitors, creating opportunities for local businesses and generating income for conservation efforts (Jenyo-Oni 2024).

The wetland-based water treatment systems offer more than just cost-effective water purification; they also generate valuable byproducts that can be monetized, contributing to the economic sustainability of these systems. The production of biomass, fish, and other marketable products from wetlands aligns with the principles of the circular economy, promoting resource efficiency and waste reduction. Studies by Turpie and Kleynhans (2010), Dey and Banerjee (2018), and Sharma et al. (2020) demonstrate the economic benefits of wetland systems and highlight the potential for wetlands to serve as multifunctional landscapes that provide environmental, social, and economic benefits. As the demand for sustainable and resilient water management solutions grows, wetland-based systems offer a promising approach that integrates ecological preservation with economic development.

5.3 Social Benefits of Wetland-Based Water Treatment

Wetlands, often referred to as the earth's kidneys, are vital ecosystems offering numerous benefits, including biodiversity conservation, agricultural energy balance, and climate regulation (Bhowmik 2022; Huryna et al. 2014; Mitsch and Gosselink 2000; Nayak and Bhushan 2022; Williams and Dodd Jr 1978). One of their most essential roles is in water treatment (Hammer and Bastian 1989), where wetlands naturally purify water through a combination of physical, chemical, and biological processes (Singh et al. 2019; Verhoeven and Meuleman 1999). This inherent purification ability is increasingly being utilized for wastewater and stormwater treatment, offering a sustainable and cost-effective alternative to traditional treatment facilities (House et al. 1999). Therefore, using wetlands for water treatment is not only an environmental and economic gain but also a significant social asset.

This section explores the social benefits of wetland-based water treatment, highlighting how these systems enhance community well-being, public health, recreational opportunities, educational initiatives, and cultural values. Understanding the social implications of wetland-based water treatment is essential, as it provides a holistic view of the diverse advantages offered by these ecosystems. Beyond their ecological functions, wetlands are invaluable resources that elevate the quality of life, promote social cohesion, and ensure equitable access to clean water. As communities worldwide face challenges such as water scarcity and pollution, recognizing and harnessing the social benefits of wetlands can significantly contribute to the development of sustainable and resilient societies.

5.3.1 Enhancing Community Well-Being

Wetland-based water treatment systems are integral to enhancing community well-being by providing essential ecosystem services that extend beyond traditional water management. The multifunctionality of wetlands contributes to various aspects of life, particularly in communities where access to clean water, sanitation, and other basic services is limited. Wetlands are not just natural water purifiers; they also serve as valuable resources that support livelihoods, public health, and environmental sustainability.

The nexus of wetlands with broader societal issues is well recognized. Horwitz et al. (2015) emphasized the importance of adopting a comprehensive approach to wetland management that addresses not only nature conservation but also crucial human activities and needs. They argue that the well-being of communities is deeply tied to the health of wetland ecosystems, as these ecosystems influence food production, water use, and extraction. The integration of wetlands into community planning can improve sanitation and water supply, support agriculture and fisheries, and enhance education and training programs. Moreover, wetlands can play a role in trade, tourism, transport, and infrastructure development, all of which are vital to community development and well-being.

Wetland-based water treatment systems also offer sustainable and cost-effective solutions for managing wastewater and stormwater. This is particularly significant in rural and underserved areas where traditional treatment facilities may be limited or unavailable. Through the natural filtration and purification of water, wetlands improve water quality and reduce pollution, creating a healthier living environment for residents. For example, in Eastern India, a study by Adhya and Banerjee (2022) revealed that wetland-dependent villagers identified 18 ecosystem services provided by the wetland under study. These services included 12 provisioning services, such as food and water, and regulatory, cultural, and supporting services, which all contributed to the community's well-being. The sustainable management of wetlands ensures that these services continue to benefit communities, enhancing their quality of life and economic resilience.

Furthermore, the implementation of wetland-based water treatment systems can lead to significant public health improvements. Communities that adopt these systems often experience a decline in waterborne diseases, which are prevalent in areas with poor water quality. Metcalfe et al. (2018) highlighted the role of wetlands in reducing healthcare costs by preventing the spread of diseases associated with contaminated water. Urban wetlands, for instance, act as natural filters that remove harmful contaminants and pathogens from the water supply, addressing the challenge of providing clean water in densely populated areas (Ribbe et al. 2024). This natural purification process is critical in preventing illnesses such as cholera, dysentery, and other gastrointestinal diseases, which are common in areas with inadequate water treatment facilities (Anthonj et al. 2019; Ezenwa et al. 2007; Rey et al. 2012).

In addition to physical health benefits, wetlands also positively impact mental health and well-being. The natural environment of wetlands, with their blue-green spaces, has been shown to reduce stress and improve mental health. Maund et al. (2019) conducted a 6-week intervention in the UK where participants were exposed to wetland environments as part of a treatment for anxiety and depression. The study found substantial improvements in mental health indicators, including reduced stress and enhanced emotional well-being. This suggests that wetlands provide more than just ecological benefits; they also offer therapeutic environments that support mental health. Similarly, Reeves et al. (2019) supported the notion that exposure to blue-green spaces aids in stress recovery, though they called for further research to strengthen the evidence. The potential for wetlands to serve as spaces for mental health interventions underscores their value in promoting holistic well-being in communities.

5.3.2 Recreational Opportunities and Quality of Life

Wetlands offer numerous recreational opportunities that significantly enhance the quality of life for community members. These ecosystems provide a unique combination of ecological, aesthetic, and recreational values, making them indispensable assets in both rural and urban settings.

In an Australian case study, Carter (2015) explored the positive health benefits associated with the recreational use of urban wetlands. The study found that access to these natural spaces led to improvements in both physical and psychological health. Regular exposure to nature, facilitated by wetlands, helped to reduce stress levels, improve mood, and promote physical activity. Activities such as walking, jogging, and cycling around wetlands contributed to better cardiovascular health and overall well-being. The study also highlighted the social benefits of wetlands, noting that they fostered increased community connection and a heightened sense of place. Wetlands often serve as communal gathering spots where people of all ages can engage in recreational activities, thereby strengthening social ties and enhancing community cohesion.

These ecosystems, teeming with biodiversity, provide unique settings for a variety of outdoor activities such as birdwatching, fishing, hiking, and photography. Birdwatching, in particular, is a popular activity in wetland areas due to the presence of diverse avian species that are often not found in other ecosystems. This biodiversity not only attracts nature enthusiasts but also provides educational opportunities for schools and environmental groups. The aesthetic appeal of wetlands, characterized by their lush vegetation and vibrant wildlife, draws photographers and artists seeking inspiration from the natural world. Moreover, the tranquil and serene environment of wetlands offers a peaceful retreat from the hustle and bustle of urban life, providing city dwellers with a much-needed escape into nature (Bacon 1987).

In addition to their recreational and aesthetic value, wetlands contribute significantly to local economies through tourism. In their study, Shah and Islam (2021) quantified the economic benefits derived from tourism in wetland areas, with a focus on Dal Lake in Jammu and Kashmir, India. They estimated that the average consumer surplus per visitor was Rs 6250 (US$96.15) and Rs 25,000 (US$384.61), depending on the type of tourism activity and the visitor's willingness to pay for recreational experiences. These figures underscore the high value that tourists place on their wetland experiences. The total recreational value of Dal Lake was assessed at Rs 1713 million (US$26 million), highlighting the substantial revenue generated from the wetland's tourism activities. This revenue supports local businesses, creates jobs, and contributes to the region's overall economic development. Wetland tourism can also drive investments in infrastructure, such as trails, visitor centers, and educational programs, further enhancing the recreational value of these ecosystems.

However, the economic benefits of wetlands are not universally recognized. In a study conducted in Uganda, Bikangaga et al. (2007) found that while local community leaders acknowledged the tourism potential of wetlands, they expressed mixed feelings about their overall value. Some community members perceived the benefits of wetland tourism as limited, citing concerns over the creation of low-paying jobs and questioning the long-term contribution of wetlands to the local economy. These concerns reflect broader challenges in balancing conservation and economic development, particularly in regions where immediate financial needs may outweigh environmental considerations.

Nonetheless, there is evidence that these perceptions can be positively influenced through education and awareness campaigns. Macharia et al. (2010) demonstrated in their research in Kenya that when local communities are informed about the ecological and economic benefits of wetlands, they are more likely to support conservation efforts and participate in sustainable tourism initiatives. Education can also help to shift perceptions by highlighting the long-term benefits of wetlands, including their role in ecosystem services such as water purification, flood control, and carbon sequestration, which ultimately contribute to the well-being of the entire community.

Summing it up, Zedler and Leach (1998) affirmed through their studies in California and Wisconsin that the 3Rs—recreation, restoration, and research—are not only compatible but also synergistic uses of urban wetlands. Their work underscores the multifaceted value of these ecosystems in enhancing urban environments through diverse and complementary applications. Recreation brings people closer to nature, fostering a sense of stewardship and appreciation for wetland conservation. Restoration efforts, in turn, improve the ecological health of wetlands, making them more resilient and attractive for recreational use. Research conducted in these environments contributes to our understanding of wetland dynamics, informing better management practices that balance human use with ecological integrity.

5.3.3 Educational and Research Opportunities

Wetlands offer a wealth of educational and research opportunities that are crucial for enhancing societal knowledge and environmental awareness. These ecosystems, with their rich biodiversity and complex ecological processes, serve as living laboratories where students, educators, and researchers can engage in hands-on learning and scientific inquiry. The dynamic and diverse nature of wetlands makes them ideal environments for studying ecological processes, species-environment interactions, and the impact of human activities on natural ecosystems.

One of the primary educational benefits of wetlands is their ability to provide experiential learning opportunities. Educational institutions can integrate wetlands into their curricula, allowing students to participate in fieldwork, observe natural processes, and engage in environmental stewardship activities. This type of experiential learning is invaluable, as it enables students to apply theoretical knowledge in real-world settings, deepening their understanding of ecological concepts and fostering a sense of environmental responsibility (Maddock 1991; Mitsch et al. 2008). For example, students can learn about hydrology by observing water flow patterns in wetlands, study plant biology by identifying wetland vegetation, or explore animal behavior by monitoring wildlife activity.

Moreover, wetlands serve as platforms for interdisciplinary learning, bridging the gap between biology, chemistry, geology, and environmental science. By studying wetlands, students can gain insights into the interconnectedness of various natural processes, such as nutrient cycling, energy flow, and food web dynamics. This interdisciplinary approach not only enhances students' scientific literacy but also prepares them to address complex environmental challenges in their future careers.

In addition to formal education, wetlands play a pivotal role in informal science learning. Research conducted in wetland environments immerses participants in scientific inquiry, encouraging them to observe, hypothesize, and experiment. This hands-on research experience is particularly valuable for fostering critical thinking and problem-solving skills. Participants can directly observe and measure environmental variables, analyze data, and draw conclusions based on empirical evidence. This active engagement in the scientific process helps to demystify science and makes it more accessible to learners of all ages (Maddock 1991).

Wetlands also provide unique opportunities for climate change research. These ecosystems are crucial for studying carbon sequestration and methane formation, two key processes that influence global climate dynamics. Wetlands store large amounts of carbon in their soils and vegetation, acting as carbon sinks that help mitigate climate change. However, they can also release methane, a potent greenhouse gas, under certain conditions. According to Mitsch et al. (2008), scientists can gain a better understanding of how these ecosystems respond to changing environmental conditions and how they can be managed to maximize their climate benefits by conducting research in wetlands. This research not only contributes to the global body of knowledge on climate change but also informs policy decisions related to wetland conservation and management.

Furthermore, wetlands offer opportunities for research that balances students' knowledge with their attitudes and values. With fieldwork and environmental monitoring, students can develop a deeper appreciation for the natural world and a stronger commitment to environmental conservation. Sukhontapatipak and Srikosamatara (2012) argued that this type of research fosters a sense of place and connection to nature, which is essential for cultivating environmental ethics and promoting sustainable behavior. Students who participate in wetland research are more likely to become advocates for environmental protection and stewards of natural resources.

The educational and research opportunities provided by wetlands extend beyond the academic settings. These ecosystems also offer valuable insights for community engagement and public education. Wetland restoration projects, for example, often involve collaboration between scientists, educators, and local communities. These projects provide opportunities for citizens to participate in environmental monitoring, habitat restoration, and conservation efforts. By involving the public in wetland research, these initiatives promote environmental awareness and encourage community-based conservation (Calhoun et al. 2003).

Moreover, wetlands offer opportunities for long-term ecological research that generates invaluable data on ecosystem dynamics and environmental change. These long-term studies are critical for understanding how wetlands respond to natural and anthropogenic disturbances, such as climate change, pollution, and habitat loss. The data collected from research sites can be used to inform conservation strategies, guide land-use planning, and assess the effectiveness of wetland management practices. Additionally, the insights gained from long-term wetland research can be applied to other ecosystems, contributing to a broader understanding of ecological processes and environmental sustainability.

Finally, wetlands inspire students to pursue advanced scientific research and careers in environmental science. Providing access to unique ecosystems and research opportunities can encourage students to explore new areas of inquiry and contribute to the advancement of scientific knowledge. The experience of conducting research in wetlands can be transformative, sparking curiosity and passion for science that extends beyond the classroom. As a result, students who engage in wetland research are more likely to pursue graduate studies and careers in environmental science, ecology, and conservation (Calhoun et al. 2003).

5.3.4 Cultural and Aesthetic Values

Wetlands also holds significant cultural and aesthetic values for many communities around the world. These ecosystems are often deeply intertwined with local traditions, history, and cultural practices. For many indigenous and local communities, wetlands serve as sacred spaces, imbued with spiritual significance and often integral to their cultural heritage. These areas may be sites for traditional ceremonies, storytelling, and community gatherings, thus playing a crucial role in maintaining cultural identity and continuity. Additionally, wetlands are often associated with local folklore and mythology, further embedding them in the cultural fabric of the societies that rely on them (Alikhani et al. 2021).

The aesthetic value of wetlands also cannot be overstated. The natural beauty of these ecosystems, characterized by their lush vegetation, serene water features, and abundant wildlife, provides inspiration for various forms of artistic expression, including painting, literature, and photography. Wetlands offer a visual and sensory experience that fosters creativity and reflection, making them valuable not only as ecological resources but also as sources of cultural enrichment as opined by Carter (2015) and Jenyo-Oni (2024). The tranquil landscapes of wetlands often serve as places of recreation and relaxation, offering spaces where people can connect with nature and find solace away from the urban hustle.

Moreover, the presence of well-maintained wetlands can significantly enhance the attractiveness of a region, contributing to its identity and making it more appealing for both residents and visitors alike. This aesthetic appeal can foster a strong sense of pride and connection to the local environment, which in turn encourages community efforts to protect and preserve these vital ecosystems. The visual and cultural significance of wetlands thus plays a crucial role in promoting environmental stewardship and sustainability, reinforcing the need to integrate these values into conservation and management strategies (Zhou et al. 2020).

5.4 Environmental Benefits of Wetland-Based Water Treatment

Wetland-based water treatment systems provide extensive environmental benefits beyond their primary function of water purification. They enhance biodiversity, create habitats for various species, and support ecosystem balance and resilience (Solano et al. 2004). The diversity of flora and fauna in wetland ecosystems aids nutrient cycling and increases environmental stability and productivity, making wetlands crucial components of a healthy landscape and water management strategy (Karia 2012).

Wetlands act as natural filtration systems, trapping suspended solids and nutrients to reduce pollution and improve water quality. These natural processes cleanse polluted waters, stabilize water supplies, and mitigate the impacts of floods and droughts (Mitsch et al. 2015). Additionally, wetlands serve as significant carbon sinks, capturing and storing atmospheric carbon dioxide, thus playing a vital role in climate regulation.

5.4.1 Enhanced Biodiversity and Habitat Creation

Wetlands are among the most biologically diverse ecosystems, providing essential habitats for various species, including many that are threatened or endangered. These ecosystems support complex food webs and overall biodiversity, making them critical for ecological health. Wetlands offer direct life support, nutrient exchange, and fish food production, which are essential for sustaining various species. The flora of wetlands varies by region, with tropical swamplands featuring palm species like Sabal jamaicensis in Jamaica and Nypa fruiticans in the Philippines, while freshwater swamps in Nigeria have more species diversity than mangrove ecosystems. Mangroves, dominated by Rhizophora species, provide highly durable wood and other resources to nearby communities (Ayodele and Jenyo-Oni 2002).

The fauna in wetlands is equally diverse, including mollusks, crustaceans, finfish, reptiles, birds, and mammals. These ecosystems serve as vital breeding and rearing grounds for aquatic resources, contributing significantly to local economies, such as in Nigeria where the bonga fish (Ethmalosa fimbriata) and shrimp fisheries (Penaeus notialis) contribute to the GDP (FAO 2020).

Wetlands provide essential services for many endangered species, offering a lifeline for freshwater species and acting as crucial stopovers for migratory birds. The rich biodiversity in wetlands, often referred to as nature's supermarkets, supports extensive food chains, making them as productive as rainforests and coral reefs (Mitsch and Gosselink 2015). Wetlands play a vital role in watershed ecology, providing habitats for a wide variety of flora and fauna, which form the base of the food web and sustain numerous species, including migratory birds and mammals.

A critical review by Li et al. (2021) emphasized the role of constructed wetland animals in water environment protection. The authors noted that animals such as fish, amphibians, and invertebrates play essential roles in maintaining the ecological balance of wetlands and enhancing their water treatment capabilities. Therefore, use of wetlands for water treatment can contribute to biodiversity conservation by providing habitat for wildlife, including migratory birds, amphibians, and fish. Constructed wetlands, when properly designed, can mimic natural habitats and support native species.

5.4.2 Reduction of Pollution and Improved Water Quality

Wetlands are dynamic systems that store water and improve its quality by acting as natural filters. They purify water, recharge groundwater, and control urban runoff, making them essential for maintaining water quality. Wetlands have been extensively researched and proven to remove or reduce pollutants such as nutrients, heavy metals, hydrocarbons, and pesticides. Their ability to improve water quality depends on various factors, including the type of vegetation, size, design, and local conditions (Clarkson et al. 2013).

Wetland-based water treatment systems contribute to water security by enhancing water quality and supporting water reuse initiatives. In regions facing water scarcity, wetlands can play a crucial role in treating and recycling water for agricultural and industrial use. For instance, Gabr et al. (2023) proposed a wetland-based economic approach for wastewater treatment in arid regions, demonstrating the potential of wetlands to address water scarcity issues. Moreso, wetlands promote water quality by absorbing excess nutrients, such as nitrogen and phosphorus, which can threaten water bodies through eutrophication. Wetland plants absorb and transform these nutrients into less harmful forms, preventing algal blooms and other harmful effects. Some plants, like water hyacinths, are particularly effective at absorbing toxic metals and chemicals, acting as natural sewage treatment systems (Jenyo-Oni 2024). Wetlands also reduce pollutants through physical, chemical, and biological processes involving vegetation, microorganisms, soil, and water.

5.4.3 Climate Regulation and Carbon Sequestration

Wetlands play a crucial role in climate regulation by acting as significant carbon sinks, storing carbon in their biomass and soils. This function helps mitigate climate change by reducing carbon dioxide emissions. Wetlands' importance in climate change mitigation is increasingly recognized, as they enhance local resilience to extreme weather events, such as floods and droughts, by providing flood storage, storm buffering, and groundwater recharge (Moomaw et al. 2018).

Peatlands and vegetated coastal wetlands are particularly effective at sequestering carbon, storing as much carbon as global forest ecosystems. Wetlands accumulate soil carbon over time due to limited decomposition in water-saturated environments, leading to deep peat deposits that can persist for thousands of years (Moomaw et al. 2018). However, when wetlands are damaged or destroyed, they can become significant sources of greenhouse gases (GHGs). Land-use changes, such as drainage and peat extraction, reduce wetlands' ability to store carbon and increase carbon emissions. For example, fires in tropical peatlands in Southeast Asia have led to massive carbon losses to the atmosphere (Moomaw et al. 2018).

Protecting and restoring wetlands is essential for maintaining their role as carbon sinks and preventing the release of stored carbon. Policies and management strategies that prioritize wetland protection are critical in the context of climate change. Additionally, wetlands enhance climate resilience by buffering the impacts of extreme weather events such as floods and droughts. Wetlands absorb excess water during heavy rainfall, reducing the risk of flooding, and release stored water during dry periods, maintaining water flow in rivers and streams. This dual function makes wetlands an essential component of climate adaptation strategies.

5.5 Conclusion and Recommendations

5.5.1 Conclusion

This chapter on the socioeconomic benefits of wetland-based water treatment systems has highlighted the diverse advantages of utilizing wetlands for water purification. Wetlands, both natural and constructed, offer substantial economic, social, and environmental benefits that are increasingly recognized in contemporary water management strategies. The natural processes occurring in wetlands—such as filtration, sedimentation, and biological uptake—allow for the efficient removal of pollutants from water, making them a cost-effective alternative to conventional water treatment methods. Furthermore, wetlands contribute to biodiversity conservation, climate regulation, and carbon sequestration, thereby supporting broader environmental goals.

Economically, wetland-based water treatment systems are attractive due to their lower construction and operational costs compared to traditional treatment plants. Their ability to generate valuable byproducts, such as biomass, which can be used for bioenergy production or other commercial purposes, further enhances their financial sustainability. The integration of wetlands into water treatment strategies also provides long-term financial savings through reduced energy consumption and maintenance costs. Moreover, wetlands play a crucial role in supporting water reuse initiatives, particularly in regions facing water scarcity, by providing an alternative source of irrigation water.

Socially, wetlands offer significant benefits to communities, particularly in rural and underserved areas where access to clean water and sanitation is limited. Wetland-based systems improve public health by reducing waterborne diseases and enhancing overall water quality. Additionally, wetlands provide recreational opportunities and serve as communal spaces that foster social cohesion and community well-being. The aesthetic and cultural values of wetlands also contribute to the quality of life for residents and can drive tourism, further supporting local economies.

Environmentally, wetlands are vital for maintaining ecological balance and resilience. They enhance biodiversity by providing habitats for various species, including many that are endangered. Wetlands also play a critical role in climate regulation by acting as carbon sinks, capturing and storing atmospheric carbon dioxide. Their ability to reduce pollution and improve water quality is well-documented, making them an essential component of sustainable water management strategies. The environmental benefits of wetlands extend beyond water purification, as they also help mitigate the impacts of floods and droughts, thereby enhancing local resilience to climate change.

5.5.2 Recommendations

1. **Promote Wetland Conservation and Restoration**: Given the extensive benefits of wetlands, it is crucial to prioritize their conservation and restoration. Governments and environmental organizations should implement policies that protect existing wetlands and encourage the restoration of degraded ones. This will not only preserve biodiversity but also enhance the ecosystem services provided by wetlands.
2. **Integrate Wetlands into Urban Planning**: Urban planners should incorporate wetlands into city landscapes to manage stormwater, reduce pollution, and provide recreational spaces. Wetlands can serve as natural infrastructure that mitigates the challenges of urbanization, such as flooding and water contamination.
3. **Encourage Sustainable Tourism**: Wetland-based tourism should be promoted as a sustainable economic activity that supports conservation efforts. Local communities should be involved in tourism initiatives to ensure that the economic benefits of wetland preservation are equitably distributed.
4. **Support Research and Education**: Continued research on wetland ecosystems is essential for understanding their full potential in water treatment and environmental management. Educational programs should also be developed to raise awareness about the importance of wetlands and to train future generations in sustainable water management practices.
5. **Foster Public-Private Partnerships**: Collaborations between the public and private sectors can drive investments in wetland-based water treatment systems. These partnerships can support the development of innovative technologies and business models that enhance the economic viability of wetland conservation efforts.

References

Adhya T, Banerjee S (2022) Impact of wetland development and degradation on the livelihoods of wetland-dependent communities: a case study from the lower Gangetic floodplains. Wetlands 42(7):65. https://doi.org/10.1007/s13157-022-01588-w

Alikhani S, Nummi P, Ojala A (2021) Urban wetlands: a review on ecological and cultural values. Water 13(22):Article 22. https://doi.org/10.3390/w13223301

Anthonj C, Diekkrüger B, Borgemeister C, Kistemann T (2019) Health risk perceptions and local knowledge of water-related infectious disease exposure among Kenyan wetland communities. Int J Hyg Environ Health 222(1):34–48. https://doi.org/10.1016/j.ijheh.2018.08.003

Ayodele IA, Jenyo-Oni A (2002) Fish resources of Lagos state coastal wetlands. Afr J Livestock Ext 1:28–31

Bacon PR (1987) Use of wetlands for tourism in the insular Caribbean. Ann Tour Res 14(1):104–117. https://doi.org/10.1016/0160-7383(87)90050-8

Bhowmik S (2022) Ecological and economic importance of wetlands and their vulnerability: a review. In: Research anthology on ecosystem conservation and preserving biodiversity. IGI Global, Hershey, pp 11–27. https://doi.org/10.4018/978-1-6684-5678-1.ch002

Bikangaga S, Picchi MP, Focardi S, Rossi C (2007) Perceived benefits of littoral wetlands in Uganda: a focus on the Nabugabo wetlands. Wetl Ecol Manag 15(6):529–535. https://doi.org/10.1007/s11273-007-9049-3

Brix H (1994) Functions of macrophytes in constructed wetlands. Water Sci Technol 29(4):71–78. https://doi.org/10.2166/wst.1994.0160

Calhoun AJK, McGarry MA, Reeve A (2003) Wetland connections: linking university researchers and high school teachers to advance science education and wetland conservation. J Geosci Educ 51(4):387–397. https://doi.org/10.5408/1089-9995-51.4.387

Carter M (2015) Wetlands and health: how do urban wetlands contribute to community wellbeing? In: Finlayson CM, Horwitz P, Weinstein P (eds) Wetlands and human health, Wetlands: ecology, conservation and management, vol 5. Springer Netherlands, Dordrecht, pp 149–167. https://doi.org/10.1007/978-94-017-9609-5_8

Clarkson BR, Ausseil AE, Gerbeaux P (2013) Wetland ecosystem services. In: Dymond JR (ed) Ecosystem services in New Zealand—conditions and trends. Manaaki Whenua Press, Lincoln. Retrieved from https://climateandnature.org.nz/wp-content/uploads/2021/06/Wetland-Ecosystem-Services-2013.pdf

Dey D, Banerjee S (2018) How expensive is the decay of East Kolkata wetlands? An estimation of opportunity cost for Kolkata. In: Mukherjee J (ed) Sustainable urbanization in India, exploring urban change in South Asia. Springer Nature Singapore Pte Ltd., Singapore, pp 181–205. https://doi.org/10.1007/978-981-10-4932-3_10

Donoso N (2018) Environmental assessment of constructed wetlands for agricultural wastewater treatment. PhD thesis. Ghent University, Ghent. ISBN: 978-94-6357-142-5

Ezenwa VO, Milheim LE, Coffey MF, Godsey MS, King RJ, Guptill SC (2007) Land cover variation and West Nile virus prevalence: patterns, processes, and implications for disease control. Vector Borne Zoonotic Dis 7(2):173–180. https://doi.org/10.1089/vbz.2006.0584

Food and Agriculture Organization (FAO) (2020) The state of the world fisheries and aquaculture. Retrieved from http://www.fao.org/state-of-fisheries-aquaculture (18.02.2024)

Gabr ME, Al-Ansari N, Salem A, Awad A (2023) Proposing a wetland-based economic approach for wastewater treatment in arid regions as an alternative irrigation water source. Hydrology 10(1):20. https://doi.org/10.3390/hydrology10010020

Griffith KL (1992) Constructed wetlands: a growing opportunity for the construction industry. Mit.edu. http://hdl.handle.net/1721.1/45719. Thesis: M.S., Massachusetts Institute of Technology, Department of Civil Engineering

Hammer DA, Bastian RK (1989) Wetland Ecosystem Natural Water Purifiers. In Constructed Wetlands for Wastewater Purifiers. In: Hammer DA, Ed., Constructed Wetlands for Waste Water

Treatment: Municipal, Industrial and Agriculture, Proceedings, First International Conference of Constructed Wetlands for Waste Water Treatment, Chattanooga, 13–17 June 1988, 508–514.
Horwitz P, Finlayson CM, Kumar R (2015) Interventions required to enhance wetlands as settings for human well-being. In: Finlayson CM, Horwitz P, Weinstein P (eds) Wetlands and human health. Springer Netherlands, Dordrecht, pp 193–225. https://doi.org/10.1007/978-94-017-9609-5_10
House CH, Bergmann BA, Stomp AM, Frederick DJ (1999) Combining constructed wetlands and aquatic and soil filters for reclamation and reuse of water. Ecol Eng 12(1):27–38. https://doi.org/10.1016/S0925-8574(98)00052-4
Huryna H, Brom J, Pokorny J (2014) The importance of wetlands in the energy balance of an agricultural landscape. Wetl Ecol Manag 22(4):363–381. https://doi.org/10.1007/s11273-013-9334-2
Jenyo-Oni A (2024) From aquatic splendor to wealth creation: towards sustainable management of Nigeria's wetland fisheries resources. An Inaugural Lecture at the University of Ibadan. Ibadan University Press Publishing House, University of Ibadan, Nigeria. ISBN: 978-978-8550-29-7
Kadlec RH, Wallace S (2008) Treatment wetlands, 2nd edn. CRC Press. https://doi.org/10.1201/9781420012514
Karia JP (2012) Floral and avifaunal diversity of Thol Lake Wildlife (Bird) Sanctuary of Gujarat State, India. In: Biodiversity enrichment in a diverse world. Retrieved from https://doi.org/10.5772/50073
Li Q, Long Z, Wang H, Zhang G (2021) Functions of constructed wetland animals in water environment protection—a critical review. Sci Total Environ 760:144038. https://doi.org/10.1016/j.scitotenv.2020.144038
Macharia JM, Thenya T, Ndiritu GG (2010) Management of highland wetlands in Central Kenya: the importance of community education, awareness and eco-tourism in biodiversity conservation. Biodiversity 11(1–2):85–90. https://doi.org/10.1080/14888386.2010.9712652
Maddock M (1991) Education, research and passive recreation: an integrated programme at the Wetlands Centre, Shortland. Int J Sci Edu 13(5):561–568. https://doi.org/10.1080/0950069910130507
Malaviya P, Singh A (2012) Constructed wetlands for management of urban stormwater runoff. Crit Rev Environ Sci Technol 42(20):2153–2214. https://doi.org/10.1080/10643389.2011.574107
Maund PR, Irvine KN, Reeves J, Strong E, Cromie R, Dallimer M, Davies ZG (2019) Wetlands for wellbeing: piloting a nature-based health intervention for the management of anxiety and depression. Int J Environ Res Public Health 16(22):Article 22. https://doi.org/10.3390/ijerph16224413
Metcalfe CD, Nagabhatla N, Fitzgerald SK (2018) Multifunctional wetlands: pollution abatement by natural and constructed wetlands. In: Nagabhatla N, Metcalfe CD (eds) Multifunctional wetlands: pollution abatement and other ecological services from natural and constructed wetlands. Springer International Publishing, Cham, pp 1–14. https://doi.org/10.1007/978-3-319-67416-2_1
Mitsch WJ, Gosselink JG (2000) The value of wetlands: importance of scale and landscape setting. Ecol Econ 35(1):25–33. https://doi.org/10.1016/S0921-8009(00)00165-8
Mitsch WJ, Gosselink JG (2015) Wetlands, 5th edn. Wiley, Hoboken. ISBN: 978-1-118-67682-0
Mitsch WJ, Tejada J, Nahlik A, Kohlmann B, Bernal B, Hernández CE (2008) Tropical wetlands for climate change research, water quality management and conservation education on a university campus in Costa Rica. Ecol Eng 34(4):276–288. https://doi.org/10.1016/j.ecoleng.2008.07.012
Mitsch WJ, Bernal B, Hernandez ME (2015) Ecosystem services of wetlands. Int J Biodivers Sci Ecosyst Serv Manag 11(1):1–4. https://doi.org/10.1080/21513732.2015.1006250
Moomaw WR, Chmura GL, Davies GT, Finlayson CM, Middleton BA, Natali SM, Perry JE, Roulet N, Sutton-Grier AE (2018) Wetlands in a changing climate: science, policy and management. Wetlands 38(2):183–205. https://doi.org/10.1007/s13157-018-1023-8
Nayak A, Bhushan B (2022) Wetland ecosystems and their relevance to the environment: importance of wetlands. In: Handbook of research on monitoring and evaluating the ecological health of wetlands. IGI Global, Hershey, pp 1–16. https://doi.org/10.4018/978-1-7998-9498-8.ch001

Reeves JP, Knight AT, Strong EA, Heng V, Neale C, Cromie R, Vercammen A (2019) The application of wearable technology to quantify health and wellbeing co-benefits from urban wetlands. Front Psychol 10. https://doi.org/10.3389/fpsyg.2019.01840

Rey JR, Walton WE, Wolfe RJ, Connelly CR, O'Connell SM, Berg J, Sakolsky-Hoopes GE, Laderman AD (2012) North American wetlands and mosquito control. Int J Environ Res Public Health 9(12):Article 12. https://doi.org/10.3390/ijerph9124537

Rhymes JM, Arnott D, Chadwick DR, Evans CD, Jones DL (2023) Assessing the effectiveness, practicality and cost effectiveness of mitigation measures to reduce greenhouse gas emissions from intensively cultivated peatlands. Land Use Policy 134:106886. https://doi.org/10.1016/j.landusepol.2023.106886

Ribbe L, Dekker G, Thapak G (2024) Chapter 6—Urban wetlands and water bodies. In: Shinde VR, Mishra RR, Bhonde U, Vaidya H (eds) Managing urban rivers. Elsevier, Amsterdam, pp 91–107. https://doi.org/10.1016/B978-0-323-85703-1.00007-9

Shah SA, Islam MS (2021) Recreational benefits of wetlands: a survey on the Dal Lake in Jammu and Kashmir of India. Int Hosp Rev 37(1):94–109. https://doi.org/10.1108/IHR-03-2021-0018

Sharma S, Lekshmi B, Sutar RS, Parikh YJ, Ranade DR, Asolekar SR (2020) Reuse of washing machine effluent using constructed wetland: the circular economy of sanitation. In: Recent trends in waste water treatment and water resource management. Springer, Singapore, pp 85–100. https://doi.org/10.1007/978-981-15-0706-9_9

Singh RP, Wu J, Fu D (2019) Purification of water contaminated with Hg using horizontal subsurface constructed wetlands. Environ Sci Pollut Res 26(10):9697–9706. https://doi.org/10.1007/s11356-019-04260-9

Snyder BF (2019) The inclusion of ecosystem service valuations in bioenergy cost analysis: a case study of constructed wetlands in the neotropics. Ecol Econ 156:196–201. https://doi.org/10.1016/j.ecolecon.2018.10.005

Solano ML, Soriano P, Ciria MP (2004) Constructed wetlands as a sustainable solution for wastewater treatment in small villages. Biosyst Eng 87(1):109–118. https://doi.org/10.1016/j.biosystemseng.2003.10.005

Sukhontapatipak C, Srikosamatara S (2012) The role of field exercises in ecological learning and values education: action research on the use of campus wetlands. J Biol Educ 46(1):36–44. https://doi.org/10.1080/00219266.2011.554574

Turpie J, Kleynhans M (2010) Wetland valuation volume IV: a protocol for the quantification and valuation of wetland. Retrieved from https://www.efdinitiative.org/sites/default/files/wetlands20vol20iv.pdf

Verhoeven JTA, Meuleman AFM (1999) Wetlands for wastewater treatment: opportunities and limitations. Ecol Eng 12(1):5–12. https://doi.org/10.1016/S0925-8574(98)00050-0

Vymazal J (2010) Constructed wetlands for wastewater treatment. Water 2(3):530–549. https://doi.org/10.3390/w2030530

Vymazal J (2011) Plants used in constructed wetlands with horizontal subsurface flow: a review. Hydrobiologia 674(1):133–156. https://doi.org/10.1007/s10750-011-0738-9

Zedler JB, Leach MK (1998) Managing urban wetlands for multiple use: research, restoration, and recreation. Urban Ecosystems 2:189–204. https://doi.org/10.1023/A:1009528505009

Williams JD, Dodd CK Jr (1978) Importance of wetlands to endangered and threatened species. In: Wetland functions and values: the state of our understanding. American Water Resources Association, USA, pp 565–575

Zhou J, Wu J, Gong Y (2020) Valuing wetland ecosystem services based on benefit transfer: a meta-analysis of China wetland studies. J Clean Prod 276:122988. https://doi.org/10.1016/j.jclepro.2020.122988

Part II
Threats to Wetland Environments

Chapter 6
Threats to Wetland Environments: Human Activities, Pollution and Climate Change

Miracle Uwa Livinus, Sunday Zeal Bala, Mustapha Abdulsalam, Musa Ojeba Innocent, Madinat Hassan, Katimu Yusuf, Joshua Oluwatobi Akinola, and Gado Mustapha Ali

Abstract Wetlands are important ecosystems that offer various utilities, including water purification, several species' home, and carbon storage. However, wetlands are progressively under pressure due to multiple anthropogenic pressures, pollution, and the alteration of climate. This chapter offers an analysis to the diverse issues that wetland face today starting with a discussion on how they are harmed by aggression such as agricultural expansion, urbanization, and industrial development. Consequences of human activities including wetlands being transformed for agricultural use, reclamation of land for development, and pollution from industrial and recreational activities is described. The chapter then turns to the major concern of pollution whereby chemical run offs from agriculture, nutrient over loads leading to eutrophication and solid waste and even micro plastics are discussed. Furthermore, the article describes thermal and noise pollution impacts on the wetland environment. The chapter advances to further discuss and present the consequences of cli-

M. U. Livinus (✉)
Department of Biochemistry, School of Sciences and Information Technology, Skyline University Nigeria, Kano, Nigeria

S. Z. Bala
Department of Education, University of Maryland Eastern Shore, USA, Princess Anne, MD, USA

M. Abdulsalam · M. Ojeba Innocent
Department of Microbiology, School of Sciences and Information Technology, Skyline University Nigeria, Kano, Nigeria

M. Hassan
Foresight Institute of Research and Translation (FIRAT), Kigali, Rwanda

K. Yusuf
Department of Microbiology, Federal University Gashua, Yobe State, Nigeria

J. O. Akinola
Department of Natural Sciences University of Maryland, College Park, MD, USA

G. M. Ali
National Biotechnology Development Agency, Abuja, Nigeria

B. R. Babaniyi et al. (eds.), *Wetland Ecosystems: Conservation Strategies, Policy Management and Applications*, Wetlands: Ecology, Conservation and Management 12, https://doi.org/10.1007/978-3-031-91982-4_6

mate change including sea level changes, temperature variations, hydrological cycles, and carbon storage. In response to these threats, this chapter stresses the need for applying multiple strategies that are aimed at land use, conservation, policy, as well as the rehabilitation of habitats and species. There is also a recognition of the significance of engaging the community while educating them on the need to support conservation. It is hoped that this chapter has offered a brief introduction to ongoing threats and possible strategies for wetland conservation, while stressing the importance of cooperation to address these threats and to preserve this important type of natural resource.

Keywords Wetlands · Human activities · Pollution · Climate change · Restoration · Conservation

6.1 Introduction

Wetlands are some of the most beneficial and 'productive' habitats economically, socially and ecologically. Wetlands are areas of land either entirely covered in water throughout the year throughout the year or during growing season or by water that is in direct contact with the earth's surface year-round or any time during the year. Such regions include marshes, swamps, bogs, and other similar regions that foster the growth of water plants and other water dwelling creatures. Wetlands have distinctive features described by hydrological, vegetation and soil characteristics, and serve important functions in ecosystems (Mitsch and Gosselink 2015). Wetlands are very crucial to the overall wellbeing of the planet for the following reasons. These perform the role of water purification, holding pollution like nitrogen, phosphorus and heavy metal before the water enters rivers, lakes or the ocean. This process of filtration also enhances the quality of water and the many ecosystems dependent on these water bodies (Zedler and Kercher 2005). They also serve as important carbon repositories where carbon dioxide is stored and which contribute to the fight against climate change. Wetlands are one subcategory of which peatlands have been shown to sequester more carbon than all of the forests on earth combined, even with the wetlands occupying far less land (Joosten et al. 2012).

Besides supporting ecological processes, wetlands can also support rich plant and animal life. These are some of the most diverse habitats, containing species which are specialized for life in wetland environments. This is important for sustaining ecosystem wing and flexibility so that wetlands can continue to perform their functions despite changes in the environment (Reddy and DeLaune 2008). They also provide social and economic values and opportunities as well throughout the society. They yield products including fish, shellfish, and timber, and they sustain human lives especially in rural and indigenous populations. Additionally, they play the roles of cultural and recreational significance by providing facilities for tourism, education, and religious beliefs. Flood control services are another important advantage since wetlands can hold rainfall and equally minimize floods affecting human settlements (Millennium Ecosystem Assessment 2005; Livinus et al. 2024).

However, wetlands continue to be among the world's most vulnerable environments despite the significant value they hold. Wetlands in the recent past have been increasingly degraded and lost due to several factors attributed to human activities. Analysis of over 2000 wetland records reveals a global average decline of 35% in natural wetland extent, with the rate of loss accelerating between 1970 and 2015. Conversely, human-made wetlands experienced a remarkable 233% increase from 1970 to 2014 (Darrah et al. 2019). In contrast, Fluet-Chouinard et al. (2023) have reported a net loss of 21% of inland wetlands from the year 1700, primarily due to conversion to croplands. Other cause of wetlands degradation is conversion to other use which include agricultural and urban development. The conversion of wetlands to farmland has been widespread, especially in areas where there is a scarcity of available land. It also results in the elimination of wetland species habitats and crashes the hydrological systems that support them (Verhoeven and Setter 2010). Urbanization is another severe threat, as people consider wetlands to be ideal for infrastructural development due to the flat and low nature of the land. However, this development often leads to the total loss of wetlands and their numerous services (Mitsch and Gosselink 2015).

Pollution is another major hazard or threat to wetlands—and this has some important implications that will be discussed later. Chemicals such as fertilizers, pesticides, industrial wastes, and untreated sewage are washed down to the wetlands from agriculture and other industrial sources. These pollutants are capable of causing eutrophication, which results from nutrient pollution and excessive growth of plants and depletion of oxygen in water bodies harming aquatic life (Reddy and DeLaune 2008; Musa et al. 2024). Moreover, the pollutants remain stored in tissues of wetland species leading to chronic impact on wetland ecosystems (Junk et al. 2013). Additionally, climate change is an issue that is relatively new and potentially catastrophic to wetlands. Due to global warming, there has been a rise in the average temperature and changes in the pattern of rainfall; this affects the wetland hydrology in terms of wetness. Global warming leading to sea level rise is causing higher levels of saltwater intrusion in the freshwater wetlands especially in coastal regions (Nicholls and Cazenave 2010). In addition, intense and more frequent storms, hurricanes, and floods detrimental to wetland flora and fauna occur and impact the wetland habitats physically (IPCC 2014). Wetlands are very important ecological assets in the world since they offer various services that promote the welfare of ecosystems as well as people. Though, they are extremely delicate to threats posed by changes in climate, and human interference. It is necessary to take collaborative steps to conserve and rehabilitate wetlands to guarantee that they deliver their precious services for the future generation.

6.2 Human Activities

Globally wetlands which are productive natural systems that offer multiple services are gradually being degraded by various human activities. Many of these activities, which are propelled by economic and developmental considerations, lead to degradation, transformation, or elimination of wetlands. The main causes of wetland loss include: the conversion of wetlands to agricultural lands, conversion to urban and industrial areas, and commercial activities such as tourism. These activities have different effects but all ultimately degrade wetlands' ecological structure and processes.

6.2.1 Agricultural Expansion

Wetland loss and degradation are well documented with agricultural expansion commonly cited as the primary cause. The development of agricultural land from wetlands normally involves the flooding of these ecosystems in order to remove water and make the soil suitable for cultivation. Traditional or Cultural perception was that wetlands were barren and unproductive land, therefore suitable for conversion to agricultural land. This has caused the elimination of large portions of wetlands and changing their hydraulic and physiognomic features drastically. Converting wetlands to oil palm plantations, for instance, in Southeast Asia has led to significant wetland elimination adding to GHG and loss of bio-diversity (Miettinen et al. 2017).

When wetlands are leached to support agricultural activities, not only are the habitats eliminated, but also the water system cycle is interfered with. Additionally, they aid in replenishing aquifers and managing river and stream flow rates during dry seasons. When wetlands are drained, these hydrological functions are altered, making the area more susceptible to droughts and floods (Junk et al. 2013). Furthermore, the changes of wetland hydrology may result in soil shrinkage, which will further amplify impacts of rising sea levels, especially in wetlands along shorelines (Taillie et al. 2020).

Irrigation and pesticides also affect wetland ecosystems because of their roles in agriculture besides drainages. Irrigation usually occurs through diverting water from wetlands or their source, thus limiting the amount of water in these systems. This decrease in the level of water could result in the loss of the wetland habitats and subsequently affect the natural vegetation and animals that could exist in those environments (Davidson et al. 2020). Chemical pesticides and fertilizers for agricultural use find their way into wetlands through runoff resulting in the introduction of toxic substances. This runoff can cause eutrophication, where algal blooms are encouraged to flourish by an abundance of nutrients like nitrogen and phosphorous. These blooms lower the water's oxygen content, which threatens fish and other marine organisms and hence affects the food chain and the overall diversity (Moss 2019).

6.2.2 Urbanization and Infrastructure Development

Wetlands are being destroyed mainly due to urbanization and infrastructural expansion due to rapid urban growth. With the growth of cities, wetlands are eliminated though filling, drainage, or modification to accommodate for residential or business sectors, or facilities like roads and airports. Wetlands are directly degraded when they are converted for use as urban spaces, which also alters the hydrologic and ecological processes of wetlands (Gómez-Baggethun et al. 2013). For instance, urbanization and development in the United States have resulted in the loss of freshwater wetlands; some areas have lost more than half of their initial wetlands (Dahl 2017).

Another aspect of wetland loss is brought about by the construction of roads, dams and other infrastructures. They can negatively impact the spatial connectivity of wetland habitats by creating barriers to dispersal for animals and by separating regions of suitable habitat. This fragmentation could lead to reduced gene pool and make populations more susceptible to changes in environment (Laurance et al. 2015). These structures, constructed with the purpose of regulating the flow of water for irrigation and power generation and flood management, can significantly impact wetlands. Through regulating the water flow, the presence of dams alters the natural flooding that many wetlands depend on which in turn alters the vegetation and creatures in the wetlands altering the total wetland species diversity (Poff et al. 2017).

Apart from their obvious effects, the growth of infrastructure and urbanization also contributes to the contamination of wetlands. Urban runoff can harm aquatic life and reduce the quantity of potable water in wetlands that is accessible to humans through contaminating the water with a mixture of heavy metals, oil, and other pollutants (Wright et al. 2020). Moreover, the heat produced by cities can raise the temperature of wetland waters, intensifying the consequences of climate change and causing species that are sensitive to temperature to become extinct (Seto et al. 2012).

6.2.3 Industrial Activities

Wetland degradation is largely caused by industrial operations, such as mining and oil production. Large volumes of dirt and rock are frequently removed during mining operations, which can result in the habitats of wetlands being destroyed. Wetlands may be destroyed during the mineral extraction process, especially in open-pit mining, and water supplies may get contaminated with heavy metals and other harmful materials when mining waste is disposed of (Cordell et al. 2021). For instance, hundreds of hectares of wetlands have been destroyed in the Canadian Athabasca oil sands region due to oil extraction, which has significantly harmed the local wildlife and water quality (Rooney et al. 2012).

Wetlands, in particular, mangroves and salt marsh habitats, are seriously threatened by oil extraction, especially in offshore and coastal areas. Wetland ecosystems may be permanently harmed by oil spills, while the building of oil rigs, pipelines, and associated infrastructure may result in the degradation of wetland habitats. Oil pollution can destroy aquatic life, harm plants and animals, and upset the wetlands' natural equilibrium. According to Mendelssohn et al. (2012), the Gulf of Mexico's Deepwater Horizon oil disaster severely damaged coastal wetlands, even years after it occurred.

Wetlands are further degraded when they are converted for industrial purposes, including building factories or storage facilities. These practices not only cause wetland habitats to disappear but also hinder wetlands' ability to carry out essential tasks including filtering water, storing carbon, and preventing flooding. Reduced water quality, a higher risk of flooding, and an upsurge in emissions of greenhouse gases are only a handful of the extensive consequences that could result from the loss of these services (Junk et al. 2013).

6.2.4 Tourism and Recreation

Despite their frequent perception as having little influence, tourism and leisure activities can seriously harm wetland ecosystems. Reclamation of wetland regions is frequently necessary for the construction of tourism infrastructure, such as hotels, resorts, and marinas. This results in the loss of habitat and the change of natural landscapes (Brander et al. 2013). For instance, wetlands have been filled or drained in many coastal locations to build beaches, golf courses, and other tourist attractions; this has resulted to the loss of important habitats for fish, birds, and other wildlife (Gössling 2018).

Wetlands may be cumulatively affected by even small-scale leisure activities. Hiking, fishing, and boating are a few examples of activities that might cause wildlife disturbance, water body contamination, and vegetation trampling. Particularly when boats are involved, propellers can seriously impair wetlands by churning up silt, lowering water clarity, and destroying aquatic vegetation. According to Duarte et al. (2021) this may result in a deterioration in the quality of the water, and fish and other marine life's habitat.

The entry of non-native species via tourism represents a noteworthy peril to the ecosystems of wetlands. Travellers may unintentionally bring non-native animals, plants, or diseases into wetland habitats, where they may outcompete or feed on local species, resulting in a decrease in biodiversity. For instance, non-native species like the Burmese python that were introduced into the Florida Everglades have had a disastrous impact on native ecosystems (Dove et al. 2011). Moreover, the development of amenities and access roads to facilitate tourism has the potential to split wetlands, worsening their biological integrity (Finlayson and Davidson 2021).

6.3 Pollution

The destruction of these vital ecosystems is a result of pollution, which is one of the most subtle threats to wetland settings. Wetlands serve as natural filters, collecting and decomposing pollutants before they reach rivers, lakes, and seas. They are frequently found near the meeting point of terrestrial and aquatic systems. Wetlands, however, can sustain serious ecological harm when the inflow of contaminants overwhelms these systems. Chemical, nutrient, plastic, solid waste, noise, and thermal pollution are the main forms of pollution that harm wetlands. Eutrophication is another name for this sort of pollution. Although the causes and effects of each of these pollution types vary, they all lead to the degradation of wetland ecosystems.

6.3.1 Chemical Pollution

Wetland ecosystems are seriously threatened by chemical pollution, which is mostly caused by runoff from industrial and agricultural discharges. Agricultural runoff is a major cause of chemical pollution in wetlands since it frequently contains high concentrations of fertilisers, herbicides, and pesticides. Rainwater and irrigation systems can carry these chemicals into wetlands, where they can result in a variety of ecological issues. Because pesticides and herbicides are intended to destroy or stop the growth of living things, they are especially dangerous. These substances have the potential to harm aquatic life, lower biodiversity, and upset food webs when they get into wetlands (Moss 2019). According to Relyea and Jones (2009), there is a correlation between the decrease in amphibian populations in wetlands and the discharge of glyphosate, a herbicide that is frequently employed in agricultural regions.

Chemical contamination in wetlands is also a result of industrial activity. Pollutants including heavy metals, hazardous chemicals, and oil are frequently released into neighbouring water bodies by factories, refineries, and other industrial sites. These contaminants can build up in the sediments of wetlands, where they may linger for years or even decades, putting at jeopardy the long-term health and wellbeing of people and wildlife. According to Vignati et al. (2013), a unique risk comes from heavy metals like mercury and lead to apex predators such as birds and mammals due to their ability to bioaccumulate in the food chain and induce harmful effects. Furthermore, wetland waters' chemical composition can be changed by industrial pollution to become more acidic or alkaline, which can upset the ecosystems' delicate balance (Gleason et al. 2011).

6.3.2 Nutrient Pollution (Eutrophication)

Eutrophication in wetlands is mostly caused by nutrient contamination, especially excess nitrogen and phosphorus. A water body experiences eutrophication when an excessive amount of nutrients are introduced, which causes an overabundance of algae and other aquatic plants. The ecosystems of wetlands may suffer greatly as a result of this process. Smith et al. (1999) state that the primary sources of pollution from nutrients are urban wastewater discharge and agricultural runoff, which transports fertilizers high in phosphate and nitrogen.

Eutrophication in wetlands can cause algal blooms, or the fast growth of algae that obstruct sunlight from reaching submerged vegetation. The absence of light can lead to the demise of aquatic plants, which are essential for preserving the composition and capabilities of wetland environments. Also, degradation and algae death lead to very high consumption of oxygen hence creating anoxic or hypoxic environments in the water which in effect leads to very low oxygen levels in the water. Fish and different marine organisms may perish as a result of these low oxygen levels, which would decrease biodiversity and change the wetland's ecological balance (Davidson et al. 2020).

Even though nutrient pollution has detrimental effects, wetlands are vital to the cycling of nutrients and can serve as natural buffers, capturing and preserving excess nutrients before they enter water bodies downstream. Nevertheless, the system is overloaded and eutrophication and its related effects result when the nutrient load surpasses the wetland's ability to handle it. Unchecked nutrient pollution in wetlands has long-term effects that include decline in biodiversity, deterioration of the quality of water, and disruption of ecosystem functions (Dodds and Smith 2016).

6.3.3 Solid Waste and Plastic Pollution

Plastic pollution in particular is a growing danger to wetland ecosystems from solid waste. Solid garbage can enter wetlands through wind, rain, and water currents, whereupon it is often sunk. Solid waste buildup in wetlands can have a number of detrimental effects. For example, wetland ecosystems can degrade due to the suffocation of plants, reduction in available habitat, and obstruction of water flow caused by abandoned plastic bags, bottles, and other waste (Klemick and Robalino 2012).

The existence of microplastics, or microscopic plastic particles that are produced as small fragments or arise from the breakdown of bigger plastic trash, is one of the most alarming elements of plastic pollution. Wetland ecosystems are rife with microplastics, which aquatic species can swallow and consume, which can have toxicological and physical consequences. Studies have demonstrated that microplastics can build up in the tissues of fish, birds, and other species, which may cause growth retardation, infertility, and higher mortality rates (Thompson et al. 2009).

Microplastics' influence on wetland ecosystems can further be exacerbated by their ability to serve as carriers of various contaminants, including as persistent organic pollutants and toxic metals (Browne et al. 2010).

Leachate from landfills and other hazardous materials may be introduced into wetlands through the disposal of solid waste in or close to these areas. Leachate, the liquid that separates from waste in landfills, frequently has a combination of dangerous materials, like organic contaminants and toxic metals. Leachate can contaminate soil and water when it reaches wetlands, endangering both human health and animals (Eggen et al. 2010). The versatility of wetlands to accomplish vital functions like flood management and water purification can be severely hampered by the combined effects of plastic pollution and solid waste (Livinus et al. 2024).

6.3.4 Thermal and Noise Pollution

Wetland habitats are also impacted by thermal pollution, which is mostly the result of industrial activity. Industries that release heated water or wastewater into wetlands can cause thermal pollution because they raise the water's temperature. High water temperatures can harm wetland habitats in several ways. Higher temperatures, for instance, have the potential to lower the water's dissolved oxygen concentration, which would exacerbate the effects of eutrophication and result in hypoxic conditions (Kaushal et al. 2010). Furthermore, a lot of wetland species have evolved to survive in particular temperature ranges, so even slight temperature rises can interfere with their ability to reproduce, feed, and survive as a whole (Woodward et al. 2010).

Despite being frequently disregarded, environmental noise can have a significant impact on organisms found in wetlands. Wetlands are frequently found close to places where people live and work, like busy cities, industrial parks, and busy roads where noise pollution is a problem. Wetland species, especially birds and amphibians, who depend on vocalisations for mating and territorial defence, can have their communication, breeding, and foraging activities disrupted by noise pollution (Slabbekoorn and Ripmeester 2008). For example, frogs and toads may find it difficult to communicate due to noise from adjacent roadways or industrial activity. This could result in less successful reproduction and population decreases (Sun and Narins 2005). Furthermore, persistent noise pollution can stress out wildlife, which over time might change their behaviour and decrease their fitness (Francis et al. 2011).

6.4 Climate Change

Wetland ecosystems are experiencing significant changes in their dynamics due to climate change, which puts their ecological integrity and vital functions in jeopardy. Wetlands, which comprise bogs, marshes, swamps, and estuaries, are among of the ecosystems most vulnerable to fluctuations in the climate. Increased sea levels, temperature swings, modified hydrological cycles, and changes in carbon absorption and release are some of the ways that climate change affects wetlands. The biodiversity of the environment, ecosystem function, and wetlands' ability to serve as organic shields against the effects of climate change are all significantly impacted by these changes.

6.4.1 Sea Level Rise

One of the most visible and thus pressing impacts of climate change on coastal wetlands is the sea level rise. Global temperatures lead to enhanced melt of polar ice and thermal expansion of seawater hence resultant rise in sea levels. It is especially common in coastal wetlands, for example those formed by mangroves and salt marshes. Flooding can occur in wetlands because of increase in sea level, this kills off vegetation, and consequently destroys the habitat. Intrinsic to wetlands, the spatial expansion witnessed with rising sea levels is strained by conventional accretionary edges, which are mainly interfered with by human developments like buildings and roads—an occurrence commonly referred to as the "coastal squeeze" (Nicholls et al. 2007).

Freshwater wetlands are seriously threatened by saltwater intrusion, another effect of sea level rise. Many plant and animal species that are not acclimated to saline conditions find these environments uninhabitable as seawater infiltrates freshwater systems, changing their salinity. This may result in a decrease in biodiversity and the disintegration of freshwater species-dependent food webs (Herbert et al. 2015). Further aggravating the socio-economic effects of climate change on coastal communities is saltwater intrusion's potential to reduce freshwater resources available for human use (Titus et al. 2009).

6.4.2 Temperature Changes

Wetland habitats are being impacted by the considerable temperature fluctuations brought on by global warming, which is being fuelled by a rise in greenhouse gas concentrations. Increased temperatures can change a species' physiology, behaviour, and distribution in wetlands. Since many marsh plants and animals, for example, are acclimated to particular temperature ranges, even minute temperature increases can cause havoc with their life cycles. Because of their heightened sensitivity to temperature fluctuations, amphibians, for instance, may experience changes in their breeding seasons as well as a decrease in the success of their reproduction (Blaustein et al. 2010).

Species dispersion is also altered by temperature elevation, since several species shift their range northward or westward in response to warmer weather. This may cause non-native species to be introduced into wetlands, where they may displace or outcompete native species, resulting in a decline in complexity (Walther et al. 2002). Warmer weather can also amplify the consequences of other stressors, such as pollution and habitat fragmentation, increasing the susceptibility of wetlands to degradation.

Temperature changes have a wide range of intricate consequences on the biodiversity of wetlands. Warming temperatures have the potential to promote the growth of specific plant species, which could alter the vegetation's structure and chemistry. For instance, rising temperatures can hasten the proliferation of alien species, such as the highly aggressive marsh grass *Phragmites australis*, which can displace native plant communities and lower the quality of wildlife habitats (Mozdzer et al. 2010). These vegetation-related modifications may have an impact on the wetlands' capacity to sustain a variety of wildlife populations and control the cycling of nutrients, among other ecological activities.

6.4.3 Altered Hydrological Cycles

Hydrological cycles are also being altered by climate change, which has important ramifications for wetland water levels and ecosystem health. Worldwide, variations in rainfall patterns are having an impact on the drainage of wetlands, which include modifications to the timing, intensity, and pattern of rainfall. Wetland areas may experience erosion and floods due to severe weather events like hurricanes and storms that are increasing in frequency in certain places due to climate change (Hunt and Watkiss 2011). Drought and waterlogging episodes may result from these changes, which may upset the delicate equilibrium of water levels in wetlands.

Altered hydrological cycles have a significant influence on wetlands through increased frequency of droughts. Wetlands may dry up as a result of droughts, which lowers the amount of water available to plants and animals and results in the loss of wetland ecosystems (Erwin 2009). Delays in precipitation can also cause peatlands—wetlands that store a lot of carbon in the form of peat—to dry out. Carbon dioxide (CO_2) is released into the atmosphere as peatlands dry out and the stored carbon is exposed to oxygen, which exacerbates global warming (Fenner and Freeman 2011).

Conversely, wetlands may also be negatively impacted by rising water levels and flooding. Overwatering can cause alterations in the chemistry of the soil, such as a decrease in the amount of oxygen present, which can impact the survival and growth of plants. Flooding can also cause vegetation to shift and wetland soils to erode, which can degrade and destroy habitat. Furthermore, wetland ecosystems can sustain physical harm from extreme weather events like hurricanes and cyclones, which can uproot trees, destroy habitats, and change the terrain (Alongi 2008).

6.4.4 *Carbon Sequestration and Release*

Wetlands are important stores and drains of carbon in the global carbon cycle. Because they have substantial carbon storage capacity in their soils and plants, wetlands—especially mangroves and peatlands—are crucial natural reservoirs for reducing the effects of climate change. For instance, peatlands store over 30% of the global soil carbon but only make up 3% of the terrestrial area on Earth (Yu et al. 2010). Numerous elements, such as temperature, vegetation type, and water levels, affect wetlands' capacity to store carbon (Aransiola et al. 2023).

On the other hand, the carbon balance in wetlands is seriously threatened by climate change. Wetland habitats may deteriorate due to changing hydrological conditions and rising temperatures, which will lessen their ability to store carbon. For example, organic matter's decay deposited in the soils of wetlands can release methane (CH_4) and CO_2 into the atmosphere when the wetlands dry out due to drainage or drought. According to Dise (2009), this process, referred to as carbon release, has the potential to accelerate global warming by converting wetlands from carbon drains to carbon stores.

The capacity of wetlands to sustain a variety of plant communities is intimately related to their function in sequestering carbon. Through photosynthesis, wetlands plants like sedges and mangroves are essential for absorbing and storing carbon. Wetlands' capacity to sequester carbon dioxide may be impacted by changes in the composition of the vegetation brought on by climate change, such as the extinction of native plant species or the introduction of invasive ones. As to Zedler and Kercher (2005), the ecosystem's overall ability to sequester carbon can be diminished when invasive species take the place of native wetland plants, as they possess a reduced capacity to store carbon.

The capacity of wetlands to sustain a variety of plant communities is intimately related to their function in sequestering carbon. Through photosynthesis, wetlands plants like sedges and mangroves are essential for absorbing and storing carbon. Wetlands' capacity to sequester carbon dioxide may be impacted by changes in the composition of the vegetation brought on by climate change, such as the extinction of native plant species or the introduction of invasive ones. As to Zedler and Kercher (2005), the ecosystem's overall ability to sequester carbon can be diminished when invasive species take the place of native wetland plants, as they possess a reduced capacity to store carbon.

6.5 Case Studies on Specific Wetlands Under Threat

1. **The Everglades, USA**

The Everglades, a distinctive and expansive wetland environment located in southern Florida, is seriously threatened by pollution, climate change, and human activity. The Everglades once covered around four million acres, but the environment has been drastically diminished as a result of increased agricultural production, urbanisation, and water management techniques (Davis and Ogden 1994). Natural plant and animal species' distribution and health have been impacted by the drastic decrease in water flow caused by the draining and redirection of water for urban and agricultural use (Kushlan 2011).

The Everglades, a distinctive and expansive wetland environment located in southern Florida, is seriously threatened by pollution, climate change, and human activity. The Everglades once covered around four million acres, but the environment has been drastically diminished as a result of increased agricultural production, urbanisation, and water management techniques (Davis and Ogden 1994). Natural plant and animal species distribution and health have been impacted by the drastic decrease in water flow caused by the draining and redirection of water for urban and agricultural use (Kushlan 2011).

2. **The Aral Sea, Central Asia**

The Aral Sea, which was once among the world's largest inland lakes, has drastically shrunk as a result of massive water diversion for irrigation over the twentieth century. Due to this human interference, the sea dried up, having detrimental effects on the environment and the economy. Fisheries, which were vital to the local inhabitants, have collapsed and wetland habitats have been lost as a result of the fall in water volume (Micklin 2007).

Pesticides and fertilisers, among other pollutants from agricultural runoff, have exacerbated the degradation of the surviving bodies of water and harmed aquatic life. According to Kemp et al., the exposed lakebed dust presents a health danger to the nearby inhabitants due to its high concentration of contaminants and salts. These problems have been made worse by climate change, which has changed local precipitation patterns and further decreased the amount of water available.

3. **The Sundarbans, India and Bangladesh**

Human activity, pollution, and climate change are posing serious challenges to the Sundarbans, a massive mangrove forest that spans the delta of the Ganges, Brahmaputra, and Meghna rivers. According to Rao et al. (2012), habitat loss and degradation are results of industrial development combined with the conversion of mangrove forests for agriculture and aquaculture. Due to pollution by chemicals from industries and fertilizers used in farming, fish and other water animals are nowhere to be found in this region. Sundarbans is also the most vulnerable to the impacts of climate change for instance; flooding, storm surges, surge frequency and intensity. According to Das and Vincent (2009), these modifications are making seawater intrusion more common, worsening coastal erosion, and endangering the fragile mangrove ecosystems.

4. **The Lake Chad Basin, Africa**

The Sahel region of Africa's Lake Chad has seen significant environmental degradation as a result of both human activity and climate change. Lake Chad, which was formerly one of Africa's largest lakes, has drastically decreased in size in recent years as a result of changes in precipitation patterns brought on by climate change combined with water extraction for agriculture. The lake's size has significantly decreased as a result of less water entering it due to irrigation projects and agricultural growth. Gao et al. (2020) report that the reduction has influenced the wetland ecosystems surrounding the area as well as the farming and fishing-dependent local communities. By changing rainfall patterns and making droughts more frequent, climate change has made these problems worse and further reduced lake levels. It has been difficult to manage water supplies and rehabilitate the lake because of the intricate interactions between socioeconomic and environmental elements (Boko et al. 2007).

6.5.1 Success Stories of Wetland Conservation and Restoration

1. **The Oostvaardersplassen Nature Reserve, Netherlands**

One noteworthy example of wetland protection and restoration achieved is the Oostvaardersplassen Nature Reserve in the Netherlands. The location was formerly developed as a polder area by land reclamation; however, rewetting and habitat development initiatives allowed the area to become a wetland. The project's objectives were to improve biodiversity and restore the natural functions of wetlands (Kiehl et al. 2010). Creating a variety of wetland habitats and reinstating seasonal flooding and other natural hydrological processes were two of the restoration attempts. Thanks to these actions, biodiversity has significantly increased, and the reserve is now home to a wide range of bird species, including the uncommon Great Egret and the Common Crane. According to Bossema and Van Wijnen (2011), the Oostvaardersplassen is a crucial illustration of how human-altered landscapes can be effectively restored to provide vital ecological functions and maintain wildlife.

2. **The Kissimmee River Restoration, USA**

In Florida, the Kissimmee River underwent a significant restoration project to undo the biological harm that channelisation for flood control and water management had caused. According to Davis and Ogden (1994), the original river was straightened and changed in the 1960s, which resulted in habitat loss and decreases in fish and bird populations. Restoring the river's gentle flow and rebuilding drained wetlands were the goals of the restoration project, which got underway in the 1990s. The project's objectives were to enhance water quality and restore natural hydrological cycles. Significant gains in ecological health have occurred during the restoration, including higher biodiversity, the resurgence of native fish species, and better water quality (Kushlan 2011). The restoration of the Kissimmee River serves as an example of how extensive initiatives can improve ecosystem services and restore damaged wetlands.

3. **The Danube Delta, Romania and Ukraine**

One of the most biodiverse wetlands in Europe, the Danube Delta, has been the target of significant conservation and restoration initiatives to counter challenges from pollution, increased agricultural production, and changing hydrology. The delta's ecological significance and biodiversity have earned it recognition as a UNESCO World Heritage Site (Cirstea and Oprea 2010). Habitat restoration initiatives, pollution control strategies, and sustainable land use practices have all been used in the efforts to preserve and restore the Danube Delta. The objectives of these initiatives are to lessen the effects of agricultural runoff, restore hydrological processes to their original state, and safeguard important habitats for fish, birds, and other animals. The recovery of native species and enhancements in water quality in the delta demonstrate the efficacy of conservation efforts and demonstrate the usefulness of integrated management strategies in safeguarding important wetland ecosystems (Romanescu and Tudor 2014).

4. **The Niger Delta, Nigeria**

Pollution, climate change, and oil exploration and production have all contributed to significant environmental deterioration in the Niger Delta, one of the largest wetland systems in the world. Gas flaring, oil spills, and habitat degradation have all had a major negative influence on the delta, which is essential to the local populations' way of life and Nigeria's biodiversity (Obi 2009). Community-driven restoration projects and government-led conservation programs are two approaches being used to solve these problems. For instance, the Niger Delta Development Commission (NDDC) has worked on several initiatives to support local people impacted by oil pollution, restore damaged wetlands, and enhance environmental conditions (NDDC 2020). NGOs and international organisations have also contributed to restoration initiatives, such as the reforestation of mangroves, to lessen the effects of environmental degradation and improve the resilience of ecosystems (Dube and Oduro 2014; Aransiola et al. 2024). These initiatives show that, despite major obstacles, it is possible to restore wetland ecosystems.

6.6 Mitigation and Conservation Strategies

6.6.1 Policy and Regulation

1. **International and National Policies**

The Convention on Wetlands of International Importance, Especially as Waterfowl Habitat (the Ramsar Convention), was a groundbreaking agreement when adopted in 1971 (Ramsar Convention 1971) and came into effect in 1975 (Stroud et al. 2022). This pivotal international agreement plays a crucial role in the protection and conservation of wetlands, focusing on their preservation and sustainable use globally, alongside several other international and national laws and regulations. The Convention encourages international collaboration to safeguard these essential places and highlights the significance of wetlands for ecosystem services and biodiversity (Ramsar Convention Secretariat 2019; Okewu et al. 2024).

At the national level, nations have created specialized laws and policies to protect wetlands. To safeguard wetlands, for instance, the Clean Water Act in the United States governs the release of pollutants into waters and wetlands (U.S. Environmental Protection Agency 2020). In a similar vein, before starting any projects, federal agencies are required by the National Environmental Policy Act (NEPA) to evaluate the effects on the environment, including those on wetlands (Council on Environmental Quality 2021).

Wetland ecosystems are fully protected in the European Union under the EU Habitats Directive and the EU Water Framework Directive, which combine habitat preservation with water quality control (European Commission 2020). As part of larger environmental conservation initiatives, these rules place a strong emphasis on the necessity for member nations to preserve and restore wetlands.

2. **Legal Frameworks**

By providing rules and procedures for enforcement to safeguard these habitats, regulatory structures are essential to the conservation of wetlands. Regulating land use, pollution management, and habitat conservation are common components of effective legal systems. For instance, Australia's Wetlands Conservation and Management Act establishes a legislative framework for the management and protection of wetlands through the creation of conservation zones and land use regulations (Australian Government 2021).

Compliance and enforcement are crucial elements of legal structures. Conservation initiatives can be upheld with the support of policies that involve monitoring, reporting, and sanctions for violators. To increase the overall efficacy of legal frameworks, the Environmental Protection Agency (EPA) in the United States, for instance, regularly performs inspections and enforces regulations of the protection of wetlands (U.S. Environmental Protection Agency 2020).

6.6.2 Sustainable Land Use Practices

1. **Sustainable Agriculture**

Reducing the amount of damage farming does to wetlands requires the use of sustainable farming methods. Wetland water quality is preserved and runoff is lessened by using practices such as buffer strips, cover crops, and sparingly applying fertiliser and pesticides (Gibbs and Manning 2008). Planting plants along waterways creates buffer strips, which serve as filters, capturing pollutants before they reach wetlands. Another sustainable method for incorporating trees and shrubs into agricultural areas is agroforestry. According to Nair (2012), this strategy protects nearby wetland ecosystems by improving soil health, lowering erosion, and improving water management.

2. **Sustainable Urban Planning**

The avoidance of environmental consequences and the preservation of natural areas are important considerations in urban development near wetlands. Rain gardens and artificial wetlands are two examples of green infrastructure solutions that assist in control of stormwater runoff and lower pollution. According to Tzoulas et al. (2007), these methods can replicate natural processes that filter pollutants and give animals habitat. In urban planning, zoning laws that limit development in delicate wetland areas are also essential. Cities can prevent wetlands from being overtaken and degraded by restricting development activity and designating special zones for conservation (Bertram and Vivier 2010).

3. **Sustainable Industry Practices**

Industry located next to wetlands need to take action to reduce their influence on the environment. According to Hsu et al. (2014), wastewater treatment methods like bioreactors and natural filtration systems can successfully lower the amount of pollutants released into wetland regions. Evaluating the possible impacts of industrial operations on wetlands and developing mitigation strategies to lessen harm require the use of environmental impact assessments (EIAs) (Glasson et al. 2012). Regulations governing EIAs contribute to ensuring that industrial projects take wetlands' environmental effects into account.

6.6.3 Restoration and Rehabilitation

1. **Techniques for Wetland Restoration**

Wetland restoration is the process of restoring degraded regions and their natural functions using a variety of techniques. According to Mitsch and Gosselink (2015), rewetting is a basic restoration approach that attempts to restore natural water levels and flow rates restore natural hydrological conditions. Planting native plants to restore vegetation cover is known as revegetation. This process can boost soil stability, support biodiversity, and improve habitat quality. Because invading plants and animals can outcompete native species and disturb wetland ecosystems, managing invasive species is also essential (Ehrenfeld 2010; Livinus et al. 2024).

2. **Rewilding**

Rewilding is a strategy that involves returning native species and letting ecosystems self-regulate in order to restore natural processes and ecosystems. This strategy can improve ecosystem resilience and biodiversity. For instance, it has been demonstrated that reintroduction animals like beavers to temperate wetlands enhances the water quality and habitat structure (Rosell et al. 2005).

6.6.4 Community Engagement and Education

1. **Involving Local Communities**

Effective wetland conservation requires community involvement. The livelihoods of the local inhabitants frequently depend on wetlands, and they possess invaluable information about these ecosystems. Communities' involvement in conservation initiatives ensures that restoration initiatives take into account local interests and needs (Danielsen et al. 2009). Communities can participate in the design and execution of conservation initiatives by being included in the decision-making process through the use of participatory methodologies. By encouraging local ownership and management, this strategy can improve the effectiveness of conservation activities (Barton and Kates 2009; Abdulsalam et al. 2024).

2. **Educational Programs and Public Awareness**

Promoting the preservation of wetlands is largely accomplished through educational initiatives and public awareness campaigns. Public outreach initiatives, community seminars, and school programs all contribute to increasing understanding of the value of wetlands and promoting eco-friendly behaviour (Chawla 1999). In addition, citizen science programs involve the general public in the observation and reporting of wetland conditions, which aids in data gathering and strengthens community participation in conservation efforts (Silvertown 2009).

6.7 Conclusion

Wetlands are vulnerable to a wide range of problems resulting from pollution, climate change, and human activity, all of which undermine the vital ecological services and functions that wetlands provide. Natural hydrological systems are altered, habitats are destroyed, and pollution results from human activities including urbanization, industrialization, and agricultural growth. Pesticides, fertilizers, and the draining of wetlands for farming are examples of agricultural practices that lead to nutrient overloads and habitat destruction. Infrastructure and land reclamation brought on by urban and industrial growth frequently disturb water flow and wetland health. These problems are made worse by pollution from solid waste, plastic contamination, and industrial discharges. Sea level rise, temperature changes, and changed hydrological cycles brought on by climate change also increase the pressure. A comprehensive conservation strategy that incorporates strong policy frameworks, sustainable land use practices, and efficient restoration techniques is needed to address these complex concerns. Wetland legal safeguards are established by national legislation and policies like the Ramsar Convention. Negative effects can be lessened by using sustainable techniques in industry, urban planning, and agriculture. Restoring degraded wetlands and increasing their resilience requires restoration initiatives such as rewetting, revegetation, and rewilding. In order to cultivate a culture of stewardship and guarantee the success of conservation efforts, community involvement and public education are essential. Protecting wetlands requires the combined efforts of local communities, governments, conservation groups, and the commercial sector. Together, we can safeguard these priceless ecosystems and guarantee that future generations will continue to benefit from vital functions like water filtering, biodiversity support, and climate regulation.

References

Abdulsalam M, Elelu SA, Musa OI, Ganiyat OI, Livinus MU, Lateefat SO, Auwal SM (2024) Women empowerment in environmental conservation. In: Prospects for soil regeneration and its impact on environmental protection, Earth and environmental sciences library. Springer Nature Switzerland, Cham. https://doi.org/10.1007/978-3-031-53270-2_9

Alongi DM (2008) Mangrove forests: resilience, protection from tsunamis, and responses to global climate change. Estuar Coast Shelf Sci 76(1):1–13

Aransiola SA, Victor-Ekwebelem MO, Leh-Togi Zobeashia SS, Maddela NR (2023) Sources and techniques for biofuel generation. In: Shah MP (ed) Green approach to alternative fuel for a sustainable future. Elsevier Inc., Amsterdam. https://doi.org/10.1016/B978-0-12-824318-3.00026-6

Aransiola SA, Leh-Togi Zobeashia SS, Ikhumetse AA, Musa OI, Abioye OP, Ijah UJJ, Maddela NR (2024) Niger delta mangrove ecosystem: biodiversity, past and present pollution, threat and mitigation. Reg Stud Mar Sci 75(2024):103568. https://doi.org/10.1016/j.rsma.2024.103568

Australian Government (2021) Wetlands conservation and management act. Retrieved from Australian Government website

Barton DN, Kates RW (2009) Participatory approaches to environmental management. Environ Manag 43(6):1020–1032

Bertram C, Vivier J (2010) Urban planning and the role of zoning. Urban Stud 47(3):527–544

Blaustein AR, Walls SC, Bancroft BA et al (2010) Direct and indirect effects of climate change on amphibian populations. Diversity 2(2):281–313

Boko M, Niang I, Nyong A (2007) Africa. In: Climate change 2007: impacts, adaptation and vulnerability. Cambridge University Press, Cambridge, pp 433–467

Bossema I, Van Wijnen H (2011) The Oostvaardersplassen nature reserve: restoration and biodiversity. Ecol Restor 29(1):18–28

Brander LM, Wagtendonk AJ, Hussain SS et al (2013) Ecosystem service values for mangroves in Southeast Asia: a meta-analysis and value transfer application. Ecosyst Serv 5:1–9
Browne MA, Galloway TS, Thompson RC (2010) Spatial patterns of plastic debris along estuarine shorelines in the United Kingdom. Estuar Coast Shelf Sci 87(1):136–142
Chawla L (1999) Growing up in an environmentally responsible way. Environ Educ Res 5(3):275–287
Cirstea C, Oprea A (2010) The Danube delta biosphere reserve: status and conservation efforts. Wetl Ecol Manag 18(6):741–753
Cordell D, White S, Drangert JO (2021) Towards sustainable phosphorus management: identifying priority areas for research and intervention. Ambio 50(4):849–859
Council on Environmental Quality (2021) National Environmental Policy Act (NEPA). Retrieved from CEQ website
Dahl TE (2017) Status and trends of wetlands in the conterminous United States 2004 to 2009. Fish and Wildlife Service, Washington, D.C.
Danielsen F, Beukema SA, Burgess ND (2009) Local participation in wetland conservation. Conserv Biol 23(5):1304–1312
Darrah SE, Shennan-Farpón Y, Loh J, Davidson NC, Finlayson CM, Gardner RC, Walpole MJ (2019) Improvements to the wetland extent trends (WET) index as a tool for monitoring natural and human-made wetlands. Ecol Indic 99:294–298. https://doi.org/10.1016/j.ecolind.2018.12.032
Das S, Vincent K (2009) Vulnerability of the Sundarbans to climate change and sea level rise. Glob Environ Chang 19(3):436–449
Davidson NC, van Dam AA, Finlayson CM, McInnes RJ (2020) Worth of wetlands: revised global monetary values of coastal and inland wetland ecosystem services. Mar Freshw Res 71(2):134–146
Davis SM, Ogden JC (1994) Everglades: the ecosystem and its restoration. St. Lucie Press, Delray Beach
Dise NB (2009) Peatland response to global change. Science 326(5954):810–811
Dodds WK, Smith VH (2016) Nitrogen, phosphorus, and eutrophication in streams. Inland Waters 6(2):155–164
Dove CJ, Snow RW, Rochford MR, Mazzotti FJ (2011) Birds consumed by the invasive Burmese python (*Python molurus bivittatus*) in Everglades National Park, Florida, USA. Wilson J Ornithol 123(1):126–131
Duarte CM, Middelburg JJ, Caraco N (2021) Major role of marine vegetation on the oceanic carbon cycle. Biogeosciences 2(1):1–8
Dube S, Oduro K (2014) Restoration and management of mangrove forests in The Niger Delta. Environ Manag 54(4):765–774
Eggen T, Moeder M, Arukwe A (2010) Municipal landfill leachates: a significant source for new and emerging pollutants. Sci Total Environ 408(21):5147–5157
Ehrenfeld JG (2010) The effects of invasive species on wetland ecosystems. Conserv Biol 24(4):1018–1027
Erwin KL (2009) Wetlands and global climate change: the role of wetland restoration in a changing world. Wetl Ecol Manag 17:71–84
European Commission (2020) EU habitats directive and water framework directive. Retrieved from European Commission website
Fenner N, Freeman C (2011) Drought-induced carbon loss in peatlands. Nat Geosci 4:895–900
Finlayson CM, Davidson NC (2021) The wetland book II: distribution, description, and conservation. Springer, Dordrecht
Fluet-Chouinard E, Stocker BD, Zhang Z, Malhotra A, Melton JR, Poulter B, Kaplan JO, Goldewijk KK, Siebert S, Minayeva T, Hugelius G, Joosten H, Barthelmes A, Prigent C, Aires F, Hoyt AM, Davidson N, Finlayson CM, Lehner B, Jackson RB et al (2023) Extensive global wetland loss over the past three centuries. Nature 614(7947):281–286. https://doi.org/10.1038/s41586-022-05572-6

Francis CD, Ortega CP, Cruz A (2011) Noise pollution filters bird communities based on vocal frequency. PLoS One 6(11):e27052
Gao J, Shalaby A, Rahman MM (2020) Impacts of climate change and human activities on the Lake Chad Basin: a review. Environ Monit Assess 192(9):560
Gibbs HK, Manning DP (2008) Buffer strips for wetlands: a review. J Environ Manag 87(2):226–233
Glasson J, Therivel R, Chadwick A (2012) Introduction to environmental impact assessment. Routledge, New York
Gleason RA, Euliss NH Jr, Tangen BA et al (2011) Agricultural conservation practices increase wetland ecosystem services in the prairie pothole region of the United States. Ecol Appl 21(3):S65–S81
Gómez-Baggethun E, De Groot R, Lomas PL, Montes C (2013) The history of ecosystem services in economic theory and practice: from early notions to markets and payment schemes. Ecol Econ 69(6):1209–1218
Gössling S (2018) Tourism and development in tropical islands: political ecology perspectives. Island Stud J 13(2):149–166
Herbert ER, Boon P, Burgin AJ et al (2015) A global perspective on wetland salinization: ecological consequences of a growing threat to freshwater wetlands. Ecosphere 6(10):1–43
Hsu S, Wu C, Chen K (2014) Wastewater treatment technologies. Environ Sci Technol 48(6):3334–3343
Hunt A, Watkiss P (2011) Climate change impacts and adaptation in cities: a review of the literature. Clim Chang 104:13–49
IPCC (2014) Climate change 2014: impacts, adaptation, and vulnerability. Cambridge University Press, Cambridge
Joosten H, Tapio-Biström M-L, Tol S (2012) Peatlands: guidance for climate change mitigation by conservation, rehabilitation and sustainable use. FAO, Rome
Junk WJ, An S, Finlayson CM, Gopal B, Kvet J, Mitchell SA, Mitsch WJ, Robarts RD (2013) Current state of knowledge regarding the world's wetlands and their future under global climate change: a synthesis. Aquat Sci 75:151–167
Kaushal SS, Likens GE, Jaworski NA et al (2010) Rising stream and river temperatures in the United States. Front Ecol Environ 8(9):461–466
Kiehl K, Bork M, Schulte A (2010) The Oostvaardersplassen: a successful wetland restoration project. J Appl Ecol 47(2):285–293
Klemick H, Robalino J (2012) Solid waste management and environmental impacts in developing countries: evidence from Ghana. World Bank Working Paper
Kushlan JA (2011) Ecological and hydrological restoration of the Kissimmee River. Wetlands 31(1):123–136
Laurance WF, Clements GR, Sloan S et al (2015) A global strategy for road building. Nature 513(7517):229–232
Livinus MU, Bala SZ, Abdulsalam M, Innocent MO, Hassan M, Kini P (2024) Natural occurrences of soil dilapidation. In: Prospects for soil regeneration and its impact on environmental protection, Earth and environmental sciences library. Springer Nature Switzerland, Cham. https://doi.org/10.1007/978-3-031-53270-2_9
Mendelssohn IA, Andersen GL, Baltz DM et al (2012) Oil impacts on coastal wetlands: implications for the Mississippi River Delta ecosystem after the deepwater horizon oil spill. Bioscience 62(6):562–574
Micklin PP (2007) The Aral sea disaster. Annu Rev Earth Planet Sci 35:47–72
Miettinen J, Hooijer A, Wang J et al (2017) Peatland degradation and conversion sequences and their impact on downstream hydrology in Southeast Asia. Glob Planet Chang 161:23–38
Millennium Ecosystem Assessment (2005) Ecosystems and human well-being: wetlands and water synthesis. World Resources Institute, Washington, DC
Mitsch WJ, Gosselink JG (2015) Wetlands, 5th edn. Wiley, Hoboken
Moss B (2019) The water crisis: prevention and cure. Philos Trans R Soc B Biol Sci 374(1764):20180416

Mozdzer TJ, Megonigal JP, McKee KL (2010) Elevating CO2 and climate change alter species composition of a tidal freshwater marsh. Glob Chang Biol 17(3):705–710

Musa IO, Samuel JO, Adams M, Abdulsalam M, Nathaniel V, Maude AM, Adedayo OA, Tiamiyu AT (2024) Soil erosion, mineral depletion and regeneration. In: Prospects for soil regeneration and its impact on environmental protection, Earth and environmental sciences library. Springer Nature Switzerland, Cham. https://doi.org/10.1007/978-3-031-53270-2_9

Nair PKR (2012) Agroforestry systems and practices. Springer, New Delhi

Nicholls RJ, Cazenave A (2010) Sea-level rise and its impact on coastal zones. Science 328(5985):1517–1520

Nicholls RJ, Wong PP, Burkett VR et al (2007) Coastal systems and low-lying areas. In: Climate change 2007: impacts, adaptation and vulnerability. Cambridge University Press, Cambridge, pp 315–356

Niger Delta Development Commission (NDDC). (2020). Annual report 2020. Retrieved from NDDC website

Obi A (2009) Oil exploration and environmental degradation in The Niger Delta: implications for environmental policy. Environ Policy Law 39(3):155–163

Okewu M, Aransiola AB, Aransiola SA, Adewoye OP, Agarry OO, Funmilayo TF, Maddela NR (2024) Roles of marine microbial products to the Nigeria economy. In: Aransiola SA, Bamisaye A, Abioye OP, Maddela NR (eds) Marine bioprospecting for sustainable blue-bioeconomy. Springer, Cham. https://doi.org/10.1007/978-3-031-68911-6_11

Poff NL, Olden JD, Merritt DM, Pepin DM (2017) Climate change and dams: An analysis of impacts and management strategies for US dam infrastructure. Front Ecol Environ 15(10):573–582

Ramsar Convention (1971). The convention on wetlands text, as originally adopted in 1971. Available at https://www.ramsar.org/document/theconvention-on-wetlands-text-as-originally-adopted-in-1971 [Verified 26 May 2021]

Ramsar Convention Secretariat (2019) The Ramsar convention on wetlands. Retrieved from Ramsar website

Rao MR, Mukherjee S, Verma P (2012) Mangrove degradation in the Sundarbans and conservation measures. Mar Pollut Bull 64(6):1102–1112

Reddy KR, DeLaune RD (2008) Biogeochemistry of wetlands: science and applications. CRC Press, Boca Raton

Relyea RA, Jones DK (2009) The toxicity of roundup original max to 13 species of larval amphibians. Environ Toxicol Chem 28(9):2004–2008

Romanescu G, Tudor V (2014) Conservation and restoration of the Danube Delta: a case study. J Environ Manag 146:234–244

Rooney RC, Bayley SE, Schindler DW (2012) Oil sands mining and reclamation cause massive loss of peatland and stored carbon. Proc Natl Acad Sci 109(13):4933–4937

Rosell F, Bozser O, Collen P (2005) The beaver: a reintroduction success story. Anim Conserv 8(3):321–328

Seto KC, Guneralp B, Hutyra LR (2012) Global forecasts of urban expansion to 2030 and direct impacts on biodiversity and carbon pools. Proc Natl Acad Sci 109(40):16083–16088

Silvertown J (2009) A new dawn for citizen science. Trends Ecol Evol 24(9):467–471

Slabbekoorn H, Ripmeester EAP (2008) Birdsong and anthropogenic noise: implications and applications for conservation. Mol Ecol 17(1):72–83

Smith VH, Tilman GD, Nekola JC (1999) Eutrophication: impacts of excess nutrient inputs on freshwater, marine, and terrestrial ecosystems. Environ Pollut 100(1-3):179–196

Stroud DA, Davidson NC, Finlayson CM, Gardner RC (2022) Development of the text of the Ramsar convention: 1965–1971. Mar Freshw Res 73(10):1107–1126. https://doi.org/10.1071/mf21312

Sun JWC, Narins PM (2005) Anthropogenic sounds differentially affect amphibian call rate. Biol Conserv 121(3):419–427

Taillie PJ, Moorman CE, Poulter B, Song C (2020) Reversing centuries of wetland drainage on a coastal landscape: patterns and implications of wetlands restoration under the clean water act. Wetl Ecol Manag 28(1):1–19

Thompson RC, Swan SH, Moore CJ, vom Saal FS (2009) Our plastic age. Philos Trans R Soc B Biol Sci 364(1526):1973–1976
Titus JG, Anderson KE, Cahoon DR et al (2009) Coastal sensitivity to sea-level rise: a focus on the mid-Atlantic region. U.S. Climate Change Science Program, Washington, DC
Tzoulas K, Korpela K, Venn S, James P (2007) Green infrastructure: the conceptual framework. J Environ Manag 84(3):1059–1076
U.S. Environmental Protection Agency (2020) Clean water act overview. Retrieved from EPA website
Verhoeven JTA, Setter TL (2010) Agricultural use of wetlands: opportunities and limitations. Ann Bot 105(1):155–163
Vignati DAL, Ferrari BJD, Dominik J (2013) Contamination of wetlands by heavy metals: the need for a more realistic approach. Water Air Soil Pollut 224:1384
Walther GR, Post E, Convey P et al (2002) Ecological responses to recent climate change. Nature 416(6879):389–395
Woodward G, Perkins DM, Brown LE (2010) Climate change and freshwater ecosystems: impacts across multiple levels of organization. Philos Trans R Soc B Biol Sci 365(1549):2093–2106
Wright AJ, Emeana CJ, Cooper AH, Holderied MW (2020) Long-term impacts of urbanisation on freshwater biodiversity: a global meta-analysis. Glob Chang Biol 26(9):4692–4705
Yu Z, Loisel J, Brosseau DP et al (2010) Global peatland dynamics since the last glacial maximum. Geophys Res Lett 37(13):L13402
Zedler JB, Kercher S (2005) Wetland resources: status, trends, ecosystem services, and restorability. Annu Rev Environ Resour 30:39–74

Chapter 7
Presence-Behaviour of Emerging Contaminants in Natural Wetlands and Strategies for their Remediation

Gema Alexandra Anchundia-Vélez, Joan Manuel Rodríguez-Díaz, and Carlos Augusto Morales-Paredes

Abstract Natural wetlands are ecosystems with multiple biotic and abiotic components in constant interaction, covering 6% of the planet's surface. These ecosystems are crucial in preserving biodiversity, regulating the water cycle, and preventing soil erosion. However, high levels of anthropogenic pollution are reported in natural wetlands, especially due to the presence of emerging pollutants. Therefore, the objective of the present work is to review the presence and behavior of emerging contaminants in natural wetlands, as well as the strategies implemented for their restoration. The presence of pesticides such as pentachlorobenzene (62 ng g^{-1}) and orisastrobin (700 ng L^{-1}), drugs such as tramadol (222 ng L^{-1}) and paracetamol (168 ng L^{-1}), and personal care products such as triclosan in concentrations in the order of 72 ng μL^{-1} have been reported in these environmental matrices. To mitigate this problem, significant progress has been made in restoration and treatment, showing improvements in water quality and habitat recovery. For example, technologies such as phytoremediation and the use of activated carbon in adsorption processes have made it possible to eliminate hydrophobic organic compounds. In addition, microbial restoration and hydrology management prevent and reduce pollution levels significantly. Finally, it is crucial to implement integrated measures that include

---

G. A. Anchundia-Vélez
Maestría en Química mención Química Ambiental, Facultad de Posgrado, Universidad Técnica de Manabí, Portoviejo, Ecuador

J. M. Rodríguez-Díaz (✉)
Laboratorio de Análisis Químicos y Biotecnológicos, Instituto de Investigación, Universidad Técnica de Manabí, Portoviejo, Ecuador

Departamento de Procesos Químicos, Biotecnología y Alimentos, Facultad de Ingeniería y Ciencias Aplicadas, Universidad Técnica de Manabí, Portoviejo, Ecuador
e-mail: joan.rodriguez@utm.edu.ec

C. A. Morales-Paredes
Laboratorio de Análisis Químicos y Biotecnológicos, Instituto de Investigación, Universidad Técnica de Manabí, Portoviejo, Ecuador

B. R. Babaniyi et al. (eds.), *Wetland Ecosystems: Conservation Strategies, Policy Management and Applications*, Wetlands: Ecology, Conservation and Management 12, https://doi.org/10.1007/978-3-031-91982-4_7

source reduction, active habitat restoration, and riparian zone protection to mitigate adverse effects and ensure the sustainability of natural wetlands.

Keywords Natural wetlands · Emerging pollutants · Removal · Restoration · Treatment technologies

7.1 Introduction

Natural wetlands are organic network ecosystems with multiple levels or various biotic and abiotic components that interact with each other (Hao et al. 2020). Wetlands cover 6% of the Earth's land surface and play a fundamental role in the preservation of biodiversity, water cycle, climate regulation and prevention of soil erosion processes (Cao et al. 2020). However, these ecosystems are susceptible to variations in terms of both composition and extent. According to Cai et al. (2021), approximately 50% of natural wetlands in the world have been degraded, which is due to high levels of pollution, overfishing, and artificial regeneration. One of the main factors that alter the composition, size, and ecosystemic interactive dynamics of these environments is environmental pollution of anthropogenic origin since the introduction of domestic, hospital, agricultural, and industrial waste is reported (Munford et al. 2023; Zhu et al. 2022), affecting their biotic and abiotic configuration.

Within the broad field of environmental chemistry, there is a specific group of contaminants that have generated increasing interest in recent years. Generally, these emerging contaminants (EC) can be subdivided into pesticides, pharmaceutical compounds, personal care products, and hormones (Lopez et al. 2022). Although they are found in low concentrations and are not yet regulated in environmental regulations, they cause harmful effects on public health and ecosystem sustainability (Morales-Paredes et al. 2022; Richardson and Ternes 2021). The occurrence of these EC in wetlands is associated with agricultural runoff, industrial discharges, and effluents from water treatment plants. Fungicides and pesticides such as tebuconazole and atrazine and more than 14 pharmaceuticals have been found in natural wetlands at concentrations exceeding 100 ng L^{-1} (Martínez-Megías et al. 2024). In addition, concentrations of caffeine (555 ng L^{-1}) and paracetamol (168 ng L^{-1}) are reported that can generate environmental risks in sensitive organisms inhabiting Mediterranean coastal wetlands (Sadutto et al. 2021), causing environmental alterations and reducing the reproductive rate of species. In turn, concentrations of the personal hygiene products triclosan and triclocarban were reported in the order of 134 and 1318 ng g^{-1} in the sediments of a natural wetland in Brazil. The vast majority of EC can present significant ecotoxicological risks, especially considering the wide variety of chemical compounds that can be found in the wetland aquatic environment and the metabolites that can be formed by chemical degradation reactions in the aqueous environment.

Ecological restoration of wetlands is a sustainable alternative to preserve these ecosystems and reduce their pollutant load (Shackelford et al. 2013), since in some cases, restoration efficiencies in coastal wetlands have been achieved in the order of 38% for seagrasses and 65% in marshes (Cadier et al. 2020). Restoration strategies include phytoextraction, where plant species are used to accumulate and metabolize pollutants, and bioremediation which employs microorganisms to degrade pollutant compounds and convert them into by-products with low toxicity. In addition, hydrological restoration plays a crucial role by restoring natural patterns of water flow and flood levels, thus improving the capacity of wetlands to filter and degrade pollutants (Sapkota and White 2020). As for indicators to monitor the aforementioned strategies, these are based on the availability of species in their environment. A study on marsh restoration considered hydrology, soil and sediment, vegetation, and nekton presence as reliable indicators (Neckles et al. 2002). Vegetation determined by leaf area index and aerial biomass as well as soil characteristics: organic matter, total nitrogen, and redox potential, have been taken as indicators to evaluate mangroves (Salmo III and Shielameh 2013). Meanwhile, microbial metrics such as enzymatic activities, bioenergetic parameters and fungal abundance were used to indicate progress in marsh restoration (Duarte 2012).

Therefore, the objective of the present work is to review the presence and behavior of EC in natural wetlands, as well as the strategies implemented for the restoration of these ecosystems.

7.2 Natural Wetlands

Natural wetlands are ecosystems partially or totally saturated with water, which fulfill vital functions for biodiversity. A wetland is formed by the spontaneous flooding of soils, either by rainfall or by seepage from subsoil where groundwater deposits exist. They are also generated by rising seawater levels, although it has been proven that most wetlands are composed of water with low salinity. They are inhabited by species with an affinity for the physical-chemical characteristics of the environment, including animal species such as birds, fish, reptiles, terrestrial and aquatic plants, and biomass-producing microorganisms (Keddy 2010). Wetlands have been classified into various categories, generally considering specific aspects such as vegetation, quality, and volume of water bodies and other hydraulic aspects, soil properties, and biotic and abiotic composition of the environment (Finlayson et al. 2018). In addition, various systems classify wetlands according to ecological and geographic criteria. For example, Cowardin's classification divides them into marine, estuarine, lacustrine, riparian, and palustrine systems, with subdivisions according to vegetation and water type. For its part, the classification of Mitsch and Gosselink is based on ecological functions and hydrological processes, distinguishing marshes, swamps, bogs, peatlands, marshes, and mixed systems. Meanwhile, the FAO classification focuses on the influence of land use and water resources. Each of these systems has specific applications in environmental management and consequent environmental restoration (Richardson et al. 2022; Wu et al. 2021). However, in the present work, we have selected a specific group of natural wetlands in which the presence of emerging contaminants and their subsequent remediation have been reported (Fig. 7.1).

Wetlands provide ecosystem services, among which the relative value of nutrient transport stands out with a 43% contribution, followed by disturbance regulation (18%), water treatment (16%), water source/supply (14%), recreation, culture, and environmental education area (6%), food production (2%) and wildlife habitat/refuge (1%) (Kingsford et al. 2016). Another relevant aspect is the biomass contribution. For example, in China, the average amount of carbon in aboveground and belowground biomass of mangroves is 254 Mg ha^{-1} and 84 Mg ha^{-1}, respectively. For their part, salt marshes and seagrasses reported average aboveground biomass of 9 and 0.36, and belowground biomass in the order of 10 and 0.41 ± 0.46 Mg ha^{-1}, respectively (Meng et al. 2019).

Wetlands also exhibit the ability to store carbon, reflecting a total of 770 billion tons of carbon, accounting for 35% of terrestrial ecosystem carbon storage, of which 500 billion tons of carbon is stored in peatland wetlands (Meng et al. 2017). Additionally, the final disposal of plant debris in coastal wetlands is mostly controlled by the activities of benthic fauna and water circulation. Crabs, bivalves, and snails remove up to 50% and 80% of the leaf litter from mangroves, this allows nutrients to be recycled within the ecosystem as part of a food chain. Birds also consume and export plant and animal biomass from the wetlands; they connect various components of the coastal wetlands and to the surrounding ecosystems.

For the reasons discussed above, the global value of wetland ecosystem services ($26.4 billion per year) is estimated to contribute more than 20% of the total value of ecosystem services globally, exceeding the contributions of terrestrial forests and coral reefs (Thorslund et al. 2017). However, areas are evident where wetland habitat once covered more than 80% of the area that is currently only 9% covered (Wang et al. 2019). Meanwhile, results based on the calculation of the wetland extent trend

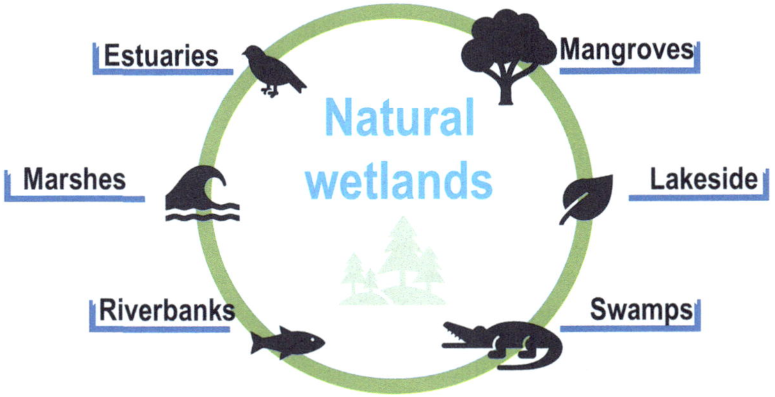

Fig. 7.1 Wetland types in which a higher presence of emerging contaminants is reported

index revealed that globally and on average, natural wetlands decreased by about 31% between 1970 and 2008 (Dixon et al. 2016), with marine/coastal type wetlands decreasing more than inland wetlands (Xue et al. 2018). Currently, factors such as agriculture (25%), urbanization (16.8%), aquaculture (10.7%), and industry (7.6%) have been found to have influenced the disappearance of natural wetlands, concluding that inland wetlands proved to be the most affected compared to coastal and marine wetlands (Ballut-Dajud et al. 2022). At a political-social level, wetlands are fundamental to the fulfillment of 17 of the United Nations Sustainable Development Goals and have therefore been declared one of nature's most threatened ecosystems (Xu et al. 2019).

7.3 Biotic and Abiotic Composition of Natural Wetlands

7.3.1 Biotic Factors

The biotic components of wetlands include a diverse range of organisms adapted to variable aquatic and terrestrial conditions. This ranges from plant species such as sedges, cattails, water lilies, and water lilies, which form the basis of primary productivity, to a wide variety of animals such as waterfowl, fish, amphibians, reptiles, and mammals adapted to wet environments. In addition, microorganisms such as bacteria, algae, and fungi play fundamental roles in the biogeochemical cycles within these ecosystems, contributing to the decomposition of organic matter and the maintenance of water quality (Alvarez-Mieles et al. 2013; He et al. 2016). Permanent wetlands present two stable states depending on water conditions: (i) clear water condition dominated by submerged aquatic macrophytes, with a low abundance of phytoplankton, and (ii) turbid water condition dominated by phytoplankton with a low abundance of submerged aquatic plants. Aquatic invertebrates play a key role in stabilizing both states and also in inducing changes from one state to the other (Zimmer et al. 2016).

In terms of vegetation, three functional groups predominate in a natural wetland. The plant functional groups are based on growth form, habitat use, and responses to environmental stress (Lichvar 2013). The predominant species are ruderal, interstitial perennials and stress-tolerant clonal plants. It has been verified that the vegetation of artificially created wetlands compared to natural wetlands does not differ significantly, however, the content of minerals and organic matter that may form over time does show differences (Campbell et al. 2002). In addition, there are species of wetland plants that can accumulate heavy metals such as nickel, cadmium, lead, copper, zinc, and chromium with a concentration range of 5–400 mg of substance per kg of macrophyte (Ramachandra et al. 2020). Plants typical of natural wetlands such as *Pistia estratiotes*, *Spirodela polirhiza*, and *Eichhornia crassipes*, allowed the removal of up to 79% of the metals present in these ecosystems (Rai 2019).

However, the presence of plants with adverse effects on other biological species is also noted. Such is the case of *Typha*, a wetland plant whose presence and growth are associated with negative ecological impacts on wetlands in terms of fauna and in turn on agricultural systems, since by invading territory and absorbing nutrients it affects wet crops such as rice (Heynes-Silerio et al. 2017); but it is also linked to ecosystem services such as bioremediation, biomass production, and traditional uses in indigenous cultures (Bansal et al. 2019). Another problem linked to impacts on biota is runoff. In Italy, a case of vegetation loss in wetlands caused by increased nutrient runoff from surrounding agricultural lands was reported, which led to intense pleustophytic algal blooms (Wu et al. 2018), inhibited the growth of submerged plants, and changed the coastal habitat from dense macrophyte meadows to an algae-dominated ecosystem (Gaglio et al. 2017).

Regarding the fauna, the trend of arthropods in some mangroves is reported as follows: 50% *Lepidoptera*, 5% *Orthoptera*, 3% *Hymenoptera*, 20% *Coleoptera*, 15% *Hemiptera*, 5% *Diptera*, 2% *Thysanoptera*. The importance of these species lies in their role as drivers of multiple ecological processes and community regulators (food chain) (Majeed et al. 2020). On the other hand, in the report by (Elliott et al. 2019) on wetlands in the United States, various birds of interest have been reported, among which waterfowl species such as ducks, grebes, skimmers, wading birds such as storks, cranes, shorebirds such as plovers, blackbirds, passerine birds, harriers, and pelicans stand out. The list of fish species found in marshes of the Caspian Sea basins (Iraq) includes fish of the order *Clupeiformes* such as the *African Ilisha*, *Cypriniformes* such as the common carp, *Siluriformes* such as the stinging catfish, *Symbrachiformes* such as the freshwater eel, *Cichliformes* such as the tilapia and Perciformes such as the perch (Finkl 2021). Regarding amphibian species, (Wassens et al. 2017) highlight certain groups of frogs and toads such as the eastern poster-bearing frog (*Crinia parinsignifera*), marsh barking frog (*Limnodynastes fletcheri*), marsh spotted frog (*Limnodynastes tasmaniensis*). On the other hand, (Tozer et al. 2018) mentions species native to northern zone wetlands such as the American toad (*Anaxyrus americanus*), bullfrog (*Lithobates catesbeianus*), green frog (*Lithobates clamitans*), gray tree frog (*Dryophytes versicolor*), and northern leopard frog (*Lithobates pipiens*).

7.3.2 Abiotic Factors

Natural wetlands are profoundly influenced by a series of abiotic factors that determine their structure, function, and biodiversity. Hydrology plays a fundamental role in establishing the flooding regime and the availability of water, both surface and groundwater. This factor not only affects the distribution of plant and animal species but also the dynamics of nutrients and the capacity of the wetland to retain water during dry periods (Brisson et al. 2020). Water chemistry in wetlands, which encompasses pH, concentration of nutrients such as nitrogen and phosphorus, salinity, and dissolved oxygen, is crucial for aquatic life and biogeochemical processes. Variations in these parameters can significantly alter the biological community and the ability of the wetland to act as a natural filter for pollutants. From this perspective, it is observed that the spatial variation of hydrological factors exerted a significant influence on the presence of the main animal species in the water. Specifically, water transparency was found to have a significant effect, as a decrease in water turbidity could reduce the number of these species (Zhao et al. 2020).

Climate, which includes variables such as temperature, precipitation and solar radiation, directly impacts evaporation, water temperature and biological productivity. Climate fluctuations can drastically alter wetland composition and structure, also influencing vital biogeochemical processes (Salimi et al. 2021). Significant trends (from 1960 to 2012) indicating climate change were found in the Biebrza Basin, located in northeastern Poland: increases in temperature, evapotranspiration and early onset of spring floods. Drier vegetation types were identified following drainage areas. However, restoration measures that were implemented such as blocking drainage, avoiding clearing of aquatic vegetation, removal of shrubs, etc., were found to have mitigated the effects of climate change by increasing soil moisture and converting existing vegetation types into wetter vegetation (Berezowski et al. 2018).

Soil in wetlands varies in texture, organic matter content, and water-holding capacity. These characteristics affect plant distribution and the availability of essential nutrients. The underlying geology determines the structure of the substrate and its permeability, which impacts the hydrology of the wetland and the chemistry of the water it harbors (Steinmuller and Chambers 2019). For example, in the Yellow River Delta region (China) the high salinity pattern in low nitrogen soils reflected in degraded wetlands in certain sectors, mainly due to seawater intrusion. In the study, there were significant negative relationships between salinity, total organic carbon content, and total soil nitrogen factors (Yu et al. 2016). In addition, the topography of the terrain defines how water accumulates and is distributed within the wetland, affecting the connectivity of the system and its response to floods or droughts. In this sense, Olsson et al. (2015) studied the methane content generated in these ecosystems, evidencing that the emission was accelerated in high water tables and decreased in water tables below the soil surface.

Taken together, these factors interact in a complex and dynamic manner to determine the health and functionality of natural wetlands, so that pollution sources can affect the composition and conservation of these ecosystems. In this sense, concentrations of EC have been determined to be progressively increasing in recent years, posing an imminent risk to the conservation of wetlands.

7.4 Presence of Emerging Contaminants in Natural Wetlands

The presence of EC in natural wetlands is of increasing concern because of their ability to affect the ecological balances and biodiversity of these aquatic ecosystems. Drugs, pesticides, cleaning products, and other chemical compounds can enter wetlands through a variety of pathways, such as inadequately treated

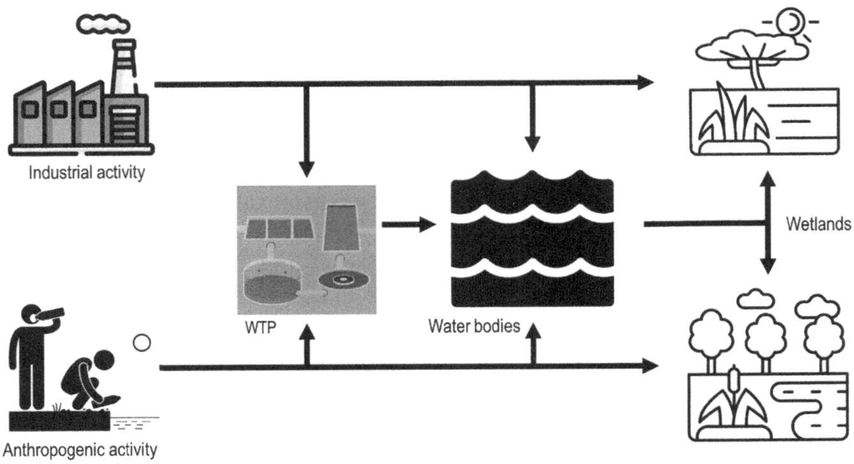

Fig. 7.2 Sources of contamination of natural wetlands. WTP water treatment plants

wastewater, agricultural runoff, and direct discharges, persisting and accumulating in water and sediments. This pollution poses potential risks to the health of biological communities that depend on wetlands, as well as to the environmental services they provide, such as water filtration and nutrient regulation. Figure 7.2 shows the possible sources of contamination of these ecosystems.

In the study conducted in the Upo wetland, South Korea (Kang et al. 2020), the presence of 35 contaminants from different types of pesticides was analyzed, where residual products from fungicide treatment were found: orysastrobin (700 ng L^{-1}), carbendazim (280 ng L^{-1}), tricyclazole (200 ng L^{-1}); insecticides: etoprop (100 ng L^{-1}), carbofuran (85 ng L^{-1}), imidacloprid (70 ng L^{-1}); herbicides: dimethenamid (100 ng L^{-1}), metolachlor (72 ng L^{-1}), prometryn (22 ng L^{-1}). By-products of the decomposition of the pesticides described above were also found, in concentrations of 4 to 60 ng L orysastrobin. Total concentrations of organochlorine contaminants ranging from 95.69 to 412.76 ng g^{-1} have been detected in sediments of natural wetlands on the east coast of China: dichlorodiphenyldichloroethylene, dichlorodiphenyldichloroethane and dichlorodiphenyltrichloroethane, hexachlorocyclohexane and pentachlorobenzene, with the latter having the highest concentrations: 61.82 ng g^{-1} (Liu et al. 2024). It should be noted that some natural wetlands have demonstrated their capacity for self-purification since significant removal of emerging organic pollutants was observed (García et al. 2020). High concentrations of pesticides such as pentachlorophenol (2.49–3.89 ng μL^{-1}) were also reported in wetlands located near urban areas in Mexico (Díaz-Torres et al. 2013).

Another type of EC detected in these environmental matrices are pharmaceuticals. This is due to the reported occurrence of illicit drugs such as cocaine, amphetamine-like compounds, cannabinoids, and opiates in concentrations of 0.03 to 78.78 ng L^{-1} in water samples taken from different sampling points in the wetland area of L'Albufera Natural Park in Valencia (Spain). The study evidences that pharmacological type pollutants in affluents of the Natural Park are related to irrigation systems and networks of wastewater treatment plants, on the other hand, the presence of these components is also attributed to the population density that occupies various recreational areas near the site (Pascual Aguilar et al. 2014). For its part, the Lutter wetland (Alsace, France) presented as results of studies in surface flows three pharmacological-type compounds detected in the affluents: ethinylestradiol (1140.62 μg L^{-1}), tramadol (222. 36 μg L^{-1}) and theophylline (31.8 μg L^{-1}), while in the outflow (effluent) the mainly detected compounds were alpha ethinylestradiol (901.62 μg L^{-1}), tramadol (193.72 μg L^{-1}) and losartan (22.87 μg L^{-1}). At both sampling points, indomethacin was detected less frequently. Finally, prednisolone, amoxicillin, and topiramate were detected only in the inflowing water; sulfadiazine, sertraline, gemfibrozil, and haloperidol in the outflowing water (Nuel et al. 2018).

In the Qinghai-Tibetan Plateau, 32 wetlands were investigated to evaluate the determinant of antibiotic resistance genes (ARGs), considering wetland type, altitude in the area, and environmental factors as variation factors. ARGs were detected in all wetlands, ranging from 1.80 × 105 to 1.35 × 107 copies per gram (viral load) of soils. Bacterial community composition was the main driver in shaping ARGs diversity and geographic distribution, as well as soil moisture and temperature factors, which were positively correlated with ARG abundance in this region (Yang et al. 2019).

In a study carried out during different climatic seasons in a wetland of the Canary Islands, Spain, the presence of 8 different types of hormones/steroids was detected, specifically in samples of affluents, whose concentration ranges were: norgestrel (20–230 ng L^{-1}), testosterone (80–240 ng L^{-1}), cortisone (0–150 ng L^{-1}), norethisterone (0–100 ng L^{-1}), boldenone (20–80 ng L^{-1}), estrone (15–20 ng L^{-1}) and estriol (0–4. 5 ng L^{-1}) (Guedes-Alonso et al. 2020). In Mexico, the presence of endocrine disrupting compounds was also determined in samples from the Xochimilco wetland, where high concentrations of bisphenol A (140.33 ng μL^{-1}), butyl benzyl phthalate (17.38 ng μL^{-1}) and estradiol (1.72 ng μL^{-1}) were obtained (Díaz-Torres et al. 2013). Whereas, other studies evaluated the presence of EC derived from personal care products. Galaxolide is a component present in personal care products: perfumes, and lotions, its concentration in the effluents of the analyzed wetland reached 263 ng L^{-1} (Matamoros and Salvadó 2012).

In addition, microplastics are reported as non-degradable pollutants in natural ecosystems. Their presence is most common in marine ecosystems, however, they are also found in freshwater ecosystems, such as lakes, ponds, river basins, wetlands, or even in wet agricultural land and groundwater. Microplastic contamination in wetlands has been reported in amounts up to 5531 particles m^{-3} and 6360 particles kg^{-1} in water and sediment samples, respectively, with a size range ranging from 0.12–9.5 mm (Kumar et al. 2021).

Table 7.1 shows the presence of various EC in natural wetlands and the concentrations detected.

Table 7.1 Emerging contaminants detected in natural wetlands

Emerging contaminant groups	Contaminant detected	Localization	Concentration	References
Pesticides	Pentachlorobenzene	China	62 ng g^{-1}	Liu et al. (2024)
	Orisastrobin	South Korea	700 ng L^{-1}	Kang et al. (2020)
	Ethoprop		100 ng L^{-1}	
	Tebuconazole	Spain	128 ng L^{-1}	Martínez-Megías et al. (2024)
	Atrazine		100 ng L^{-1}	
Drugs	Tramadol	France	222 ng L^{-1}	Nuel et al. (2018)
	Theophylline		32 ng L^{-1}	
	Testosterone	Spain	240 ng L^{-1}	Guedes-Alonso et al. (2020)
	Norgestrel		230 ng L^{-1}	
	Cortisone		150 ng L^{-1}	
	Paracetamol	Spain	168 ng L-1	Sadutto et al. (2021)
Disinfectants-personal care products	Triclosan	Mexico	72 ng uL^{-1}	Díaz-Torres et al. (2013)
	Galaxolide	Spain	263 ng L^{-1}	Matamoros and Salvadó (2012)
Plastic consumables	Bisphenol A	Mexico	140 ng uL^{-1}	Díaz-Torres et al. (2013)
	Butyl Benzyl Phthalate		17 ng uL^{-1}	

It is evident that natural wetlands are experiencing an increasing threat due to the occurrence and accumulation of ECs such as pesticides, pharmaceuticals, cleaning products and hormones. These pollutants, often introduced through agricultural activities, urban discharges and industrial processes, can persist in the environment and affect the quality of water, soil and the health of aquatic and terrestrial organisms. Pesticides and pharmaceuticals, for example, can alter the biological cycles of native species, affect reproduction and cause hormonal disruptions. Cleaning products and hormones, on the other hand, can contribute to eutrophication and alter microbial communities. The accumulation of these pollutants in wetlands not only threatens biodiversity and ecosystem functionality, but also has implications for human health through the food chain and water supply (Fig. 7.3). Therefore, it is essential to discuss restoration strategies to mitigate and protect the integrity of these vital ecosystems and public health.

7.5 Natural Wetland Restoration Strategies

Restoration strategies in wetlands can focus on restoring habitats in all functional aspects or on certain characteristics that help maintain the services that naturally predominate (Bertolini and da Mosto 2021). It is observed that the success of restoration in coastal wetlands ranges in the order of 38% for seagrasses and 65% in salt

Fig. 7.3 Presence of emerging contaminants in wetlands

marshes, furthermore in the economic scope, expenditures are projected between $80,000 and 1,600,000 USD, so projects may not appear cost-effective (Rains et al. 2016). Therefore, the restoration of coastal ecosystems suggests then, choosing the most influential indicators in the environment, eliminating biological threats, and performing continuous monitoring at the sites with specific periods (Cadier et al. 2020).

Liu et al. (2021) suggests implementing structures such as check dams and breakwaters to protect sediment availability against sea level interference in tidal marshes. In addition, he proposes strategic planting of marsh grasses or mangroves to prevent erosion, capture more sediment, and promote accumulation through subway roots, although their economic feasibility is yet to be determined. Marsh restoration can take a variety of approaches, such as managing hydrology, managing invasive vegetation, regulating plant species intrusion, mitigating disturbances to wildlife, implementing controlled grazing practices, and reducing pollution, as well as reintroducing angiosperms. Mitigation of wetland degradation caused by *Spartina alterniflora* invasion has been investigated in the Yangtze River, especially in Chongming Dongtan, using techniques such as cutting and flooding, digging and tilling, rhizome breaking, mowing, and biological replacement with *Phragmitas australis*, demonstrating high efficacy in that study (Zhao et al. 2016).

Likewise, in situ sediment treatment using activated carbon is effective for the removal of hydrophobic organic compounds in water bodies, such as oils, dioxins, and furans, among others. Although factors are mentioned that may vary the success of the result such as porosity and adsorption capacity of the soil (Patmont et al.

2015). Other authors report that soil replacement achieves nutrient restoration with the presence of fungi and bacteria, although there are losses in native microbial communities. Regarding soil restoration, it was obtained that the diversity of soil microorganisms represents the stability of microbial communities and reflects the effects of ecological mechanism (Ding 2023). The degree of restoration of bacterial community diversity and richness after different periods of natural restoration has been compared, showing that the role of the dominant bacterial community is crucial in the recovery of soil quality (Inglett et al. 2011).

Phytoremediation is reported to be one of the best alternatives for the removal, decontamination, and primary isolation of toxic substances in wetlands such as EC since it does not destroy soil structure or fertility and allows the elimination of various contaminants. This method also has a positive cost-benefit ratio and favorable engineering-economy ratios, including environmentally friendly technologies (Oyuela Leguizamo et al. 2017). Phytoremediation technologies include phytostabilization, which prevents erosion and reduces the bioavailability of contaminants without the need for soil removal through revegetation; phytodegradation, which uses plant enzymes to break down contaminants in sterile soils where natural biodegradation is limited; rhizofiltration, which can be applied to both terrestrial and aquatic plants to remove contaminants from water or soils, being more effective in terrestrial plants; and phytoextraction, which relies on the ability of plants to absorb contaminants, where the collected biomass is generally incinerated or composted. Under this approach, 90% of studies that used hydrophytic plants for wetland restoration in countries such as Japan and New Zealand showed success, as improvements in water quality parameters were proven after their introductory planting (Rodrigo 2021).

It is worth noting that some of the approaches to manage and treat EC are good manufacturing practices, installation of buffer strips, riparian zone protection and wetlands (Talib and Randhir 2017). In the work of (Zhou et al. 2020), the proportion and incidence of conventional technologies used for wetland restoration are observed, where techniques such as water-soil restoration engineering (including sediment dredging), microbial restoration, and soil fauna restoration are mentioned. However, in China, they rely on the principles of substrate design, and biological population selection, to delineate the main ecological wetland restoration techniques: lakeside wetland engineering technique, aquatic vegetation remediation technique, artificial floating island technique, natural imitation dam engineering technique and artificial media purification, combined grassland and forest ecosystem engineering technique, river corridor bioremediation technique, and riparian zone wastewater treatment engineering technique.

In terms of actual cases of implementation of wetland restoration strategies, there have been major contributions over the last few years. A case study is the rewetting of the Hedwige Polder (The Netherlands) in the Western Scheldt estuary to facilitate navigation. In Europe, there are restoration programs supported by the EU-LIFE Program. Another example of collaboration on this topic was the restoration of a floodplain on the Rhine River within a wetland, the objective of which was to reconnect floodplains, side channels, ridge and ditch complexes. Initiatives to restore or create riparian wetlands in open water systems in wetlands improved catchment, water quality, and biodiversity at a landscape scale. In southern Sweden, hundreds of wetlands were created in agricultural catchments that discharge effluent into the Baltic Sea. These wetlands have proven to be effective in removing and retaining nitrates while enhancing aquatic plant and macrofaunal species diversity (Verhoeven 2014). On the other hand, (Mitsch and Day 2006) states that it is necessary to regulate the discharge of nitrate-nitrogen-based compounds (fertilizers) in riverbank-type wetlands since these cause episodic and persistent eutrophication and hypoxia (dissolved oxygen <2 mg L^{-1}) in the joint wetlands, and therefore hinder the application of treatment technologies. Table 7.2 below summarizes the main strategies for natural wetland restoration.

Table 7.2 Restoration strategies for different types of wetlands and their main contributions

Type of wetland	Localization	Restoration strategy	Contributions	References
Estuaries	Colombia	Phytoremediation with allochthonous and autochthonous plants.	Removal of contaminants.	Oyuela Leguizamo et al. (2017)
	United States	Application of activated carbon.	Elimination of hydrophobic organic compounds.	Patmont et al. (2015)
Mangroves	China	Implementation of barriers/containment walls.	Erosion prevention, sediment trapping, and subway root accumulation.	Liu et al. (2021)
Lakeside	United States	Restoration of soil microbial community.	Increased abundance of fungal biomarkers.	Inglett et al. (2011)
	China	Restoration of soil microbial community.	Degradation of polymers from plant residues.	Ding (2023)
Marshes	China	Hydrology management and phytoremediation against invasive species.	Pollution and eutrophication prevention.	Zhao et al. (2016)
Swamps	Japan and New Zealand	Phytoremediation with hydrophytic plants.	Improvement of water quality in 90% of the cases.	Rodrigo (2021)
Riverbanks	Germany	Creation of artificial wetlands.	Improving catchment, water quality and biodiversity.	Verhoeven (2014)
	United States	Regulation of discharge of N and P-based pollutants.	Prevention of episodic and persistent eutrophication and hypoxia.	Mitsch and Day (2006)

7.6 Conclusions

The presence of ECs in natural wetlands has generated significant concern due to their negative impact on the ecological health and functionality of these ecosystems. It is noted that these contaminants come from a variety of sources, including industrial discharges, agricultural runoff, and inadequate wastewater treatment systems, and accumulate in water and sediments, affecting biodiversity and ecosystem services. The presence of emerging pollutants such as pesticides, pharmaceuticals and hormones can alter the biological balance of wetlands, representing a persistent threat to these environmental spaces. However, important advances are evident in wetland restoration strategies, which have shown potential for improving water quality and habitat recovery. Finally, it is crucial to implement integrated measures that include source reduction of pollutants, active habitat restoration, and riparian zone protection to mitigate adverse effects and ensure the resilience and sustainability of natural wetlands.

References

Alvarez-Mieles G, Irvine K, Griensven AV, Arias-Hidalgo M, Torres A, Mynett AE (2013) Relationships between aquatic biotic communities and water quality in a tropical river–wetland system (Ecuador). Environ Sci Pol 34:115–127. https://doi.org/10.1016/J.ENVSCI.2013.01.011

Ballut-Dajud GA, Herazo LCS, Fernández-Lambert G, Marín-Muñiz JL, Méndez MCL, Betanzo-Torres EA (2022) Factors affecting wetland loss: a review. Land 11(3). https://doi.org/10.3390/land11030434

Bansal S, Lishawa SC, Newman S, Tangen BA, Wilcox D, Albert D, Anteau MJ, Chimney MJ, Cressey RL, DeKeyser E, Elgersma KJ, Finkelstein SA, Freeland J, Grosshans R, Klug PE, Larkin DJ, Lawrence BA, Linz G, Marburger J et al (2019) Typha (Cattail) invasion in North American wetlands: biology, regional problems, impacts, ecosystem services, and management. Wetlands 39(4):645–684). Springer Netherlands. https://doi.org/10.1007/s13157-019-01174-7

Berezowski T, Wassen M, Szatyłowicz J, Chormański J, Ignar S, Batelaan O, Okruszko T (2018) Wetlands in flux: looking for the drivers in a central European case. Wetl Ecol Manag 26(5):849–863. https://doi.org/10.1007/s11273-018-9613-z

Bertolini C, da Mosto J (2021) Restoring for the climate: a review of coastal wetland restoration research in the last 30 years. Restor Ecol 29(7). Wiley. https://doi.org/10.1111/rec.13438

Brisson J, Rodriguez M, Martin CA, Proulx R (2020) Plant diversity effect on water quality in wetlands: a meta-analysis based on experimental systems. Ecol Appl 30(4):e02074. https://doi.org/10.1002/EAP.2074

Cadier C, Bayraktarov E, Piccolo R, Adame MF (2020) Indicators of coastal wetlands restoration success: a systematic review. In: Frontiers in marine science, vol 7. Frontiers Media S.A., Lausanne. https://doi.org/10.3389/fmars.2020.600220

Cai Y, Liang J, Zhang P, Wang Q, Wu Y, Ding Y, Wang H, Fu C, Sun J (2021) Review on strategies of close-to-natural wetland restoration and a brief case plan for a typical wetland in northern China. Chemosphere 285:131534. https://doi.org/10.1016/J.CHEMOSPHERE.2021.131534

Campbell DA, Cole CA, Brooks RP (2002) A comparison of created and natural wetlands in Pennsylvania, USA. Wetlands Ecol Manag 10(1):41–49

Cao T, Yi Y, Liu H, Yang Z (2020) Integrated ecosystem services-based calculation of ecological water demand for a macrophyte-dominated shallow lake. Glob Ecol Conserv 21:e00858. https://doi.org/10.1016/J.GECCO.2019.E00858

Díaz-Torres E, Gibson R, González-Farías F, Zarco-Arista AE, Mazari-Hiriart M (2013) Endocrine disruptors in the Xochimilco wetland, Mexico City. Water Air Soil Pollut 224(6). https://doi.org/10.1007/s11270-013-1586-1

Ding JN (2023) Effect of cultivation and natural restoration on soil microbial functional structure in coldregion wetlands. Appl Ecol Environ Res 21(2):1471–1484. https://doi.org/10.15666/aeer/2102_14711484

Dixon MJR, Loh J, Davidson NC, Beltrame C, Freeman R, Walpole M (2016) Tracking global change in ecosystem area: the wetland extent trends index. Biol Conserv 193:27–35. https://doi.org/10.1016/j.biocon.2015.10.023

C. M. Duarte. (2012). Evaluación de los ecosistemas del milenio en España

Elliott LH, Igl LD, Johnson DH (2019) The relative importance of wetland area versus habitat heterogeneity for promoting species richness and abundance of wetland birds in the prairie pothole region, USA. Condor 122(1). https://doi.org/10.1093/condor/duz060

Finkl CW (2021) Checklist of freshwater fishes of southwestern wetlands of Iran. http://www.springer.com/series/8795

Finlayson CM, Milton GR, Prentice RC (2018) Wetland types and distribution. In: The wetland book II: distribution, description, and conservation, vol 1. Springer Netherlands, Dordrecht, pp 19–35. https://doi.org/10.1007/978-94-007-4001-3_186/TABLES/4

Gaglio M, Aschonitis VG, Gissi E, Castaldelli G, Fano EA (2017) Land use change effects on ecosystem services of river deltas and coastal wetlands: case study in Volano–Mesola–Goro in Po river delta (Italy). Wetl Ecol Manag 25(1):67–86. https://doi.org/10.1007/s11273-016-9503-1

García J, García-Galán MJ, Day JW, Boopathy R, White JR, Wallace S, Hunter RG (2020) A review of emerging organic contaminants (EOCs), antibiotic resistant bacteria (ARB), and antibiotic resistance genes (ARGs) in the environment: increasing removal with wetlands and reducing environmental impacts. In: Bioresource technology, vol 307. Elsevier Ltd., New York. https://doi.org/10.1016/j.biortech.2020.123228

Guedes-Alonso R, Montesdeoca-Esponda S, Herrera-Melián JA, Rodríguez-Rodríguez R, Ojeda-González Z, Landívar-Andrade V, Sosa-Ferrera Z, Santana-Rodríguez JJ (2020) Pharmaceutical and personal care product residues in a macrophyte pond-constructed wetland treating wastewater from a university campus: presence, removal and ecological risk assessment. Sci Total Environ 703. https://doi.org/10.1016/j.scitotenv.2019.135596

Hao J, Xu G, Luo L, Zhang Z, Yang H, Li H (2020) Quantifying the relative contribution of natural and human factors to vegetation coverage variation in coastal wetlands in China. CATENA 188:104429. https://doi.org/10.1016/J.CATENA.2019.104429

He W, Choi I, Lee JJ, Hur J (2016) Coupling effects of abiotic and biotic factors on molecular composition of dissolved organic matter in a freshwater wetland. Sci Total Environ 544:525–534. https://doi.org/10.1016/J.SCITOTENV.2015.12.008

Heynes-Silerio SA, González-Elizondo M d S, Ruacho-González L, González-Elizondo M, López-Enríquez IL (2017) Vegetación de humedales del municipio de Durango, Durango, México. Revista Mexicana de Biodiversidad 88(2):358–364. https://doi.org/10.1016/j.rmb.2017.03.005

Inglett KS, Inglett PW, Reddy KR (2011) Soil microbial community composition in a restored calcareous subtropical wetland. Soil Sci Soc Am J 75(5):1731–1740. https://doi.org/10.2136/sssaj2010.0424

Kang D, Doudrick K, Park N, Choi Y, Kim K, Jeon J (2020) Identification of transformation products to characterize the ability of a natural wetland to degrade synthetic organic pollutants. Water Res 187. https://doi.org/10.1016/j.watres.2020.116425

Keddy PA (2010) Wetlands: an overview. In: Wetland ecology. Cambridge University Press, pp 1–41. https://doi.org/10.1017/cbo9780511778179.003

Kingsford RT, Basset A, Jackson L (2016) Wetlands: conservation's poor cousins. Aquat Conserv Marine and Freshw Ecosyst 26(5):892–916). John Wiley and Sons Ltd. https://doi.org/10.1002/aqc.2709

Kumar R, Sharma P, Bandyopadhyay S (2021) Evidence of microplastics in wetlands: extraction and quantification in freshwater and coastal ecosystems. J Water Proc Engineering 40. https://doi.org/10.1016/j.jwpe.2021.101966

Lichvar RW (2013) The national wetland plant list: 2013 wetland ratings. Phyton 49:1–241. http://www.bonap.org
Liu Z, Fagherazzi S, Cui B (2021) Success of coastal wetlands restoration is driven by sediment availability. Commun Earth Environ 2(1). https://doi.org/10.1038/s43247-021-00117-7
Liu M, Su X, Yuan J, Chen Y, Huang X, Yang X, Zheng J, Li Q, Xu J, He Y (2024) Residual effects of chlorinated organic pollutants on microbial community and natural redox processes in coastal wetlands. J Hazard Mater 465:133364. https://doi.org/10.1016/J.JHAZMAT.2023.133364
Lopez FJ, Pitarch E, Botero-Coy AM, Fabregat-Safont D, Ibáñez M, Marin JM, Peruga A, Ontañón N, Martínez-Morcillo S, Olalla A, Valcárcel Y, Varó I, Hernández F (2022) Removal efficiency for emerging contaminants in a WWTP from Madrid (Spain) after secondary and tertiary treatment and environmental impact on the Manzanares River. Sci Total Environ 812:152567. https://doi.org/10.1016/J.SCITOTENV.2021.152567
Majeed W, Rana N, de Azevedo Koch EB, Nargis S (2020) Seasonality and climatic factors affect diversity and distribution of arthropods around wetlands. Pak J Zool 52(6):2135–2144. https://doi.org/10.17582/journal.pjz/20200112020107
Martínez-Megías C, Arenas-Sánchez A, Manjarrés-López D, Pérez S, Soriano Y, Picó Y, Rico A (2024) Pharmaceutical and pesticide mixtures in a Mediterranean coastal wetland: comparison of sampling methods, ecological risks, and removal by a constructed wetland. Environ Sci Pollut Res 31(10):14593–14609. https://doi.org/10.1007/S11356-024-31968-0/FIGURES/5
Matamoros V, Salvadó V (2012) Evaluation of the seasonal performance of a water reclamation pond-constructed wetland system for removing emerging contaminants. Chemosphere 86(2):111–117. https://doi.org/10.1016/j.chemosphere.2011.09.020
Meng W, He M, Hu B, Mo X, Li H, Liu B, Wang Z (2017) Status of wetlands in China: a review of extent, degradation, issues and recommendations for improvement. Ocean Coastal Manag 146:50–59). Elsevier Ltd. https://doi.org/10.1016/j.ocecoaman.2017.06.003
Meng W, Feagin RA, Hu B, He M, Li H (2019) The spatial distribution of blue carbon in the coastal wetlands of China. Estuarine Coastal Shelf Sci 222:13–20). Academic. https://doi.org/10.1016/j.ecss.2019.03.010
Mitsch WJ, Day JW (2006) Restoration of wetlands in the Mississippi-Ohio-Missouri (MOM) river basin: experience and needed research. Ecol Eng 26(1):55–69. https://doi.org/10.1016/j.ecoleng.2005.09.005
Morales-Paredes CA, Rodríguez-Díaz JM, Boluda-Botella N (2022) Pharmaceutical compounds used in the COVID-19 pandemic: a review of their presence in water and treatment techniques for their elimination. Sci Total Environ 814:152691. https://doi.org/10.1016/J.SCITOTENV.2021.152691
Munford KE, Gilbert-Parkes S, Mykytczuk NCS, Basiliko N, Yakimovich KM, Poulain A, Watmough SA (2023) How arsenic contamination influences downslope wetland plant and microbial community structure and function. Sci Total Environ 876:162839. https://doi.org/10.1016/J.SCITOTENV.2023.162839
Neckles HA, Dionne M, Burdick DM, Roman CT, Buchsbaum R, Hutchins E (2002) A monitoring protocol to assess tidal restoration of salt marshes on local and regional scales. Restor Ecol 10(3):556–563
Nuel M, Laurent J, Bois P, Heintz D, Wanko A (2018) Seasonal and ageing effect on the behaviour of 86 drugs in a full-scale surface treatment wetland: removal efficiencies and distribution in plants and sediments. Sci Total Environ 615:1099–1109. https://doi.org/10.1016/j.scitotenv.2017.10.061
Olsson L, Ye S, Yu X, Wei M, Krauss KW, Brix H (2015) Factors influencing CO2 and CH4 emissions from coastal wetlands in the Liaohe Delta, Northeast China. Biogeosciences 12(16):4965–4977. https://doi.org/10.5194/bg-12-4965-2015
Oyuela Leguizamo MA, Fernández Gómez WD, Sarmiento MCG (2017) Native herbaceous plant species with potential use in phytoremediation of heavy metals, spotlight on wetlands — a review. Chemosphere 168:1230–1247). Elsevier Ltd. https://doi.org/10.1016/j.chemosphere.2016.10.075

Pascual Aguilar JA, Andreu V, Vázquez P, Picó Y (2014) Presence and spatial distribution of emerging contaminants (drugs of abuse) in protected agroecological systems (L'Albufera de Valencia Coastal Wetland, Spain). Environ Earth Sci 71(1):31–37. https://doi.org/10.1007/s12665-013-2688-1

Patmont CR, Ghosh U, LaRosa P, Menzie CA, Luthy RG, Greenberg MS, Cornelissen G, Eek E, Collins J, Hull J, Hjartland T, Glaza E, Bleiler J, Quadrini J (2015) In situ sediment treatment using activated carbon: a demonstrated sediment cleanup technology. Integr Environ Assess Manag 11(2):195–207). Wiley-Blackwell. https://doi.org/10.1002/ieam.1589

Rai PK (2019) Heavy metals/metalloids remediation from wastewater using free floating macrophytes of a natural wetland. Environ Technol Innov 15. https://doi.org/10.1016/j.eti.2019.100393

Rains MC, Leibowitz SG, Cohen MJ, Creed IF, Golden HE, Jawitz JW, Kalla P, Lane CR, Lang MW, Mclaughlin DL (2016) Geographically isolated wetlands are part of the hydrological landscape. Hydrol Process 30(1):153–160. https://doi.org/10.1002/hyp.10610

Ramachandra TV, Sudarshan P, Vinay S, Asulabha KS, Varghese S (2020) Nutrient and heavy metal composition in select biotic and abiotic components of Varthur wetlands, Bangalore, India. SN Appl Sci 2(8). https://doi.org/10.1007/s42452-020-03228-6

Richardson SD, Ternes TA (2021) Water analysis: emerging contaminants and current issues. Anal Chem 94(1):382–416. https://doi.org/10.1021/ACS.ANALCHEM.1C04640

Richardson DC, Holgerson MA, Farragher MJ, Hoffman KK, King KBS, Alfonso MB, Andersen MR, Cheruveil KS, Coleman KA, Farruggia MJ, Fernandez RL, Hondula KL, López Moreira Mazacotte GA, Paul K, Peierls BL, Rabaey JS, Sadro S, Sánchez ML, Smyth RL, Sweetman JN (2022) A functional definition to distinguish ponds from lakes and wetlands. Sci Rep 12(1):1–13. https://doi.org/10.1038/s41598-022-14569-0

Rodrigo MA (2021) Wetland restoration with hydrophytes: a review. Plants 10(6). https://doi.org/10.3390/plants10061035

Sadutto D, Andreu V, Ilo T, Akkanen J, Picó Y (2021) Pharmaceuticals and personal care products in a Mediterranean coastal wetland: impact of anthropogenic and spatial factors and environmental risk assessment. Environ Pollut 271:116353. https://doi.org/10.1016/J.ENVPOL.2020.116353

Salimi S, Almuktar SAAAN, Scholz M (2021) Impact of climate change on wetland ecosystems: a critical review of experimental wetlands. J Environ Manag 286:112160. https://doi.org/10.1016/J.JENVMAN.2021.112160

Salmo SG III, Shielameh P-M (2013) Evaluating patterns of fish assemblage changes from different-aged reforested mangroves in Lingayen gulf. J Environ Sci Manag 16(1):11

Sapkota Y, White JR (2020) Carbon offset market methodologies applicable for coastal wetland restoration and conservation in the United States: a review. Sci Total Environ 701:134497. https://doi.org/10.1016/J.SCITOTENV.2019.134497

Shackelford N, Hobbs RJ, Burgar JM, Erickson TE, Fontaine JB, Laliberté E, Ramalho CE, Perring MP, Standish RJ (2013) Primed for change: developing ecological restoration for the 21st century. Restor Ecol 21(3):297–304. https://doi.org/10.1111/rec.12012

Steinmuller HE, Chambers LG (2019) Characterization of coastal wetland soil organic matter: implications for wetland submergence. Sci Total Environ 677:648–659. https://doi.org/10.1016/J.SCITOTENV.2019.04.405

Talib A, Randhir TO (2017) Managing emerging contaminants in watersheds: need for comprehensive, systems-based strategies. In: Sustainability of water quality and ecology, vol 9–10, pp 1–8). Elsevier B.V. https://doi.org/10.1016/j.swaqe.2016.05.002

Thorslund J, Jarsjo J, Jaramillo F, Jawitz JW, Manzoni S, Basu NB, Chalov SR, Cohen MJ, Creed IF, Goldenberg R, Hylin A, Kalantari Z, Koussis AD, Lyon SW, Mazi K, Mard J, Persson K, Pietro J, Prieto C et al (2017) Wetlands as large-scale nature-based solutions: status and challenges for research, engineering and management. Ecol Eng 108:489–497. https://doi.org/10.1016/j.ecoleng.2017.07.012

Tozer DC, Steele O, Gloutney M (2018) Multispecies benefits of wetland conservation for marsh birds, frogs, and species at risk. J Environ Manag 212:160–168. https://doi.org/10.1016/j.jenvman.2018.01.055

Verhoeven JTA (2014) Wetlands in Europe: perspectives for restoration of a lost paradise. Ecol Eng 66:6–9. https://doi.org/10.1016/j.ecoleng.2013.03.006

Wang G, Otte ML, Jiang M, Wang M, Yuan Y, Xue Z (2019) Does the element composition of soils of restored wetlands resemble natural wetlands? Geoderma 351:174–179. https://doi.org/10.1016/j.geoderma.2019.05.032

Wassens S, Hall A, Spencer J (2017) The effect of survey method on the detection probabilities of frogs and tadpoles in large wetland complexes. Mar Freshw Res 68(4):686–696. https://doi.org/10.1071/MF15183

Wu Y, Liu J, Rene ER (2018) Periphytic biofilms: a promising nutrient utilization regulator in wetlands. Bioresour Technol 248:44–48). Elsevier Ltd. https://doi.org/10.1016/j.biortech.2017.07.081

Wu Y, Xu N, Wang H, Li J, Zhong H, Dong H, Zeng Z, Zong C (2021) Variations in the diversity of the soil microbial community and structure under various categories of degraded wetland in Sanjiang Plain, northeastern China. Land Degrad Dev 32(6):2143–2156. https://doi.org/10.1002/LDR.3872

Xu W, Fan X, Ma J, Pimm SL, Kong L, Zeng Y, Li X, Xiao Y, Zheng H, Liu J, Wu B, An L, Zhang L, Wang X, Ouyang Z (2019) Hidden loss of wetlands in China. Curr Biol 29(18):3065–3071.e2. https://doi.org/10.1016/j.cub.2019.07.053

Xue Z, Zou Y, Zhang Z, Lyu X, Jiang M, Wu H, Liu X, Tong S (2018) Reconstruction and future prediction of the distribution of wetlands in China. Earth's Future 6(11):1508–1517. https://doi.org/10.1029/2017EF000807

Yang Y, Liu G, Ye C, Liu W (2019) Bacterial community and climate change implication affected the diversity and abundance of antibiotic resistance genes in wetlands on the Qinghai-Tibetan Plateau. J Hazard Mater 361:283–293. https://doi.org/10.1016/j.jhazmat.2018.09.002

Yu J, Zhan C, Li Y, Zhou D, Fu Y, Chu X, Xing Q, Han G, Wang G, Guan B, Wang Q (2016) Distribution of carbon, nitrogen and phosphorus in coastal wetland soil related land use in the modern Yellow River Delta. Sci Rep 6. https://doi.org/10.1038/srep37940

Zhao Q, Bai J, Huang L, Gu B, Lu Q, Gao Z (2016) A review of methodologies and success indicators for coastal wetland restoration. In: Ecological Indicators, vol 60, pp 442–452). Elsevier B.V. https://doi.org/10.1016/j.ecolind.2015.07.003

Zhao CS, Yang Y, Yang ST, Xiang H, Ge YR, Zhang ZS, Zhao Y, Yu Q (2020) Effects of spatial variation in water quality and hydrological factors on environmental flows. Sci Total Environ 728:138695. https://doi.org/10.1016/J.SCITOTENV.2020.138695

Zhou D, Yu J, Guan B, Li Y, Yu M, Qu F, Zhan C, Lv Z, Wu H, Wang Q, Yang J (2020) A comparison of the development of wetland restoration techniques in China and other nations. Wetlands 40(6):2755–2764. https://doi.org/10.1007/s13157-020-01305-5

Zhu X, Wang L, Zhang X, He M, Wang D, Ren Y, Yao H, net Victoria Ngegla J, Pan H (2022) Effects of different types of anthropogenic disturbances and natural wetlands on water quality and microbial communities in a typical black-odor river. Ecol Indic 136:108613. https://doi.org/10.1016/J.ECOLIND.2022.108613

Zimmer KD, Hanson MA, Wrubleski DA (2016) Invertebrates in permanent wetlands (long-hydroperiod marshes and shallow lakes). In: Invertebrates in freshwater wetlands: an international perspective on their ecology. Springer International Publishing, Cham, pp 251–286. https://doi.org/10.1007/978-3-319-24978-0_8

Chapter 8
Heat Waves in Wetlands Ecosystem Biodiversity and Bioresources

Faith Ann Miracle, Madaki Enyojo Nathan, Adejoke Blessing Aransiola, Paul Ojo Adewoye, Ayodeji Adams Dahunsi, Sesan Abiodun Aransiola, and Naga Raju Maddela

Abstract Heatwaves pose a threat to wetlands, which are important ecosystems that offer essential functions like carbon sequestration, water filtration, and habitats for a variety of species. Because of their sensitivity to temperature changes, these wetlands are at risk of biodiversity loss as heatwaves become more frequent and upset their delicate equilibrium. This may result in local extinctions, disturbed migration patterns, and altered habitats. The consequences extend beyond wetlands, impacting human livelihoods in agriculture, fishing, and tourism, with potential economic losses from declining fish populations, lower crop yields, and reduced tourism revenue. Heatwaves also reduce wetlands' resilience to other environmental stresses and disrupt hydrology and nutrient cycling, further degrading water quality and ecosystem function. This chapter examines the impacts of heatwaves on wetland ecosystems, including changes in species composition, population dynamics, and ecosystem functions, as well as the effects on human economies and ways of life. Understanding these impacts is crucial for developing effective conservation and management strategies to protect these vital ecosystems and their services.

Keywords Heatwaves · Wetlands · Environment · Ecosystem · Water

F. A. Miracle · M. E. Nathan (✉) · A. A. Dahunsi · S. A. Aransiola
Department of Microbiology, Faculty of Science, University of Abuja, Abuja, Nigeria

A. B. Aransiola
Department of Surveying and Geoinformatics, Faculty of Environmental Science, University of Abuja,
Abuja, Nigeria

P. O. Adewoye
Bioresources Development Centre, National Biotechnology Research and Development Agency (NBRDA), Ogbomoso, Nigeria

N. R. Maddela
Departamento de Ciencias Biológicas, Facultad de Ciencias de la Salud, Universidad Técnica de Manabí, Portoviejo, Ecuador

B. R. Babaniyi et al. (eds.), *Wetland Ecosystems: Conservation Strategies, Policy Management and Applications*, Wetlands: Ecology, Conservation and Management 12, https://doi.org/10.1007/978-3-031-91982-4_8

8.1 Introduction

Heatwaves are becoming more common and powerful, endangering ecosystems all over the world (IPCC 2020). Due to their distinct features and purposes, wetlands in particular are susceptible to the effects of heatwaves (Mitsch and Gosselink 2015). Often called "ecological supermarkets," these ecosystems offer vital functions like carbon sequestration, water filtration, climate regulation, running water storage and habitat provision for a variety of plant and animal species (Barbier et al. 2011). Heatwaves, however, have the potential to interfere with these processes, which could have a significant impact on biodiversity and bio resources (Heino et al. 2015).

According to Keddy (2010), wetlands are intricate ecosystems that are home to a diverse array of flora and fauna species including microorganisms, many of which are adapted to particular temperature ranges for their optimum survival. These species may experience population declines, alterations in the makeup of their communities, or even local extinctions if heatwaves force them beyond their threshold for heat (Cahill et al. 2013). These heatwaves have a variety of effects on wetland biodiversity including habitat loss, increased mortality, behavioral adaptations, disrupted migration patterns, increased competition etc. influencing not just individual species but also the complex interactions that exist within ecosystems (Poff et al. 2016).

Heatwaves can have an effect on wetlands' bio resources in addition to their effects on biodiversity. Among the sectors that depend on robust and productive wetlands are the fishing, agricultural, and tourism sectors (Barbier et al. 2011). Heatwaves have the potential to negatively impact human livelihoods and local economies by reducing fish populations, crop yields, and tourism revenue. Heatwaves have broad repercussions on wetland ecosystems and may affect ecosystem resilience over the long run (IPCC 2020). Heatwaves that occur frequently can weaken wetlands' resistance to other shocks like floods, droughts, and invasive species (Heino et al. 2015). Furthermore, according to Mitsch et al. (2013), heatwaves can affect wetlands' hydrology and nutrient cycling, which can affect water quality and ecosystem function. Barbier et al. (2011) assert that although heatwaves have a substantial impact on wetland ecosystems, more research and comprehension of these effects are required. The value of estuarine and coastal ecosystem services. This chapter aims to address this knowledge gap by exploring the impacts of heatwaves on wetland ecosystem biodiversity and bio resources. According to Cahill et al. (2013), we will look at how heatwaves affect species composition, population dynamics, and ecosystem function, As well as the consequences for human livelihoods and local economies (Poff et al. 2016). Furthermore, in light of the growing frequency and intensity of heatwaves, we will talk about adaptation and mitigation strategies to support the management and conservation of wetland ecosystems (Keddy et al. 2017). Understanding how heatwaves affect wetland ecosystems can help us create practical plans to save and maintain these vital ecosystems (Millennium Ecosystem Assessment 2005).

8.2 Biodiversity in Wetlands

Wetlands are covered by about 6% of the earth surface providing a number of ecosystem services and maintaining biodiversity on earth (OMRF 2017), fishes, mollusks, reptiles, amphibians, insects, mammal's and even plants sum up the diversities in a wetland. Over 450 species of wetland plants are found in the province of Ontario, Canada alone, (United Nations Environment Programme 2021). Consequently, wetlands also supported essential services for humans such as water filtration, flood and erosion prevention, recharge ground water aquifers, serves as sinks and climate moderations. Wetlands are actually biologically diverse from primary consumers like crustaceans, aquatic insects, molluscs to varieties of carnivores like osprey, alligators, otters and dragonfly (Bull et al. 2017).

Several communities of ecological and economic values are maintained by wetlands around the world. Wetlands diversity is also in types as all wetlands share common features like waterlogged or wet soils (hydrology), soils capable of anaerobic process, rooted vascular plants, according to Linh et al. (2023), biodiversity is a very complex system made up of the relationships that organisms have with their external environment, which are expressed in genetic, species, and ecosystem diversities. Water purification, nutrient cycling, and biodiversity maintenance are among the functions of wetlands that can be stabilized by high biodiversity (Thomaz 2023). In addition, the biodiversity of wetlands enhances the stability and resistance to external disturbances of the food chain, preventing external disturbances in wetlands and achieving this by developing complex food web structures. Rich diversity can reduce competitive pressures among organisms by offering more ecological niches (Peel et al. 2019; Hatten et al. 2024). Numerous microbial colonies can cooperate to break down complex organic matter and transform it into harmless products, and wetlands with a high microbial diversity typically have a higher capacity to degrade organic pollutants (Berrier et al. 2022). Microbial aggregates consisting of *pseudomonas*, *rhinococus*, and Nocardia are formed to enhance the elimination of n-alkanes and polycyclic aromatic hydrocarbons (PAHs).

The process of denitrification in wetlands is carried out by anaerobic ammonia-oxidizing bacteria working in tandem with specific archea (*nitrate archea*) (Wang et al. 2019). Wetlands come in a variety of sizes, which further contributes to their diversity. They can be either freshwater or saltwater swamps, home to a wide range of insects, reptiles, and amphibians, including frogs, as well as thick canopy trees like bubinga and ovangkol, as well as pricey luxury woods. In more temperate climates, cypress trees grow alongside willows and other shrubs, tiny water plants called duckweed, and some hydrophytes like mangrove trees. Marshes are among the most biologically diverse ecosystems in the world. They are primarily made up of grasses, which give animals a place to live and food. Other types of wetlands that are recognized by scientists as major contributors to biodiversity worldwide include sloughs, muskegs, bogs, fens, and mires. Also, because of their special high altitude features, alpine wetlands, play a unique ecological role in purifying and conserving water, maintaining biodiversity, climate and nutrient cycling regulations (Cao et al.

2017; Yin et al. 2022). Alpine wetlands display rich biodiversity worldwide enhancing ecosystem reliability and providing ecological services, functionality and stability (Qiu et al. 2024). However, factors such as anthropological factors, global climate change, and geological disasters like earthquakes, have led to biodiversity loss and even severe degradation of global wetlands (He et al. 2021; Jiang et al. 2023).

8.3 Wetland Bioresources

These materials are found to be naturally occurring and capable of sustaining renewable biodegradable materials, they are also known as bio resources. Resources obtained from biological origin or nature wetlands can be in different forms (Fig. 8.1).

8.3.1 Food

Aquatic lives obtained from the wetland ecosystems provides food in sufficient nutrition to human beings and animals. Bird species are being hunted and gotten as sources of food from barnacles, crabs, meats from alligators, crocodile's, salmons, ducks and geese, honey collectors also get bee hives on these mangrove trees canopy.

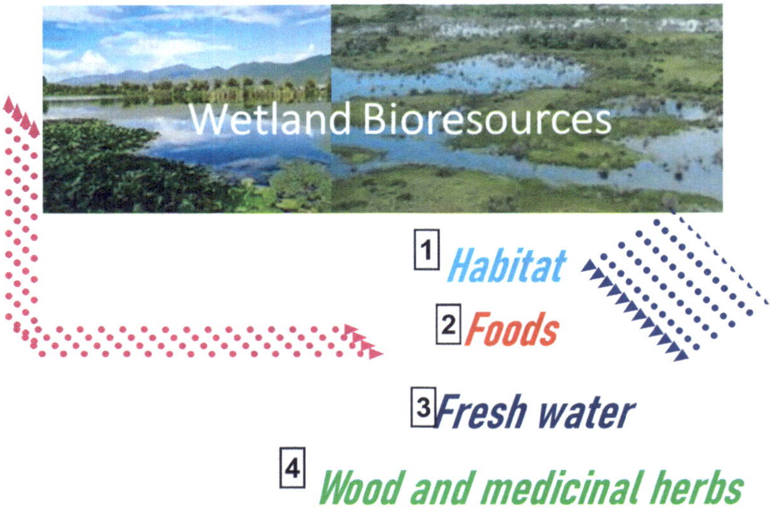

Fig. 8.1 Bioresources in the wetlands

8.3.2 *Sources of Wood and Medicinal Herbs*

Mangrove leaves and other woods like bubinga and ovangkol are pricey luxury woods used to make furniture and violins, among other musical instruments (National Geographic); some of these woods can also be used for research and medicinal purposes (timber products, saw logs, transmission poles, building poles (McGinley 2008).

8.3.3 *Habitat*

Hundreds of insects, reptiles, and amphibians can be found living on the muddy floors of these wetlands. Various organisms are supported by the roots of mangroves. The root system of mangrove swamps provides cover and a place for fallen leaves and other materials to be eaten, and it is home to a variety of shellfish, including crabs and conchs.

8.3.4 *Fresh Water*

Wetlands help maintain water quality and can be used for irrigation farming. Rice is always cultivated in wetlands to aid the growth of the plants these only occurs due to the presence of water and these in turn supports biodiversity in our ecosystem and creating a balance.

8.4 Causes of Heatwave in Wetlands Ecosystem

8.4.1 *Climate Change*

The change from numbers of high heat days to severe storms, increased flooding and drought all has an effect on wetlands. Wetlands are vulnerable to change in hydrology as they exist between full aquatic and full terrestrial ecosystem. These slight changes can overtime lead to shift in species community, warming temperatures, biochemical soil, increased carbon dioxide levels and change in the rainfall frequency. Urbanization, pollution along with other stressors all bring about heat waves in wetlands (Aransiola et al. 2024).

8.4.2 Land use Changes

The conversion of wetlands to different uses can alter regional microclimates and local hydrological cycles (Bai et al. 2013). Wetlands converted into built-up areas increase impervious surfaces, which improve surface runoff generation. When wetlands are converted to bare surfaces, soil water changes, which may affect latent heat fluxes (Kueppers et al. 2008). Increased heat emissions are the primary cause of urban heat island (UHI) effects, which have been linked to elevated temperatures. These lands use most times reduces vegetative covers leading to higher surface temperatures and high heat.

8.4.3 Methane Emissions

The planet's main source of methane, which is a powerful greenhouse gas, is wetlands. The global methane pledge was signed in 2021 by the US, EU, Indonesia, Canada, Brazil, UK, and other countries. It called for a 30% reduction in methane emissions over 2020–2030. Nevertheless, 40% of methane emissions originate from wetlands—wet surfaces that release methane. Permafrost wetlands are made of soil that is partially frozen and wet in cold climates found at high latitudes. A portion of these emissions cause the climate to warm and the permafrost to thaw. Additionally, dormant microbes awaken and release more methane into the atmosphere, leading to more weather fluctuations and a 25% increase in global temperatures.

8.4.4 Human Activities (Anthropogenic)

The world has lost 57% of it's wetlands since 1900 AD (Davidson and Davidson 2014). Wetlands due to their hydrological and chemical functions are regarded as the kidneys of the earth (Keddy 2010). Threats from wetlands degradation resulting from human disturbance and climate change are present in many countries across continents (Gong et al. 2010; Na et al. 2008). Wetlands have also been significantly impacted by human activity (per capita GDP, population densities, and agricultural population, for example). Human policies, economics are part of the major driving force, lands like transportation lands, canals, residential lands are on the increase whereas areas of wetlands are significantly degraded and these are actually as a result of increase in population, economic activities, these summing up to cause temperature decline in wetlands even extermination of wetlands ecosystem. Policies should enact to prevent the encroachment of human activities across these wetlands activities like fishing, hunting and damming (Fig. 8.2) Millennium Ecosystem Assessment (2005).

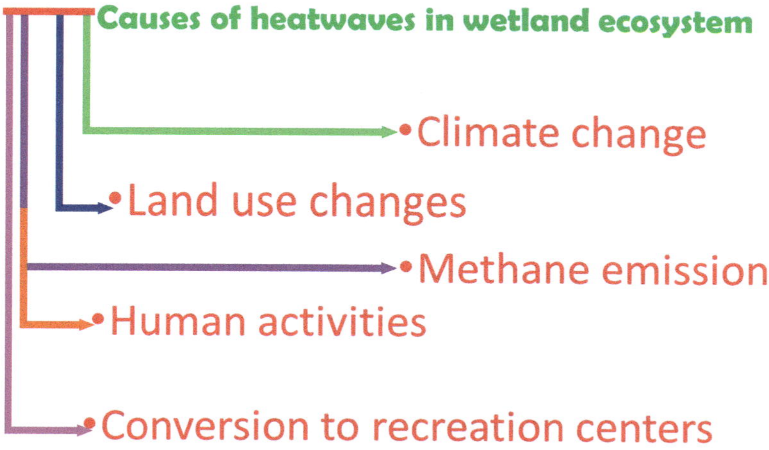

Fig. 8.2 Causes of heatwave in wetlands ecosystem

8.4.5 Reclamation of Land, Conversion to Ports, Urban Areas, Agriculture, and Aquaculture

Mangrove wetlands disappear as a result of various human activities that modify natural hydrology, including dredging, poldering, and the building of dams, dykes, and drainage channels. Consequently, maintaining coastal wetlands can aid in mitigating and adapting to climate change (Neubauer and Verhoeven 2019; Serrano and de la Hera 2019).

8.5 Benefits of Wetlands Ecosystem During Heat Waves

Wetlands are essential for mitigating the effects of climate change and promoting adaptation, according to the 2005 Millennium Ecosystem Assessment. Vegetated and healthy wetlands are among the most effective sinks for carbon on the planet, temperature regulators, and bio filters and serve a whole lot of other benefits during heat wave.

8.5.1 Running Water Retention and Groundwater Recharge

Wetlands serve as natural sponges, trapping and retaining water during floods, snowmelt, and even rain which helps to recharge groundwater aquifers, maintain water tables, retain soil moisture, and reduce drought stress during heat waves.

8.5.2 Carbon Sequestration

Sanderson et al. (2016) suggested that sequestration is used to refer to the photosynthetic removal of CO_2 from the atmosphere and its conversion into cellulose and other carbon compounds in plants and its conversion from decaying plants into soil organic matter. Wetlands serve as carbon sinks, absorbing carbon from the atmosphere and keeping it trapped in soils and vegetation for thousands of years. Wetland flora absorb carbon from the environment as they grow and when they die, due to the low oxygen levels in water compared to land, decomposition is slower. This carbon does not easily get to be released back into the environment.

8.5.3 Bio Filtering and Elimination of Particulate Matter

Wetlands serve the purpose of Biofilters that eliminate particulate matter, collect nutrients, and transform them into varying gaseous forms of nitrogen, this helps to improve air quality and develop cool breezes during heat waves.

8.5.4 Weather Pattern Influence

Wetlands influence local weather patterns such as precipitation and fog formation.

8.5.5 Climate Regulation

Wetlands also mitigate climate change by regulating greenhouse gases, temperature, and other climatic compositions of the atmosphere.

8.5.6 Recreation

Wetlands serve as recreational spaces, reducing the heat stress on the human race and improving their mental well-being during heat waves.

8.5.7 Habitation and Shades for Species

Wetlands' unique combination of water, vegetation, and shade creates a haven for species during heat waves (Elmqvist et al. 2015).

8.5.8 Environmental Coolant

Wetlands act as natural coolants through evaporation and evapotranspiration, regulating temperature by transferring water from land and vegetation to the atmosphere in the form of water vapor thereby cooling the surrounding atmosphere, especially in urban areas where heat islands are massive during heat waves.

8.6 Effect of Heats Waves on Wetland Ecosystem (Effects on Plants, Microorganism, Animals etc)

Heatwaves can have very significant effects on wetlands ecosystems. Wetlands are delicate ecosystems that are highly susceptible to temperature variations, including heat waves. During heatwaves, the increased temperatures can lead to elevated water temperatures in wetlands, affecting the delicate balance of these ecosystems. The flora, fauna, and microorganisms that live in wetlands may be impacted by this temperature rise, which may also change the wetlands' biodiversity. The extension of these heatwaves can now also result in wetlands experiencing a shortage of water, which could disrupt the availability of water for different species and lead to habitat loss.

In addition, heat waves have the potential to worsen conditions like water contamination in wetlands. Warmer temperatures have the potential to hasten the growth of algae, resulting in algal blooms that can lower the oxygen content of the water and endanger aquatic life. These severe weather conditions can also dry up wetlands, which results in soil erosion and a reduction in the amount of vegetation. Heatwaves can cause wetlands to dry up, raising the possibility of wildfires and putting the ecosystem and its inhabitants in even greater danger. Extreme heat waves would put the wetlands at risk for soil erosion, loss of vegetation cover, increased wildfire danger, disturbances to food chains, heightened competition for scarce resources, species migration, and general ecosystem imbalance.

8.6.1 Heat Waves Can Affect Wetlands' Flora in a Number of Ways

8.6.1.1 Phenology Modification

Heatwaves can cause irreversible changes in the timing of the life cycle events in plants, such as flowering, fruiting, and leaf senescence. This can disrupt the synchrony between plants and their pollinators or seed dispersers, thereby altering the phenology of wetlands flora. Heatwaves cause phenological stages to speed up, which causes timing errors that impact plant reproduction and the health of the ecosystem in wetland areas.

8.6.1.2 Variations in the Composition of Species

Changes in temperature, nutrient availability, and water levels are a few examples of environmental conditions that can cause this. The diversity and distribution of plants within the wetland ecosystem may fluctuate as a result of these changes, which may favor some plant species over others. This may have a domino effect on wetlands' resilience and general ability to function.

8.6.1.3 Decreased Growth of Plants

Heatwaves can cause stress to plants, which can impair their capacity for photosynthetic processes and efficient growth. A decrease in plant growth may affect the availability of food, the structure of the habitat, and the general health of the wetland ecosystem.

8.6.1.4 Decrease in Aquatic Vegetation

Reduced water levels brought on by the severe weather or heat wave will expose aquatic plants to the air and result in die-offs.

8.6.1.5 Variations in the Availability of Nutrients

Heatwaves cause changes in the cycling of nutrients, which impact plant uptake of nutrients and heighten competition.

8.6.1.6 A Rise in Death Rates

The severe weather, such as heat waves, increases evaporation and decreases water availability. This exacerbates drought conditions and results in a high mortality rate for wetlands flora.

8.6.2 Effects of Heat Waves on Wetland Fauna

The following list includes the effects that the animals (fauna) living in wetlands experience during heat waves, which range from mild inconveniences to fatalities;

8.6.2.1 Variations in the Dynamics of Population

Changes in the distribution of species, fluctuations in their abundance, and adjustments to how they interact with other species are examples of this. These disturbances may have an effect on food webs, predator-prey dynamics, and biodiversity in the wetland ecosystem, all of which could upset the ecosystem's delicate equilibrium.

8.6.2.2 Abruptions in the Patterns of Migration

In response to the shifting environmental conditions brought on by heatwaves, wetland fauna may modify their customary migration paths or timing. This has an effect on species distribution, interactions between species, and the general health of the wetland ecosystem.

8.6.2.3 Physiological Strain

Fauna may exhibit elevated body temperatures, dehydration, and altered metabolic rates during heatwaves. The organisms' survival and well-being may be impacted by these stressors, which may result in behavioral changes and possible changes in the ecosystem's population dynamics.

8.6.2.4 Adaptations in Behavior

In wetlands fauna, heatwaves can lead to a variety of behavioral adaptations, such as activity shifts, in which animals change their activity patterns to mitigate heat stress by being more active at night or during cooler months, Seeking shade by avoiding direct sunlight to reduce heat stress, Water-seeking behavior to cool down or rehydrate, Animals may migrate, engage in burrowing behavior, or enter a state akin to hibernation in which they reduce their activity or enter a state of dormancy in order to conserve energy. Adapting their foraging techniques to mitigate heat stress, such as gathering food during chilly times of the day, or even changing their social behavior by gathering more often or dispersing widely as a result of heat intolerance.

8.6.3 Heatwaves' Effects on Microorganisms in Wetlands

Heat waves and elevated temperatures also have a significant impact on the microorganisms in the wetlands; They undergo changes in their metabolic pathways, increased microbial mortality, altered nutrient cycling, altered community composition, altered enzyme activity, broken symbiotic relationships, and altered methane production and emission, all of which have a significant impact on the ecosystem as a whole (Maddela et al. 2022).

8.7 Wetlands Preservative Strategies

In china the national wetland conservation plan (2022–2030) a sectorial policy document aims to increase wetlands protection to 55% by 2025, while increasing international and national importance to 20 and 50 respectively by 2030 in China.

8.7.1 Addition of Wetland Actions into Climate Mitigation and Resiliency Plans

Policy makers should include the goals for restoration and protecting wetlands in national policies with plans for climate actions, adaptation and disaster risk reduction (convention om wetlands 2021) with the diverse benefits from commercial, food, agriculture and other benefits and importance of wetlands policies should be enacted to protect these waterlogged areas.

8.7.2 Support of Wetlands and Conservative Efforts

Donations to wetlands conservation organizations, volunteer activities, and advocating for wetlands protection along support organizations, communal groups, and local support organizations advocating and donations to ensure the preservation of wetlands.

8.7.3 Education, Research and Enlightenment

Educating local farmers and hunters and stakeholders on the importance of ecological benefits of wetlands and how they help balance our planet. Educating them on the threats faced by wetlands and the consequences to the environment, especially causing heatwaves leading to greenhouse effects and several other consequences. Getting them involved in the conservation strategies and efforts to curb the degradation of wetlands. Research can also be done to come up with plans to help improve and protect wetlands and reduce emissions and additional knowledge these gotten will be used to educate communities on the need to protect wetlands and it's surrounding.

8.7.4 Restoration Programs

Degraded and damaged wetlands can be aided towards the restoration of these bodies by either replanting the vegetation, stopping of human activities that have posed a threat to the wetlands preventing deforestation and reconstruction of artificial wetlands. In Balancing the planets ecosystem these bodies need to be restored and by these both climate changes and other ecological effects will be maintained. In Louisiana's Burrataria Basin, (NOAA 2022) Began construction of one of the largest habitat restoration efforts till date, a 1200-acre marsh in hopes to restore and create new habitat by these, wetlands are preserved along with species associated with the wetlands.

Several other preservation strategies include; water management, vegetation changes, ecotourism, economic incentives, wetlands friendly infrastructure and climate change mitigation along with as discussed above are strategies to prevent the extinction or degradation of wetlands ecosystem.

8.8 Conclusion

It is imperative that we respond to the threat posed to our wetlands by increasingly intense heatwaves. These ecosystems are more than just geographic areas; they are functioning life support systems that clean our water, control our temperature, and sustain a vast number of species, including ourselves. Wetlands are becoming less resilient as temperatures rise and weather patterns change, endangering the delicate balance that keeps them intact. However, this difficulty also offers a chance to reconsider how we relate to the natural world. Our comprehension of the effects of heatwaves on wetlands will increase, enabling us to create plans for their preservation and restoration. This is about protecting future generations' well-being as much as it is about preserving the environment.

References

Aransiola SA, Babaniyi BR, Aransiola AB, Maddela NR (eds) (2024) Prospects for soil regeneration and its impact on environmental protection, 1st edn. Springer, Cham. https://link.springer.com/book/9783031532696

Bai J, Chen X, Li J, Luo G (2013) Changes in wetland area and landscape pattern in the Yellow River Delta from 1976 to 2010. J Coast Res 29(4):831–841

Barbier EB, Hacker SD, Kennedy C, Koch EW, Stier AC, Silliman BR (2011) The value of estuarine and coastal ecosystem services. Ecol Monogr 81(2):169–193

Berrier QE, Frazier TG, Guntenspergen GR (2022) Microbial diversity and ecosystem functioning in wetlands. Environ Microbiol 24(1):1–14

Bull JW, Lloyd SP, Strange N (2017) Implementation gap between the theory and practice of biodiversity offset multipliers. Conserv Lett 10(6):656–669. https://doi.org/10.1111/conl.12335

Cahill AE, Aiello-Lammens ME, Fisher-Reid MC, Hua X, Karanewsky CJ, Yeong Ryu H et al (2013) How does climate change cause extinction? Proc R Soc B Biol Sci 280(1765):20121890

Cao S, Cao G, Feng Q, Han G, Lin Y, Yuan J, Wu F, Cheng S (2017) Alpine wetland ecosystem carbon sink and its controls at the Qinghai Lake. Environ Earth Sci 76:210

Davidson NC, and Davidson MA (2014) Wetland loss and the role of wetland restoration in the United Kingdom. Hydrobiologia, 725(1), Publisher: Springer, Dordrecht, Netherlands, 1–14

Elmqvist T, Setälä H, Handel SN, van der Ploeg S, Aronson J, Blignaut JN et al (2015) Benefits of restoring ecosystem services in urban areas. Curr Opin Environ Sustain 14:101–108

Gong Z, Kaina R, Gao C (2010) Wetland degradation and its ecological consequences in the Yellow River Delta, China. J Environ Manag 91(11):2364–2373

Hatten JA, Guntenspergen GR, Nurminen L (2024) Wetland biodiversity and ecosystem services: a synthesis. Wetlands 44(1):1–15

He M, Xin C, Baskin CC, Li J, Zhao Y, An H, Sheng X, Zhao L, Zhao Y, Ma M (2021) Different response of transient and persistent seed bank of alpine wetland to grazing disturbance on the Tibetan plateau. Plant Soil 459:93–107

Heino J, Melo AS, Siqueira T (2015) Climate change and freshwater biodiversity: a review. Biol Rev 90(3):761–777

Intergovernmental Panel on Climate Change (2020) Climate change 2020: impacts, vulnerability, and adaptation. Part A: global and sectoral aspects contribution of working group II to the fifth assessment report of the Intergovernmental Panel on Climate Change

Jiang R, Liu J, Liu W, Zhang D, Hu W (2023) Changes and driving forces analysis of alpine wetlands in the first meander of the Yellow River based on long-term time series remote sensing data. Front Ecol Evol 11:1193059

Keddy PA (2010) Wetland ecology: principles and conservation, 2nd edn. Cambridge University Press, Cambridge

Keddy PA, Fraser LH, Solomeshch AI, Junk WJ, Campbell DR, Arroyo MTK, Cox CJ (2017) Wetland ecology: principles and conservation, 2nd edn. Cambridge University Press, Cambridge

Kueppers LM, Snyder MA, Sloan LC (2008) Impacts of climate change on wetlands in the California Central Valley. Clim Chang 87(1):137–154

Linh NH, Hien NT, Giang LT (2023) Biodiversity and ecosystem services in wetlands. J Wetland Ecol 13(1):1–12

Maddela NR, Aransiola SA, Prasad R (eds) (2022) Ecological interplays in microbial enzymology, 1st edn. Springer Nature Singapore Pte Ltd. ISSN: 2662-1681. https://www.springer.com/series/16324

McGinley M (2008) Wetlands restoration and creation: A review of the state of the science. Wetlands, 28(3), Publisher: Springer, Dordrecht, Netherlands, 1–14

Millennium Ecosystem Assessment (2005) Ecosystems and human well-being: synthesis. Publisher: Island Press, Washington, D.C., USA

Mitsch WJ, Bernal B, Nahlik AM, Mander Ü, Zhang L, Anderson CJ et al (2013) Wetlands, carbo and climate change. Landsc Ecol 28(1):1–21

Mitsch WJ, and Gosselink JG (2015) wetlands (5th ed). wiley. Publisher location: Wiley, Hoboken, New Jersey, USA

Na C, Zhang S, Wang Z (2008) Impacts of human activities on wetland ecosystems in the Yangtze River Delta, China. J Environ Sci 20(11):1423–1430

National Oceanic and Atmospheric Administration (2022) Louisiana's Burrataria Basin wetland restoration project

Neubauer SC and Verhoeven JTA (2019) Wetlands biogeochemistry and greenhouse gas emissions. In Wetlands and Natural Resource management. publisher location: Springer, Cham, Switzerland. (pp 1–22)

Ontario Ministry of Natural Resources and Forestry (2017) Wetlands conservation in Ontario

Peel MC, Finlayson CM, McMahon TA (2019) Wetland ecology and management: a review. J Environ Manag 235:345–355

Poff NL, Brown CM, Grantham TE, Matthews JH, Palmer MA, Spence CM et al (2016) Sustainable water management under future uncertainty with eco-engineering decision scaling. Nat Clim Chang 6(1):25–34

Qiu D, Zhang H, Ren Y, Zhu Y (2024) The lost biodiversity and degraded alpine wetlands caused by strong earthquake on the Qinghai–Tibet Plateau did not self-restore in the short term. Glob Ecol Conserv 50(2024):e02830

Sanderson MG, Burton CM, Shine KP (2016) Spatial and temporal distribution of methane emissions from wetlands. Atmos Environ 143:247–256

Serrano L, de la Hera A (2019) Wetlands and climate change: a review. J Hydrol 569:345–355

Thomaz SM (2023) Wetland biodiversity and ecosystem services. Wetl Ecol Manag 31(1):1–13

United Nations Environment Programme (2021) The importance of wetlands

Wang Y, Liu J, Liu X (2019) Anaerobic ammonia-oxidizing bacteria and nitrate archea in wetland denitrification. Environ Sci Technol 53(19):10714–10723

Yin T, Feng Y, Liang S, Wang G, Yuan N, Yan D, Wu P, Kuang X, Wan L (2022) A suitable method for alpine wetland delineation: An example for the headwater area of the Yellow River, Tibetan Plateau. Front Environ Sci 10:1062954

Chapter 9
Climate Change Impacts on Wetlands: Vulnerabilities, Adaptation, and Mitigation Strategies

Gabriel Gbenga Babaniyi, Ulelu Jessica Akor, Olaniran Victor Olagoke, and Oluwatosin Emmanuel Daramola

Abstract In the complex relationship between wetlands and climate change, understanding vulnerabilities and identifying adaptation strategies are essential for ensuring the resilience of these important ecosystems. Wetlands, encompassing diverse habitats like marshes, swamps, and mangroves, face significant threats from climate change-induced phenomena such as sea-level rise, altered precipitation patterns, and temperature extremes. Also, wetland ecosystem sensitivity varies, influenced by factors like hydrological changes and biodiversity loss. Changes in precipitation patterns can disrupt water cycles, while temperature extremes pose risks to wetland vegetation and wildlife. Sea-level rise exacerbates coastal erosion and inundation, threatening wetland habitats. In addition, adaptation strategies include wetland restoration and rehabilitation, enhancing resilience, sustainable land use planning, and managed retreat. Restoring degraded wetlands and implementing nature-based solutions can bolster their ability to withstand climate impacts. Engaging communities through education and awareness campaigns is vital for fostering stewardship and driving conservation efforts. Mitigation approaches like carbon sequestration and reducing anthropogenic stressors aim to mitigate climate impacts on wetlands. Policy interventions and governance play a pivotal role in coordinating global efforts to protect wetlands and mitigate climate change. However, despite progress, challenges remain, including technological limitations and the need for political will and international cooperation. Overcoming these challenges requires innovative solutions and collective action. In conclusion, securing a resilient future for wetlands necessitates a holistic approach encompass-

G. G. Babaniyi (✉) · O. V. Olagoke
Department of Agricultural Development and Management, Agricultural and Rural Management Training Institute (ARMTI), Ilorin, Nigeria

U. J. Akor
Department of Animal Production, University of Ilorin, Ilorin, Nigeria

O. E. Daramola
Department of Chemistry and Biochemistry, Texas Tech University, Lubbock, TX, USA

B. R. Babaniyi et al. (eds.), *Wetland Ecosystems: Conservation Strategies, Policy Management and Applications*, Wetlands: Ecology, Conservation and Management 12, https://doi.org/10.1007/978-3-031-91982-4_9

ing adaptation and mitigation measures, community engagement, and policy support. By prioritizing wetland conservation and integrating climate resilience into decision-making processes, we can safeguard these invaluable ecosystems for future generations.

Keyword Climate change · Wetlands · Hydrological changes · Sea level rise

9.1 Introduction to Wetlands and Climate Change

Wetlands, crucial ecosystems teeming with water, wield significant influence over global biogeochemical cycles, biodiversity upkeep, and climate stabilization. Their susceptibility to the burgeoning impacts of climate change is inimitable. The consequences of human-induced climate alterations present formidable trials for wetland ecosystems across the globe. Wetlands stand acutely vulnerable to the repercussions of climate change (DCCEEW 2023; Wardhana and Prawira 2024). As temperatures climb, rainfall patterns shift, evapotranspiration rates soar, and water dynamics undergo transformations. Many permanent wetlands face the prospect of transitioning into seasonal habitats, while seasonal one's risk becoming ephemeral. Alpine wetlands face species composition shifts as rising temperatures thaw snow cover (Giri et al. 2020; Salimi et al. 2021). Intertidal wetlands confront the prospect of migrating inland with rising sea levels, where feasible. Additionally, frequent, high-intensity fires pose threats to wetlands, including valuable peat systems.

The impacts of climate change are poised to intersect with prevailing land use patterns. Wetlands already grappling with the effects of water management practices, agricultural encroachments, and urbanization are less resilient and face heightened vulnerability. Climate change will also intertwine with natural fluctuations and extreme climate events like droughts, heatwaves, high tides, and storm surges (Troxler et al. 2018; Salimi et al. 2021; DCCEEW 2023; Gimba et al. 2024). Wetlands, encompassing diverse habitats like marshes, swamps, and bogs, serve as vital reservoirs of biodiversity, establishing complex webs of life (Giri et al. 2020). Nonetheless, the onslaught of climate change, marked by rising temperatures, erratic precipitation patterns, and intensified storms, imperils the delicate equilibrium within wetland ecosystems (Moomaw et al. 2018; Meehl 2023; Solecki et al. 2024; Wardhana and Prawira 2024).

Surges in temperature, propelled by greenhouse gas emissions, elevate water temperatures, intensifying stress on aquatic life and disrupting critical ecological processes (Naskar et al. 2022). Furthermore, altered precipitation patterns, characterized by prolonged droughts and intensified rainfall events, trigger fluctuations in water availability, compromising wetland hydrology and imperiling the survival of specialized flora and fauna (Karimi et al. 2022; Roy et al. 2023). Moreover, the specter of rising sea levels looms ominously over coastal wetlands, threatening to inundate vast expanses of vital habitat and precipitating irreversible biodiversity loss. The convergence of rising seas and human encroachments exacerbates the vulnerability of these ecosystems, exacerbating habitat fragmentation and obstructing the migratory routes of numerous species (Karimi et al. 2022).

Comprehending the underlying dynamics is imperative for devising effective mitigation and adaptation strategies. Integrating interdisciplinary research efforts encompassing hydrological modeling, ecological assessments, and climatological projections is crucial for unraveling the complex feedback mechanisms shaping wetland resilience in the face of escalating climatic disruptions (Wang and Gu 2021; Duffield et al. 2021). Moreover, fostering collaborative endeavors among diverse stakeholders, including policymakers, scientists, and local communities, is paramount for catalyzing proactive measures aimed at safeguarding wetland ecosystems (Wang and Gu 2021). Embracing innovative conservation approaches such as ecosystem-based adaptation and nature-based solutions holds promise for enhancing wetland resilience and nurturing sustainable coexistence amidst the tumultuous challenges posed by climate change (Duffield et al. 2021; Roy et al. 2023). In conclusion, the burgeoning impacts of climate change present formidable challenges to wetland ecosystems, imperiling their invaluable ecological services and global biodiversity. Nonetheless, concerted efforts, grounded in scientific rigor and collective stewardship, offer a glimmer of hope in navigating the precarious intersection of wetlands and climate change, thereby heralding a resilient future for these indispensable ecosystems.

9.2 Understanding Vulnerabilities

The comprehension of vulnerabilities entails an examination of the susceptibilities inherent within wetland ecosystems. These vulnerabilities, woven into the fabric of wetland dynamics, stem from complex factors influenced by anthropogenic and climatic perturbations. Comprehending the weaknesses of wetlands necessitates a diverse methodology, drawing upon insights from hydrology, ecology, climatology, and geomorphology (Yang et al. 2021a; Bhowmik 2022). Central to the understanding of vulnerabilities is the assessment of wetland ecosystem sensitivity to external stressors. Wetlands, characterized by dynamic hydrological regimes and biotic interactions, exhibit varying degrees of sensitivity to climatic fluctuations and human-induced disturbances (Bhowmik 2022). Unraveling the relationship between these stressors and wetland resilience mandates examination of hydrological dynamics, sediment transport mechanisms, and biogeochemical cycling processes.

Hydrological changes, precipitated by shifting rainfall patterns, alterations in river flow regimes, and sea level rise, emerge as primary drivers of vulnerability within wetland ecosystems. Fluctuations in water availability, compounded by anthropogenic water extraction and land-use alterations, disrupt the delicate balance of wetland hydrology, exacerbating stress on resident flora and fauna (Hoq et al. 2021). Moreover, the specter of biodiversity loss looms large over vulnerable wetland ecosystems. The interdependencies among species, woven into the fabric of wetland food webs, render them particularly susceptible to disruptions induced by climatic perturbations and habitat alterations. Elevated temperatures, prolonged droughts, and habitat fragmentation precipitate cascading impacts on wetland biodiversity, imperiling the persistence of specialized species and compromising ecosystem functioning (Nielsen et al. 2020).

Furthermore, wetland vulnerabilities are exacerbated by anthropogenic encroachments and land-use alterations. Urbanization, industrial development, and agricultural expansion impose pressures on wetland ecosystems, fragmenting habitats, altering hydrological regimes, and degrading water quality (Sarkar et al. 2020). These anthropogenic stressors compound the vulnerabilities of wetlands, rendering them less resilient to the impacts of climate change. Understanding wetland vulnerabilities demands efforts to decipher the interplay of climatic, hydrological, ecological, and anthropogenic factors shaping their resilience (Twinomuhangi et al. 2021). Embracing an understanding of vulnerabilities is imperative for informing conservation and management strategies aimed at safeguarding the services rendered by wetland ecosystems. In conclusion, unraveling wetland vulnerabilities reveals interactions, underscoring the need for measures to mitigate the impacts of climate change and anthropogenic disturbances on these ecosystems (Table 9.1).

9.2.1 Wetland Ecosystem Sensitivity

Wetland ecosystem sensitivity encapsulates the responsiveness of wetlands to external stressors, reflecting their dynamic nature and complex interdependencies. It takes investigation to fully grasp this sensitivity, considering the factors that influence wetland dynamics (Su et al. 2021; Malekmohammadi et al. 2023). Climatic fluctuations, anthropogenic disturbances, habitat fragmentation, and land-use changes shape the susceptibility of wetlands to perturbations. The assessment of wetland sensitivity entails deciphering shifts in vegetation composition, alterations in species distribution, changes in hydrological patterns, and variations in ecosystem productivity (Malekmohammadi et al. 2023). These indicators provide insights into the responses of wetlands to environmental changes.

Impacts of heightened sensitivity encompass increased vulnerability to stressors, loss of ecosystem services, changes in wetland functions, and reduced resilience to disturbances. To address these challenges, enhanced monitoring and assessment efforts, adaptive management approaches, restoration and rehabilitation initiatives, and promotion of wetland conservation are paramount (Su et al. 2021). However, further research is needed to unravel the mechanisms underlying wetland sensitivity, develop robust monitoring frameworks, and integrate socio-ecological perspectives into management strategies (Koffi et al. 2020). Embracing the complexities of wetland ecosystem sensitivity is crucial for fostering resilience and sustainability in the face of evolving environmental pressures.

Table 9.1 Differences between wetland ecosystem sensitivity, hydrological changes, and biodiversity loss

Aspect	Wetland ecosystem sensitivity	Hydrological changes	Biodiversity loss
Definition	The degree to which wetlands respond to external stressors, influenced by their dynamic nature and complex interactions.	Alterations in water flow, availability, and distribution within wetland ecosystems, driven by climatic fluctuations and human activities.	The decline in the variety and abundance of species within wetland habitats due to disruptions in ecological processes and habitat degradation.
Influencing factors	Climatic fluctuations, anthropogenic disturbances, habitat fragmentation, and land-use changes.	Shifting precipitation patterns, sea level rise, alterations in river flow regimes, and anthropogenic water extraction.	Climate change, habitat degradation, habitat fragmentation, invasive species, and human disturbances.
Indicators	Changes in wetland vegetation composition, shifts in species distribution, alterations in hydrological patterns, and variations in ecosystem productivity.	Changes in water levels, frequency and intensity of flooding/drought events, alterations in sediment deposition, and shifts in wetland boundaries.	Decline in species richness, loss of keystone species, disruptions in trophic interactions, and changes in ecosystem stability.
Impacts	Increased vulnerability to stressors, loss of ecosystem services, changes in wetland functions, and reduced resilience to disturbances.	Disruption of wetland hydrology, alterations in sediment transport, loss of habitat connectivity, and increased risk of wetland degradation.	Reduction in ecosystem resilience, loss of genetic diversity, decline in ecosystem services, and diminished capacity for ecosystem functioning.
Management strategies	Enhanced monitoring and assessment, adaptive management approaches, restoration and rehabilitation efforts, and promoting wetland conservation initiatives.	Implementation of water management strategies, restoration of natural hydrological processes, sustainable land-use planning, and integrated watershed management practices.	Habitat restoration and creation, invasive species management, establishment of protected areas, and promotion of sustainable resource use practices.
Research needs	Further investigation into the mechanisms underlying wetland sensitivity, development of robust monitoring frameworks, and integration of socio-ecological perspectives.	Improved understanding of hydrological processes in wetlands, development of predictive models, and assessment of long-term hydrological trends.	Enhanced research on biodiversity dynamics, identification of key drivers of biodiversity loss, and assessment of the effectiveness of conservation interventions.

9.2.2 Hydrological Changes

Hydrological changes wield significant influence over the dynamics of wetland ecosystems, eliciting diverse responses and challenging dynamics. These changes, stemming from climatic fluctuations and human interventions, disrupt the balance of water flow, availability, and distribution within wetland habitats (Shen et al. 2020; Yang et al. 2021b). Shifts in precipitation patterns, alterations in river flow regimes, and sea level rise act as primary drivers of hydrological changes, triggering cascading impacts on wetland ecosystems (Shen et al. 2020). However, the ramifications of hydrological changes include disruptions to sediment transport mechanisms, changes in water levels, habitat fragmentation, and heightened vulnerability to wetland degradation (Yang et al. 2021b). These alterations stress resident flora and fauna, compromise ecosystem functions, and reduce the provision of crucial ecosystem services.

More so, hydrological changes stem from various drivers that induce stress and alterations in hydrological systems and the water cycle at large. Primary stressors on hydrological systems comprise extensive modifications in land cover, urban expansion, industrial growth, and substantial engineering interventions (Ng et al. 2020). Dealing with hydrological changes in wetland ecosystems requires efforts to decipher underlying mechanisms, anticipate future trends, and implement proactive management interventions (Duethmann et al. 2020; Cheng et al. 2020). Embracing this challenge is essential for safeguarding the resilience and sustainability of wetland ecosystems amidst the turbulent dynamics of hydrological perturbations.

9.2.2.1 Hydrological Cycle

A foundational attribute of the hydrological cycle is its perpetual nature, devoid of a distinct beginning or end. It can be scrutinized by commencing at any of the subsequent stages: evaporation, condensation, precipitation, interception, infiltration, percolation, transpiration, runoff, and storage (Ng et al. 2020). Meanwhile, the water cycle denotes a natural phenomenon characterized by the continuous recycling of water within the atmosphere, also referred to as the hydrologic cycle. Throughout this cyclical process between Earth and the atmosphere, water undergoes transitions across three states of matter: solid, liquid, and gas (BYJUS 2022) (Fig. 9.1).

9.2.3 Biodiversity Loss

Biodiversity loss entails the diminishing or vanishing of biological diversity, encompassing the variety of organisms inhabiting Earth, their diverse biological levels, and genetic variability, alongside the natural patterns found in ecosystems. It occurs

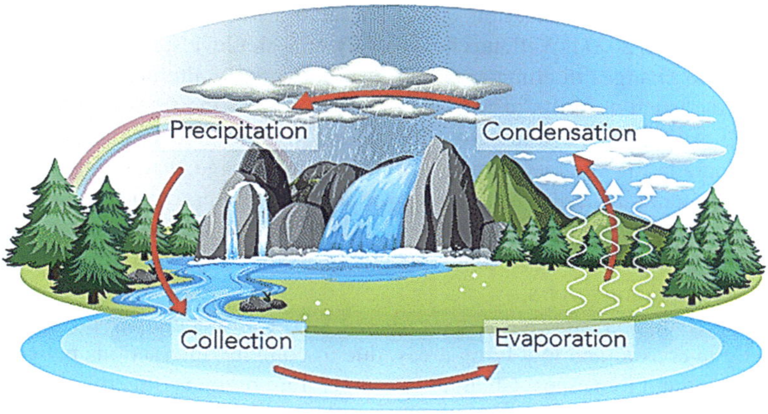

Fig. 9.1 Water cycle

with the complete disappearance of plant or animal species or a notable reduction or extinction of species in specific regions. This phenomenon denotes a reduction in biological diversity within a defined locale, with potential for temporary or permanent repercussions (Peixoto et al. 2022). There are five major primary drivers of habitat loss according to Singh et al. (2021), invasive species, overexploitation, pollution, and climate change are responsible for this loss, all attributable to human activities.

Furthermore, biodiversity loss signifies the decrease in the diversity and abundance of living organisms within ecosystems, impacting their crucial services like food provision, climate regulation, and disease mitigation. The complex and interconnected causes encompass habitat destruction, pollution, resource overexploitation, introduction of invasive species, and climate change, which disrupt ecosystems, leading to species decline or extinction (Singh et al. 2021; Peixoto et al. 2022). This loss can diminish ecosystem resilience, heightening vulnerability to disturbances like disease outbreaks and extreme weather events, while disrupting vital ecological processes such as pollination and nutrient cycling (Shin et al. 2022). Combatting biodiversity loss necessitates concerted action at various scales, including conservation strategies like protected areas, habitat restoration, and sustainable resource management, alongside efforts to raise awareness and promote biodiversity-friendly practices, essential for safeguarding Earth's diverse life forms.

9.3 Impacts of Climate Change on Wetlands

Wetlands encompass a diverse range of aquatic habitats, serving as critical biodiversity hotspots and significant components of the global carbon budget. They offer various ecosystem goods, like food and drinking water, and provide essential services such as water purification, climate regulation, and flood protection (Lloréns 2008; Salimi et al. 2021). Wetlands heavily rely on water levels, making them highly susceptible to changes in climatic conditions, particularly alterations in the hydrological cycle, such as shifts in wet and dry seasons and the frequency of extreme events (Xi et al. 2021). While both climatic and non-climatic factors contribute to wetland degradation and loss, human activities like drainage, water withdrawal, and habitat fragmentation have been more pronounced drivers than climate change (Mahdian et al. 2023). The combined effects of climatic and non-climatic drivers can lead to abrupt and irreversible changes in wetlands, posing significant challenges for restoration efforts.

Climate change is expected to exacerbate the loss and degradation of wetlands, especially in coastal and low-lying areas, due to complex interactions between climate and wetland dynamics. Changes in temperature regimes directly affect physiological traits and behaviors of wetland organisms, influencing their growth rates, phenology, and habitat preferences (Xi et al. 2021; Mahdian et al. 2023). Additionally, shifts in temperature can alter species compositions, impacting the overall metabolism and productivity of wetlands (Lloréns 2008). However, global warming is projected to shift the geographic ranges of species towards higher latitudes or elevations, rendering certain areas unsuitable for certain species. Addressing the impacts of climate change on wetlands requires integrated approaches, including habitat restoration, sustainable water management, and conservation efforts aimed at enhancing wetland resilience and mitigating future climate risks (Taillardat et al. 2020; Wang et al. 2020). Thus, collaboration among scientists, policymakers, and local communities is crucial for developing effective adaptation strategies to safeguard these valuable ecosystems. By implementing proactive measures and fostering interdisciplinary partnerships, we can enhance the resilience of wetlands and ensure their continued provision of vital services for both humans and biodiversity.

9.3.1 Sea Level Rise

Sea level rise denotes the elevation of the Earth's ocean levels attributed to global warming's repercussions. The combustion of fossil fuels, releasing carbon dioxide and other heat-retaining gases into the atmosphere, is a prominent contributor to global warming (Hauer et al. 2020; Durand et al. 2022). Consequently, the oceans absorb a substantial portion of this heat, resulting in a gradual increase in their average level over time. This escalation is primarily driven by thermal expansion and the thawing of polar ice caps and glaciers (Frederikse et al. 2020; Isiaka et al. 2021). With rising temperatures due to climate change, oceanic waters undergo expansion, contributing to this phenomenon. Simultaneously, the melting of ice formations infuses additional water into the oceans, further propelling sea level rise (Oloyede et al. 2022).

The ramifications of sea level rise are profound, posing significant threats to coastal communities, ecosystems, and infrastructure. It precipitates coastal erosion, heightened flooding, land loss, and the intrusion of saltwater into freshwater reservoirs (Hauer et al. 2020). Especially vulnerable are coastal habitats like wetlands and mangroves, prone to submersion or loss of protective functions (Durand et al. 2022). Among the myriad causes, glaciers and ice sheets worldwide are diminishing, contributing water to the ocean, while the ocean's volume expands with rising temperatures. A lesser contributor is the depletion of liquid water on land, encompassing aquifers, lakes, rivers, and soil moisture (Dada et al. 2020; Frederikse et al. 2020).

In the coming decades, projections suggest a substantial increase in sea levels, with U.S. coastlines expecting an average rise of 10–12 inches (0.25–0.30 m) by 2050, comparable to the ascent observed over the past century (1920–2020). This escalation poses significant threats to southern cities such as Lagos and coastal regions, heightening susceptibility to flooding and waterborne diseases (Adegun 2023). Mitigating sea level rise necessitates efforts to curtail greenhouse gas emissions and implement adaptive measures to safeguard coastal areas and communities. These measures encompass the construction of seawalls, restoration of natural protective features like dunes and marshes, and the enactment of land use policies to mitigate risks posed by escalating sea levels.

9.3.2 Changes in Precipitation Patterns

Changes in precipitation patterns refer to alterations in the frequency, intensity, duration, and spatial distribution of rainfall and snowfall over time. These changes are primarily driven by factors such as global warming, atmospheric circulation patterns, and land use changes. As the Earth's climate warms due to the accumulation of greenhouse gases in the atmosphere, it affects the dynamics of the water cycle, leading to shifts in precipitation patterns. These alterations can manifest in various ways, including changes in the timing and magnitude of rainfall events, more frequent or intense storms, prolonged droughts, and shifts in the geographical distribution of precipitation (Rahman and Islam 2019; Pepin et al. 2022). For example, some regions may experience increased rainfall and flooding, while others may face decreased precipitation and drought conditions. Precipitation according to UCMP (2022), exerts considerable influence over various Earth system phenomena, including:

1. Regional temperature and humidity modulation.
2. Terrestrial ecosystem vitality and biomass dynamics, impacting species distributions and population sizes as freshwater availability dictates growth and survival.
3. Organismal life cycle regulation, where precipitation patterns serve as environmental cues guiding developmental transitions.
4. Pollutant dispersion and atmospheric particulate concentration modulation, contributing to air quality maintenance through precipitation-mediated removal or dissolution of airborne particles.

5. Run-off water volume and sediment erosion mitigation, crucial for preserving aquatic ecosystems' integrity and water quality in oceans, rivers, and lakes.
6. Fire incidence frequency and magnitude regulation, particularly in regions prone to dry seasons. Climate change-induced elongation of dry periods exacerbates fire susceptibility in affected areas.

Various human activities and environmental phenomena wield influence over precipitation patterns. Fossil fuel combustion, agricultural practices, and deforestation escalate greenhouse gas levels, consequently elevating Earth's average temperature. This temperature surge alters atmospheric circulation, impacting water distribution across the planet. Such shifts amplify extreme weather events like hurricanes and droughts (Jiang et al. 2020). Trees play an important role in precipitation dynamics through transpiration, where they release moisture into the atmosphere. Deforestation curtails this water cycle, whereas reforestation bolsters it. Pepin et al. (2022) opined that mountains intercept warm, moisture-laden air, prompting condensation and rainfall on their windward slopes. Conversely, the leeward side, or rain shadow, experiences diminished precipitation due to descending, drier air. Continental drift over millennia reshapes solar radiation absorption and reflection, dictating wind patterns and precipitation. This redistribution affects atmospheric and oceanic circulation, dictating precipitation intensity, timing, and location (Mendez et al. 2020; Wise and Dannenberg 2022). Hence, collaboration among scientists, policymakers, and communities is key for mitigating these impacts and fostering sustainable environmental stewardship.

Furthermore, changes in precipitation patterns have significant implications for ecosystems, agriculture, water resources, and human communities (Wise and Dannenberg 2022). They can influence the availability of water for drinking, irrigation, and industrial purposes, as well as impact the health of ecosystems and biodiversity. Additionally, extreme precipitation events can lead to flooding, landslides, and other natural disasters, causing damage to infrastructure and posing risks to human safety (Mendez et al. 2020). Addressing changes in precipitation patterns requires comprehensive strategies, including improved water management practices, conservation efforts, and adaptation measures. This may involve implementing measures such as water conservation and efficiency measures, developing drought-resistant crops, and improving flood preparedness and infrastructure resilience. Also, collaboration among scientists, policymakers, and local communities is essential for developing effective strategies to mitigate the impacts of changing precipitation patterns and build resilience to future climate challenges.

9.3.3 Temperature Extremes

In periods of extreme temperature, the body undergoes extra exertion to maintain its normal temperature, a strain that can lead to fatalities. Hot weather can prompt dehydration and muscle weariness, especially when coupled with high humidity.

Conversely, cold temperatures stiffen muscles, heightening the risk of strains and pulls. Symptoms of extreme temperatures include difficulty breathing, fatigue, reduced mobility, heightened sensitivity to sensations, and weakened grip strength (Harvey et al. 2020; Cheung et al. 2021). According to National Center for Environmental Health (NCEH) (2019), the highest officially recorded air temperature on Earth stands at 56.7 °C (134.1 °F), noted on July 10, 1913, at Furnace Creek Ranch in Death Valley, USA. Heatwaves have long posed health hazards globally, with many countries and cities witnessing surges in mortality rates during such periods. Deaths often arise from heat stroke and related conditions, as well as cardiovascular, respiratory, and cerebrovascular ailments. Heatwaves also correlate with spikes in hospitalizations for cardiovascular, renal, and respiratory issues. Extreme heat incidents are on the rise in some nations, including the United States, with climate forecasts predicting even more frequent and intense occurrences in the future (Harvey et al. 2020). While advancements like improved forecasting and heat-health warning systems, along with increased access to air conditioning, have helped reduce heat-related risks, such events still pose a significant threat to life. Urban heat islands, coupled with an aging populace and expanding urbanization, are expected to heighten the vulnerability of urban residents to heat-related health challenges in the coming years (Cheung et al. 2021).

However, temperature extremes play a crucial role in shaping the impacts of climate change on wetlands, exerting significant influence on their ecological dynamics. These extremes encompass both heatwaves and cold spells, characterized by unusually high or low temperatures compared to historical averages (NCEH 2019; Mendez et al. 2020). With climate change, the frequency, intensity, and duration of temperature extremes are expected to escalate, posing multifaceted challenges for wetland ecosystems. High temperatures can trigger heat stress in wetland vegetation and wildlife, disrupting critical ecological processes such as photosynthesis and nutrient cycling (Wang et al. 2021). Heatwaves can also exacerbate water scarcity by accelerating evaporation rates, leading to decreased water levels in wetlands and exacerbating habitat loss (Dunn et al. 2020). Conversely, cold spells can result in frost damage to wetland vegetation and disrupt the reproductive cycles of aquatic species, impacting population dynamics.

Furthermore, temperature extremes can exacerbate other climate change-related stressors on wetlands, such as altered precipitation patterns and sea level rise (Ma et al. 2021). For example, intense heatwaves can amplify the impacts of droughts, further compromising wetland hydrology and biodiversity. Similarly, cold spells can exacerbate the effects of sea level rise by accelerating ice melt in polar regions, contributing to increased flooding and habitat loss in coastal wetlands. Addressing the impacts of temperature extremes on wetlands requires comprehensive strategies that integrate mitigation and adaptation measures. Mitigation efforts aimed at reducing greenhouse gas emissions are essential for curbing the frequency and intensity of temperature extremes in the long term. Meanwhile, adaptation strategies focus on enhancing the resilience of wetland ecosystems to withstand and recover from temperature-related stressors. This may involve measures such as restoring wetland habitats, implementing sustainable water management practices, and promoting ecosystem-based approaches to conservation. Thus, temperature extremes are integral components of climate change that profoundly influence the resilience and vulnerability of wetland ecosystems.

9.3.4 Storm Intensity and Frequency

Storm intensity and frequency play essential roles in shaping the impacts of climate change on wetlands, exerting significant influence on their ecological dynamics. Storms, characterized by intense precipitation, strong winds, and sometimes accompanied by thunder and lightning, are integral components of the Earth's climate system (Cortese et al. 2024). Storm intensity is measured by the maximum surface wind speed or the minimum sea level pressure. Climate change affects many of the factors that can result in greater storm intensity. With climate change, there is growing evidence suggesting an increase in the intensity and frequency of storms. Warmer ocean temperatures fuel the formation of more powerful storms, such as hurricanes, cyclones, and typhoons (Moron et al. 2021; Cortese et al. 2024). The intensity of a hurricane is closely related to its central pressure with the lower the pressure the more intense the storm and its maximum sustained winds. These storms can bring torrential rainfall, leading to flooding in wetland areas. Additionally, strong winds associated with storms can uproot trees, destroy vegetation, and alter wetland habitats (Harvey et al. 2020; Cheung et al. 2021; Tamm et al. 2023). The increased frequency of storms can also disrupt wetland ecosystems. Repeated storm events can erode wetland soils, leading to loss of habitat and sedimentation of water bodies. This sedimentation can negatively impact aquatic species and alter nutrient cycling processes within the wetland (Priestley and Catto 2022).

Moreover, storms can introduce pollutants and debris into wetland ecosystems, further compromising their ecological integrity. Runoff from urban areas and agricultural lands, exacerbated by storm events, can carry pollutants such as pesticides, fertilizers, and heavy metals into wetlands, contaminating water bodies and affecting the health of resident flora and fauna (Moron et al. 2021). Comprehensive solutions that incorporate both adaptation and mitigation measures are needed to address the effects of storm strength and frequency on wetlands. Mitigation efforts aimed at reducing greenhouse gas emissions are essential for curbing the intensity and frequency of storms in the long term (Priestley and Catto 2022; Cortese et al. 2024). In addition, adaptation strategies focus on enhancing the resilience of wetland ecosystems to withstand and recover from storm-related impacts. This may involve measures such as wetland restoration, floodplain management, and stormwater management practices. Storm intensity and frequency are key drivers of change in wetland ecosystems under climate change.

9.4 Adaptation Strategies

Adaptation strategies are essential plans and actions designed to help individuals, communities, and ecosystems cope with the impacts of climate change. These strategies aim to reduce vulnerability and enhance resilience to changing environmental conditions. Adaptation strategies in the context of climate change refer to methods aimed at responding to and mitigating its adverse effects rather than halting or reversing the phenomenon itself (Wong et al. 2021; Ogundeji and Okolie 2022). These practices encompass a wide array of proactive measures, including diversifying crops and livelihoods, utilizing seasonal climate forecasts, implementing community-based disaster risk reduction initiatives, establishing early warning systems for famines, adopting insurance schemes, and enhancing water storage and supplementary irrigation methods (Borg et al. 2021). It's important to recognize that there is no universal solution to adaptation, as it varies greatly depending on the specific circumstances and challenges faced. Adaptation efforts can range from constructing flood defenses and implementing cyclone early warning systems to transitioning to drought-resistant crops and reconfiguring communication networks, business strategies, and governmental policies (Ogundeji and Okolie 2022).

The typology of adaptation encompasses five primary strategies according to Borg et al. (2021) listed them as follow: foraging, horticulture, agriculture, pastoralism, and industrialism. Historically, foraging was the predominant mode of sustenance until approximately 10,000 years ago when environmental variations led to divergent practices among foraging societies (Khan et al. 2020; Ojo et al. 2022). One adaptation strategy is ecosystem-based adaptation, which involves restoring and conserving natural habitats, such as wetlands, forests, and coral reefs, to enhance their ability to withstand climate change impacts. For example, restoring degraded wetlands can help buffer coastal areas from storm surges and flooding by absorbing excess water and reducing erosion (Ojo et al. 2022). Another strategy is infrastructure adaptation, which involves designing and implementing measures to protect built environments from climate-related hazards. This can include building seawalls, levees, and flood barriers to mitigate the impacts of sea level rise and storm surges on coastal communities (Khan et al. 2020). Furthermore, community-based adaptation involves empowering local communities to develop and implement strategies that address their specific climate change challenges. This can include initiatives such as early warning systems, community-based disaster risk reduction programs, and sustainable land management practices.

Additionally, policy adaptation involves developing and implementing laws, regulations, and policies that promote climate resilience at local, national, and international levels. This can include measures such as zoning regulations, building codes, and incentives for renewable energy development (Kabir et al. 2022). Likewise, technological adaptation involves developing and deploying innovative technologies to help mitigate and adapt to climate change impacts. This can include technologies such as drought-resistant crops, water-efficient irrigation systems, and renewable energy technologies (Fuchs et al. 2022). Therefore, adaptation strategies are crucial for building resilience and ensuring the long-term sustainability of communities and ecosystems in the face of climate change. By implementing a combination of ecosystem-based, infrastructure, community-based, policy, and technological adaptation measures, societies can better prepare for and respond to the challenges posed by a changing climate.

9.4.1 Wetland Restoration and Rehabilitation

Saltmarshes and mangroves are frequently restored as wetland ecosystems for coastal protection purposes, while seagrasses are less commonly used for this purpose alone (USACE 1989; Linham and Nicholls 2017). These habitats play vital roles in managing coastal flood and erosion by dissipating wave and tidal energy (Brampton 1992) and trapping sediment materials to build land seawards. The dense root systems of wetland plants also aid in stabilizing shore sediments, mitigating erosion (USACE 1989). However, many wetlands worldwide have undergone degradation due to natural and human-induced factors, necessitating restoration efforts (Linham and Nicholls 2017). Techniques have been developed to reintroduce wetlands to their original locations or suitable new areas. Various methods are employed depending on the specific habitat being restored (Fig. 9.2).

Wetland restoration aims to restore the original hydrology, topography, and native plant cover of the area. Wetland enhancement, on the other hand, focuses on managing existing wetlands to maintain or improve their functions or values (Linham and Nicholls 2017; Spieles 2022). Human activities such as changes in water quality, quantity, and flow rates, as well as the introduction of non-native species, contribute to wetland degradation and loss. The main objectives of wetland restoration or creation projects typically include providing wildlife habitat, improving water quality, and offering flood protection. While restoration aims to return an

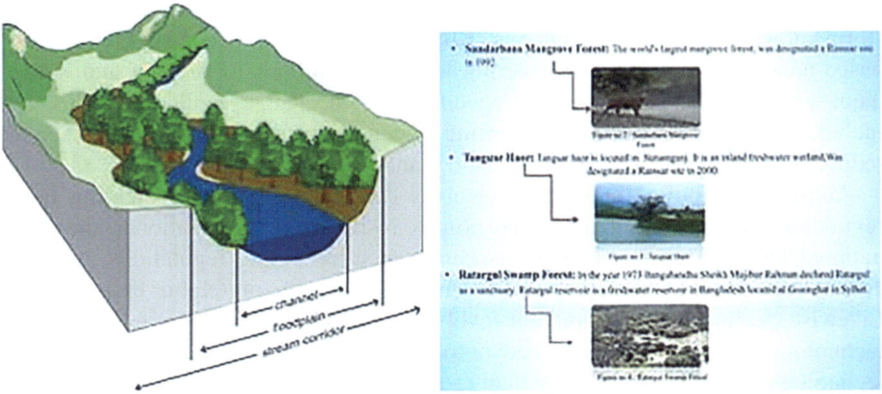

Fig. 9.2 Wetlands restoration

ecosystem to its original state, rehabilitation involves adapting the landscape to serve a specific human purpose. Equally, wetland restoration and rehabilitation are critical adaptation strategies aimed at bolstering the resilience of wetland ecosystems in the face of climate change (Spieles 2022). These strategies involve efforts to revive degraded or damaged wetland areas and enhance their ability to withstand environmental stressors (Rodrigo 2021). Here's a detailed look at wetland restoration and rehabilitation:

Restoring Hydrological Processes: Wetland restoration focuses on re-establishing natural hydrological processes, such as water flow and retention. This may involve restoring natural watercourses, constructing levees or dams to control water levels, and removing artificial drainage systems that disrupt wetland hydrology (Klaus and Kiehl 2021).

Replanting Native Vegetation: Rehabilitating wetlands often involves replanting native vegetation species that are well-adapted to local conditions. Native plants help stabilize soil, improve water quality, and provide habitat for wildlife (Klaus and Kiehl 2021; Spieles 2022). Additionally, restoring vegetation helps enhance biodiversity and ecosystem resilience.

Enhancing Habitat Connectivity: Restoring wetland connectivity with surrounding habitats is crucial for supporting species migration and maintaining ecological processes. This may involve creating wildlife corridors, constructing fish ladders, or removing barriers that impede natural movement between wetland and upland areas (Klaus and Kiehl 2021; Spieles 2022).

Controlling Invasive Species: Rehabilitating wetlands requires controlling invasive plant and animal species that can outcompete native species and disrupt ecosystem functioning. This may involve mechanical removal, herbicide application, or biological control methods to manage invasive populations (Klaus and Kiehl 2021; Spieles 2022).

Implementing Soil Conservation Measures: Wetland restoration often involves implementing soil conservation measures to prevent erosion and sedimentation (Klaus and Kiehl 2021; Spieles 2022). This may include re-vegetating eroded areas, stabilizing stream banks, and installing erosion control structures such as sediment traps or silt fences.

Monitoring and Adaptive Management: Monitoring the progress of wetland restoration efforts is essential for evaluating success and identifying areas for improvement. Adaptive management approaches allow for ongoing adjustments based on monitoring data and stakeholder feedback to ensure the effectiveness of restoration activities (Klaus and Kiehl 2021; Spieles 2022).

Stakeholder Engagement and Capacity Building: Engaging local communities, landowners, and stakeholders in wetland restoration efforts is critical for long-term success. Building capacity through education, training, and outreach initiatives helps foster stewardship and support for wetland conservation (Spieles 2022).

By prioritizing wetland restoration and rehabilitation as adaptation strategies, we can enhance the resilience of these valuable ecosystems to climate change impacts, protect biodiversity, and ensure the continued provision of ecosystem services for future generations.

9.4.1.1 Enhancing Wetland Resilience

Enhancing wetland resilience involves implementing strategies to improve the ability of wetland ecosystems to withstand and recover from environmental stressors, such as climate change and human disturbances. This is crucial for maintaining the valuable services provided by wetlands, such as flood regulation, water purification, and habitat provision (Cahoon et al. 2021; Khelifa et al. 2022). Coastal wetlands ecosystems provide a wide range of goods and services, including timber/wood; fish/shellfish; protection of the inlands from high waves, storm surges, and tsunamis; and carbon stored in the ecosystem. Low-lying coastal wetlands are among the most vulnerable landscape to climate change/sea level rise and subsequent land-use change for economic development (Cahoon et al. 2021). Finding the balance between conservation and development including restoration, requires innovative actions to enhance the resilience of the ecosystems and minimize loss and damage for the society. However, one key strategy for enhancing wetland resilience is restoring natural hydrological processes. This involves ensuring that wetlands receive adequate water inputs and maintaining their natural water flow patterns (Hidayat et al. 2022). By restoring hydrological connectivity and mimicking natural water regimes, wetlands can better adapt to changing environmental conditions and support diverse plant and animal communities.

Another important approach is promoting biodiversity within wetland ecosystems. High biodiversity increases the resilience of wetlands by enhancing their ability to adapt to disturbances and changes in environmental conditions. This can be achieved through habitat restoration, reintroducing native species, and creating diverse microhabitats to support a wide range of plants and animals (Sun et al. 2022a). Additionally, managing human activities around wetlands is essential for enhancing their resilience. This includes reducing pollution inputs, minimizing habitat fragmentation, and implementing sustainable land-use practices. By reducing anthropogenic pressures on wetlands, their ability to recover from disturbances and maintain ecosystem functions is improved (Törnqvist et al. 2021). Furthermore, incorporating climate-smart management practices can enhance wetland resilience to climate change. This may involve implementing adaptive management strategies, such as monitoring environmental conditions and adjusting management actions accordingly. It also includes incorporating climate change projections into wetland conservation and restoration plans to ensure they remain effective in the face of future environmental changes (Suir et al. 2020). Hence, enhancing wetland resilience requires a holistic approach that integrates ecological restoration, biodiversity conservation, sustainable management practices, and climate adaptation strategies. By implementing these measures, we can help ensure that wetlands continue to provide vital ecosystem services and support biodiversity in the face of ongoing environmental challenges.

9.4.1.2 Sustainable Land Use Planning

Sustainable land use planning is a strategic approach to managing land resources in a way that meets the needs of present and future generations while minimizing negative impacts on the environment. This involves making decisions about how to use land for various purposes, such as agriculture, urban development, conservation, and recreation, in a manner that balances economic, social, and environmental considerations (Carlos Loures 2019). Also, sustainable land-management practices such as conservation tillage, cover cropping, and crop rotation can improve soil health by reducing erosion, increasing organic matter, and enhancing soil structure. These practices increase soil fertility, improve water infiltration, and reduce soil degradation (Domingo et al. 2021). Nevertheless, one key aspect of sustainable land use planning is preserving and protecting natural habitats, including wetlands, through measures such as zoning regulations, protected areas, and conservation easements. By designating areas for conservation and limiting development in ecologically sensitive areas, we can help maintain biodiversity and ecosystem services provided by wetlands (Sun et al. 2022b). Another important component is promoting smart growth and compact development patterns to reduce urban sprawl and preserve agricultural and natural landscapes. This includes encouraging mixed land uses, compact development, and transportation options that reduce reliance on private vehicles and promote walking, cycling, and public transit.

Additionally, sustainable land use planning involves integrating climate change considerations into land use decisions to enhance resilience to climate impacts. This may include avoiding development in areas prone to flooding or erosion, protecting natural buffers such as coastal wetlands and dunes, and incorporating green infrastructure features like rain gardens and permeable pavement to manage stormwater runoff (Zhang et al. 2023). Furthermore, sustainable land use planning seeks to promote sustainable agriculture practices that minimize environmental degradation while ensuring food security (Liang et al. 2021). This may involve supporting agroecological approaches, conserving soil and water resources, and promoting the use of organic farming methods. Therefore, sustainable land use planning is essential for adapting to climate change and ensuring the long-term health and resilience of ecosystems, including wetlands. By integrating ecological, economic, and social considerations into land use decisions, we can help protect natural habitats, conserve biodiversity, mitigate climate impacts, and promote sustainable development for future generations.

9.4.1.3 Managed Retreat

Managed retreat is a strategic approach to coping with the impacts of climate change, particularly sea level rise and coastal erosion. It involves the deliberate and organized movement of people, infrastructure, and assets away from vulnerable coastal areas to safer locations. This proactive strategy recognizes the inevitability of ongoing environmental changes and aims to reduce risks to human lives and property while promoting the long-term sustainability of coastal ecosystems (Liang et al. 2021; Zhang et al. 2023). In managed retreat, decisions are made collaboratively among governments, communities, and stakeholders to identify areas at high risk of coastal hazards and develop plans for relocating or abandoning vulnerable assets. This may include homes, businesses, infrastructure such as roads and utilities, and even entire communities (Sun et al. 2022a, 2022b). Scientific assessments and modeling are used to evaluate current and future coastal hazards, such as storm surges, flooding, and erosion, and to identify areas where retreat may be necessary. Social, economic, and environmental factors are also considered to minimize negative impacts on affected communities and ecosystems (Carlos Loures 2019; Domingo et al. 2021).

Managed retreat can take various forms, including voluntary buyouts of properties in high-risk areas, relocation of critical infrastructure to safer locations, and restoration of natural buffers such as dunes, wetlands, and mangroves to provide natural coastal defenses (Carlos Loures 2019; Törnqvist et al. 2021). While managed retreat can be challenging and complex, it offers several benefits. By moving away from vulnerable coastal areas, communities can reduce their exposure to future risks and avoid the high costs associated with repeated damage from coastal hazards (Suir et al. 2020). It also allows for the restoration and protection of coastal ecosystems, which provide valuable services such as storm protection, habitat for wildlife, and recreational opportunities. However, managed retreat also raises important ethical, social, and economic considerations. It can be emotionally difficult for residents to leave their homes and communities, and there may be concerns about equity, fairness, and access to resources in the relocation process (Suir et al. 2020; Klaus and Kiehl 2021). Additionally, managing the transition to new locations requires careful planning, coordination, and investment in infrastructure and services to support relocated communities. Therefore, managed retreat is an important adaptation strategy for addressing the impacts of climate change on coastal areas. By proactively planning for retreat and considering the needs of both people and nature, we can help build resilient and sustainable coastal communities in the face of environmental uncertainty.

9.4.2 Mitigation Approaches

Preserving important wetlands and fostering their restoration stand not only as pivotal measures to combat climate change but also as effective strategies to mitigate escalating flood risks. By expanding the natural capacity of wetlands to absorb and retain water, we can curtail flood damage, which is poised to amplify in magnitude. Coastal wetlands, in particular, excel in carbon sequestration while emitting minimal methane (Taillardat et al. 2020). Their soil stores twice as much carbon as all tropical forests combined, underscoring their significance in climate mitigation efforts. Reducing climate change entails staunching the flow of greenhouse gases from prominent sources like power plants, factories, vehicles, and agricultural operations. Additionally, forests, oceans, and soil serve as vital sinks, absorbing and storing these gases, thus playing a critical role in combating climate change. The loss of wetlands not only diminishes habitat availability but also disrupts landscape diversity and connectivity among aquatic ecosystems (Priestley and Catto 2022). Alterations such as transitioning from forested wetlands to other types can substantially alter habitat structures and community compositions, exerting profound ecological impacts.

Mitigation approaches aim to reduce the causes and effects of climate change on wetlands. They involve taking actions to minimize greenhouse gas emissions and enhance the resilience of wetland ecosystems to changing environmental conditions. One key mitigation approach is reducing greenhouse gas emissions, which are major contributors to climate change (Grieger et al. 2020). This involves decreasing the use of fossil fuels, which release carbon dioxide and other greenhouse gases when burned. Additionally, efforts to transition to renewable energy sources, such as solar and wind power, can help reduce emissions and mitigate climate change impacts on wetlands. Another mitigation approach is carbon sequestration, which involves capturing and storing carbon dioxide from the atmosphere. (Lovelock et al. 2023) Wetlands are natural carbon sinks, meaning they absorb and store large amounts of carbon. Protecting and restoring wetlands can enhance their ability to sequester carbon, helping to mitigate climate change. Wetland conservation and restoration are also important mitigation approaches. By preserving existing wetlands and restoring degraded ones, we can maintain and enhance their ability to provide important ecosystem services, such as carbon storage, water filtration, and habitat for biodiversity (Fairchild et al. 2021). This helps to mitigate the impacts of climate change on wetlands and the communities that depend on them.

Adaptive management is another key mitigation approach. This involves continuously monitoring wetland ecosystems, assessing their vulnerability to climate change, and adjusting management strategies accordingly (Lovelock et al. 2023). By being proactive and responsive to changing conditions, we can better mitigate the impacts of climate change on wetlands and improve their long-term resilience. Hence, international cooperation and policy initiatives play a crucial role in mitigation efforts. By working together to set emissions targets, implement climate policies, and support wetland conservation and restoration projects, countries can collectively mitigate the impacts of climate change on wetlands and protect these valuable ecosystems for future generations.

9.4.2.1 Carbon Sequestration in Wetland Ecosystems

Wetlands serve as critical natural resources, adept at absorbing atmospheric carbon and preventing its subsequent loss, thereby enabling long-term storage. Purposeful management of wetlands offers a natural solution to combat climate change and mitigate the direct impacts of wetland loss due to diverse land-use changes and natural factors. All types of wetlands contribute to carbon sequestration by absorbing carbon dioxide through photosynthesis and trapping sediments from runoff. Carbon is retained in living vegetation, litter, peat, organic soils, and accumulated sediments, some of which have accumulated over millennia (Krauss et al. 2023). Compared to forests and agroecosystems, wetlands exhibit lower carbon sequestration rates, albeit still significant. Wetland plants, like all vegetation, absorb carbon dioxide from the air during photosynthesis, storing it in plant biomass. While much of this carbon is stored aboveground in tree biomass, a substantial portion is also stored belowground in intricate root systems. In non-wetland ecosystems, decomposing roots release carbon dioxide back into the atmosphere due to readily available oxygen (Ye et al. 2022). However, in wetland soils where oxygen availability is limited, root decomposition is slowed, leading to the accumulation of organic matter in the soil, with carbon constituting approximately 50% of this organic matter.

Wetland ecosystems play a crucial role in the fight against climate change through a process called carbon sequestration. This process involves the capture and storage of carbon dioxide from the atmosphere, helping to mitigate the effects of greenhouse gas emissions. Wetlands are particularly efficient at sequestering carbon due to their unique environmental conditions. In wetlands, carbon is stored in various forms, including in the soil, vegetation, and water (Yan et al. 2021). Plant growth in wetlands removes carbon dioxide from the atmosphere during photosynthesis and stores it in plant tissues. When plants die and decompose, some of this carbon is transferred to the soil, where it can be stored for long periods. Additionally, waterlogged conditions in wetlands slow down the decomposition of organic matter, further enhancing carbon storage in the soil (Valach et al. 2021). Moreover, wetlands act as sinks for carbon dioxide by preventing the release of methane, another potent greenhouse gas, into the atmosphere. Methane production is reduced in waterlogged conditions typical of wetlands, helping to limit its impact on climate change (Ballut-Dajud et al. 2022). Thus, the carbon sequestration capacity of wetlands makes them invaluable in efforts to combat climate change. Protecting and restoring wetland ecosystems can enhance their ability to sequester carbon and contribute to global climate mitigation strategies.

9.4.2.2 Reducing Anthropogenic Stressors

To safeguard wetlands from the impacts of climate change, it's crucial to reduce anthropogenic stressors, which are human-caused pressures on these ecosystems. These stressors include various activities like urbanization, deforestation, pollution, and unsustainable resource extraction. Urbanization, for instance, involves the development of cities and infrastructure, which often leads to habitat loss and fragmentation, disrupting wetland ecosystems (Setter et al. 2022). Deforestation removes trees and vegetation that help regulate water flow and stabilize soil, leaving wetlands vulnerable to erosion and degradation. Pollution from industrial runoff, agricultural runoff, and sewage can contaminate wetland waters, harming aquatic life and altering ecosystem dynamics. Additionally, unsustainable resource extraction, such as overfishing or excessive water withdrawal, can disrupt the balance of wetland ecosystems, affecting biodiversity and ecosystem services (Firmiano et al. 2021). By reducing these anthropogenic stressors, we can help wetlands better cope with the impacts of climate change, such as sea level rise, extreme weather events, and shifts in temperature and precipitation patterns. Therefore, implementing measures like sustainable land use planning, habitat restoration, and pollution control can mitigate these stressors and promote the resilience of wetland ecosystems. Collaboration among policymakers, scientists, local communities, and stakeholders is essential to develop and implement effective strategies for reducing anthropogenic pressures on wetlands and ensuring their long-term survival in the face of climate change.

9.4.2.3 Policy Interventions and Governance

Multilevel climate governance involves ongoing discussions and negotiations among a diverse array of stakeholders, including national and local governments, international organizations, businesses, NGOs, and other social actors. Its primary aim is to foster opportunities and spur action to tackle climate change. The Adaptation Policy Framework is a structured approach for crafting adaptation strategies, policies, and measures to bolster human development in light of climate change and variability (Hossen et al. 2022; Babaniyi et al. 2024b). The National Climate Change Policy for Nigeria 2021–2030 supports the country's efforts to significantly reduce greenhouse gas emissions and mitigate the socio-economic impacts of climate change. To address the impacts of climate change on wetlands, effective policy interventions and governance mechanisms are essential (Obasuyi and Sabbo 2022; Echendu 2023). These measures involve the development and implementation of regulations, policies, and strategies at local, national, and international levels to protect and sustainably manage wetland ecosystems.

One key aspect of policy interventions is the establishment of protected areas and conservation measures to safeguard wetland habitats from further degradation and loss. These protected areas can include national parks, wildlife reserves, and marine protected areas, which serve as sanctuaries for diverse flora and fauna found in wetland ecosystems (Graversgaard et al. 2021; Babaniyi et al. 2024a, 2024b). Additionally, policies aimed at promoting sustainable land use practices, such as land zoning regulations and environmental impact assessments, can help prevent the conversion of wetlands for agricultural or urban development purposes. These measures ensure that wetlands are conserved for their ecological value and the services they provide, such as flood regulation, water purification, and carbon sequestration (Bell-James et al. 2023).

More so, governance mechanisms play a crucial role in coordinating and implementing policy interventions effectively. This involves collaboration among government agencies, non-governmental organizations, local communities, and stakeholders to develop and enforce regulations, monitor compliance, and address emerging challenges. Integrated management approaches that consider the social, economic, and ecological dimensions of wetland conservation are essential for sustainable governance (Dixon et al. 2021). By incorporating scientific research, traditional knowledge, and stakeholder engagement into decision-making processes, policymakers can develop adaptive strategies that enhance the resilience of wetland ecosystems to climate change impacts. Therefore, robust policy interventions and effective governance mechanisms are critical for mitigating the impacts of climate change on wetlands and ensuring their long-term sustainability (Okon et al. 2021; Edo 2023). By adopting proactive measures and fostering collaboration among various stakeholders, policymakers can help preserve these invaluable ecosystems for future generations.

9.4.2.4 Community Engagement and Education

Community engagement and education play crucial roles in addressing the impacts of climate change on wetlands. Involving local communities in decision-making processes and providing them with education about climate change can empower them to become active participants in conservation efforts. This engagement fosters a sense of ownership and responsibility, leading to more effective conservation outcomes (Uzoaru and Ijah 2021). Educating communities about the importance of wetlands in mitigating climate change and protecting biodiversity helps raise awareness and garner support for conservation initiatives. Scientific concepts related to climate change, such as greenhouse gas emissions and carbon sequestration, can be explained in simple terms to make them accessible to a wide audience (Bukar et al. 2021; Babaniyi et al. 2024a).

Furthermore, community engagement initiatives can encourage the adoption of sustainable practices that reduce anthropogenic stressors on wetlands. For example, promoting eco-friendly agricultural techniques and advocating for responsible waste management practices can help minimize pollution and habitat degradation (Greenwalt et al. 2020; Akpabio and Umoh 2021). In addition to education, community engagement efforts should involve collaboration with local leaders, NGOs, and other stakeholders to develop and implement adaptation strategies tailored to the specific needs and circumstances of each community (Effiong et al. 2024). By fostering a sense of collective responsibility and promoting environmental stewardship, community engagement and education contribute to building resilience and ensuring the long-term conservation of wetland ecosystems in the face of climate change.

9.4.2.5 Public Awareness Campaigns

Addressing issues related to climate change, such as air pollution and water contamination, can be tackled through heightened awareness of environmental concerns. By taking simple steps, individuals can play a role in preserving the environment and mitigating these challenges. Awareness campaigns, often organized on specific days, weeks, or months, aim to educate and raise public consciousness about various philanthropic causes, medical conditions, organizations, or topics (Akinola et al. 2020; Okon et al. 2021). The primary objective of such campaigns is to enhance brand recognition, ensuring that target audiences are familiar with the brand and informed about its latest initiatives and products.

Increasing awareness and understanding of the health impacts of climate change can facilitate changes in behavior and garner societal support for efforts to reduce greenhouse gas emissions. By spreading environmental awareness, we can protect natural resources and mitigate environmental degradation caused by rising temperatures. For instance, glacial melting is a consequence of global warming (Adeleke 2022). Well-crafted campaigns have the potential to inspire individuals to adopt energy-saving practices, potentially leading to significant reductions in residential energy consumption. In the United States, it is estimated that behavioral changes in the residential sector could save up to 20% of home energy demand (Galadima et al. 2020; Madaki et al. 2023).

Public awareness campaigns are essential for educating people about the impacts of climate change on wetlands and promoting action to mitigate these effects. These campaigns aim to raise awareness about the importance of wetlands in regulating the climate, supporting biodiversity, and providing valuable ecosystem services (Benson 2020; Madaki et al. 2023). Through various communication channels such as social media, television, radio, and community events, public awareness campaigns disseminate information about the role of wetlands in carbon sequestration, flood control, and water purification. Scientific concepts related to climate change, such as greenhouse gas emissions, sea level rise, and extreme weather events, are explained in simple terms to make them accessible to a wide audience (Akpan et al. 2020; Adeleke 2022).

Public awareness campaigns also highlight the potential consequences of wetland degradation and loss, such as increased flooding, loss of habitat, and declines in water quality. By illustrating the direct link between human activities and the health of wetland ecosystems, these campaigns seek to mobilize individuals and communities to take action to protect and restore wetlands (Benson 2020). Furthermore, public awareness campaigns often provide practical tips and guidelines for sustainable living practices that reduce environmental impact, such as conserving water, reducing energy consumption, and supporting wetland conservation initiatives. By empowering individuals to make informed choices and adopt environmentally friendly behaviors, these campaigns contribute to building resilience and fostering a culture of environmental stewardship in society (Akpan et al. 2020; Madaki et al. 2023). However, public awareness campaigns play a vital role in mobilizing public support for wetland conservation efforts and driving policy change at local, national, and international levels. By engaging and empowering individuals to take action, these campaigns contribute to the collective effort to address the challenges posed by climate change and ensure the long-term health and sustainability of wetland ecosystems.

9.4.2.6 Challenges and Future Directions

Addressing the challenges posed by climate change on wetlands requires concerted efforts and innovative approaches. One of the primary challenges is the loss and degradation of wetlands due to human activities and environmental changes. Urbanization, agricultural expansion, and industrial development continue to encroach upon wetland habitats, fragmenting ecosystems and disrupting vital ecological processes (Gxokwe et al. 2020). Moreover, climate change exacerbates these challenges by altering precipitation patterns, increasing the frequency and intensity of extreme weather events, and causing sea-level rise. Common direct impacts to wetlands include filling, grading, removal of vegetation, building construction and changes in water levels and drainage patterns. However, according to Shiru et al. (2020), 80% of our global wastewater is released into wetlands untreated. Pollution from factories, fertilisers, pesticides or from major spills, all pose serious threats to wetlands. That is why wetlands are considered especially vulnerable to climate change in short, medium or long term. Additionally, wetlands are also strongly vulnerable to land use changes and anthropic disturbances such as pollution and sediment loads carried out to their water bodies.

Likewise, another significant challenge is the limited understanding of wetland dynamics and their responses to climate change. Wetlands are complex ecosystems with intricate interactions between biotic and abiotic factors. Further research is needed to elucidate these dynamics and develop effective strategies for wetland conservation and management (Vushe 2021; Iwuchukwu et al. 2023). Additionally, there are socio-economic challenges associated with wetland conservation and adaptation. Balancing the needs of local communities with conservation goals can be challenging, especially in developing countries where wetlands are crucial for livelihoods and food security (Shiru et al. 2020; Vushe 2021). Looking ahead, future directions in wetland management and conservation will require interdisciplinary collaboration, innovative technologies, and adaptive governance frameworks (Gxokwe et al. 2020; Iwuchukwu et al. 2023). Embracing nature-based solutions, such as wetland restoration and sustainable land use planning, can enhance the resilience of wetland ecosystems to climate change impacts.

Furthermore, there is a need for increased public awareness and engagement in wetland conservation efforts. Education and outreach initiatives can empower communities to take action and support policy changes that promote wetland conservation and climate resilience (Isfat and Raihan 2022; Hernandez and Park 2024). In all, focusing on the challenges posed by climate change on wetlands requires a multiple approach that integrates scientific research, policy interventions, community engagement, and adaptive management strategies. By working together and taking proactive measures, we can safeguard these invaluable ecosystems for future generations.

9.5 Conclusion: Toward a Resilient Future for Wetlands

In conclusion, safeguarding the future of wetlands in the face of climate change requires collective action and innovative solutions. Wetlands play a critical role in maintaining ecological balance, supporting biodiversity, and providing essential ecosystem services. However, they are increasingly vulnerable to the impacts of climate change, including sea-level rise, altered precipitation patterns, and extreme weather events. To ensure a resilient future for wetlands, it is imperative to prioritize conservation efforts, restore degraded habitats, and implement adaptive management strategies. This includes enhancing wetland resilience through nature-based solutions such as wetland restoration, sustainable land use planning, and reducing anthropogenic stressors.

Moreover, fostering interdisciplinary collaboration and engaging local communities in conservation efforts are essential for long-term success. Education and awareness initiatives can empower individuals to take action and advocate for policy changes that support wetland conservation and climate resilience. By working together and embracing innovative approaches, we can protect and preserve wetlands for future generations. Investing in the resilience of wetland ecosystems not only benefits biodiversity and ecosystem health but also enhances our ability to adapt to the challenges of a changing climate. Together, we can build a resilient future where wetlands thrive and continue to provide invaluable services to society.

References

Adegun OB (2023) Flood-related challenges and impacts within coastal informal settlements: a case from Lagos, Nigeria. Int J Urban Sustain Dev 15(1):1–13

Adeleke BO (2022) Need for emergency response to wetlands loss in southwestern Nigeria: a review. Ghana J Geogr 14(3):90–114

Akinola AO, Opoko AP, Ibem EO, Okagbue HI, Afolabi AO (2020) Climate change adaptation and mitigation strategies in Lagos, Nigeria: built environment professionals' perspective. Int J Eng Adv Technol 9(3):1273–1282

Akpabio EM, Umoh GS (2021) The practical challenges of achieving sustainable wetland agriculture in Nigeria's Cross River basin. Water Int 46(1):83–97

Akpan UF, Akpan FS, Nsien IB, Suleiman RA (2020) Public perception of climatic risks and use of legislative measures in mitigating climatic variation in Rivers State, Nigeria. J Res For Wildl Environ 12(3):10–19

Babaniyi GG, Akor UJ, Adebomi JI (2024a) Legislation and programs supporting environmental protection. In: Aransiola SA, Babaniyi BR, Aransiola AB, Maddela NR (eds) Prospects for soil regeneration and its impact on environmental protection. Springer, Singapore, pp 3–14. https://doi.org/10.1007/978-3-031-53270-2_1

Babaniyi GG, Akor UJ, Bisi-Omotosho A (2024b) Application of marine greens in aquaculture. In: Aransiola SA, Babaniyi BR, Maddela NR, Bello OS (eds) Marine greens. CRC Press, Boca Raton, pp 57–65. https://doi.org/10.1201/9781003369738-8

Ballut-Dajud GA, Sandoval Herazo LC, Fernández-Lambert G, Marín-Muñiz JL, López Méndez MC, Betanzo-Torres EA (2022) Factors affecting wetland loss: a review. Land 11(3):434

Bell-James J, Foster R, Lovelock CE (2023) Identifying priorities for reform to integrate coastal wetland ecosystem services into law and policy. Environ Sci Pol 142:164–172

Benson A (2020) Social and environmental drivers of climate change vulnerability in the Niger Delta region, Nigeria. Eur J Environ Earth Sci 1(5):1–9

Bhowmik S (2022) Ecological and economic importance of wetlands and their vulnerability: a review. In: Research anthology on ecosystem conservation and preserving biodiversity. IGI Global, Hershey, pp 11–27

Borg FH, Greibe Andersen J, Karekezi C, Yonga G, Furu P, Kallestrup P, Kraef C (2021) Climate change and health in urban informal settlements in low-and middle-income countries – a scoping review of health impacts and adaptation strategies. Glob Health Action 14(1):1908064

Brampton AH (1992) Engineering significance of British saltmarshes. In: Allen JRL, Pye K (eds) Saltmarshes: morphodynamics, conservation and engineering significance. Cambridge University Press, Cambridge, pp 115–122

Bukar Y, Monguno AK, AbdulRahman AT (2021) Environmental change and livelihood activities in Hadejia-Nguru wetlands of Yobe State, north east Nigeria. J Geogr Geol 12(3):1–12

BYJUS (2022) Water cycle – a diagram of water cycle with detailed explanations. 18 May 2022. Retrieved April 20, 2024, from https://byjus.com/biology/diagram-of-water-cycle/#:~:text=The%20water%20cycle%20is%20defined,–%20solid%2C%20liquid%20and%20gas

Cahoon DR, McKee KL, Morris JT (2021) How plants influence resilience of salt marsh and mangrove wetlands to sea-level rise. Estuar Coasts 44(4):883–898

Carlos Loures L (2019) Introductory chapter: land-use planning and land-use change as catalysts of sustainable development. In: Land use – assessing the past, envisioning the future. IntechOpen. https://doi.org/10.5772/intechopen.84520

Cheng L, Trenberth KE, Gruber N, Abraham JP, Fasullo JT, Li G et al (2020) Improved estimates of changes in upper ocean salinity and the hydrological cycle. J Clim 33(23):10357–10381

Cheung WW, Frölicher TL, Lam VW, Oyinlola MA, Reygondeau G, Sumaila UR et al (2021) Marine high temperature extremes amplify the impacts of climate change on fish and fisheries. Sci Adv 7(40):eabh0895

Cortese L, Zhang X, Simard M, Fagherazzi S (2024) Storm impacts on mineral mass accumulation rates of coastal marshes. J Geophys Res Earth 129(3):e2023JF007065

Dada OA, Almar R, Oladapo MI (2020) Recent coastal sea-level variations and flooding events in the Nigerian Transgressive Mud coast of Gulf of Guinea. J Afr Earth Sci 161:103668

Department of Climate Change, Energy, the Environment and Water (DCCEEW) (2023) Wetlands and climate change – information resources. Australian Government. 20 October 2023. Retrieved April 15, 2024, from https://www.dcceew.gov.au/water/wetlands/climate-change-resources#daff-page-main

Dixon A, Wood A, Hailu A (2021) Wetlands in Ethiopia: lessons from 20 years of research, policy and practice. Wetlands 41(2):20

Domingo D, Palka G, Hersperger AM (2021) Effect of zoning plans on urban land-use change: a multi-scenario simulation for supporting sustainable urban growth. Sustain Cities Soc 69:102833

Duethmann D, Blöschl G, Parajka J (2020) Why does a conceptual hydrological model fail to correctly predict discharge changes in response to climate change? Hydrol Earth Syst Sci 24(7):3493–3511

Duffield SJ, Le Bas B, Morecroft MD (2021) Climate change vulnerability and the state of adaptation on England's National Nature Reserves. Biol Conserv 254:108938

Dunn RJ, Alexander LV, Donat MG, Zhang X, Bador M, Herold N et al (2020) Development of an updated global land in situ-based data set of temperature and precipitation extremes: HadEX3. J Geophys Res Atmos 125(16):e2019JD032263

Durand G, van den Broeke MR, Le Cozannet G, Edwards TL, Holland PR, Jourdain NC et al (2022) Sea-level rise: from global perspectives to local services. Front Mar Sci 8:709595

Echendu AJ (2023) Human factors vs climate change; experts' view of drivers of flooding in Nigeria. Nat Hazard Res 3(2):240–246

Edo I (2023) Environmental laws as tools for effective management of Niger-Delta wetland: a case study of Upper Orashi forest. Doctoral dissertation, BTU Cottbus-Senftenberg

Effiong C, Ngang E, Ekott I (2024) Land use planning and climate change adaptation in river-dependent communities in Nigeria. Environ Dev 49:100970

Fairchild TP, Bennett WG, Smith G, Day B, Skov MW, Möller I et al (2021) Coastal wetlands mitigate storm flooding and associated costs in estuaries. Environ Res Lett 16(7):074034

Firmiano KR, Castro DM, Linares MS, Callisto M (2021) Functional responses of aquatic invertebrates to anthropogenic stressors in riparian zones of Neotropical savanna streams. Sci Total Environ 753:141865

Frederikse T, Landerer F, Caron L, Adhikari S, Parkes D, Humphrey VW et al (2020) The causes of sea-level rise since 1900. Nature 584(7821):393–397

Fuchs JM, Hittenbeck A, Brandl S, Schmidt M, Paul C (2022) Adaptation strategies for spruce forests—economic potential of bark beetle management and Douglas fir cultivation in future tree species portfolios. Forestry 95(2):229–246

Galadima A, Moyi SI, Lawal AM (2020) Awareness of climate change and its dangers among local farmers of Katsina State, Nigeria. Int J Sci Glob Sustain 6(3):7–7

Gimba, U.A., Aransiola, S.A., Oyewole, O.A., Maddela, N.R. (2024). Marine greens: roles in climate change and global warming mitigations. In Sesan Abiodun Aransiola, Oluwafemi Adebayo Oyewole, Naga Raju Maddela, Udeme Joshua Josiah Ijah, Bangeppagari Manjunatha (eds.) Marine greens: environmental, agricultural, industrial and biomedical applications. CRC Press, Boca Raton. ISBN: 9781032438856. https://www.routledge.com/Marine-Greens-Environmental-Agricultural-Industrial-and-Biomedical-Applications/Aransiola-Oyewole-Maddela-Josiah-Ijah-Manjunatha/p/book/9781032438856

Giri S, Lathrop RG, Obropta CC (2020) Climate change vulnerability assessment and adaptation strategies through best management practices. J Hydrol 580:124311

Graversgaard M, Jacobsen BH, Hoffmann CC, Dalgaard T, Odgaard MV, Kjaergaard C et al (2021) Policies for wetlands implementation in Denmark and Sweden – historical lessons and emerging issues. Land Use Policy 101:105206

Greenwalt J, Dede M, Johnson I, Nosa P, Precious A, Summers B (2020) Climate change adaptation and community development in Port Harcourt, Nigeria. In: African handbook of climate change adaptation. Springer, Cham, pp 1–28

Grieger R, Capon SJ, Hadwen WL, Mackey B (2020) Between a bog and a hard place: a global review of climate change effects on coastal freshwater wetlands. Clim Chang 163(1):161–179

Gxokwe S, Dube T, Mazvimavi D (2020) Multispectral remote sensing of wetlands in semi-arid and arid areas: a review on applications, challenges and possible future research directions. Remote Sens 12(24):4190

Harvey JA, Heinen R, Gols R, Thakur MP (2020) Climate change-mediated temperature extremes and insects: from outbreaks to breakdowns. Glob Chang Biol 26(12):6685–6701

Hauer ME, Fussell E, Mueller V, Burkett M, Call M, Abel K et al (2020) Sea-level rise and human migration. Nat Rev Earth Environ 1(1):28–39

Hernandez JO, Park BB (2024) Litterfall production and decomposition in tropical and subtropical mangroves: research trends and interacting effects of biophysical, chemical, and anthropogenic factors. Wetlands 44(2):23

Hidayat AS, Rajiani I, Arisanty D (2022) Sustainability of floodplain wetland fisheries of rural Indonesia: does culture enhance livelihood resilience? Sustainability 14(21):14461

Hoq MS, Raha SK, Hossain MI (2021) Livelihood vulnerability to flood hazard: understanding from the flood-prone haor ecosystem of Bangladesh. Environ Manag 67(3):532–552

Hossen MA, Netherton C, Benson D, Rahman MR, Salehin M (2022) A governance perspective for climate change adaptation: conceptualizing the policy-community interface in Bangladesh. Environ Sci Pol 137:174–184

Isfat M, Raihan A (2022) Current practices, challenges, and future directions of climate change adaptation in Bangladesh. Int J Res Publ Rev. 3:3429–3437

Isiaka IO, Ndukwe KO, Chibuike US (2021) Mean sea level: the effect of the rise in the environment. J Geoinform Environ Res 2(2):92–102

Iwuchukwu FU, Ewuzie U, Ajala OJ, Ojukwu VE, Nnorom IC, Egbueri JC et al (2023) A consideration of the climatic drivers, focal points and challenges of soil erosion, land degradation, landslides and landscapes in Nigeria. In: Climate change impacts on Nigeria: environment and sustainable development. Springer, Cham, pp 449–477

Jiang J, Zhou T, Chen X, Zhang L (2020) Future changes in precipitation over Central Asia based on CMIP6 projections. Environ Res Lett 15(5):054009

Kabir KH, Khan MA, Hasan K, Aftab S (2022) Driving forces of adaptation decision and strategies to climate-related events: case on farming households in south–west coastal Bangladesh. JEAPM 24(1):2250019

Karimi V, Valizadeh N, Rahmani S, Bijani M, Karimi M (2022) Beyond climate change: impacts, adaptation strategies, and influencing factors. In: Climate change: the social and scientific construct. Springer, Cham, pp 49–70

Khan I, Lei H, Shah IA, Ali I, Khan I, Muhammad I et al (2020) Farm households' risk perception, attitude and adaptation strategies in dealing with climate change: promise and perils from rural Pakistan. Land Use Policy 91:104395

Khelifa R, Mahdjoub H, Samways MJ (2022) Combined climatic and anthropogenic stress threaten resilience of important wetland sites in an arid region. Sci Total Environ 806:150806

Klaus VH, Kiehl K (2021) A conceptual framework for urban ecological restoration and rehabilitation. Basic Appl Ecol 52:82–94

Koffi EN, Bergamaschi P, Alkama R, Cescatti A (2020) An observation-constrained assessment of the climate sensitivity and future trajectories of wetland methane emissions. Sci Adv 6(15):eaay4444

Krauss KW, Zhu Z, Stagg CL, Krauss KW, Zhu Z, Stagg CL (2023) Managing wetlands to improve carbon sequestration. Eos. 1 July 2023. https://eos.org/editors-vox/managing-wetlands-to-improve-carbon-sequestration

Liang X, Guan Q, Clarke KC, Liu S, Wang B, Yao Y (2021) Understanding the drivers of sustainable land expansion using a patch-generating land use simulation (PLUS) model: a case study in Wuhan, China. Comput Environ Urban Syst 85:101569

Linham MM, Nicholls RJ (2017) Wetland restoration and rehabilitation. UNEP Copenhagen Climate Centre (UNEP-CCC). 4 May 2017. Retrieved May 2, 2024, from https://www.ctc-n.org/technologies/wetland-restoration-and-rehabilitation

Lloréns JLP (2008) Impacts of climate change on wetland ecosystems. Water Supply 7(2.117):8

Lovelock CE, Adame MF, Bradley J, Dittmann S, Hagger V, Hickey SM et al (2023) An Australian blue carbon method to estimate climate change mitigation benefits of coastal wetland restoration. Restor Ecol 31(7):e13739

Ma CS, Ma G, Pincebourde S (2021) Survive a warming climate: insect responses to extreme high temperatures. Annu Rev Entomol 66:163–184

Madaki MY, Muench S, Kaechele H, Bavorova M (2023) Climate change knowledge and perception among farming households in Nigeria. Climate 11(6):115
Mahdian M, Hosseinzadeh M, Siadatmousavi SM, Chalipa Z, Delavar M, Guo M et al (2023) Modelling impacts of climate change and anthropogenic activities on inflows and sediment loads of wetlands: case study of the Anzali wetland. Sci Rep 13(1):5399
Malekmohammadi B, Uvo CB, Moghadam NT, Noori R, Abolfathi S (2023) Environmental risk assessment of wetland ecosystems using Bayesian belief networks. Hydrology 10(1):16
Meehl GA (2023) The role of the IPCC in climate science. In: Oxford research encyclopedia of climate science. Oxford University Press, Oxford
Mendez M, Maathuis B, Hein-Griggs D, Alvarado-Gamboa LF (2020) Performance evaluation of bias correction methods for climate change monthly precipitation projections over Costa Rica. Water 12(2):482
Moomaw WR, Chmura GL, Davies GT, Finlayson CM, Middleton BA, Natali SM et al (2018) Wetlands in a changing climate: science, policy and management. Wetlands 38(2):183–205
Moron V, Barbero R, Fowler HJ, Mishra V (2021) Storm types in India: linking rainfall duration, spatial extent and intensity. Phil Trans R Soc A 379(2195):20200137
Naskar M, Sarkar UK, Mishal P, Karnatak G, Saha S, Bandopadhyay A et al (2022) Assessing vulnerability of wetland fisheries to climate change: a stakeholders' perception-based approach. Clim Dev 14(7):600–616
National Center for Environmental Health (2019) Climate effects on health. Centers for Disease Control and Prevention; U.S. Department of Health & Human Services. https://www.cdc.gov/climateandhealth/effects/temperature_extremes.htm
Ng CWW, Ni JJ, Leung AK (2020) Effects of plant growth and spacing on soil hydrological changes: a field study. Géotechnique 70(10):867–881
Nielsen DL, Merrin LE, Pollino CA, Karim F, Stratford D, O'Sullivan J (2020) Climate change and dam development: effects on wetland connectivity and ecological habitat in tropical wetlands. Ecohydrology 13(6):e2228
Obasuyi A, Sabbo JC (2022) The effectiveness of the law and environmental policies towards protecting Upper Niger Lake wetlands in West Africa. African Journal of Emerging Issues 4(12):10–20
Ogundeji AA, Okolie CC (2022) Perception and adaptation strategies of smallholder farmers to drought risk: a scientometric analysis. Agriculture 12(8):1129
Ojo TO, Ogundeji AA, Emenike CU (2022) Does adoption of climate change adaptation strategy improve food security? A case of rice farmers in Ogun State, Nigeria. Land 11(11):1875
Okon EM, Falana BM, Solaja SO, Yakubu SO, Alabi OO, Okikiola BT et al (2021) Systematic review of climate change impact research in Nigeria: implication for sustainable development. Heliyon 7(9):e07941
Oloyede MO, Williams AB, Ode GO, Benson NU (2022) Coastal vulnerability assessment: a case study of the Nigerian coastline. Sustainability 14(4):2097
Peixoto RS, Voolstra CR, Sweet M, Duarte CM, Carvalho S, Villela H et al (2022) Harnessing the microbiome to prevent global biodiversity loss. Nat Microbiol 7(11):1726–1735
Pepin NC, Arnone E, Gobiet A, Haslinger K, Kotlarski S, Notarnicola C et al (2022) Climate changes and their elevational patterns in the mountains of the world. Rev Geophys 60(1):e2020RG000730
Priestley MD, Catto JL (2022) Future changes in the extratropical storm tracks and cyclone intensity, wind speed, and structure. Weather Clim Dyn 3(1):337–360
Rahman MS, Islam ARMT (2019) Are precipitation concentration and intensity changing in Bangladesh overtimes? Analysis of the possible causes of changes in precipitation systems. Sci Total Environ 690:370–387
Rodrigo MA (2021) Wetland restoration with hydrophytes: a review. Plants 10(6):1035
Roy P, Pal SC, Chakrabortty R, Chowdhuri I, Saha A, Shit M (2023) Effects of climate change and sea-level rise on coastal habitat: vulnerability assessment, adaptation strategies and policy recommendations. J Environ Manag 330:117187

Salimi S, Almuktar SA, Scholz M (2021) Impact of climate change on wetland ecosystems: a critical review of experimental wetlands. J Environ Manag 286:112160

Sarkar UK, Bakshi S, Lianthuamluaia L, Mishal P, Das Ghosh B, Saha S, Karnatak G (2020) Understanding enviro-climatological impact on fish biodiversity of the tropical floodplain wetlands for their sustainable management. Sustain Water Resour Manag 6:1–12

Setter RO, Franklin EC, Mora C (2022) Co-occurring anthropogenic stressors reduce the timeframe of environmental viability for the world's coral reefs. PLoS Biol 20(10):e3001821

Shen YJ, Shen Y, Guo Y, Zhang Y, Pei H, Brenning A (2020) Review of historical and projected future climatic and hydrological changes in mountainous semiarid Xinjiang (northwestern China), central Asia. Catena 187:104343

Shin YJ, Midgley GF, Archer ER, Arneth A, Barnes DK, Chan L et al (2022) Actions to halt biodiversity loss generally benefit the climate. Glob Chang Biol 28(9):2846–2874

Shiru MS, Shahid S, Shiru S, Chung ES, Alias N, Ahmed K et al (2020) Challenges in water resources of Lagos mega city of Nigeria in the context of climate change. J Water Clim Chang 11(4):1067–1083

Singh V, Shukla S, Singh A (2021) The principal factors responsible for biodiversity loss. Open J Plant Sci 6(1):11–14

Solecki W, Roberts D, Seto KC (2024) Strategies to improve the impact of the IPCC Special Report on Climate Change and Cities. Nat Clim Chang 14(7):685–691

Spieles DJ (2022) Wetland construction, restoration, and integration: a comparative review. Land 11(4):554

Su X, Zhou Y, Li Q (2021) Designing ecological security patterns based on the framework of ecological quality and ecological sensitivity: a case study of Jianghan Plain, China. Int J Environ Res Public Health 18(16):8383

Suir GM, Sasser CE, Harris JM (2020) Use of remote sensing and field data to quantify the performance and resilience of restored Louisiana wetlands. Wetlands 40(6):2643–2658

Sun B, Jiang M, Han G, Zhang L, Zhou J, Bian C et al (2022a) Experimental warming reduces ecosystem resistance and resilience to severe flooding in a wetland. Sci Adv 8(4):eabl9526

Sun X, Wu J, Tang H, Yang P (2022b) An urban hierarchy-based approach integrating ecosystem services into multiscale sustainable land use planning: the case of China. Resour Conserv Recycl 178:106097

Taillardat P, Thompson BS, Garneau M, Trottier K, Friess DA (2020) Climate change mitigation potential of wetlands and the cost-effectiveness of their restoration. Interface Focus 10(5):20190129

Tamm O, Saaremäe E, Rahkema K, Jaagus J, Tamm T (2023) The intensification of short-duration rainfall extremes due to climate change – need for a frequent update of intensity–duration–frequency curves. Clim Serv 30:100349

Törnqvist TE, Cahoon DR, Morris JT, Day JW (2021) Coastal wetland resilience, accelerated sea-level rise, and the importance of timescale. AGU Adv 2(1):e2020AV000334

Troxler TG, Kennedy HA, Crooks S, Sutton-Grier AE (2018) Introduction of coastal wetlands into the IPCC greenhouse gas inventory methodological guidance. In: A blue carbon primer. CRC Press, Boca Raton, pp 217–234

Twinomuhangi R, Sseviiri H, Mulinde C, Mukwaya PI, Nimusiima A, Kato AM (2021) Perceptions and vulnerability to climate change among the urban poor in Kampala City, Uganda. Reg Environ Chang 21:1–13

United States Army Corps of Engineers (USACE) (1989) Environmental engineering for coastal shore protection. USACE, Washington, DC. http://140.194.76.129/publications/eng-manuals/em1110-2-1100/PartI/PartI.htm. Accessed 29 Aug 2010

University of California Museum of Paleontology (UCMP) (2022) Precipitation – understanding global change. University of California Regents. 9 March 2022. Retrieved April 30, 2024, from https://ugc.berkeley.edu/background-content/precipitation/#:~:text=Various%20human%20activities%20and%20environmental,thus%20the%20Earth%27s%20average%20temperature

Uzoaru OC, Ijah CN (2021) Community based environmental education a strategy for mitigating impacts of climate change on livelihood of riverine communities in rivers state. Int J Weather Clim Change Conserv Res 7:45–54

Valach AC, Kasak K, Hemes KS, Anthony TL, Dronova I, Taddeo S et al (2021) Productive wetlands restored for carbon sequestration quickly become net CO2 sinks with site-level factors driving uptake variability. PLoS One 16(3):e0248398

Vushe A (2021) Proposed research, science, technology, and innovation to address current and future challenges of climate change and water resource management in Africa. In: Climate change and water resources in Africa: perspectives and solutions towards an imminent water crisis. Springer, Cham, pp 489–518

Wang YS, Gu JD (2021) Ecological responses, adaptation and mechanisms of mangrove wetland ecosystem to global climate change and anthropogenic activities. Int Biodeterior Biodegradation 162:105248

Wang J, He G, Fang H, Han Y (2020) Climate change impacts on the topography and ecological environment of the wetlands in the middle reaches of the Yarlung Zangbo-Brahmaputra River. J Hydrol 590:125419

Wang J, Chen Y, Liao W, He G, Tett SF, Yan Z et al (2021) Anthropogenic emissions and urbanization increase risk of compound hot extremes in cities. Nat Clim Chang 11(12):1084–1089

Wardhana DHA, Prawira MR (2024) The analysis of Indonesia's climate change policies in response to the 2021 Intergovernmental Panel on Climate Change (IPCC) Assessment Report/AR6 Group 1 (2021–2023). PROIROFONIC 1(1):42–53

Wise EK, Dannenberg MP (2022) Simulating the impacts of changes in precipitation timing and intensity on tree growth. Geophys Res Lett 49(22):e2022GL100863

Wong NH, Tan CL, Kolokotsa DD, Takebayashi H (2021) Greenery as a mitigation and adaptation strategy to urban heat. Nat Rev Earth Environ 2(3):166–181

Xi Y, Peng S, Ciais P, Chen Y (2021) Future impacts of climate change on inland Ramsar wetlands. Nat Clim Chang 11(1):45–51

Yan Z, Shen T, Li W, Cheng W, Wang X, Zhu M et al (2021) Contribution of microalgae to carbon sequestration in a natural karst wetland aquatic ecosystem: an in-situ mesocosm study. Sci Total Environ 768:144387

Yang X, Liu S, Jia C, Liu Y, Yu C (2021a) Vulnerability assessment and management planning for the ecological environment in urban wetlands. J Environ Manag 298:113540

Yang D, Yang Y, Xia J (2021b) Hydrological cycle and water resources in a changing world: a review. Geogr Sustain 2(2):115–122

Ye S, Pei L, He L, Xie L, Zhao G, Yuan H et al (2022) Wetlands in China: evolution, carbon sequestrations and services, threats, and preservation/restoration. Water 14(7):1152

Zhang T, Xin X, He F, Wang X, Chen K (2023) How to promote sustainable land use in Hangzhou Bay, China? A decision framework based on fuzzy multiobjective optimization and spatial simulation. J Clean Prod 414:137576

Part III
Ecosystem Services, Revitalization and Benefits for Human Well-being

Chapter 10
Phytoremediation in Wetlands

O. D. Ogundele, D. T. Ogundele, O. J. Popoola, A. A. Adelekun, and V. A. Olagunju

Abstract This chapter explores the role of phytoremediation in wetlands, focusing on the use of wetland plants to absorb, detoxify, and mitigate contaminants present in water. Wetlands serve as natural filtration systems, and the integration of phytoremediation techniques enhances their ability to treat various pollutants, including heavy metals, organic compounds, and nutrients. The chapter begins by providing an overview of wetland ecosystems, emphasizing their ecological importance and their function as buffers against environmental degradation. It then delves into the principles of phytoremediation, highlighting the mechanisms by which plants in wetland environments absorb and metabolize contaminants. The various types of phytoremediation; phytoextraction, phytodegradation, phytostabilization, phytovolatilization, and rhizofiltration are discussed in detail, with a focus on their applications in wetland management. Case studies of constructed wetlands designed for industrial and agricultural wastewater treatment are presented, demonstrating the effectiveness of phytoremediation in real-world scenarios. The chapter also examines factors influencing the efficiency of phytoremediation, such as plant species selection, environmental conditions, and plant-microbe interactions. Future perspectives and opportunities in the field are explored, including innovations in biotechnology, the integration of phytoremediation with other treatment methods, and policy considerations. Overall, this chapter underscores the potential of wetland phytoremediation as a sustainable and cost-effective approach to environmental remediation, contributing to the preservation and restoration of wetland ecosystems.

O. D. Ogundele (✉) · A. A. Adelekun
Department of Chemical Sciences, Achievers University, Owo, Nigeria
e-mail: olusoladavidogundele@gmail.com

D. T. Ogundele
Department of Chemistry and Industrial Chemistry, Kwara State University, Malete, Nigeria

O. J. Popoola
Department of Geological Sciences, Achievers University, Owo, Nigeria

V. A. Olagunju
Department of Chemistry and Biochemistry, Miami University, Oxford, OH, USA

B. R. Babaniyi et al. (eds.), *Wetland Ecosystems: Conservation Strategies, Policy Management and Applications*, Wetlands: Ecology, Conservation and Management 12, https://doi.org/10.1007/978-3-031-91982-4_10

Keywords Marshes · Invasive species · Phytodegradation · Rhizofiltration · Nanotechnology

10.1 Introduction

Wetlands are among the most productive ecosystems in the world, playing a crucial role in maintaining ecological balance and supporting biodiversity. These ecosystems are characterized by the presence of water, either permanently or seasonally, which creates conditions that support the growth of hydrophytic vegetation (Mitsch and Gosselink 2015). Wetlands can be found on every continent and encompass a wide variety of habitats, from mangrove swamps and peat bogs to floodplains and marshes. They act as natural buffers, mitigating the impacts of floods, controlling water quality, and sequestering carbon, which helps combat climate change (Gell et al. 2023). Despite their ecological importance, wetlands are increasingly under threat from human activities, leading to the degradation and loss of these vital ecosystems (Davidson 2014).

Wetlands are among the most vital ecosystems on Earth, providing essential services such as water purification, flood control, carbon sequestration, and biodiversity support. Despite covering only about 6% of the Earth's surface, wetlands are estimated to host nearly 40% of global biodiversity, making them crucial for maintaining ecological balance and supporting human livelihoods (Mitsch and Gosselink 2015). However, these ecosystems are increasingly under threat due to various anthropogenic activities, leading to the degradation of water quality and loss of biodiversity. Industrial discharges, agricultural runoff, and urban development have introduced a wide range of contaminants, including heavy metals, nutrients, and organic pollutants, into wetland environments, thereby necessitating effective remediation strategies (Mitsch and Gosselink 2015).

Wetlands are typically defined as areas where water covers the soil or is present at or near the surface for varying periods, leading to conditions that favor water-tolerant vegetation (Cowardin 1979). This definition includes a wide range of environments, from coastal estuaries and tidal marshes to inland lakes, rivers, and peatlands. Wetlands are highly dynamic systems, with their hydrology and vegetation changing over time due to natural processes such as flooding, sedimentation, and succession (Mitsch and Gosselink 2015).

Wetlands can be classified into several types based on their hydrology, vegetation, and location. The primary types of wetlands include marshes, swamps, bogs, and fens, each with distinct characteristics and ecological functions.

Marshes are wetlands dominated by herbaceous plants, such as grasses, sedges, and reeds. They are typically found in areas with shallow, standing water and are divided into two main types: freshwater marshes and saltwater marshes (Tiner

2002). Freshwater marshes occur along rivers, lakes, and floodplains, while saltwater marshes are found along coastlines, particularly in estuaries where freshwater meets saltwater (Mitsch and Gosselink 2015). Marshes play a critical role in water purification by trapping sediments and filtering pollutants, as well as providing habitat for a wide range of wildlife, including birds, amphibians, and fish (Tiner 2002).

Swamps are wetlands characterized by the presence of woody vegetation, such as trees and shrubs. They can be further divided into two categories: forested swamps and shrub swamps. Forested swamps are dominated by trees, while shrub swamps are dominated by shrubs (Keddy 2010). Swamps are typically found in areas with slow-moving or stagnant water, such as river floodplains, lake edges, and coastal regions (Mitsch and Gosselink 2015). Swamps are important for flood control, as they can absorb excess water during storms and release it slowly over time, reducing the risk of downstream flooding (Keddy 2010).

Bogs are wetlands that receive water primarily from precipitation, resulting in low nutrient levels and acidic conditions. They are characterized by the accumulation of peat, a type of organic soil formed from the slow decomposition of plant material, particularly mosses (Rydin et al. 2013). Bogs are typically found in cool, temperate regions and are home to unique plant species, such as sphagnum moss, cranberries, and carnivorous plants (Mitsch and Gosselink 2015). Bogs play a vital role in carbon sequestration, as the slow decomposition of plant material in these environments leads to the accumulation of large amounts of carbon in the form of peat (Rydin et al. 2013).

Fens are wetlands that receive water from both precipitation and groundwater, resulting in higher nutrient levels compared to bogs. They are characterized by the presence of peat and are typically found in areas with a mix of grasses, sedges, and woody vegetation (Amon et al. 2002). Fens are important for maintaining water quality, as they can filter and store nutrients, preventing them from entering nearby water bodies (Mitsch and Gosselink 2015). Fens also provide critical habitat for a variety of plant and animal species, including rare and endangered species (Amon et al. 2002).

10.1.1 Importance of Wetlands in the Global Ecosystem

Wetlands are often referred to as the "kidneys of the landscape" because of their ability to filter and cleanse water, removing pollutants and sediments from surface and groundwater (Mitsch and Gosselink 2015). They play a critical role in maintaining water quality by trapping and transforming pollutants, such as nitrogen, phosphorus, and heavy metals, through processes like sedimentation, adsorption, and microbial degradation (Verhoeven et al. 2006). Wetlands also act as buffers between land and water, absorbing excess nutrients and preventing them from reaching rivers, lakes, and oceans, where they can cause harmful algal blooms and other forms of pollution (Mitsch and Gosselink 2015).

In addition to their role in water purification, wetlands are essential for flood control. By storing excess water during periods of heavy rainfall and slowly releasing it over time, wetlands reduce the risk of flooding downstream and help to maintain stable water levels in rivers and lakes (Bullock and Acreman 2003). This function is particularly important in urban areas, where the loss of wetlands has led to increased flood risks and costly damage to infrastructure (Mitsch and Gosselink 2015).

Wetlands also play a crucial role in climate regulation by acting as carbon sinks. Peatlands, in particular, store large amounts of carbon in the form of peat, which accumulates over thousands of years (Parish et al. 2008). By sequestering carbon, wetlands help to mitigate climate change by reducing the amount of carbon dioxide in the atmosphere (Joosten et al. 2012). However, when wetlands are drained or degraded, the stored carbon is released back into the atmosphere, contributing to global warming (Parish et al. 2008).

Biodiversity is another key aspect of wetland ecosystems. Wetlands provide habitat for a wide variety of species, including many that are rare or endangered (Davidson 2014). They support diverse communities of plants, animals, and microorganisms, and are particularly important for migratory birds, which rely on wetlands for breeding, feeding, and resting during their long journeys (Kingsford et al. 2017). Wetlands also provide critical habitat for fish and amphibians, many of which are important for commercial and recreational fishing (Mitsch and Gosselink 2015).

In addition to their ecological functions, wetlands provide numerous cultural, recreational, and economic benefits. Many wetlands are important sites for tourism and recreation, offering, opportunities for birdwatching, fishing, and hiking. Wetlands also provide valuable resources, such as fish, timber, and medicinal plants, which support the livelihoods of millions of people around the world (Barbier et al. 1997).

Despite their importance, wetlands are increasingly threatened by human activities, which have led to significant loss and degradation of these ecosystems. Understanding the environmental threats to wetlands is essential for developing effective conservation strategies and ensuring the continued provision of the ecological services they provide (Davidson 2014).

10.1.2 Environmental Threats to Wetland Ecosystems

Wetlands are among the most threatened ecosystems globally, with an estimated 64% of the world's wetlands having been lost since 1900 (Davidson 2014). The primary drivers of wetland loss and degradation include land conversion, pollution, water diversion, and climate change.

10.1.2.1 Land Conversion

One of the most significant threats to wetlands is land conversion for agriculture, urban development, and infrastructure projects. Wetlands are often drained or filled in to create space for farming, housing, roads, and industrial facilities (Mitsch and Gosselink 2015). This process not only destroys the wetland habitat but also disrupts the hydrological processes that sustain wetland ecosystems (Davidson 2014). For example, the conversion of wetlands for agriculture has led to the loss of important floodplains and the degradation of water quality in many regions.

10.1.2.2 Pollution

Pollution from agricultural runoff, industrial discharges, and urban stormwater is another major threat to wetlands. Nutrient pollution, in particular, can lead to eutrophication, a process in which excess nutrients, such as nitrogen and phosphorus, stimulate the growth of algae and other aquatic plants, depleting oxygen levels in the water and leading to the death of fish and other aquatic organisms (Verhoeven et al. 2006; Adewumi et al. 2023). Heavy metals, pesticides, and other toxic chemicals can also accumulate in wetland sediments and plants, posing risks to wildlife and human health (Mitsch and Gosselink 2015).

10.1.2.3 Water Diversion

Water diversion for irrigation, hydropower, and urban water supply has significantly altered the hydrology of many wetland ecosystems, leading to changes in water levels, flow patterns, and water quality (Kingsford 2000). Dams and levees, for example, can block the natural flow of water to wetlands, reducing their ability to recharge and maintain healthy water levels (Mitsch and Gosselink 2015). In some cases, water diversion has led to the complete drying up of wetlands, resulting in the loss of critical habitat for wildlife and the degradation of ecosystem services (Kingsford 2000).

10.1.2.4 Climate Change

Climate change is expected to have profound impacts on wetland ecosystems, altering temperature and precipitation patterns, sea levels, and the frequency and intensity of extreme weather events (Erwin 2009). Rising sea levels, for example, threaten coastal wetlands, such as salt marshes and mangroves, which may be inundated by seawater or eroded by increased storm surges (Erwin 2009). Changes in temperature and precipitation patterns can also affect the hydrology of inland wetlands, leading to shifts in vegetation and wildlife communities (Erwin 2009).

10.1.2.5 Invasive Species

Invasive species pose a significant threat to wetland ecosystems by outcompeting native species, altering habitat structure, and disrupting ecological processes (Zedler and Kercher 2004). For example, invasive plants, such as Phragmites and water hyacinth, can rapidly colonize wetlands, displacing native vegetation and reducing biodiversity (Zedler and Kercher 2004). Invasive animals, such as non-native fish and amphibians, can also have negative impacts on wetland ecosystems by preying on native species and altering food webs (Zedler and Kercher 2004).

The cumulative effects of these threats have led to significant declines in wetland extent and quality, with many wetlands now degraded or lost entirely (Davidson 2014). Addressing these threats requires coordinated efforts at local, national, and international levels, including the implementation of policies and practices that promote wetland conservation, restoration, and sustainable management.

10.2 Phytoremediation

One promising approach to addressing water contamination in wetlands is phytoremediation, a green technology that utilizes plants to absorb, detoxify, or stabilize contaminants from the environment (Salt et al. 1998). Phytoremediation is gaining popularity due to its cost-effectiveness, environmental sustainability, and ability to enhance ecosystem services (Ogundele and Anaun 2022). Wetland plants, in particular, are uniquely suited for phytoremediation because of their adaptive traits, such as tolerance to waterlogged conditions, efficient nutrient uptake, and symbiotic relationships with microbial communities that enhance contaminant degradation (Wang and Que 2013; Aransiola et al. 2024).

Phytoremediation in wetlands involves several mechanisms by which plants remove or neutralize contaminants. These include phytoextraction, where plants absorb contaminants into their tissues; phytodegradation, in which plants metabolize pollutants into less toxic forms; phytostabilization, where contaminants are immobilized in the soil or sediments by plant roots; and phytovolatilization, where volatile contaminants are released into the atmosphere through transpiration (Ali et al. 2013). Wetland plants such as cattails (*Typha spp.*), common reed (*Phragmites australis*), and bulrushes (*Schoenoplectus spp.*) have been widely studied for their phytoremediation potential (Sarwar et al. 2017).

Phytoremediation is an innovative and environmentally friendly approach that utilizes plants to remove, detoxify, or stabilize environmental pollutants from soil, water, and air. Derived from the Greek word "phyto" (meaning plant) and the Latin word "remedium" (meaning to restore balance), phytoremediation represents a natural process through which plants, in association with their symbiotic microorganisms, mitigate contamination in the environment (Salt et al. 1998). This technique has gained considerable attention over recent decades as a cost-effective and sustainable alternative to conventional remediation methods, which are often expensive, invasive, and disruptive to the ecosystem (Pulford and Watson 2003).

Phytoremediation exploits the unique abilities of certain plants, known as hyperaccumulators, to tolerate and accumulate high concentrations of pollutants in their tissues. These plants, either naturally occurring or genetically modified, can absorb contaminants such as heavy metals, pesticides, petroleum hydrocarbons, and radionuclides, reducing their bioavailability in the environment (Pilon-Smits 2005). The effectiveness of phytoremediation depends on several factors, including the type of pollutant, the characteristics of the contaminated site, and the plant species used. Understanding the principles underlying this process is essential for optimizing its application and maximizing its benefits (Ali et al. 2013).

The principles of phytoremediation are grounded in plant physiology, soil science, and environmental chemistry. At the core of this process is the plant's ability to interact with contaminants through various mechanisms, including uptake, accumulation, degradation, and stabilization. These mechanisms are supported by the plant's root system, which plays a crucial role in accessing and transforming pollutants in the soil and water (Macek et al. 2000). Additionally, the rhizosphere, the area surrounding the plant roots, is a hotspot of microbial activity, where beneficial microorganisms enhance phytoremediation by breaking down contaminants and facilitating their uptake by plants (Sarwar et al. 2017).

One of the primary advantages of phytoremediation is its ability to remediate large areas of contaminated land with minimal disturbance to the environment. Unlike traditional methods that often involve the excavation and disposal of contaminated soil, phytoremediation works in situ, meaning that the remediation occurs directly at the site of contamination (Macek et al. 2000). This reduces the risk of secondary contamination and helps preserve the integrity of the ecosystem. Moreover, phytoremediation can improve soil health and promote biodiversity by enhancing soil structure, increasing organic matter content, and providing habitat for wildlife (Pulford and Watson 2003).

Phytoremediation is also a visually appealing remediation strategy, as it transforms contaminated sites into green spaces that can be used for recreational purposes, wildlife habitat restoration, or even agricultural production after successful remediation (Salt et al. 1998; Aransiola et al. 2013). Additionally, the plants used in phytoremediation can provide economic benefits, such as the production of biomass for bioenergy or the recovery of valuable metals through a process known as phytomining (Ghosh and Singh 2005). However, the success of phytoremediation depends on the careful selection of plant species and the appropriate management of the site to ensure that the contaminants are effectively removed or neutralized (Pilon-Smits 2005).

Phytoremediation can be categorized into several types, each targeting specific pollutants and environmental matrices. The selection of the appropriate phytoremediation strategy depends on the nature of the contamination and the desired outcome of the remediation process (Ali et al. 2013). Table 10.1 contains potential of some species of plant for phytoremediation.

Table 10.1 Potential of some species of plant for phytoremediation

S/n	Plant species	Pollutant	Remediation results	References
1	*Brassica juncea*	Heavy metals (Lead, Cadmium)	Significant reduction of lead and cadmium in contaminated soil.	Salt et al. (1995)
2	*Phragmites australis*	Nutrients (Nitrogen, Phosphorus)	Reduced nutrient levels in wastewater	Vymazal (2007)
3	*Helianthus annuus*	Arsenic	Accumulation and reduction of arsenic in contaminated soil	Tangahu et al. (2011)
4	*Salix viminalis*	Heavy metals (Cadmium, Zinc)	Successful accumulation of cadmium and zinc from contaminated soil.	Keller et al. (2003)
5	*Lemna minor*	Organic pollutants (PCBs)	Degradation of PCBs in contaminated water	Reinhold et al. (2010)
6	*Spartina alterniflora*	Hydrocarbons	Degradation and reduction of hydrocarbons in wetland soils	Lin and Mendelssohn (1998)
7	*Arundo donax*	Heavy metals (Cadmium, Lead)	Significant uptake and reduction of cadmium and lead from soils	Pilu et al. (2012)

10.2.1 Phytoextraction

Phytoextraction is a specialized form of phytoremediation, a green technology that uses plants to clean up contaminated environments. This process involves the uptake of contaminants, particularly heavy metals, from the soil and their accumulation in the above-ground parts of plants, which are then harvested and disposed of or recycled (Ali et al. 2013). Over the past few decades, phytoextraction has emerged as a promising, cost-effective, and environmentally friendly method for remediating contaminated soils, particularly those polluted by heavy metals from industrial activities, mining, and agricultural practices (Salt et al. 1998). The process not only reduces the bioavailability of harmful substances in the environment but also offers the potential for recovering valuable metals through phytomining (Ghosh and Singh 2005; Aransiola et al. 2019).

Phytoextraction is based on the ability of plants to absorb contaminants from the soil through their root systems and translocate them to the aerial parts of the plant, such as stems, leaves, and shoots. This process is influenced by several key mechanisms, including root uptake, translocation, and accumulation of contaminants in the plant tissues (Rascio and Navari-Izzo 2011; Aransiola et al. 2024). The uptake of contaminants by plant roots is the first step in phytoextraction. This process occurs through active or passive transport mechanisms, depending on the nature of the contaminant and the plant species. Active uptake involves the use of energy to transport ions across cell membranes, while passive uptake relies on diffusion and osmosis (Verbruggen et al. 2009). The efficiency of root uptake is influenced by factors such as soil pH, contaminant concentration, and the presence of chelating agents, which can increase the solubility and mobility of contaminants in the soil (Evangelou et al. 2007).

After contaminants are absorbed by the roots, they are translocated to the above-ground parts of the plant via the plant's vascular system, primarily through the xylem (Kramer 2010). Translocation depends on the plant's ability to move contaminants efficiently from the roots to the shoots. The efficiency of this process is influenced by the plant's physiological characteristics, such as transpiration rate and the presence of specific transport proteins that facilitate the movement of contaminants (Verbruggen et al. 2009). Once contaminants reach the aerial parts of the plant, they are stored in vacuoles or bound to cell wall components in the leaves, stems, and shoots (Kramer 2010). The ability of a plant to accumulate high levels of contaminants without suffering toxicity is a critical factor in the success of phytoextraction. Hyperaccumulator plants, which have developed unique physiological and biochemical mechanisms to tolerate and store large amounts of contaminants, are often used in phytoextraction (Baker et al. 2020).

Several factors influence the efficiency of phytoextraction, including plant species, soil properties, and the characteristics of the contaminant. The selection of appropriate plant species is one of the most important factors in phytoextraction. Hyperaccumulator plants, which can tolerate and accumulate high levels of contaminants, are often preferred for this purpose. Examples of hyperaccumulators include *Brassica juncea* (Indian mustard), *Thlaspi caerulescens* (Alpine pennycress), and *Pteris vittata* (Chinese brake fern) (Baker et al. 2020). These plants have developed unique adaptations, such as enhanced root uptake, efficient translocation, and the ability to sequester contaminants in vacuoles, which make them ideal for phytoextraction (Verbruggen et al. 2009).

Soil properties, such as pH, texture, organic matter content, and nutrient levels, play a significant role in the phytoextraction process. Soil pH, for example, affects the solubility and mobility of contaminants, with acidic soils generally increasing the availability of heavy metals for plant uptake (Evangelou et al. 2007). Soil texture influences the ease with which roots can penetrate the soil and access contaminants, while organic matter content affects the binding and mobility of contaminants in the soil (Ali et al. 2013). The nature of the contaminant itself also affects the efficiency of phytoextraction. Heavy metals, for instance, differ in their bioavailability and mobility in the soil, which influences their uptake by plants. Metals such as cadmium (Cd) and zinc (Zn) are more readily absorbed by plants compared to lead (Pb) and chromium (Cr), which tend to be more strongly bound to soil particles (Rascio and Navari-Izzo 2011). The chemical form of the contaminant, as well as its concentration in the soil, also plays a crucial role in determining its phytoextraction potential (Verbruggen et al. 2009).

To improve the efficiency of phytoextraction, soil amendments can be used to increase the bioavailability of contaminants and enhance their uptake by plants. Chelating agents, for example, are compounds that bind to metal ions, increasing their solubility and mobility in the soil (Evangelou et al. 2007). Common chelating agents used in phytoextraction include ethylenediaminetetraacetic acid (EDTA) and ethylenediaminedisuccinic acid (EDDS) (Ali et al. 2013). While chelating agents can significantly enhance metal uptake, their use must be carefully managed to prevent leaching and contamination of groundwater. Other soil amendments, such as organic matter and fertilizers, can also improve the effectiveness of phytoextraction. Organic matter can enhance soil structure, increase microbial activity, and improve nutrient availability, all of which can promote plant growth and contaminant uptake (Ghosh and Singh 2005). Fertilizers, particularly those containing nitrogen, phosphorus, and potassium, can support plant growth and increase biomass production, leading to higher contaminant removal (Pulford and Watson 2003).

Phytomining is a specialized form of phytoextraction that involves the use of plants to recover economically valuable metals from contaminated soils. This process is particularly relevant for metals such as nickel (Ni), cobalt (Co), and gold (Au), which can be accumulated in high concentrations by hyperaccumulator plants (Baker et al. 2020). After harvesting, the plant biomass is processed to extract the metals, which can then be sold for profit. Phytomining offers a unique opportunity to combine environmental remediation with resource recovery, making it an attractive option for contaminated sites where valuable metals are present (Ghosh and Singh 2005).

Phytoextraction offers several advantages over traditional remediation methods. One of the primary benefits is its cost-effectiveness. Compared to physical and chemical remediation techniques, which often require significant investment in equipment, labor, and disposal, phytoextraction is relatively inexpensive and can be implemented with minimal disruption to the environment (Ali et al. 2013). Another advantage of phytoextraction is its environmental sustainability. Unlike traditional methods that may involve the excavation and disposal of contaminated soil, phytoextraction works in situ, meaning that the contaminants are removed directly from the soil without disturbing the site. This reduces the risk of secondary contamination and helps preserve the integrity of the ecosystem (Salt et al. 1998).

Phytoextraction also has the potential to improve soil health and promote biodiversity. By increasing organic matter content, enhancing soil structure, and providing habitat for wildlife, phytoextraction can contribute to the restoration of degraded ecosystems (Pulford and Watson 2003). Additionally, the plants used in phytoextraction can provide economic benefits, such as the production of biomass for bioenergy or the recovery of valuable metals through phytomining (Ghosh and Singh 2005).

Despite its potential, phytoextraction has several limitations that can restrict its applicability. One of the main challenges is the time required for effective remediation. Phytoextraction is generally a slow process, and it may take several years or even decades to achieve significant reductions in contaminant levels, especially in highly contaminated sites (Ali et al. 2013). This extended time frame can be a drawback for sites that require immediate remediation. Another limitation is the depth of contamination. Most plants have shallow root systems, which limits the effectiveness of phytoextraction to the upper layers of soil. Contaminants located at greater depths may not be accessible to the plants, reducing the overall effectiveness of the remediation process (Kramer 2010). The potential for contaminant leaching and volatilization is another concern. In some cases, contaminants may be released back into the environment during the phytoextraction process, leading to the redistribution of pollutants rather than their removal. Careful site management and the use of appropriate soil amendments can help mitigate these risks, but they cannot be entirely eliminated (Evangelou et al. 2007). Additionally, the disposal of contaminated plant biomass presents a challenge. After harvesting, the plant material must be safely disposed of or processed to recover the contaminants. This can be a complex and costly process, particularly for plants that have accumulated high levels of toxic metals (Rascio and Navari-Izzo 2011).

10.2.2 Phytodegradation

Phytodegradation is a specialized form of phytoremediation that involves the use of plants to degrade organic contaminants in the environment. Unlike other forms of phytoremediation, which focus on the removal or stabilization of contaminants, phytodegradation specifically targets the breakdown of organic pollutants through biological and chemical processes facilitated by plant systems (Ghosh and Singh 2005). This process leverages the natural metabolic activities of plants and their associated microorganisms to transform harmful substances into less toxic or non-toxic forms, ultimately leading to the remediation of contaminated sites (Moosavi and Seghatoleslami 2013).

Phytodegradation encompasses several key mechanisms through which plants and their associated microorganisms contribute to the degradation of organic contaminants. These mechanisms include the uptake and accumulation of contaminants, the enzymatic breakdown of pollutants, and the transformation of contaminants into less harmful products (Kumar et al. 2019). The initial step in phytodegradation involves the uptake of contaminants by plant roots from the soil or water. This uptake is often facilitated by plant root systems, which can absorb organic pollutants from the surrounding environment (Ghosh and Singh 2005). Once inside the plant, contaminants may be transported to different tissues and accumulate in various cellular compartments, where they can be further metabolized or transformed. One of the central mechanisms of phytodegradation is the enzymatic breakdown of contaminants. Plants produce a range of enzymes, such as peroxidases, laccases, and dioxygenases, that can degrade organic pollutants (Wei et al. 2021). These enzymes are involved in various biochemical reactions that break down complex organic molecules into simpler, less toxic forms. For example, peroxidases can catalyze the oxidation of aromatic hydrocarbons, while laccases can degrade lignin and other phenolic compounds (Kumar et al. 2019).

Plants also support microbial communities in their rhizospheres, which play a crucial role in the degradation of organic contaminants. These microorganisms, including bacteria and fungi, can further metabolize contaminants that are taken up by the plant or present in the soil (Miller and Jastrow 2013). The interactions between plants and microbes create a synergistic effect, enhancing the overall degradation process. After contaminants are taken up by plants and broken down by enzymes, they may undergo further transformation into less toxic forms. For instance, some contaminants can be converted into non-toxic or less toxic metabolites through processes such as hydroxylation, methylation, or conjugation (Wei et al. 2021). These transformed products can be stored in plant tissues or further degraded by soil microorganisms.

Phytodegradation is effective against a wide range of organic contaminants, including pesticides, solvents, petroleum hydrocarbons, and polycyclic aromatic hydrocarbons (PAHs). Understanding the types of contaminants that can be degraded by plants is essential for selecting appropriate species and optimizing remediation strategies (Moosavi and Seghatoleslami 2013). Phytodegradation has been successfully used to remediate soils contaminated with pesticides, such as atrazine, chlorpyrifos, and parathion. Plants can degrade these compounds through enzymatic processes or by promoting microbial activity in the rhizosphere (Wei et al. 2021). For example, Panicum maximum and Brassica juncea have been shown to effectively degrade atrazine in contaminated soils (Ghosh and Singh 2005). Organic solvents, such as trichloroethylene (TCE) and tetrachloroethylene (PCE), are common environmental contaminants resulting from industrial activities. Plants can degrade these solvents through processes such as phytotransformation and microbial degradation. Populus species, for instance, have demonstrated the ability to degrade TCE and PCE in contaminated groundwater (Miller and Jastrow 2013). Petroleum hydrocarbons, including benzene, toluene, ethylbenzene, and xylenes (BTEX), are major pollutants from oil spills and industrial discharges. Phytodegradation can effectively remediate petroleum-contaminated soils by promoting the microbial degradation of these hydrocarbons or by direct plant uptake and metabolism (Moosavi and Seghatoleslami 2013). Canna indica and Helianthus annuus are examples of plants used for this purpose (Ghosh and Singh 2005). Polycyclic Aromatic Hydrocarbons (PAHs) are a group of organic compounds with multiple aromatic rings that are resistant to degradation. Phytodegradation can target PAHs by enhancing microbial activity in the rhizosphere or by direct enzymatic breakdown within plant tissues (Kumar et al. 2019). Plants such as Phragmites australis and Salix species have shown promise in degrading PAHs in contaminated soils (Wei et al. 2021).

Several factors influence the efficiency of phytodegradation, including plant species, environmental conditions, and contaminant characteristics. Optimizing these factors is crucial for achieving successful remediation outcomes (Moosavi and Seghatoleslami 2013). The selection of appropriate plant species is critical for effective phytodegradation. Plants with high metabolic activity and the ability to produce specific enzymes required for contaminant degradation are preferred. Hyperaccumulator plants, which have evolved mechanisms to tolerate and degrade high levels of contaminants, are often used in phytoremediation (Ghosh and Singh 2005). Additionally, plant species with extensive root systems and high biomass production can enhance contaminant uptake and degradation. Environmental conditions, such as soil pH, temperature, and moisture, can significantly impact the efficiency of phytodegradation. Soil pH affects the availability and mobility of contaminants, with acidic or alkaline conditions potentially influencing the activity of plant enzymes and microbial communities (Miller and Jastrow 2013). Temperature and moisture levels also affect plant growth and metabolic activity, which can in turn impact the rate of contaminant degradation (Wei et al. 2021).

The chemical properties of contaminants, such as their solubility, persistence, and toxicity, play a role in their degradation. Contaminants that are highly soluble and easily taken up by plants are generally more amenable to phytodegradation. In contrast, contaminants with low solubility or high persistence may require additional strategies, such as soil amendments or microbial inoculation, to enhance their degradation (Kumar et al. 2019).

To improve the efficiency of phytodegradation, soil amendments and microbial inoculation can be employed. These strategies aim to enhance the bioavailability of contaminants, promote plant growth, and support the activity of degrading microorganisms (Moosavi and Seghatoleslami 2013). Soil amendments, such as organic matter, fertilizers, and chelating agents, can be used to improve phytodegradation outcomes. Organic matter can enhance soil structure, increase microbial activity, and improve nutrient availability, all of which contribute to better plant growth and contaminant degradation (Wei et al. 2021). Fertilizers can support plant growth and increase biomass production, leading to higher rates of contaminant uptake and degradation. Chelating agents can increase the solubility and mobility of contaminants, making them more accessible to plants (Ghosh and Singh 2005).

The introduction of specific microorganisms into the rhizosphere can enhance the degradation of organic contaminants. These microorganisms, including bacteria and fungi, can work in synergy with plants to break down contaminants through their metabolic activities (Miller and Jastrow 2013). Microbial inoculation can be particularly effective for contaminants that are resistant to plant-mediated degradation. For example, the addition of specific bacterial strains capable of degrading chlorinated solvents or PAHs can improve the overall remediation process (Kumar et al. 2019).

10.2.3 Phytostabilization

Phytostabilization is a method of phytoremediation designed to stabilize contaminants in soil and water, preventing their migration and reducing their bioavailability. Unlike other phytoremediation techniques that focus on the removal or degradation of pollutants, phytostabilization aims to immobilize contaminants, thereby mitigating their environmental and health risks (Salt et al. 1998). This process involves the use of plants to reduce the mobility and bioavailability of contaminants, thus preventing their spread into the ecosystem. Phytostabilization operates through several mechanisms that collectively reduce the movement and availability of contaminants.

Plants can uptake contaminants through their root systems. While not all contaminants are translocated to aerial parts, their immobilization in root tissues can reduce their availability to soil and water. This process is particularly relevant for heavy metals and metalloids (McGrath and Zhao 2003). Plants with extensive root systems can trap and retain contaminants in the rhizosphere, thus preventing their further migration. Plants release various organic compounds into the soil through their roots, known as root exudates. These exudates can form complexes with contaminants, particularly heavy metals, thus reducing their solubility and mobility. For example, organic acids and phosphates released by plants can bind with metal ions, effectively immobilizing them (Kumpiene et al. 2008). Phytostabilization can involve the formation of metal-organic complexes within plant tissues. Plants often produce chelating agents such as phytochelatins and metallothioneins that bind metal ions, thereby reducing their availability and toxicity. This mechanism is crucial for stabilizing heavy metals in contaminated soils (Pilon-Smits 2005).

Plant root systems can improve soil structure by increasing soil aggregation and reducing soil erosion. This physical stabilization helps prevent the dispersion of contaminants into surrounding areas. Additionally, the presence of vegetation can reduce surface runoff and erosion, which further limits the movement of contaminants (Pulford and Watson 2003). The rhizosphere is home to various microorganisms that can assist in the stabilization of contaminants. These microbes can interact with plant roots and contribute to the immobilization and transformation of contaminants. For instance, certain bacteria can precipitate metal ions as insoluble compounds, further reducing their mobility (Xu et al. 2021).

Phytostabilization is particularly effective for dealing with contaminants that are persistent and have low mobility, such as heavy metals and metalloids. Key contaminants addressed by phytostabilization includes heavy metals like lead (Pb), cadmium (Cd), zinc (Zn), and arsenic (As) are commonly targeted by phytostabilization efforts. These metals are often present in contaminated soils and can pose significant health risks. Plants used in phytostabilization can immobilize these metals through various mechanisms, including root uptake, complexation with root exudates, and soil amendments (McGrath and Zhao 2003). Metalloids such as arsenic and selenium are also of concern in phytostabilization. Arsenic, for example, can be stabilized in soil through the formation of insoluble arsenic compounds or by reducing its bioavailability through root exudates (Xu et al. 2021). Selenium can be stabilized by forming selenides or selenites in the soil, reducing its mobility and bioavailability. While phytostabilization is less commonly used for organic contaminants compared to metals, it can still play a role in immobilizing certain persistent organic pollutants. For instance, plants can stabilize some organic compounds by enhancing microbial degradation in the rhizosphere or by forming complexes with these compounds (Pilon-Smits 2005).

10.2.4 Phytovolatilization

Phytovolatilization is a specialized phytoremediation technique that involves the use of plants to remove contaminants from the soil or water and subsequently release them into the atmosphere in a less harmful or transformed form (Burken and Ma 2006). This process primarily targets volatile organic compounds (VOCs), such as certain solvents and halogenated compounds, which can be transported through the plant and emitted into the air via transpiration and volatilization processes (Cristaldi et al. 2017). By harnessing the natural metabolic capabilities of plants, phytovolatilization offers an innovative approach to managing and mitigating environmental contaminants.

Phytovolatilization operates through several key mechanisms, including the uptake of contaminants, their transport through the plant, and their release into the atmosphere. The initial stage of phytovolatilization involves the uptake of volatile contaminants by plant roots from contaminated soil or water. This process relies on the plant's ability to absorb contaminants through its root system. The uptake efficiency can be influenced by factors such as root density, plant species, and contaminant concentration (Burken and Ma 2006). Once contaminants are absorbed by the roots, they are transported through the plant's vascular system to the aerial parts. This transport occurs via the xylem and phloem, which move water and nutrients throughout the plant. The efficiency of this transport process depends on plant physiology, including root structure, stem vascularization, and leaf surface area (Cristaldi et al. 2017).

The final stage of phytovolatilization involves the release of contaminants into the atmosphere. This occurs primarily through transpiration, where water vapor and volatile compounds are expelled from the plant's leaves. Volatilization can also occur through direct evaporation from plant surfaces. The rate of emission is influenced by environmental factors such as temperature, humidity, and wind speed (Abdullah et al. 2020). Some contaminants may undergo biochemical transformation within the plant before being volatilized. Plants can metabolize volatile compounds into less harmful forms, which are then emitted. This detoxification process can involve various enzymatic reactions, such as hydroxylation or conjugation, that modify the chemical structure of contaminants (Cristaldi et al. 2017).

Phytovolatilization is particularly effective for certain types of volatile contaminants, including:

Halogenated Compounds: Halogenated VOCs, such as trichloroethylene (TCE), tetrachloroethylene (PCE), and carbon tetrachloride, are commonly targeted by phytovolatilization. These compounds are often used in industrial processes and can be persistent environmental pollutants. Plants such as Populus spp. and Salix spp. have shown the ability to volatilize these halogenated compounds effectively (Abdullah et al. 2020).

Solvents: Organic solvents like benzene, toluene, ethylbenzene, and xylenes (BTEX) are also amenable to phytovolatilization. These compounds are frequently found in petroleum products and industrial waste. Plants capable of volatilizing these solvents can help manage contamination in areas affected by oil spills and industrial activities (Burken and Ma 2006).

Pesticides: Certain pesticides, particularly those with volatile properties, can be targeted by phytovolatilization. For example, herbicides such as atrazine and glyphosate have been studied for their potential volatilization from plant tissues.

The ability of plants to volatilize these pesticides can aid in the management of agricultural residues (Cristaldi et al. 2017).

Aromatic Compounds: Aromatic compounds, such as toluene and naphthalene, are another class of VOCs that can be managed through phytovolatilization. These compounds are commonly found in industrial effluents and can pose environmental and health risks. Plants that can volatilize these aromatic compounds contribute to the reduction of pollution in affected areas (Abdullah et al. 2020).

10.2.5 Rhizofiltration

Rhizofiltration is a phytoremediation technology that utilizes plant roots to absorb, concentrate, and remove contaminants from aqueous environments, such as contaminated groundwater or wastewater (Laghlimi et al. 2015). This technique primarily targets heavy metals, metalloids, and organic pollutants present in water. By leveraging the natural processes of plant root systems, rhizofiltration offers a sustainable and cost-effective approach to managing and mitigating environmental pollution (Salt et al. 1998).

Rhizofiltration operates through several key mechanisms, including the uptake of contaminants, their accumulation in root tissues, and the subsequent removal of contaminants from the water. The initial stage of rhizofiltration involves the absorption of contaminants from the surrounding water by plant roots. This process relies on the plant's root system to interact with the aqueous environment and take up contaminants through mechanisms such as passive diffusion and active transport (Laghlimi et al. 2015). The effectiveness of uptake depends on factors such as root surface area, plant species, and contaminant concentration. Once contaminants are absorbed by the roots, they are concentrated within the root tissues. This accumulation can occur through various mechanisms, including binding to root cell walls, sequestration in vacuoles, or complexation with root exudates (Salt et al. 1998). The ability of plants to accumulate contaminants is influenced by the chemical properties of the contaminants, such as their solubility and reactivity. The final stage of rhizofiltration involves the removal of accumulated contaminants from the root system. This can occur through various methods, including periodic harvesting of plant biomass, treatment of root tissues, or disposal of contaminated plant material (Laghlimi et al. 2015). Effective management of plant biomass is essential to prevent the re-release of contaminants into the environment. In addition to physical accumulation, some plants can detoxify contaminants through biochemical processes. For example, plants may convert toxic substances into less harmful forms through enzymatic reactions or metabolic pathways. This detoxification can further enhance the effectiveness of rhizofiltration by reducing the toxicity of contaminants (Salt et al. 1998).

Rhizofiltration is particularly effective for addressing a range of contaminants, including heavy metals, metalloids, and organic pollutants.

Heavy Metals Heavy metals such as lead (Pb), cadmium (Cd), zinc (Zn), and copper (Cu) are commonly targeted by rhizofiltration. These metals can be toxic to plants, animals, and humans, making their removal from contaminated water a priority (Babaniyi et al. 2023). Plants such as Helianthus annuus (sunflower), Brassica juncea (Indian mustard), and Pteris vittata (Chinese brake fern) have shown effectiveness in removing heavy metals through rhizofiltration (Gavrilescu 2022).

Metalloids Metalloids like arsenic (As) and selenium (Se) are also amenable to rhizofiltration. These elements can pose significant health risks when present in high concentrations in water. Plants can take up and accumulate metalloids, thereby reducing their levels in contaminated water sources (Laghlimi et al. 2015).

Organic Pollutants Organic pollutants, including pesticides and solvents, can be targeted by rhizofiltration. Plants that can uptake and accumulate organic contaminants from water help in managing pollution from agricultural and industrial sources. While less common than heavy metals, the rhizofiltration of organic pollutants is an area of ongoing research (Salt et al. 1998).

10.2.6 Effectiveness of Phytoremediation in Wetland Management

Phytoremediation is an effective strategy for managing contaminants in wetlands due to the specific characteristics of wetland plants and their interactions with the contaminated environment. Wetlands are naturally equipped to handle various pollutants through biological, chemical, and physical processes. The effectiveness of phytoremediation in these ecosystems can be attributed to several factors, including plant adaptability, the natural filtering capacity of wetlands, and the integration of plant processes into the overall remediation strategy (Mitsch and Gosselink 2015).

10.2.6.1 Plant Adaptability

Wetland plants are uniquely adapted to their environments, which often include challenging conditions such as high moisture levels and variable nutrient availability. This adaptability extends to their ability to interact with contaminants in their surroundings. Plants such as Typha spp. (cattails) and Phragmites australis (common reed) have evolved mechanisms to cope with waterlogged soils and can effectively uptake and accumulate pollutants, including heavy metals and organic compounds (Veselá et al. 2021). Their tolerance to extreme conditions makes them suitable candidates for phytoremediation in wetland environments.

10.2.6.2 Natural Filtering Capacity

Wetlands act as natural filters, capturing sediments and pollutants from water before it flows to other ecosystems. This filtering capability is enhanced by the presence of wetland plants, which can further improve water quality through phytoremediation processes. For instance, Schoenoplectus californicus (California bulrush) can remove excess nutrients and heavy metals from water, thus contributing to the overall health of the wetland ecosystem (Ghavzan and Trivedy 2005). The integration of plant-based remediation with natural wetland functions results in a more comprehensive approach to managing contamination.

10.2.6.3 Integration of Plant Processes

Phytoremediation involves various plant processes that work synergistically to address contamination. These processes include the uptake, accumulation, and detoxification of contaminants. Wetland plants can absorb pollutants from the water and soil, concentrate them in their tissues, and either detoxify them or facilitate their removal from the environment. This integration of plant processes into wetland management enhances the overall effectiveness of phytoremediation strategies (Ansari et al. 2014).

10.3 Applications of Wetland Phytoremediation

10.3.1 Treatment of Industrial Wastewater

Industrial wastewater treatment is a critical concern in environmental management, given the potential for various pollutants to degrade ecosystems and harm human health. Traditional wastewater treatment methods, such as chemical precipitation, activated carbon adsorption, and ion exchange, can be expensive and generate secondary waste. In contrast, phytoremediation, particularly in the context of wetlands, offers a sustainable, cost-effective, and ecologically sound alternative for treating industrial wastewater. Wetland phytoremediation harnesses the natural abilities of plants to absorb, accumulate, and detoxify pollutants, making it a valuable tool for industries such as mining, agriculture, and textiles (Vymazal 2011).

10.3.1.1 Treatment of Industrial Wastewater in Mining

Mining activities often result in the production of large volumes of wastewater, contaminated with heavy metals, metalloids, and other toxic substances. These pollutants, if not properly managed, can lead to severe environmental degradation, including soil and water contamination, biodiversity loss, and adverse health impacts on local communities (Sheoran and Sheoran 2006). Wetland phytoremediation offers a promising solution for treating mining wastewater by utilizing the natural capabilities of wetland plants to remove or stabilize contaminants (Thompson et al. 2019).

Heavy metals such as arsenic (As), cadmium (Cd), lead (Pb), and mercury (Hg) are common pollutants in mining wastewater. Wetland plants, including Phragmites australis (common reed) and Typha latifolia (broadleaf cattail), have been extensively studied for their ability to uptake and accumulate these metals in their tissues, thereby reducing their concentration in the water (Vymazal 2010). For instance, Phragmites australis has been shown to remove up to 85% of lead and cadmium from contaminated water, making it an effective tool for mitigating heavy metal pollution in mining-affected wetlands (Yang and Shen 2020).

Acid mine drainage (AMD) is a major environmental concern associated with mining activities, particularly in coal and metal mining. AMD occurs when sulfide minerals in exposed rock surfaces oxidize upon contact with water and oxygen, producing sulfuric acid and dissolving heavy metals into the surrounding water. The acidic and metal-laden water can devastate aquatic ecosystems and contaminate drinking water sources (Fitamo and Leta 2010). Wetland phytoremediation, through the use of plants such as Juncus effusus (common rush) and Carex rostrata (beaked sedge), has been employed to neutralize acidity and remove metals from AMD (Sheoran and Sheoran 2006). These plants can raise the pH of the water and precipitate metals as less soluble compounds, reducing their mobility and toxicity.

10.3.1.2 Treatment of Agricultural Wastewater

Agricultural activities contribute significantly to water pollution through the runoff of fertilizers, pesticides, and animal waste into nearby water bodies. These pollutants can cause eutrophication, algal blooms, and contamination of drinking water sources. Wetland phytoremediation offers a natural and sustainable method for treating agricultural wastewater by using plants to absorb and transform pollutants before they reach sensitive aquatic ecosystems.

Excessive nitrogen (N) and phosphorus (P) from fertilizers are primary contributors to nutrient pollution in agricultural wastewater. Wetland plants such as Typha spp. (cattails) and Scirpus spp. (bulrushes) have demonstrated the ability to uptake and assimilate these nutrients into their biomass, thereby reducing nutrient levels in the water (Vymazal 2011). Studies have shown that constructed wetlands planted with these species can remove up to 70–90% of nitrogen and phosphorus from agricultural runoff, significantly improving water quality (Kadlec 2009).

Pesticides used in agriculture can contaminate water bodies through surface runoff, leading to harmful effects on aquatic life and human health. Wetland phytoremediation can degrade and detoxify these pesticides through processes such as phytodegradation and rhizodegradation. Plants like Spartina alterniflora (smooth cordgrass) and Phragmites australis have been found to enhance the microbial breakdown of pesticides in the rhizosphere, reducing their persistence in the environment (Dzantor 2007). The roots of these plants provide a habitat for microbial communities that can degrade pesticide molecules into less toxic forms, further mitigating their environmental impact.

Animal waste from livestock operations is another source of agricultural pollution, particularly through the release of pathogens and nutrients into water bodies. Constructed wetlands planted with species such as Juncus effusus and Scirpus validus (softstem bulrush) have been used to treat wastewater from animal farms by filtering out solids, absorbing nutrients, and reducing pathogen loads (Vymazal 2010). These wetlands can effectively reduce biochemical oxygen demand (BOD), total suspended solids (TSS), and fecal coliform bacteria in agricultural wastewater, contributing to the overall health of downstream ecosystems.

10.3.1.3 Treatment of Textile Industry Wastewater

The textile industry is one of the most water-intensive industries globally, generating large volumes of wastewater contaminated with dyes, chemicals, and heavy metals. The complex composition of textile wastewater, which includes both organic and inorganic pollutants, poses significant challenges for traditional treatment methods. Wetland phytoremediation offers an alternative approach to treating textile wastewater, utilizing the ability of plants to absorb, degrade, and stabilize contaminants.

Dyes are a major pollutant in textile wastewater, often resistant to degradation and toxic to aquatic life. Wetland plants, such as Phragmites australis and Typha latifolia, have been investigated for their ability to remove and degrade dyes from wastewater (Vymazal 2010). The rhizosphere of these plants supports microbial communities that can break down dye molecules, while the plants themselves can absorb and accumulate dyes in their tissues. Studies have shown that constructed wetlands can remove up to 90% of certain textile dyes, depending on the plant species and environmental conditions (Hussein and Scholz 2017).

Textile wastewater often contains heavy metals such as chromium (Cr), zinc (Zn), and copper (Cu), which are used in dyeing and finishing processes. These metals can be toxic to aquatic organisms and pose risks to human health if they enter drinking water supplies. Wetland plants like Scirpus validus and Juncus effusus have been shown to stabilize heavy metals in the soil and sediments of constructed wetlands, reducing their bioavailability and preventing their movement into water bodies (Vymazal 2011). This stabilization process helps mitigate the environmental impact of heavy metals from textile wastewater.

The textile industry uses a wide range of chemicals, including surfactants, solvents, and bleaching agents, which can contaminate wastewater. Wetland phytoremediation can degrade these chemicals through processes such as phytodegradation and rhizodegradation. Plants like Spartina alterniflora and Phragmites australis have been found to facilitate the breakdown of organic chemicals in the rhizosphere, reducing their toxicity and persistence in the environment (Dzantor 2007). The root systems of these plants provide a habitat for microorganisms that can degrade complex chemical compounds into simpler, less harmful substances.

10.3.2 Remediation of Agricultural Runoff

Agricultural runoff, rich in nutrients, pesticides, and herbicides, poses significant threats to water quality and ecosystem health. As agricultural activities expand, the challenges of managing the environmental impacts of runoff have become increasingly pressing. Traditional approaches to mitigate these effects, such as chemical treatments and structural barriers, often come with limitations, including high costs and the risk of secondary pollution. Wetland phytoremediation, which uses plants to absorb, degrade, or stabilize contaminants, presents an effective and sustainable alternative for managing agricultural runoff (Vymazal 2011).

10.3.2.1 Nutrient Removal in Agricultural Runoff

Agricultural runoff is a major source of nutrient pollution, particularly nitrogen (N) and phosphorus (P), which are introduced into the environment through the use of fertilizers. These nutrients, when present in excessive amounts, can lead to eutrophication, a process characterized by excessive plant and algal growth in water bodies, followed by oxygen depletion and the decline of aquatic life (Smith and Schindler 2009). Wetland phytoremediation offers a natural and effective means of removing these nutrients from runoff before they reach sensitive aquatic ecosystems.

Nitrates, a common form of nitrogen in agricultural runoff, pose serious environmental risks. High nitrate levels can lead to eutrophication and pose health risks to humans, such as methemoglobinemia or "blue baby syndrome," particularly in infants (Camargo and Alonso 2006). Wetland plants, such as Typha spp. (cattails) and Phragmites australis (common reed), play a critical role in nitrate removal through processes such as uptake, microbial denitrification, and sedimentation (Vymazal 2007).

Wetland plants absorb nitrates through their roots and incorporate them into their biomass, effectively reducing nitrate concentrations in the water. Additionally, the anaerobic conditions often found in wetlands promote microbial denitrification, a process in which nitrate is converted into nitrogen gas (N_2) and released into the atmosphere, thus permanently removing nitrogen from the water system (Reddy et al. 1999). Studies have shown that constructed wetlands can achieve nitrate removal efficiencies of up to 90%, depending on the plant species, wetland design, and environmental conditions (Vymazal 2011).

Phosphorus, primarily in the form of phosphates, is another key nutrient contributing to eutrophication in water bodies (Ogundele et al. 2023). Unlike nitrogen, phosphorus does not have a gaseous form that can be released into the atmosphere, making its removal more challenging. Wetland phytoremediation addresses this challenge through plant uptake, adsorption to soil particles, and incorporation into sediments (Kadlec 2009). Plants like Scirpus spp. (bulrushes) and Juncus effusus (soft rush) have been found to be effective in removing phosphates from agricultural runoff (Vymazal 2007). These plants absorb phosphates through their root systems and store them in their tissues. Additionally, phosphates can bind to soil particles and organic matter within the wetland, where they are less likely to be released back into the water column. Constructed wetlands have been shown to reduce phosphate concentrations by 60–80%, depending on the specific conditions and management practices (Kadlec 2009).

10.3.2.2 Pesticide and Herbicide Degradation in Agricultural Wetlands

Pesticides and herbicides used in agriculture can contaminate surface and groundwater through runoff, posing risks to aquatic ecosystems and human health (Babaniyi et al. 2024). These chemicals are designed to be toxic to specific organisms, but their persistence in the environment can lead to unintended consequences, including the disruption of non-target species and the contamination of drinking water sources (Arias-Estévez et al. 2008). Wetland phytoremediation offers a promising approach to mitigating these risks by degrading and transforming pesticides and herbicides into less harmful substances.

Phytodegradation refers to the breakdown of organic pollutants, such as pesticides, by plants and associated microbial communities. In wetland environments, this process is facilitated by the unique conditions present in the rhizosphere, the zone of soil surrounding plant roots. The rhizosphere hosts a diverse microbial community that can degrade pesticides into less toxic metabolites (Dzantor 2007). Plants such as Spartina alterniflora (smooth cordgrass) and Phragmites australis have been shown to enhance the degradation of pesticides in wetland systems. These plants exude root exudates that stimulate microbial activity, leading to increased breakdown of pesticide molecules (Gleba et al. 1999). For example, wetlands planted with Phragmites australis have demonstrated significant reductions in concentrations of organophosphate pesticides, such as malathion and chlorpyrifos, with removal efficiencies exceeding 80% in some cases (Dzantor 2007).

Rhizodegradation, a subset of phytodegradation, involves the breakdown of contaminants in the rhizosphere, driven primarily by microbial activity. Wetland plants create favorable conditions for rhizodegradation by releasing organic compounds, such as sugars and amino acids, which serve as energy sources for soil microbes (Amora-Lazcano et al. 2010). These microbes, in turn, break down pesticides and herbicides into less harmful compounds, reducing their persistence in the environment.

The effectiveness of rhizodegradation depends on several factors, including the type of pesticide, the microbial community present, and the specific plant species involved. Wetland plants with extensive root systems, such as Scirpus spp. and Typha spp., provide a larger surface area for microbial colonization, enhancing the degradation process (Dzantor 2007). Moreover, the anaerobic conditions often found in wetland sediments can promote the breakdown of certain pesticides that are resistant to degradation under aerobic conditions (Mitsch et al. 2015).

Herbicides, which are used to control unwanted plant species in agriculture, can also be effectively managed through wetland phytoremediation. Wetland plants can transform herbicides into less toxic forms through metabolic processes, reducing their environmental impact. For instance, studies have shown that plants like Eichhornia crassipes (water hyacinth) can uptake and metabolize herbicides such as atrazine, converting them into less harmful compounds that are either stored in plant tissues or volatilized into the atmosphere (Gleba et al. 1999).

Phytovolatilization, the process by which plants uptake contaminants and release them into the atmosphere as volatile compounds, is another important mechanism in the degradation of certain herbicides. While this process does not remove contaminants from the environment entirely, it can reduce their concentration in water and soil, thereby mitigating their impact on aquatic ecosystems (Amora-Lazcano et al. 2010).

10.3.3 Removal of Heavy Metals from Contaminated Water

Wetland ecosystems are recognized for their potential in treating contaminated water, particularly in the removal of heavy metals. Heavy metals, including lead (Pb), mercury (Hg), cadmium (Cd), chromium (Cr), and arsenic (As), pose significant environmental and public health risks due to their toxicity, persistence, and tendency to bioaccumulate in the food chain (Alloway 2012). Traditional methods for heavy metal remediation, such as chemical precipitation, ion exchange, and membrane filtration, can be costly and may generate secondary pollutants. Wetland phytoremediation offers a cost-effective and sustainable alternative, leveraging the natural processes of wetland plants to absorb, accumulate, and stabilize heavy metals in contaminated water and sediments (Vymazal 2010).

10.3.3.1 Role of Wetland Plants in Bioaccumulation of Heavy Metals

Wetland plants play a crucial role in the removal of heavy metals from contaminated water through processes such as bioaccumulation, phytostabilization, and phytovolatilization. Bioaccumulation refers to the uptake and accumulation of heavy metals in plant tissues, where they can be sequestered and rendered less bioavailable (Salt et al. 1998). The effectiveness of wetland plants in bioaccumulating heavy metals depends on factors such as plant species, metal type, environmental conditions, and the presence of associated microbial communities.

Lead (Pb) Bioaccumulation Lead is one of the most common heavy metals found in contaminated water bodies, primarily due to industrial activities, mining, and the use of leaded gasoline (Nriagu 1990). Wetland plants such as Typha latifolia (common cattail) and Phragmites australis (common reed) have demonstrated a high capacity for lead uptake and accumulation in their roots and shoots. These plants can bioaccumulate lead from contaminated sediments and water, thereby reducing its concentration in the environment. The mechanism of lead uptake involves the absorption of Pb^{2+} ions by plant roots, where they are often sequestered in the vacuoles or cell walls, minimizing their toxicity to the plant (Sharma and Dubey 2005). In some cases, lead can be translocated to the above-ground parts of the plant, where it can be harvested and removed from the site. Studies have shown that Typha latifolia can accumulate lead concentrations up to 1000 mg/kg in its roots, making it an effective species for phytoremediation of lead-contaminated sites.

Mercury (Hg) Bioaccumulation Mercury is a highly toxic heavy metal that poses significant environmental and health risks, particularly in its methylated form (methylmercury), which can bioaccumulate in the food chain (Boening 2000). Wetland plants, such as Juncus effusus (soft rush) and Scirpus spp. (bulrushes), have been studied for their ability to uptake and accumulate mercury from contaminated water and sediments (Weis and Weis 2004). The bioaccumulation of mercury in wetland plants occurs through root uptake, where mercury binds to thiol groups in proteins, forming stable complexes that are stored in the plant tissues (Patra and Sharma 2000). In some instances, mercury can be volatilized by the plant through phytovolatilization, where it is transformed into a less toxic form (elemental mercury) and released into the atmosphere (Meagher and Heaton 2005). However, the efficiency of phytovolatilization depends on the plant species and environmental conditions.Studies have demonstrated that Juncus effusus can accumulate mercury concentrations up to 100 mg/kg in its roots, highlighting its potential for mercury phytoremediation (Weis and Weis 2004). The presence of microbial communities in the rhizosphere also plays a significant role in mercury transformation and uptake, as certain bacteria can methylate or demethylate mercury, influencing its bioavailability and toxicity (Patra and Sharma 2000).

Cadmium (Cd) Bioaccumulation Cadmium is a highly toxic heavy metal commonly found in industrial effluents, mining runoff, and agricultural soils contaminated by phosphate fertilizers (Alloway 2012). Wetland plants such as Phragmites australis and Scirpus validus (soft-stem bulrush) have been shown to bioaccumulate cadmium in their tissues, offering a natural solution for the remediation of cadmium-contaminated water bodies (Liao ShaoWei and Chang WenLian 2004). The uptake of cadmium by wetland plants involves the absorption of Cd^{2+} ions by the roots, where they are often stored in vacuoles or bound to metallothioneins, a group of proteins that help to detoxify heavy metals within the plant (Clemens 2006). Cadmium can also be translocated to the shoots, where it can be harvested and removed from the site. Research has shown that Phragmites australis can accumulate cadmium concentrations up to 500 mg/kg in its roots, making it a suitable candidate for the phytoremediation of cadmium-contaminated wetlands (Liao ShaoWei and Chang WenLian 2004). The efficiency of cadmium bioaccumulation can be enhanced by the presence of mycorrhizal fungi in the rhizosphere, which can increase the surface area for metal uptake and improve plant tolerance to cadmium stress (Göhre and Paszkowski 2006).

Chromium (Cr) Bioaccumulation Chromium, particularly in its hexavalent form (Cr^{6+}), is a highly toxic and carcinogenic heavy metal commonly associated with industrial activities such as leather tanning, electroplating, and textile manufacturing (Alloway 2012). Wetland plants such as Typha angustifolia (narrowleaf cattail) and Phragmites australis have been studied for their ability to bioaccumulate and reduce chromium in contaminated water and sediments (Panda and Choudhury 2005). The mechanism of chromium uptake involves the reduction of Cr^{6+} to Cr^{3+} by plant roots or associated microbial communities, followed by the sequestration of Cr^{3+} in the plant tissues (Shanker et al. 2005). Cr^{3+} is less toxic and less mobile than Cr^{6+}, making its accumulation in plants a key strategy for reducing chromium toxicity in the environment. Studies have shown that Typha angustifolia can accumulate chromium concentrations up to 1200 mg/kg in its roots, demonstrating its potential for chromium phytoremediation (Panda and Choudhury 2005). The presence of certain bacteria in the rhizosphere, such as Pseudomonas spp., can enhance chromium reduction and uptake by producing enzymes that catalyze the reduction of Cr^{6+} to Cr^{3+} (Shanker et al. 2005).

Arsenic (As) Bioaccumulation Arsenic is a toxic metalloid commonly found in groundwater and soils contaminated by mining activities, industrial processes, and the use of arsenic-based pesticides (Smith et al. 1998; Adewumi and Ogundele 2024). Wetland plants such as Pteris vittata (Chinese brake fern) and Eleocharis acicularis (needle spikerush) have been identified as hyperaccumulators of arsenic, capable of accumulating high concentrations of arsenic in their tissues (Ma et al. 2001). The bioaccumulation of arsenic by wetland plants involves the uptake of arsenate (As^{5+}) and arsenite (As^{3+}) ions by the roots, followed by sequestration in the vacuoles or complexation with phytochelatins, which are small peptides that bind heavy metals (Ma et al. 2001). Arsenic can be translocated to the shoots, where it can be harvested and removed from the site. Research has shown that Pteris vittata can accumulate arsenic concentrations up to 20,000 mg/kg in its fronds, making it one of the most effective plants for arsenic phytoremediation (Ma et al. 2001). The presence of certain soil microbes, such as Bacillus spp., can enhance arsenic uptake and tolerance by promoting the reduction of arsenate to arsenite, which is more easily absorbed by plants (Smith et al. 1998).

10.4 Factors Affecting the Efficiency of Phytoremediation in Wetlands

Phytoremediation in wetlands, which utilizes plants to absorb, detoxify, or mitigate contaminants from water, is a promising technique for environmental remediation. However, the efficiency of this process is influenced by various factors, including the selection of plant species, environmental conditions, plant-microbe interactions, and seasonal and climatic variability. Understanding these factors is crucial for optimizing phytoremediation practices in wetland ecosystems.

10.4.1 Plant Species Selection

The selection of appropriate plant species is a critical factor in the efficiency of phytoremediation. Different plants exhibit varying abilities to uptake, accumulate, and degrade pollutants, depending on their physiological and biochemical properties (Salt et al. 1995). The efficiency of phytoremediation can be enhanced by choosing plants that are well-suited to the specific contaminants present in the wetland.

Hyperaccumulators Certain plants, known as hyperaccumulators, have the ability to absorb and concentrate high levels of heavy metals or organic pollutants in their tissues. For instance, Phragmites australis (common reed) and Typha latifolia (cattail) are widely used in wetlands for their ability to accumulate heavy metals such as lead, cadmium, and arsenic (Sas-Nowosielska et al. 2004). The use of hyperaccumulators in phytoremediation can significantly enhance the removal of contaminants from wetland ecosystems.

Root Architecture and Depth The root system of the plant also plays a vital role in phytoremediation. Plants with extensive and deep root systems, such as Scirpus spp. (bulrushes), can access a larger volume of contaminated water and soil, thereby increasing the efficiency of contaminant uptake (Pulford and Watson 2003). Additionally, plants with fibrous root systems can provide a greater surface area for the adsorption and absorption of pollutants.

Tolerance to Contaminants The tolerance of plants to contaminants is another important consideration. Some plants may suffer from phytotoxicity when exposed to high levels of pollutants, leading to reduced growth and efficiency in phytoremediation. Selecting plant species that are tolerant to the specific contaminants in the wetland is essential for maintaining the effectiveness of the remediation process (Cunningham and Ow 1996).

Plant Growth and Biomass Production The growth rate and biomass production of plants also affect the efficiency of phytoremediation. Fast-growing plants with high biomass production, such as Eichhornia crassipes (water hyacinth), can uptake and store larger quantities of contaminants in a shorter time frame (Reddy and DeBusk 1985). However, the disposal of contaminated plant biomass needs to be managed carefully to prevent secondary pollution.

10.4.2 Environmental Conditions

Environmental conditions, including water quality, soil characteristics, and hydrology, play a significant role in determining the success of phytoremediation in wetlands. These factors can influence the availability, mobility, and transformation of contaminants, as well as the growth and health of wetland plants.

Water Quality The quality of water in the wetland, including parameters such as pH, dissolved oxygen, and nutrient levels, affects the efficiency of phytoremediation. For instance, the availability of heavy metals for plant uptake can be influenced by the pH of the water, with certain metals becoming more soluble and bioavailable at lower pH levels (McBride 1995). Maintaining optimal water quality conditions is essential for maximizing the uptake of contaminants by wetland plants.

Soil Characteristics The physical and chemical properties of the wetland soil, including texture, organic matter content, and cation exchange capacity, can impact the efficiency of phytoremediation. Soils with high organic matter content can enhance the sorption of organic pollutants, thereby reducing their bioavailability for plant uptake (Jordahl et al. 1997). Additionally, soils with good drainage and aeration can promote root growth and microbial activity, further enhancing the phytoremediation process.

Hydrology The hydrology of the wetland, including water depth, flow rate, and water retention time, influences the contact time between contaminants and plants, as well as the overall efficiency of phytoremediation (Kadlec and Wallace 2009). Wetlands with slow-moving or stagnant water allow for greater interaction between plant roots and contaminants, thereby increasing the potential for uptake and degradation. Conversely, fast-flowing water can reduce contact time and limit the effectiveness of phytoremediation.

Salinity In coastal wetlands or areas affected by saline intrusion, salinity levels can influence the efficiency of phytoremediation. High salinity can reduce plant growth and inhibit the uptake of contaminants, particularly in salt-sensitive species (Munns and Tester 2008). Selecting salt-tolerant plants, such as Spartina alterniflora (smooth cordgrass), can help mitigate the negative effects of salinity on phytoremediation.

10.4.3 Plant-Microbe Interactions

The interactions between wetland plants and microorganisms, particularly those in the rhizosphere, are crucial for the success of phytoremediation. These interactions can enhance the breakdown and transformation of contaminants, as well as improve the overall health and growth of the plants involved in the remediation process.

Rhizosphere Microbial Communities The rhizosphere is the zone of soil directly influenced by plant roots, where microbial activity is often higher than in the surrounding soil. Microorganisms in the rhizosphere, including bacteria, fungi, and archaea, play a key role in the degradation of organic pollutants and the transformation of inorganic contaminants (Johnson et al. 2005). Wetland plants release root exudates, such as sugars, amino acids, and organic acids, which can stimulate the growth and activity of these microorganisms (Glick 2012).

Biodegradation of Organic Pollutants Rhizosphere microorganisms can enhance the breakdown of organic pollutants, such as hydrocarbons, pesticides, and pharmaceuticals, through processes such as co-metabolism and enzymatic degradation (Siciliano et al. 2003). For instance, certain rhizosphere bacteria possess enzymes such as dehalogenases and oxygenases that can degrade complex organic molecules into simpler, less toxic compounds. The presence of these microorganisms can significantly increase the efficiency of phytoremediation in wetlands.

Transformation of Inorganic Contaminants Microbial communities in the rhizosphere can also influence the transformation and mobility of inorganic contaminants, such as heavy metals and metalloids. For example, certain bacteria can reduce toxic metal ions to less soluble and less bioavailable forms, thereby reducing their mobility and potential for uptake by plants (Gadd 2004). The interactions between plants and these beneficial microorganisms are essential for enhancing the efficiency of phytoremediation.

Mycorrhizal Associations Mycorrhizal fungi form symbiotic associations with the roots of many wetland plants, enhancing nutrient uptake and improving plant tolerance to contaminants (Smith and Read 2010). These fungi can increase the surface area of plant roots, allowing for greater absorption of water and nutrients, as well as the uptake of certain pollutants. Mycorrhizal associations can also protect plants from the toxic effects of heavy metals by sequestering these metals in fungal structures, such as vesicles and spores (Leyval et al. 2002).

10.4.4 Seasonal and Climatic Variability

Seasonal and climatic variability can have a profound impact on the efficiency of phytoremediation in wetlands. Factors such as temperature, precipitation, and photoperiod can influence plant growth, microbial activity, and the availability of contaminants, leading to fluctuations in the effectiveness of the remediation process.

Temperature Temperature is a key factor influencing the growth and metabolism of wetland plants and microorganisms. In temperate regions, phytoremediation efficiency may be reduced during the colder months due to lower temperatures, which can slow down plant growth and microbial activity (Truu et al. 2017). Conversely, higher temperatures during the growing season can enhance the uptake and degradation of contaminants, leading to improved phytoremediation outcomes.

Photoperiod The length of daylight, or photoperiod, can also affect the growth and physiological processes of wetland plants. Longer photoperiods during the summer months can promote greater photosynthetic activity and biomass production, thereby increasing the capacity of plants to uptake and store contaminants (Migliaccio et al. 2010). However, shorter photoperiods during the winter can lead to reduced plant growth and a corresponding decrease in phytoremediation efficiency.

Precipitation and Hydrology Seasonal variations in precipitation can influence the hydrology of wetlands, affecting water levels, flow rates, and the availability of contaminants. During periods of high precipitation, increased water flow can lead to the dilution and dispersion of contaminants, potentially reducing their bioavailability for plant uptake (Kadlec and Wallace 2009). Conversely, low water levels during dry periods can concentrate contaminants, increasing the potential for phytoremediation but also posing a risk of phytotoxicity to wetland plants.

Seasonal Growth Patterns Many wetland plants exhibit seasonal growth patterns, with peak growth occurring during the spring and summer months. This seasonal variation in plant biomass can influence the efficiency of phytoremediation, with higher contaminant uptake and degradation typically observed during the growing season (Vymazal 2011). Managing wetlands to align phytoremediation efforts with these seasonal growth patterns can enhance the overall effectiveness of the process.

Climatic Events Extreme climatic events, such as floods, droughts, and storms, can disrupt phytoremediation processes in wetlands. For example, flooding can lead to the rapid transport of contaminants out of the wetland, reducing the opportunity for plant uptake and microbial degradation (Gill et al. 2007). Drought conditions, on the other hand, can stress wetland plants, reducing their growth and contaminant uptake capacity. Preparing for and mitigating the impacts of these climatic events is essential for maintaining the efficiency of phytoremediation in wetlands.

10.5 Future Perspectives and Researches in Wetland Phytoremediation

Wetland phytoremediation has emerged as a sustainable and effective approach to mitigate environmental contaminants. As the world faces increasing environmental challenges, the potential of phytoremediation in wetland ecosystems continues to grow. Future perspectives and researches in this field lie in innovations in wetland phytoremediation, integrating phytoremediation with other treatment methods, and addressing policy and economic considerations. These aspects are crucial for advancing the application of phytoremediation and ensuring its long-term sustainability and effectiveness.

10.5.1 Innovations in Wetland Phytoremediation

The future of wetland phytoremediation will be shaped by a range of innovative approaches aimed at enhancing the efficiency and scope of this technology. Advances in biotechnology, genetic engineering, and nanotechnology offer promising opportunities to overcome current limitations and expand the application of phytoremediation.

Biotechnology and Genetic Engineering Genetic engineering holds significant potential for enhancing the phytoremediation capabilities of wetland plants. By introducing genes that encode for specific enzymes involved in contaminant degradation, plants can be engineered to metabolize a wider range of pollutants more effectively (Pilon-Smits 2005). For example, transgenic plants with enhanced tolerance to heavy metals or organic pollutants can be developed, enabling them to thrive in highly contaminated environments and remove contaminants more efficiently (Doty 2008). Additionally, advances in synthetic biology could allow for the design of custom-made plants tailored to target specific contaminants, thus improving the precision and effectiveness of phytoremediation efforts.

Nanotechnology Nanotechnology presents another innovative avenue for enhancing phytoremediation in wetlands. The use of nanoparticles in conjunction with phytoremediation can improve the uptake, transport, and degradation of contaminants by plants (Karn et al. 2009). For instance, engineered nanoparticles can be used to increase the bioavailability of hydrophobic pollutants, making them more accessible for plant uptake and degradation. Furthermore, nanoparticles can be functionalized to carry specific enzymes or chemical agents that facilitate the breakdown of complex contaminants within plant tissues (Mustafa et al. 2022). The integration of nanotechnology with phytoremediation offers a promising approach to address the limitations of traditional methods and enhance the overall efficiency of contaminant removal.

Phycoremediation The use of algae in wetland phytoremediation, known as phycoremediation, is an emerging field with great potential. Algae are capable of accumulating heavy metals, nutrients, and organic pollutants, making them valuable agents in the remediation of contaminated water (Olguín 2003; Ogundele et al. 2023). Algae can also be used in combination with higher plants in constructed wetlands to enhance the overall efficiency of phytoremediation. The development of algal-based systems for wetland remediation offers a novel approach to address water pollution, particularly in scenarios where traditional plant-based systems may be less effective.

Microbial-Assisted Phytoremediation The role of microbial communities in enhancing phytoremediation is increasingly recognized. The future of wetland phytoremediation may see greater emphasis on harnessing plant-microbe interactions to improve contaminant degradation. For instance, the use of rhizosphere bacteria and mycorrhizal fungi to stimulate plant growth and enhance contaminant uptake is an area of active research (Glick 2012). The development of microbial inoculants and biofertilizers specifically designed to support wetland plants in phytoremediation applications could lead to significant improvements in the effectiveness of this technology.

10.5.2 Integrating Phytoremediation with Other Treatment Methods

The integration of phytoremediation with other treatment methods is a promising strategy to enhance the overall effectiveness of wetland remediation. By combining phytoremediation with physical, chemical, and biological treatments, it is possible to achieve more comprehensive and efficient contaminant removal.

Phytoremediation and Constructed Wetlands Constructed wetlands are engineered systems designed to mimic the natural processes of wetland ecosystems. The integration of phytoremediation into constructed wetlands can enhance the removal of a wide range of contaminants, including nutrients, heavy metals, and organic pollutants (Kadlec 2009). By selecting appropriate plant species and optimizing the design of the wetland, it is possible to create a synergistic system where plants, microorganisms, and hydrological processes work together to achieve effective contaminant removal. This approach has been successfully applied in the treatment of municipal wastewater, industrial effluents, and agricultural runoff.

Phytoremediation and Bioaugmentation Bioaugmentation involves the introduction of specific microorganisms or microbial consortia into a contaminated environment to enhance the degradation of pollutants. When combined with phytoremediation, bioaugmentation can improve the breakdown of complex organic compounds and enhance the overall efficiency of the remediation process (Thompson et al. 2005). For instance, the introduction of hydrocarbon-degrading bacteria into a wetland system can accelerate the degradation of petroleum pollutants, while the presence of plants provides additional habitat and nutrients to support microbial activity.

Phytoremediation and Phytomining Phytomining is a process that uses plants to extract valuable metals from contaminated soils or water. By integrating phytomining with phytoremediation, it is possible to simultaneously remove contaminants and recover economically valuable metals, such as nickel, zinc, and gold (Chaney et al. 2005). This approach offers a sustainable solution for the remediation of metal-contaminated wetlands, while also providing an economic incentive for the recovery of valuable resources.

Phytoremediation and Bioreactors Bioreactors are engineered systems that use biological processes to treat contaminated water or soil. The integration of phytoremediation with bioreactors can enhance the removal of contaminants, particularly in scenarios where the natural attenuation capacity of wetland systems is insufficient (Rittmann and McCarty 2001). For example, the use of plant roots as a biofilm support in bioreactors can enhance the degradation of organic pollutants, while the plants themselves contribute to the removal of nutrients and heavy metals.

Phytoremediation and Chemical Amendments The use of chemical amendments, such as chelating agents or surfactants, can enhance the bioavailability of contaminants for plant uptake. When combined with phytoremediation, these amendments can improve the efficiency of contaminant removal, particularly for hydrophobic organic pollutants and heavy metals (Evangelou et al. 2007). However, the use of chemical amendments must be carefully managed to avoid potential environmental side effects and ensure the long-term sustainability of the remediation process (Akinyemi et al. 2023).

10.5.3 Policy and Economic Considerations

The widespread adoption of wetland phytoremediation as a viable environmental remediation strategy requires supportive policies and economic incentives. Addressing policy and economic considerations is essential for ensuring the long-term sustainability and effectiveness of phytoremediation efforts.

Regulatory Frameworks Effective regulatory frameworks are crucial for promoting the use of phytoremediation in wetland management. Governments and environmental agencies must establish clear guidelines and standards for the implementation of phytoremediation projects, including the selection of plant species, monitoring and evaluation protocols, and the management of contaminated plant biomass (Moosavi and Seghatoleslami 2013). These regulations should be based on sound scientific principles and take into account the specific characteristics of the wetland ecosystem and the contaminants present.

Incentives for Adoption Economic incentives can play a key role in encouraging the adoption of phytoremediation practices. For example, subsidies or tax credits for landowners and businesses that implement phytoremediation projects can help offset the costs of establishing and maintaining wetland systems (Sas-Nowosielska et al. 2004). Additionally, the development of markets for the sale of phytomined metals or the production of bioenergy from contaminated biomass can provide further economic motivation for the adoption of phytoremediation.

Cost-Effectiveness The cost-effectiveness of phytoremediation compared to traditional remediation methods is a significant consideration for policymakers and stakeholders. Phytoremediation is generally less expensive than conventional methods, such as excavation and landfilling, due to its reliance on natural processes and minimal energy inputs (Pilon-Smits 2005). However, the long-term sustainability of phytoremediation must also be considered, as the process may require extended time frames to achieve the desired levels of contaminant removal. Cost-benefit analyses can help determine the most appropriate remediation strategy for a given site, taking into account both economic and environmental factors.

Public Awareness and Education Public awareness and education are essential for gaining support for phytoremediation projects. Stakeholders, including local communities, landowners, and businesses, must be informed about the benefits and limitations of phytoremediation, as well as the specific requirements for successful implementation (Dietz and Schnoor 2001). Educational programs and outreach efforts can help build public trust and support for phytoremediation initiatives, particularly in areas where the technology is relatively unknown.

Environmental Justice Environmental justice considerations are important in the context of phytoremediation, particularly in communities disproportionately affected by environmental contamination. Policies and programs should ensure that all communities have access to the benefits of phytoremediation and are not disproportionately burdened by the risks associated with contaminated wetlands (Bullard 2018). This includes providing resources and support for the implementation of phytoremediation projects in underserved or marginalized communities.

10.6 Conclusion

In conclusion, phytoremediation in wetlands represents a promising and sustainable approach to managing environmental contamination. The ability of wetland plants to absorb, detoxify, and mitigate pollutants such as heavy metals, organic compounds, and excess nutrients highlights the critical role these ecosystems play in natural water purification. By leveraging the inherent capabilities of wetland plants, phytoremediation offers a cost-effective and environmentally friendly solution for addressing pollution in both natural and constructed wetland systems. The successful application of various phytoremediation techniques, including phytoextraction, phytodegradation, and rhizofiltration, underscores the versatility of this approach across different contaminant types and environmental conditions. As global environmental challenges continue to escalate, the integration of phytoremediation into wetland management practices becomes increasingly important. Future advancements in biotechnology, along with supportive policy frameworks, will further enhance the effectiveness of phytoremediation strategies. This chapter reinforces the potential of wetland phytoremediation not only as a tool for environmental cleanup but also as a critical component in the broader effort to restore and protect vital wetland ecosystems for future generations.

References

Abdullah SRS, Al-Baldawi IA, Almansoory AF, Purwanti IF, Al-Sbani NH, Sharuddin SSN (2020) Plant-assisted remediation of hydrocarbons in water and soil: application, mechanisms, challenges and opportunities. Chemosphere 247:125932

Adewumi AJ, Ogundele OD (2024) Hidden hazards in urban soils: a meta-analysis review of global heavy metal contamination (2010–2022), sources and its ecological and health consequences. Sustain Environ 10(1):2293239

Adewumi AJ, Ogundele OD, Emumejakpor IS (2023) Potentially toxic metals in Africa, fresh and marine environment: marine green contamination, ecological and human health risk. In: Marine greens. CRC Press, pp 213–225

Akinyemi BT, Ogundele OD, Afolabi AB (2023) Advancements in sustainable membrane technologies for enhanced remediation and wastewater treatment: a comprehensive review. Acadlore Trans Geosci 2(4):196–207

Ali H, Khan E, Sajad MA (2013) Phytoremediation of heavy metals—concepts and applications. Chemosphere 91(7):869–881

Alloway BJ (ed) (2012) Heavy metals in soils: trace metals and metalloids in soils and their bioavailability, vol 22. Springer Science & Business Media

Amon JP, Thompson CA, Carpenter QJ, Miner J (2002) Temperate zone fens of the glaciated Midwestern USA. Wetlands 22(2):301–317

Amora-Lazcano E, Guerrero-Zuniga LA, Rodriguez-Tovar A, Rodriguez-Dorantes A, Vasquez-Murrieta MS (2010) Rhizospheric plant-microbe interactions that enhance the remediation of contaminated soils. In: Current research, technology and education topics in applied microbiology and microbial biotechnology, vol 1. Formatex Research Center, pp 251–256

Ansari AA, Gill SS, Khan FA, Naeem M (2014) Phytoremediation systems for the recovery of nutrients from eutrophic waters. In: Eutrophication: causes, consequences and control, vol 2. Springer Netherlands, pp 239–248

Aransiola SA, Ijah UJJ, Abioye OP (2013) Phytoremediation of lead polluted soil by glycine max L. Appl Environ Soil Sci 2013:631619. https://doi.org/10.1155/2013/631619. https://www.hindawi.com/journals/aess/2013/631619/abs/

Aransiola SA, Ijah UJJ, Abioye OP, Bala JD (2019) Microbial-aided phytoremediation of heavy metals contaminated soil: a review. Eur J Biol Res 9(2):104–125. http://www.journals.tmkarpinski.com/index.php/ejbr/article/view/157

Aransiola SA, Ijah UJJ, Abioye OP, Bala JD, Rivadeneira-Mendoza RF, Luque R, Rodríguez-Díaz JM, Maddela NR (2024) Micro and vermicompost assisted remediation of heavy metal contaminated soils using phytoextractors. Case Stud Chem Environ Eng 9:100755. https://doi.org/10.1016/j.cscee.2024.100755

Arias-Estévez M, López-Periago E, Martínez-Carballo E, Simal-Gándara J, Mejuto JC, García-Río L (2008) The mobility and degradation of pesticides in soils and the pollution of groundwater resources. Agric Ecosyst Environ 123(4):247–260

Babaniyi BR, Ogundele OD, Bisi-Omotosho A, Babaniyi EE, Aransiola SA (2023) Remediation approaches in environmental sustainability. In: Microbiology for cleaner production and environmental sustainability. CRC Press, pp 321–346

Babaniyi BR, Ogundele OD, Abe TO, Olowoyeye BR, Jayeola JO, Oyegoke DA et al (2024) Bioenergy: the environmentalist's perspectives. In: Microbial biotechnology for bioenergy. Elsevier, pp 97–113

Baker AJ, McGrath SP, Reeves RD, Smith JAC (2020) Metal hyperaccumulator plants: a review of the ecology and physiology of a biological resource for phytoremediation of metal-polluted soils. In: Phytoremediation of contaminated soil and water. CRC Press, pp 85–107

Barbier EB, Acreman M, Knowler D (1997) Economic valuation of wetlands: a guide for policy makers and planners. Ramsar Convention Bureau, Gland

Boening DW (2000) Ecological effects, transport, and fate of mercury: a general review. Chemosphere 40(12):1335–1351

Bullard RD (2018) Dumping in Dixie: race, class, and environmental quality. Routledge

Bullock A, Acreman M (2003) The role of wetlands in the hydrological cycle. Hydrol Earth Syst Sci 7(3):358–389

Burken JG, Ma X (2006) Phytoremediation of volatile organic compounds. In: Phytoremediation rhizoremediation. Springer Netherlands, Dordrecht, pp 199–216

Camargo JA, Alonso Á (2006) Ecological and toxicological effects of inorganic nitrogen pollution in aquatic ecosystems: a global assessment. Environ Int 32(6):831–849

Chaney RL, Angle JS, McIntosh MS, Reeves RD, Li YM, Brewer EP et al (2005) Using hyperaccumulator plants to phytoextract soil Ni and Cd. Z Naturforsch C 60(3–4):190–198

Clemens S (2006) Toxic metal accumulation, responses to exposure and mechanisms of tolerance in plants. Biochimie 88(11):1707–1719

Cowardin LM (1979) Classification of wetlands and deepwater habitats of the United States. Fish and Wildlife Service, US Department of the Interior

Cristaldi A, Conti GO, Jho EH, Zuccarello P, Grasso A, Copat C, Ferrante M (2017) Phytoremediation of contaminated soils by heavy metals and PAHs. A brief review. Environ Technol Innov 8:309–326

Cunningham SD and Ow DW (1996) Promises and prospects of phytoremediation. Plant physiology 110(3):715

Davidson NC (2014) How much wetland has the world lost? Long-term and recent trends in global wetland area. Mar Freshw Res 65(10):934–941

Dietz AC, Schnoor JL (2001) Advances in phytoremediation. Environ Health Perspect 109(suppl 1):163–168

Doty SL (2008) Enhancing phytoremediation through the use of transgenics and endophytes. New Phytol 179(2):318–333

Dzantor EK (2007) Phytoremediation: the state of rhizosphere 'engineering'for accelerated rhizodegradation of xenobiotic contaminants. J Chem Technol Biotechnol 82(3):228–232

Erwin KL (2009) Wetlands and global climate change: the role of wetland restoration in a changing world. Wetl Ecol Manag 17(1):71–84

Evangelou MW, Ebel M, Schaeffer A (2007) Chelate assisted phytoextraction of heavy metals from soil. Effect, mechanism, toxicity, and fate of chelating agents. Chemosphere 68(6):989–1003

Fitamo D, Leta S (2010) Assessment of plants growing on gold mine wastes for their potential to remove heavy metals from contaminated soils. Int J Environ Stud 67(5):705–724

Gadd GM (2004) Mycotransformation of organic and inorganic substrates. Mycol 18(2):60–70

Gavrilescu M (2022) Enhancing phytoremediation of soils polluted with heavy metals. Curr Opin Biotechnol 74:21–31
Gell PA, Finlayson CM, Davidson NC (2023) An introduction to the Ramsar Convention on wetlands. In: Ramsar wetlands. Elsevier, pp 1–36
Gill RB, Mayewski PA, Nyberg J, Haug GH, Peterson LC (2007) Drought and the Maya collapse. Ancient Mesoamerica 18(2):283–302
Ghavzan NJ, Trivedy RK (2005) Environmental pollution control by using phytoremediation technology. Pollut Res 24(4):875
Ghosh M, Singh SP (2005) A review on phytoremediation of heavy metals and utilization of it's by products. Asian J Energy Environ 6(4):18
Gleba D, Borisjuk NV, Borisjuk LG, Kneer R, Poulev A, Skarzhinskaya M et al (1999) Use of plant roots for phytoremediation and molecular farming. Proc Natl Acad Sci 96(11):5973–5977
Glick BR (2012) Plant growth-promoting bacteria: mechanisms and applications. Scientifica 2012(1):963401
Göhre V, Paszkowski U (2006) Contribution of the arbuscular mycorrhizal symbiosis to heavy metal phytoremediation. Planta 223:1115–1122
Johnson DL, Anderson DR, McGrath SP (2005) Soil microbial response during the phytoremediation of a PAH contaminated soil. Soil Biol Biochem 37(12):2334–2336
Hussein A, Scholz M (2017) Dye wastewater treatment by vertical-flow constructed wetlands. Ecol Eng 101:28–38
Jordahl JL, Foster L, Schnoor JL, Alvarez PJ (1997) Effect of hybrid poplar trees on microbial populations important to hazardous waste bioremediation. Environ Toxicol Chem 16(6):1318–1321
Joosten H, Tapio-Biström ML, Tol S (2012) Peatlands: guidance for climate change mitigation through conservation, rehabilitation and sustainable use. Food and Agriculture Organization of the United Nations, Rome
Kadlec RH (2009) Comparison of free water and horizontal subsurface treatment wetlands. Ecol Eng 35(2):159–174
Kadlec RH, Wallace SD (2009) Treatment wetlands, 2nd edn. CRC Press
Karn B, Kuiken T, Otto M (2009) Nanotechnology and in situ remediation: a review of the benefits and potential risks. Environ Health Perspect 117(12):1813–1831
Keddy PA (2010) Wetland ecology: principles and conservation. Cambridge University Press
Keller C, Hammer D, Kayser A, Richner W, Brodbeck M, Sennhauser M (2003) Root development and heavy metal phytoextraction efficiency: comparison of different plant species in the field. Plant Soil 249:67–81
Kingsford RT (2000) Ecological impacts of dams, water diversions and river management on floodplain wetlands in Australia. Austral Ecol 25(2):109–127
Kingsford RT, Bino G, Porter JL (2017) Continental impacts of water development on waterbirds, contrasting two Australian river basins: global implications for sustainable water use. Glob Chang Biol 23(11):4958–4969
Kramer U (2010) Metal hyperaccumulation in plants. Annu Rev Plant Biol 61(1):517–534
Kumar N, Jeena N, Gangola S, Singh H (2019) Phytoremediation facilitating enzymes: an enzymatic approach for enhancing remediation process. In: Smart bioremediation technologies. Academic, pp 289–306
Kumpiene J, Lagerkvist A, Maurice C (2008) Stabilization of As, Cr, Cu, Pb and Zn in soil using amendments–a review. Waste Manag 28(1):215–225
Laghlimi M, Baghdad B, El Hadi H, Bouabdli A (2015) Phytoremediation mechanisms of heavy metal contaminated soils: a review. Open J Ecol 5(8):375–388
Leyval C, Joner EJ, Del Val C, Haselwandter K (2002) Potential of arbuscular mycorrhizal fungi for bioremediation. Mycorrhizal technology in agriculture: From genes to bioproducts, 175–186
Liao ShaoWei LS, Chang WenLian CW (2004) Heavy metal phytoremediation by water hyacinth at constructed wetlands in Taiwan. Photogramm Eng Remote Sens 54:177–185

Lin Q, Mendelssohn IA (1998) The combined effects of phytoremediation and biostimulation in enhancing habitat restoration and oil degradation of petroleum contaminated wetlands. Ecol Eng 10(3):263–274
Ma LQ, Komar KM, Tu C, Zhang W, Cai Y, Kennelley ED (2001) A fern that hyperaccumulates arsenic. Nature 409(6820):579–579
Macek T, Macková M, Káš J (2000) Exploitation of plants for the removal of organics in environmental remediation. Biotechnol Adv 18(1):23–34
McBride M B (1995) Toxic metal accumulation from agricultural use of sludge: are USEPA regulations protective?. J Environ Qual 24(1):5–18
McGrath SP, Zhao FJ (2003) Phytoextraction of metals and metalloids from contaminated soils. Curr Opin Biotechnol 14(3):277–282
Meagher RB and Heaton AC (2005) Strategies for the engineered phytoremediation of toxic element pollution: mercury and arsenic. J Ind Microbiol Biotechnol 32(11–12):502–513
Migliaccio KW, Schaffer B, Crane JH, Davies FS (2010) Plant response to evapotranspiration and soil water sensor irrigation scheduling methods for papaya production in south Florida. Agric Water Manag 97(10):1452–1460
Miller RM, Jastrow JD (2013) The role of plant roots and soil microorganisms in organic contaminant degradation. Biorem J 17(4):177–193
Mitsch WJ, Gosselink JG (2015) Wetlands. John Wiley & Sons
Mitsch WJ, Bernal B, Hernandez ME (2015) Ecosystem services of wetlands. Int J Biodivers Sci Ecosyst Serv Manag 11(1):1–4
Moosavi SG, Seghatoleslami MJ (2013) Phytoremediation: a review. Adv Agric Biol 1(1):5–11
Munns R and Tester M (2008) Mechanisms of salinity tolerance. Annu Rev Plant Biol 59(1):651–681
Mustafa K, Kanwal J, Farrukh S, Mussaddiq S, Saddiq N, Younas M (2022) Nano-phytoremediation technology in environmental remediation. In: Phytoremediation technology for the removal of heavy metals and other contaminants from soil and water. Elsevier, pp 433–459
Nriagu JO (1990) Global metal pollution: poisoning the biosphere? Environ Sci Policy Sustain Dev 32(7):7–33
Olguín EJ (2003) Phycoremediation: key issues for cost-effective nutrient removal processes. Biotechnol Adv 22(1–2):81–91
Ogundele OD, Anaun TE (2022) Phytoremediation: a green approach for pollution cleanup. In: Phytoremediation technology for the removal of heavy metals and other contaminants from soil and water. Elsevier, pp 49–74
Ogundele OD, Adewumi AJ, Oyegoke DA (2023) Phycoremediation: algae as an effective agent for sustainable remediation and waste water treatment. Environ Earth Sci Res J 10(1):7–17
Panda SK, Choudhury S (2005) Chromium stress in plants. Braz J Plant Physiol 17:95–102
Parish F, Sirin AA, Charman D, Joosten H, Minaeva TY, Silvius M (2008) Assessment on peatlands, biodiversity and climate change
Patra M, Sharma A (2000) Mercury toxicity in plants. Bot Rev 66:379–422
Pilon-Smits E (2005) Phytoremediation. Annu Rev Plant Biol 56(1):15–39
Pilu R, Bucci A, Badone FC, Landoni M (2012) Giant reed (Arundo donax L.): a weed plant or a promising energy crop. Afr J Biotechnol 11(38):9163–9174
Pulford ID, Watson C (2003) Phytoremediation of heavy metal-contaminated land by trees—a review. Environ Int 29(4):529–540
Rascio N, Navari-Izzo F (2011) Heavy metal hyperaccumulating plants: how and why do they do it? And what makes them so interesting? Plant Sci 180(2):169–181
Reddy KR and DeBusk WF (1985) Growth characteristics of aquatic macrophytes cultured in nutrient-enriched water: II. Azolla, Duckweed, and Salvinia. Economic Botany 39(2):200–208
Reddy KR, Kadlec RH, Flaig E, Gale PM (1999) Phosphorus retention in streams and wetlands: a review. Crit Rev Environ Sci Technol 29(1):83–146
Reinhold D, Vishwanathan S, Park JJ, Oh D, Saunders FM (2010) Assessment of plant-driven removal of emerging organic pollutants by duckweed. Chemosphere 80(7):687–692

Rittmann BE, McCarty PL (2001) Environmental biotechnology: principles and applications. McGraw-Hill Education

Rydin H, Jeglum JK, Bennett KD (2013) The biology of peatlands, 2e. OUP Oxford

Salt DE, Blaylock M, Kumar NP, Dushenkov V, Ensley BD, Chet I, Raskin I (1995) Phytoremediation: a novel strategy for the removal of toxic metals from the environment using plants. Bio/Technology 13(5):468–474

Salt DE, Smith RD, Raskin I (1998) Phytoremediation. Annu Rev Plant Biol 49(1):643–668

Sarwar N, Imran M, Shaheen MR, Ishaque W, Kamran MA, Matloob A et al (2017) Phytoremediation strategies for soils contaminated with heavy metals: modifications and future perspectives. Chemosphere 171:710–721

Sas-Nowosielska A, Kucharski R, Małkowski E, Pogrzeba M, Kuperberg JM, Kryński KJEP (2004) Phytoextraction crop disposal—an unsolved problem. Environ Pollut 128(3):373–379

Siciliano SD, Germida JJ, Banks K, Greer C W (2003) Changes in microbial community composition and function during a polyaromatic hydrocarbon phytoremediation field trial. Appl Environ Microbiol 69(1):483–489

Shanker AK, Cervantes C, Loza-Tavera H, Avudainayagam S (2005) Chromium toxicity in plants. Environ Int 31(5):739–753

Sharma P, Dubey RS (2005) Lead toxicity in plants. Braz J Plant Physiol 17:35–52

Sheoran AS, Sheoran V (2006) Heavy metal removal mechanism of acid mine drainage in wetlands: a critical review. Miner Eng 19(2):105–116

Smith VH, Schindler DW (2009) Eutrophication science: where do we go from here? Trends Ecol Evol 24(4):201–207

Smith SE and Read DJ (2010) Mycorrhizal symbiosis. Academic press

Smith ERG, Naidu R, Alston AM (1998) Arsenic in the soil environment: a review. Adv Agron 64(149195):60504-0

Tangahu BV, Sheikh Abdullah SR, Basri H, Idris M, Anuar N, Mukhlisin M (2011) A review on heavy metals (As, Pb, and Hg) uptake by plants through phytoremediation. Int J Chem Eng 2011(1):939161

Thompson IP, Van Der Gast CJ, Ciric L, Singer AC (2005) Bioaugmentation for bioremediation: the challenge of strain selection. Environ Microbiol 7(7):909–915

Thompson SO, Ogundele OD, Abata EO, Ajayi OM (2019) Heavy metals distribution and pollution indices of scrapyards soils. Int J Curr Res Appl Chem Chem Eng 3(1):9–19

Tiner RW (2002) Wetland indicators: a guide to wetland identification, delineation, classification, and mapping. Lewis Publishers, CRC Press, Boca Raton, pp 1–397

Truu M, Ostonen I, Preem JK, Lõhmus K, Nõlvak H, Ligi T, et al (2017) Elevated air humidity changes soil bacterial community structure in the silver birch stand. Front Microbiol 8:557

Verbruggen N, Hermans C, Schat H (2009) Molecular mechanisms of metal hyperaccumulation in plants. New Phytol 181(4):759–776

Verhoeven JT, Arheimer B, Yin C, Hefting MM (2006) Regional and global concerns over wetlands and water quality. Trends Ecol Evol 21(2):96–103

Veselá A, Hadincová V, Vandvik V, Münzbergová Z (2021) Maternal effects strengthen interactions of temperature and precipitation, determining seed germination of dominant alpine grass species. Am J Bot 108(5):798–810

Vymazal J (2007) Removal of nutrients in various types of constructed wetlands. Sci Total Environ 380(1–3):48–65

Vymazal J (2010) Constructed wetlands for wastewater treatment. Water 2(3):530–549

Vymazal J (2011) Plants used in constructed wetlands with horizontal subsurface flow: a review. Hydrobiologia 674(1):133–156

Wang QH, Que X (2013) Phytoremediation-a green approach to environmental clean-up. Chin J Eco-Agric 21(2):261–266

Wei Z, Van Le Q, Peng W, Yang Y, Yang H, Gu H et al (2021) A review on phytoremediation of contaminants in air, water and soil. J Hazard Mater 403:123658

Weis JS, Weis P (2004) Metal uptake, transport and release by wetland plants: implications for phytoremediation and restoration. Environ Int 30(5):685–700
Xu DM, Fu RB, Wang JX, Shi YX, Guo XP (2021) Chemical stabilization remediation for heavy metals in contaminated soils on the latest decade: available stabilizing materials and associated evaluation methods-a critical review. J Clean Prod 321:128730
Yang Y, Shen Q (2020) Phytoremediation of cadmium-contaminated wetland soil with Typha latifolia L. and the underlying mechanisms involved in the heavy-metal uptake and removal. Environ Sci Pollut Res 27(5):4905–4916
Zedler JB, Kercher S (2004) Causes and consequences of invasive plants in wetlands: opportunities, opportunists, and outcomes. Crit Rev Plant Sci 23(5):431–452

Chapter 11
Wetland-Based Solutions for Urban Water Management

Orimisan Emmanuel Abata, Kemi Catherine Adesorioye, Kehinde Oluwasiji Olorunfemi, and T. J. Popoola

Abstract Urbanization presents significant challenges for water management, including increased flooding, degraded water quality, and loss of natural habitats. Wetland-based solutions (WBS) offer a promising approach to address these challenges by integrating natural wetland processes into urban landscapes. This paper provides an overview of wetlands-based solution for urban water management, discussing their principles, applications, benefits, and challenges.

Wetlands based solution utilizes constructed wetlands, green infrastructure, and riparian zones to mimic the functions of natural wetlands in urban settings. These systems effectively capture, store, and treat storm water runoff, reducing peak flows, mitigating flooding, and improving water quality through natural filtration processes. Moreover, Wetlands-based solution provide habitat for wildlife, enhance biodiversity, and offer recreational opportunities, contributing to urban resilience and quality of life. As wetlands-based solution offers numerous benefits, their implementation faces challenges such as land availability, maintenance requirements, and regulatory constraints. Successful integration of wetlands-based solution into urban planning requires interdisciplinary collaboration, community engagement, and adaptive management approaches. Case studies from cities worldwide demonstrate the effectiveness of wetlands-based solution in mitigating urban water issues and enhancing urban environments. Wetland-based solution represent a sustainable and holistic approach to urban water management, aligning with principles of green infrastructure, ecosystem-based adaptation, and nature-based solutions. By harnessing the natural functions of wetlands, cities can build resilience to

O. E. Abata · K. C. Adesorioye · K. O. Olorunfemi (✉)
Bioresources Development Centre, National Biotechnology Development Agency (NABDA), Abuja, Nigeria
e-mail: olorunfemiko@futa.edu.ng

T. J. Popoola
Department of Environmental Management, School of Health and Life Sciences, Teesside University, Middlesbrough, UK

B. R. Babaniyi et al. (eds.), *Wetland Ecosystems: Conservation Strategies, Policy Management and Applications*, Wetlands: Ecology, Conservation and Management 12, https://doi.org/10.1007/978-3-031-91982-4_11

climate change, enhance water security, and create healthier and more livable urban spaces.

Keywords Hydrology · Retention ponds · Vegetation · Flood control

11.1 Introduction

In an era of rapid urbanization and climate change, cities worldwide are facing pressing challenges in managing water resources sustainably. Urban water management must address issues such as water pollution, flooding, and degradation of ecosystem to ensure the long-term health and resilience of cities and their inhabitants. In response to these issues, wetland-based solutions (WBS) have emerged as a promising approach for sustainable urban water management. Wetlands-based solution leverage the natural functions of wetlands to mitigate urban water issues while enhancing ecosystem services and biodiversity within urban landscapes.

According to recent studies, urban areas are experiencing growing pressures on water resources due to population growth, land development, and climate change (Mitsch and Gosselink 2021). Traditional water management infrastructure often struggles to cope with these pressures, leading to increased flood vulnerability, degraded water quality, and loss of natural habitats. In this context, wetlands-based solutionoffers an alternative paradigm for managing urban water that integrates ecological principles with engineering solutions.

Wetland according US EPA, wetlands are area where water covers the soil, or is present near the surface of the soil all year or for varying periods of time in the year. Wetlands serves as a bridge between the land and the water, they are transition areas where biogeochemical interactions occur. In wetlands, nutrient cycling, water and energy from the sun interacts to produce a distinct ecosystem with features of water, soil and vegetation. This distinct eco-system can be called a watershed. Utilizing watershed approach to wetland protection ensures that the whole environmental matrix including air, soil and water resources is adequately preserved (Tiner 2016).

Wetlands-based solution encompass a range of nature-based interventions, including constructed wetlands, green infrastructure, and riparian buffers, designed to mimic the functions of natural wetlands within urban settings (Davies and Day 2020). These solutions effectively capture, store, and treat stormwater runoff, reducing peak flows, mitigating flooding, and improving water quality through natural filtration processes. Furthermore, wetlands-based solution provide habitat for wildlife, enhance biodiversity, and offer recreational opportunities, contributing to urban resilience and quality of life.

Despite the growing recognition of wetlands-based solution potential, their widespread adoption in urban water management remains limited. Challenges such as land availability, regulatory constraints, and funding barriers hinder the implementation of wetlands-based solution projects in many urban areas (Fletcher et al. 2015). Moreover, public awareness and stakeholder engagement are crucial for overcoming barriers and promoting the adoption of WBS as integral components of urban planning and development strategies.

In light of these considerations, this paper aims to explore the principles, applications, benefits, and challenges of Wetland based solution for urban water management. By examining case studies and examples from around the world, we seek to demonstrate the effectiveness of wetlands-based solution in addressing urban water challenges while enhancing urban resilience and sustainability.

11.1.1 Wetland-Based Solutions for Urban Water Management

Wetland is an example of a green infrastructure, a viable technological approach for comprehensive challenges relating to water hazard management. Natural disaster, for instance flood, associated with improper and inadequate water management facilities in urban settlement is a big threat to the rate of urbanization and urban sustainability. Certain places, such as Bangladesh, Vietnam, Netherland, Egypt, Nigeria etc., around the world are being threatened with the risk of flooding. In the face of this threat, wetland forms a basis for urban water-related disaster control and largely provide efficient means for water management. Wetland-based solutions therefore involve utilizing natural or constructed wetlands to address various water-related challenges in urban environments. These solutions take advantage of the diverse and complex natural biochemical processes of wetlands to achieve contaminant's remediation, manage stormwater, enhance water quality, improve eco-system biodiversity, and provide recreational and aesthetic benefits to mention a few.

11.1.2 Types of Wetland-Based Solutions

11.1.2.1 Natural Wetlands

Protection and Restoration: Preserving existing natural wetlands and restoring degraded ones can provide immediate benefits for urban water management. Restoration activities may include replanting native vegetation, removing invasive species, and reestablishing natural hydrology.

11.1.2.2 Constructed Wetlands

Surface Flow Wetlands These mimic natural wetlands with water flowing over the surface through a dense stand of aquatic plants. They are effective for treating wastewater and stormwater runoff.

Subsurface Flow Wetlands Water flows horizontally or vertically through a gravel or soil medium planted with vegetation. These systems are particularly effective for removing pathogens and organic matter.

Retention Ponds Designed to hold stormwater temporarily and allow sedimentation of particulates. They often have a shallow wetland area around the perimeter to further treat water before it is discharged.

Bio-swales Vegetated, shallow, landscaped depressions that manage stormwater runoff. They slow the flow of water, allowing infiltration and removal of pollutants through plant uptake and soil filtration.

11.1.3 Implementation and Design Considerations

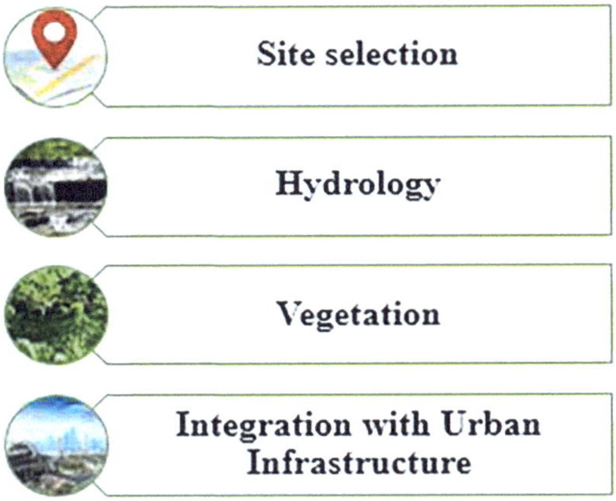

Site Selection The most important factor to consider in the implementation and design of wetlands is land use and access. The desired site should be accessible to urban runoff and/or wastewater to be treated. The site must also be accessible to construction equipment, operation personnel and chemical delivery vehicles. On the other hand, public access or unauthorized access must be prevented to protect the wetland. Therefore, choosing appropriate locations for wetland construction or restoration is important. Ideal sites include areas prone to flooding, low-lying areas, and places with existing water quality issues. Baseline information of water, fauna and flora should be performed on the selected site.

Hydrology Hydrology is the scientific study of water, its distribution, movement, and properties on Earth. It focuses on the water cycle, which includes processes such as precipitation, evaporation, infiltration, runoff, and groundwater flow.

Hydrologists study how water interacts with the environment, including rivers, lakes, aquifers, and the atmosphere. This field is crucial for understanding water resources, managing water supply, predicting floods, and addressing environmental challenges like droughts and water pollution. Hydrology plays a vital role in fields such as agriculture, urban planning, and environmental conservation (Wang and Lu 2007).

Hydrological parameters to consider includes surface water, groundwater, their flow direction, quantity present at different seasons of the year and their chemistry. A poor assessment of the hydrological characteristics of an area can jeopardize significantly the operation and general efficiency of the wetland. Ensuring the wetland has the right hydrological conditions (water depth, flow patterns, etc.) is essential for its functionality. Wetland should be sited away from major streams because of potential flooding, scouring, sedimentation, erosion, high ground water tables, saturated grounds, and variable water quantity and quality (Wang and Lu 2007).

Vegetation Selecting native wetland plants that are adapted to local conditions enhances the wetland's ability to filter pollutants and support wildlife. To select vegetation for the project, vegetation existing on or near a site and on similar nearby areas which have revegetated naturally are the best indicators of the plant species to use. If commercial wetland plant sources are not available, then on- or off-site harvesting can be considered. When nourishing plants, care must be given to local or federal laws prohibiting such plant acquisition and decimating the natural stands of wetland plants must be avoided (USDA, Natural Resources Conservation Service 2007). Additionally, care must be taken to assure that pest species, such as purple loosestrife (Lythrum salicaria), are not collected and transferred to tie project site. The availability of plants of the appropriate species, size, and quality is often a limiting factor in the final selection and plant procurement process. Some native plant species are very difficult to propagate and grow and many desirable species are not commonly available in commerce, or not available as good quality plants. As demand increases and nurserymen gain more experience in growing natives, this limitation should become less important.

Integration with Urban Infrastructure There is the necessity for new infrastructure that will meet the requirements for water treatment and sanitation and, at the same time, create green spaces in the urban environment, which could also partially compensate the lack of sufficient green areas. Also, wetlands should be integrated with other urban water management systems, such as green roofs, permeable pavements, and rain gardens, to create a holistic approach to managing storm water. Nature-based solutions and, specifically, the eco-friendly technology of constructed wetlands appears as an ideal option that can provide the desired added values to ecosystem services and promote water circularity in the urban context (Masi et al. 2018).

11.1.4 Principles of Wetland-Based Solutions for Urban Water Management

Mimic natural wetlands Constructed wetlands should be designed to replicate the structure and function of natural wetlands as closely as possible. This includes considering factors like water depth, substrate composition, and plant diversity to ensure the system supports the necessary physical, biological, and chemical processes for effective pollutant removal and habitat provision.

Integrate into the urban landscape Wetlands should be strategically integrated into the urban landscape to maximize their benefits. This may involve constructing wetlands in parks, along waterways, or as part of green infrastructure networks to ensure they effectively intercept and treat urban runoff, while also providing valuable green space for residents.

Multifunctionality WBS aim to provide multiple benefits beyond water management alone. In addition to stormwater retention and treatment, these solutions contribute to biodiversity conservation, habitat creation, carbon sequestration, and recreational opportunities within urban areas.

Adaptability and Resilience Wetland-based solution are designed to be adaptable and resilient to changing environmental conditions, including climate variability and urban development pressures. Flexible design approaches and adaptive management strategies ensure the long-term effectiveness and sustainability of WBS in dynamic urban environments.

Community Engagement and Stakeholder Participation Successful implementation of Wetland based solution requires active engagement and collaboration among various stakeholders, including local communities, government agencies, NGOs, and private sector actors. Community involvement in planning, design, and maintenance enhances ownership and promotes the long-term success of wetland-based solution projects.

Monitoring and Evaluation Regular monitoring and evaluation of wetland-based solution performance are essential to assess effectiveness, identify potential issues, and inform adaptive management decisions. Monitoring efforts may include water quality sampling, biodiversity surveys, hydrological measurements, and community feedback mechanisms.

Policy Support and Institutional Frameworks Wetland-based solution implementation relies on supportive policy frameworks, regulations, and institutional arrangements at the local, regional, and national levels. Policy incentives, funding mechanisms, and capacity-building initiatives facilitate the adoption and mainstreaming of wetland-based solution in urban planning and development processes.

11.1.5 Benefits of Wetland-Based Solutions for Urban Water Management

Natural Water Treatment Wetlands act as natural filters, removing pollutants, sediments, and nutrients from urban runoff and wastewater, thus improving water quality before it enters water bodies or groundwater reservoirs.

Flood Mitigation Wetlands can absorb and store excess water during heavy rainfall or floods, reducing the risk of urban flooding by providing natural floodplains and retention areas.

Biodiversity Conservation Wetlands support diverse ecosystems, providing habitats for various plant and animal species, including endangered or migratory species. By preserving wetlands in urban areas, biodiversity can be protected and enhanced.

Recreation and Aesthetics Well-designed wetland areas can serve as green spaces within urban landscapes, providing recreational opportunities for residents, such as hiking, birdwatching, and nature appreciation, while enhancing the aesthetic value of the environment.

Climate Resilience Wetlands contribute to climate resilience by sequestering carbon, mitigating the urban heat island effect, and providing buffer zones against extreme weather events, thus enhancing the overall resilience of urban areas to climate change impacts.

Economic benefits Wetland-based solutions can enhance property values and generate revenue through tourism and recreational activities.

11.1.6 Challenges of Wetland-Based Solutions for Urban Water Management

Design and operational challenges Ensuring optimal performance of constructed wetlands requires careful consideration of factors like plant selection, substrate composition, and water flow, as well as ongoing monitoring and maintenance.

Regulatory and policy issues Navigating complex regulatory frameworks and policies can pose challenges for implementing wetland-based solutions, particularly in jurisdictions with limited experience in nature-based approaches to water management.

Long-term sustainability Ensuring the long-term sustainability of wetland-based solutions requires careful planning, adaptation to changing conditions, and adequate resources for ongoing monitoring and maintenance.

Land Use Conflicts and cost Wetland restoration or creation projects often face challenges related to land availability and competing land uses, such as urban development, agriculture, or infrastructure expansion, which can hinder the implementation of wetland-based solution in urban areas. Constructing wetlands in urban areas particularly in densely populated cities can be expensive.

Hydrological Impacts Altering the hydrology of urban areas through wetland construction or restoration may have unintended consequences, such as changes in groundwater recharge rates, streamflow patterns, or stormwater drainage, requiring careful planning and hydrological modeling.

Maintenance and Management Wetlands require regular maintenance and management to ensure their effectiveness in water treatment, flood control, and habitat preservation. This involves ongoing monitoring, vegetation management, sediment removal, and invasive species control, which can be resource-intensive.

Community Engagement Engaging local communities and stakeholders in the planning, design, and implementation of wetland projects is crucial for their success. However, achieving consensus among diverse stakeholders with different interests, values, and priorities can be challenging and may require extensive outreach and participatory processes.

11.1.7 Application of Wetland Based-Solution

Stormwater management Wetland-based solutions are increasingly being integrated into urban stormwater management strategies to capture, store, and treat urban runoff. These systems help to reduce peak flow rates and mitigate flooding, while also improving water quality by filtering pollutants. For example, constructed wetlands in urban areas like Portland, Oregon, effectively manage stormwater while enhancing local biodiversity (Weiss et al. 2007). The Sustainable Urban Drainage Systems (SUDS) approach in the UK emphasizes the use of wetlands and other nature-based solutions for managing runoff and improving water quality (Woods-Ballard et al. 2007). The Augustenborg Eco-City project in Malmö, Sweden, is a successful example of implementing such strategies (Emilsson et al. 2002).

Wastewater Treatment Constructed wetlands are employed as natural wastewater treatment systems, providing a cost-effective and sustainable alternative to conventional treatment plants. They utilize natural processes involving vegetation, soil, and microbial communities to remove contaminants from wastewater. An example is the

East Kolkata Wetlands in India, which treat urban sewage while supporting agriculture and fisheries (Ghosh 2016). They have been widely used for treating municipal wastewater, with numerous studies demonstrating their effectiveness in removing pollutants such as nutrients, organic matter, and pathogens (Vymazal 2018). A well-known example is the Arcata Marsh and Wildlife Sanctuary in California, which has been treating wastewater while providing valuable habitat for wildlife since the 1980s (Gersberg et al. 1986).

Habitat Restoration Wetland restoration projects can revitalize degraded urban areas, creating habitats for wildlife and enhancing biodiversity. These restored wetlands also offer recreational opportunities and aesthetic benefits to urban residents. For instance, the restoration of the Cheonggyecheon Stream in Seoul, South Korea, transformed a polluted waterway into a vibrant urban park with restored wetland habitats (Lee and Anderson 2013).

Flood Control Wetlands act as natural sponges, absorbing and slowing down floodwaters, thereby reducing the impact of flooding on urban infrastructure. The use of wetlands for flood control has been demonstrated in the Netherlands, where the "Room for the River" program incorporates wetland restoration to manage high water levels and protect urban areas (Rijke et al. 2012). The Bishan-Ang Mo Kio Park in Singapore, for instance, has implemented a network of wetlands and bio-retention areas to enhance flood resilience while providing recreational spaces and wildlife habitat (PUB n.d.).

Climate Change Adaptation Wetlands contribute to climate resilience by buffering against extreme weather events, sequestering carbon, and regulating microclimates. Coastal wetlands, in particular, play a critical role in protecting urban areas from storm surges and sea-level rise. The restoration of coastal wetlands in New Orleans, Louisiana, is part of the city's strategy to enhance resilience against hurricanes and climate change (Day et al. 2007).

Water Quality Improvement Wetlands help improve water quality by removing nutrients, sediments, and other pollutants from urban runoff and wastewater. The use of wetlands for water quality improvement is exemplified by the use of the Prado Wetlands in Southern California, which treat agricultural and urban runoff, significantly reducing nitrate levels before the water enters the Santa Ana River (Horne 1995).

Urban river restoration Wetland-based solutions have been applied to restore degraded urban rivers and improve water quality. The River Skerne restoration project in Darlington, UK, used constructed wetlands to remove pollutants from urban runoff before entering the river, resulting in significant improvements in water quality (D'Arcy et al. 2006).

11.2 Integration of Wetlands into Urban Landscapes for Managing and Treating Stormwater Runoff

The successful integration of wetlands into urban landscapes requires careful design and planning.

1. **Planning and site selection**
   Identify suitable sites within the urban landscape for wetland integration.

   Factors to be considered:
   Hydrology: Evaluate the site's hydrological conditions, such as water table levels, soil moisture, and drainage patterns, to ensure they are suitable for wetland establishment and function.
   Topography: Assess the site's topography to determine the potential for water retention and flow control. Areas with gentle slopes and natural depressions may be more suitable for wetland systems.
   Soil characteristics: Analyze soil properties, such as texture, permeability, and organic matter content, to ensure they can support wetland vegetation and facilitate pollutant removal processes.
   Runoff sources: Identify potential sources of stormwater runoff, such as impervious surfaces, to ensure the wetland system can effectively capture and treat polluted runoff.
   Land availability and ownership: Consider land availability, ownership, and potential land use conflicts. Access to publicly owned land or green spaces may facilitate wetland establishment and management.
   Ecological context: Assess the site's proximity to existing wetlands, green spaces, and wildlife habitats. Integration with existing ecological networks can enhance biodiversity benefits and support habitat connectivity.
   Public access and visibility: Evaluate the potential for public access and visibility to promote educational and recreational opportunities and raise awareness about the benefits of wetland-based solutions.
   Local regulations and policies: Review local regulations, policies, and zoning requirements that may impact site selection and the implementation of wetland-based solutions.
   Climate change impacts: Consider potential climate change impacts on the site, such as changes in precipitation patterns or sea-level rise, to ensure the long-term resilience and effectiveness of wetland systems.
   Stakeholder input: Engage local stakeholders, such as residents, community groups, and municipal agencies, in the site selection process to gather input, address concerns, and build support for wetland-based solutions.

2. **Design and Engineering**:
   The design and engineering of wetland-based solutions for urban stormwater management require a multidisciplinary approach that considers hydrology,

ecology, and landscape architecture principles. (Vymazal 2018; Woods-Ballard et al. 2007).Factor tobe considered include:

Wetland Type: Decide on the type of wetland that best suits the urban area—constructed wetlands, retention ponds, bio-swales, or natural wetland restoration.
Hydraulic design: Determine the volume and flow rate of stormwater runoff to be treated and design the wetland system accordingly. Consider factors such as retention time, water depth, and flow patterns to ensure effective pollutant removal (Kadlec and Knight 1996).
Vegetation selection: Choose appropriate native wetland plant species to enhance pollutant removal, support biodiversity, and promote resilience to local environmental conditions. Consider the plants' growth habits, nutrient uptake capabilities, and tolerance to flooding and drought (Tanner 2001).
Substrate and media selection: Select appropriate substrates and media to support plant growth and facilitate pollutant removal processes, such as filtration and sorption. Materials like gravel, sand, and organic matter can be used to create an aerobic environment for efficient pollutant treatment (Vymazal 2011).
Water level control: Design and install water level control structures, such as weirs and culverts, to maintain the desired water depth within the wetland system and ensure optimal plant growth and pollutant removal (Woods-Ballard et al. 2007).
Erosion control: Incorporate erosion control measures, such as revegetation and bioengineering techniques, to prevent sedimentation and maintain the structural integrity of the wetland system (Stark et al. 2012).
Inlet and outlet structures: Design inlet and outlet structures to distribute stormwater runoff evenly throughout the wetland system and prevent short-circuiting, which can compromise pollutant removal efficiency (Kadlec and Wallace 2009).
Integration with existing infrastructure: Consider how the wetland system can be integrated with existing urban infrastructure, such as drainage networks, green spaces, and transportation systems, to maximize its benefits and minimize potential conflicts (Woods-Ballard et al. 2007).

3. **Construction and Implementation**

   Construct the wetland system in accordance with the design specifications, paying close attention to proper grading, planting, and installation of any necessary infrastructure.

   Factor to be considered include:

   Site preparation: Clear the site of any debris or vegetation that may hinder construction and implement erosion control measures to minimize sediment runoff during the construction process.

Excavation and grading: Excavate the site to create the desired depth and shape of the wetland system, and grade the site to promote even water distribution and prevent short-circuiting (Kadlec and Knight 1996).
Installation of water control structures: Install inlet and outlet structures, weirs, culverts, or other water level control devices to maintain the desired water depth within the wetland system (Woods-Ballard et al. 2007).
Installation of substrates and media: Add the appropriate substrates and media, such as gravel, sand, or organic matter, to create an aerobic environment for pollutant treatment and support plant growth (Vymazal 2011).
Planting wetland vegetation: Plant native wetland plant species that are well-suited to local conditions and have a high capacity for pollutant uptake and removal (Tanner 2001). Consider the plants' growth habits, nutrient uptake capabilities, and tolerance to flooding and drought.
Temporary irrigation: Provide temporary irrigation, if necessary, to support plant establishment and ensure the survival of wetland vegetation during the establishment phase.
Monitoring and maintenance: Monitor the construction process to ensure adherence to the design specifications and minimize environmental impact. Once construction is complete, perform regular maintenance tasks, such as sediment removal, vegetation management, and infrastructure repair, to ensure the long-term performance of the wetland system (Woods-Ballard et al. 2007).

4. **Environmental and Ecological Impact**
   The integration of wetlands into urban landscapes for managing and treating stormwater runoff can have significant environmental and ecological impacts:

Positive Impacts
Water quality improvement: Wetlands are effective in removing pollutants from stormwater runoff, such as nutrients, heavy metals, and pathogens, thus improving the overall water quality of urban water bodies (Vymazal 2011).
Flood control: Wetland systems can help mitigate flood risks by temporarily storing stormwater runoff and reducing peak flow rates (Dixon 2012).
Habitat provision: Urban wetlands provide habitat for a diverse array of plant and animal species, supporting local biodiversity and contributing to regional conservation efforts (Stark et al. 2012).
Carbon sequestration: Wetland vegetation can sequester carbon, thus contributing to climate change mitigation efforts (Mitsch and Gosselink 2015).
Aesthetic and recreational benefits: Incorporating wetlands into urban landscapes can enhance the aesthetic appeal of green spaces, provide opportunities for recreational activities, and promote overall human well-being (Woods-Ballard et al. 2007).
Potential Negative Impacts
Habitat alteration: The construction of wetland systems may alter existing habitats, potentially impacting local flora and fauna (Kadlec and Wallace 2009).

Mosquito breeding: Standing water in wetland systems may provide breeding habitats for mosquitoes, which can be a public health concern (Dale and Knight 2012).
Greenhouse gas emissions: Under certain conditions, wetlands may emit greenhouse gases like methane and nitrous oxide, which contribute to climate change (Moomaw et al. 2018).

5. **Integration with Urban Infrastructure**

   Integrating wetlands into urban landscapes for managing and treating stormwater runoff requires thoughtful planning and collaboration among various stakeholders to ensure seamless integration with existing urban infrastructure.
   Integration with stormwater drainage systems: Design wetland systems to effectively intercept stormwater runoff from impervious surfaces, such as roads, rooftops, and parking lots. Work with municipal engineers and planners to identify opportunities for connecting wetland systems to existing drainage networks, such as pipes, culverts, and swales (Woods-Ballard et al. 2007).
   Connection with green infrastructure networks: Incorporate wetlands into broader green infrastructure networks, including parks, greenways, and urban forests, to maximize ecological and recreational benefits. Collaborate with landscape architects and urban planners to identify strategic locations for wetland integration (Stark et al. 2012).
   Integration with transportation infrastructure: Consider opportunities for integrating wetlands into transportation corridors, such as roadside ditches, bioswales, and buffer zones around parking lots. Collaborate with transportation agencies to design and implement wetland systems that enhance stormwater management while minimizing potential conflicts with traffic safety and maintenance (Dixon 2012).
   Utilization of public spaces and vacant land: Identify underutilized public spaces, such as parks, schoolyards, and vacant lots that could be repurposed for wetland systems. Engage community members and local organizations in the planning and implementation process to build support and promote stewardship (Kadlec and Wallace 2009).
   Connection with water reuse systems: Explore opportunities for integrating wetlands with water reuse systems, such as rainwater harvesting, greywater reuse, and irrigation systems. This can help maximize water conservation and efficiency in urban environments (Vymazal 2011).

6. **Community and Stakeholder Engagement**
   Engaging local residents, community groups, and stakeholders can help build public support, promote stewardship, and ensure the long-term success of wetland-based solutions.

   Public participation: Involve local residents and community groups in the planning, design, and implementation phases of wetland projects. Conduct public meetings, workshops, and surveys to gather input, address concerns, and ensure the project meets local needs and priorities (Woods-Ballard et al. 2007).

Education and awareness campaigns: Develop educational materials and conduct outreach activities to raise awareness about the benefits of wetland-based solutions for stormwater management, water quality improvement, and biodiversity enhancement. Use various communication channels, such as social media, local events, and workshops, to reach diverse audiences (Kadlec and Wallace 2009).

Partnerships with local organizations: Collaborate with local environmental organizations, schools, and community groups to build support for wetland projects, share resources, and promote volunteer stewardship opportunities (Stark et al. 2012).

Incentives for participation: Offer incentives, such as grants, tax rebates, or technical assistance, to encourage homeowners, businesses, and developers to incorporate wetland systems into their properties (Dixon 2012).

Community stewardship programs: Establish community-based monitoring, maintenance, and restoration programs to involve local residents and stakeholders in the long-term stewardship of urban wetlands. Provide training and resources to support these efforts and build local capacity for wetland management (Vymazal 2011).

7. **Regulatory and Policy Compliance**

   Ensuring regulatory and policy compliance is an essential aspect of integrating wetlands into urban landscapes for managing and treating stormwater runoff. Local, state, and federal regulations and policies must be considered during the planning, design, and implementation phases of wetland-based solutions.

Permitting requirements: Obtain all necessary permits and approvals from relevant regulatory agencies, such as the U.S. Army Corps of Engineers, state environmental agencies, and local planning and zoning departments (Kadlec and Wallace 2009).

Stormwater regulations: Ensure that wetland systems comply with local and state stormwater management regulations, including requirements for stormwater detention, infiltration, and pollutant removal (Dixon 2012).

Wetland protection laws: Follow federal laws, such as the Clean Water Act and the Endangered Species Act, that protect wetlands and wetland habitats. This may involve conducting wetland delineations, assessing potential impacts, and implementing mitigation measures (Vymazal 2011).

Water quality standards

8. **Maintenance and Sustainability**

Management Plan

Vegetation management: Maintain appropriate vegetation cover in wetland systems by removing invasive species, planting native species, and managing plant growth to optimize pollutant removal and habitat quality (Tanner 2001).

Sediment and debris removal: Remove accumulated sediment and debris from wetland systems to prevent clogging and ensure continued hydraulic performance. This may involve periodic dredging, excavation, or the use of floating debris collectors (Vymazal 2011).

Monitoring and evaluation: Conduct regular monitoring and evaluation of wetland systems to assess their performance in managing stormwater runoff and treating pollutants. Use the data collected to inform adaptive management strategies and identify areas for improvement (Kadlec and Wallace 2009).

Structural maintenance: Inspect and maintain wetland infrastructure, such as inlet and outlet structures, weirs, and culverts, to ensure proper functioning and prevent erosion or structural failure (Woods-Ballard et al. 2007).

9. **Funding and Resources**

   Secure funding for both the initial construction and ongoing maintenance. Explore opportunities for public-private partnerships, grants, and community involvement in maintenance efforts.

Monitoring and Adaptation

By tracking the performance of these systems and making necessary adjustments over time, cities can ensure their continued effectiveness and adapt to changing conditions.

Performance indicators: Identify key performance indicators for wetland systems, such as pollutant removal efficiency, water quality improvements, and habitat quality. Establish baseline conditions and set target performance goals to guide monitoring efforts (Kadlec and Wallace 2009).

Data collection and analysis: Collect data on relevant hydrological, water quality, and ecological parameters through field monitoring, laboratory analyses, and remote sensing techniques. Analyze the data to assess system performance, identify trends, and detect potential issues (Vymazal 2011).

Adaptive management strategies: Develop and implement adaptive management strategies to address identified issues and improve system performance. This may involve adjusting vegetation management practices, modifying water control structures, or implementing additional pollutant removal technologies (Woods-Ballard et al. 2007).

Climate change adaptation: Consider the potential impacts of climate change on the performance of wetland systems, such as changes in precipitation patterns, temperature, and sea-level rise. Incorporate climate change projections into monitoring and adaptation strategies to ensure the long-term resilience of wetland-based solutions (Stark et al. 2012).

11.3 Wetland System for Urban Pollution Treatment

Wetland systems are increasingly being integrated into urban landscapes to treat pollution and manage stormwater runoff. These systems utilize the natural processes of wetlands to filter pollutants, improve water quality, and provide ecological benefits. Factors that need to be considered include:

Different Types of Wetland Systems for Urban Water Management (Table 11.1)

11.3.1 Design Principles and Considerations for a Wetland System for Treating Urban Pollution

Designing a wetland system for urban pollution treatment involves integrating ecological principles with engineering practices to create effective, sustainable, and resilient treatment systems.

1. Site Selection and Analysis

Table 11.1 This table highlights some key features, functions, and pollutant removal mechanisms of various wetland-based solutions that can be integrated into urban landscapes for effective water management and treatment

Wetland type	Main features	Primary functions	Pollutant removal mechanisms
Constructed wetlands	Engineered sequences of water bodies and vegetated areas, designed to mimic natural wetlands	Stormwater detention, filtration, and biological uptake	Sedimentation, adsorption, plant uptake, microbial degradation
Stormwater ponds	Shallow ponds with aquatic vegetation, often with an extended detention time	Stormwater storage and pollutant settling	Sedimentation, filtration, biological uptake
Floating treatment wetlands	Artificial platforms with wetland plants, floated in urban water bodies	Nutrient uptake, shadingand habitat provision	Plant uptake, nutrient sequestration
Biofiltration systems	Vegetated swales or trenches, often with an engineered filter media	Stormwater filtration and infiltration	Adsorption, filtration, microbial degradation
Riparian buffer zones	Natural or restored vegetated areas along waterways	Runoff interception and pollutant uptake	Sediment trapping, plant uptake, biological degradation
Urban green spaces	Multifunctional green infrastructure, including parks, green roofs, and rain gardens	Stormwater management, temperature regulation, and recreation	Evapotranspiration, infiltration, plant uptake

Hydrology: Understand the local hydrology, including water sources, flow patterns, and seasonal variations. The site should be capable of sustaining the wetland's hydrological requirements (Vymazal 2011).
Soil Characteristics: Assess soil types and permeability to ensure they are suitable for wetland construction. Soils should support plant growth and facilitate filtration (Kadlec and Wallace 2009).
Topography: Select low-lying areas that naturally collect water, facilitating gravity-fed flow and reducing the need for pumping (Mitsch and Gosselink 2015).

2. **Water Flow Management**

   Inflow and Outflow Control: Design structures to manage water inflow and outflow, ensuring consistent water levels and flow rates that maximize pollutant removal. Use weirs, gates, and other control structures as necessary (Kadlec and Wallace 2009).
   Retention Time: Ensure adequate water retention time to allow for effective sedimentation, filtration, and biological processes. This typically ranges from several days to a few weeks, depending on the pollutant load and treatment goals (Vymazal 2011).
   Flow Distribution: Use spreaders or diffusers to evenly distribute water across the wetland surface, preventing channeling and ensuring uniform treatment (Mitsch and Gosselink 2015).

3. **Wetland Configuration**

   Size and Shape: Design the wetland size based on the volume and pollutant load of incoming water. Larger wetlands generally offer more treatment capacity. Shape should promote even water flow and maximize edge habitat (Vymazal 2011).
   Depth: Vary depths within the wetland to create diverse habitats and enhance treatment processes. Shallow zones (0.5–1.5 feet) are effective for emergent vegetation, while deeper zones (up to 6 feet) support sedimentation and anaerobic processes (Kadlec and Wallace 2009).
   Multiple Cells: Use a series of wetland cells or compartments to allow staged treatment, improving efficiency and ease of maintenance (Mitsch and Gosselink 2015).

4. **Vegetation**

   Plant Selection: Choose native plant species that are adapted to local conditions and effective at pollutant uptake. Include a mix of emergent, submergent, and floating plants to enhance biodiversity and treatment capacity (Vymazal 2011).
   Planting Zones: Establish different planting zones based on water depth and plant requirements. Emergent plants should dominate shallow areas, while floating and submergent plants are suitable for deeper zones (Kadlec and Wallace 2009).

Plant Density: Ensure adequate plant density to provide effective pollutant uptake and habitat. Dense planting helps prevent erosion and stabilize sediments (Mitsch and Gosselink 2015).

5. **Pollutant Removal Mechanisms**

   Sedimentation: Design areas for sedimentation of suspended solids near the inflow points, such as sedimentation basins or forebays, to reduce sediment load on the main wetland (Vymazal 2011).
   Filtration: Use a substrate layer of gravel, sand, or soil to filter water as it flows through the wetland. The substrate also supports microbial communities that aid in pollutant degradation (Kadlec and Wallace 2009).
   Biological Uptake: Rely on wetland plants and microbial communities to uptake and transform pollutants, including nutrients (nitrogen and phosphorus), heavy metals, and organic matter (Mitsch and Gosselink 2015).
   Chemical Processes: Incorporate design elements that promote chemical processes such as adsorption and precipitation, which help remove pollutants from the water (Vymazal 2011).

6. **Habitat and Biodiversity**

   Habitat Diversity: Create a variety of habitats within the wetland, including open water, vegetated shallows, and upland buffers. This supports diverse wildlife and enhances ecological functions (Kadlec and Wallace 2009).
   Wildlife Corridors: Design the wetland to connect with other green spaces or natural habitats, promoting wildlife movement and biodiversity (Mitsch and Gosselink 2015).

7. **Aesthetics and Public Use**

   Visual Appeal: Design the wetland to be aesthetically pleasing, incorporating features such as walking paths, viewing platforms, and educational signage to encourage public interaction and awareness (Vymazal 2011).
   Recreational Space: Integrate recreational areas around the wetland, providing spaces for activities such as birdwatching, picnicking, and environmental education (Kadlec and Wallace 2009).

8. **Maintenance and Monitoring**

   Access for Maintenance: Design for easy access to key areas for maintenance activities, such as vegetation management and sediment removal (Mitsch and Gosselink 2015).
   Monitoring Systems: Implement monitoring systems to track water quality, vegetation health, and overall wetland performance. Use data to inform adaptive management practices (Vymazal 2011).
   Adaptive Management: Be prepared to adjust management strategies based on monitoring results and changing conditions to ensure long-term effectiveness and sustainability (Kadlec and Wallace 2009).

9. **Regulatory Compliance and Community Engagement**

   Permitting: Ensure compliance with local, state, and federal regulations regarding wetland construction, water quality standards, and land use (Mitsch and Gosselink 2015).
   Stakeholder Involvement: Engage local communities, stakeholders, and authorities in the planning and design process. This fosters support and ensures the wetland meets community needs and expectations (Vymazal 2011).

11.3.2 Mechanisms of Pollution Treatment

Sedimentation
Suspended solids settle out as water slows down in the wetland, reducing turbidity and removing pollutants attached to particles.

Filtration
The wetland substrate (gravel, soil, or sand) acts as a physical filter, trapping particles and associated pollutants.

Microbial Degradation
Wetlands provide a habitat for microbes that degrade organic pollutants, nutrients, and some types of heavy metals through biochemical processes.

Plant Uptake
Aquatic plants absorb nutrients (such as nitrogen and phosphorus) and heavy metals from the water, reducing their concentration.

Adsorption and Precipitation
Pollutants can adhere to soil particles (adsorption) or form solid precipitates that settle out.

11.4 Challenges of Wetland in Urban Water Treatment

While wetlands offer many benefits for urban water treatment, they also come with several challenges. These challenges need to be carefully managed to ensure that wetland systems function effectively and sustainably in urban environments.

1. **Space Constraints**

   Urban areas often have limited space available for large-scale wetland systems. Finding suitable sites that can accommodate the necessary wetland area while still fitting within the urban fabric can be difficult (Kadlec and Wallace 2009).

2. **Pollutant Load Variability**
   Urban runoff can carry highly variable pollutant loads, including heavy metals, oils, and organic contaminants. This variability can make it challenging to design wetlands that can consistently treat all types of pollutants effectively (Vymazal 2011).
3. **Hydrological Fluctuations**
   Urban wetlands must handle significant fluctuations in water flow, especially during storm events. Managing these fluctuations without causing erosion, sediment resuspension, or plant damage requires careful hydrological design (Mitsch and Gosselink 2015).
4. **Maintenance Requirements**
   Wetlands require ongoing maintenance to function effectively. This includes managing plant health, removing accumulated sediments, and controlling invasive species. Urban settings may complicate maintenance activities due to restricted access and potential interference from human activities (Kadlec and Wallace 2009).
5. **Mosquito Breeding**
   Standing water in wetlands can become a breeding ground for mosquitoes, which poses a public health concern in urban areas. Effective mosquito control strategies must be integrated into the wetland design and management plan.
6. **Public Perception and Acceptance**
   Urban residents may have concerns about the aesthetics, safety, and potential nuisances (such as odors or pests) associated with wetlands. Gaining public support and ensuring community acceptance can be a significant challenge (Mitsch and Gosselink 2015).
7. **Contaminant Accumulation**
   Over time, wetlands can accumulate contaminants in their sediments and vegetation. This can lead to long-term issues of pollutant buildup, which may require periodic remediation efforts to prevent negative environmental impacts (Vymazal 2011).
8. **Climate Change Impacts**
   Climate change can affect the performance of wetlands through changes in precipitation patterns, temperature fluctuations, and extreme weather events. These impacts can alter wetland hydrology and the effectiveness of pollutant removal processes (Kadlec and Wallace 2009).
9. **Regulatory and Policy Challenges**
   Compliance with local, state, and federal regulations can be complex and time-consuming. Ensuring that wetland projects meet all necessary regulatory requirements and obtaining the required permits can delay implementation (Mitsch and Gosselink 2015).
10. **Integration with Existing Infrastructure**
    Integrating wetlands into existing urban infrastructure, such as stormwater systems and green spaces, can be technically challenging. Ensuring that wet-

lands complement and enhance current systems rather than causing conflicts requires careful planning and design.

11. **Economic Constraints**

    The initial cost of designing and constructing wetlands, as well as the ongoing maintenance costs, can be significant. Securing funding and demonstrating cost-effectiveness compared to traditional infrastructure can be a hurdle (Kadlec and Wallace 2009).

11.5 Mitigation to the Challenges of Wetland System of Treating Urban Water

1. **Space Constraints**

   Vertical Flow Systems: Utilize vertical flow wetlands that require less horizontal space compared to surface flow wetlands (Kadlec and Wallace 2009). Integration with Green Spaces: Incorporate wetlands into existing parks, green roofs, and other multifunctional urban green spaces to maximize land use (Mitsch and Gosselink 2015).
   Modular Design: Design modular wetland units that can be distributed throughout the urban landscape in smaller, connected patches.

2. **Pollutant Load Variability**

   Pre-treatment: Implement pre-treatment systems such as sedimentation basins or vegetated swales to capture coarse sediments and reduce pollutant loads before water enters the wetland (Vymazal 2011).
   Adaptive Design: Design wetlands with flexible features that can adapt to varying pollutant loads, such as adjustable flow paths and multiple treatment cells (Kadlec and Wallace 2009).

3. **Hydrological Fluctuations**

   Stormwater Detention Features: Incorporate stormwater detention basins upstream of the wetland to manage peak flows and reduce the impact of heavy rainfall events (Mitsch and Gosselink 2015).
   Robust Inflow/Outflow Structures: Design robust inflow and outflow structures that can handle significant flow variations and prevent erosion (Vymazal 2011).

4. **Maintenance Requirements**

   Access Points: Design easy access points for maintenance vehicles and personnel to reach key areas for sediment removal and vegetation management (Kadlec and Wallace 2009).

Community Involvement: Engage local communities in maintenance activities through volunteer programs and stewardship initiatives, which can reduce costs and enhance local support (Mitsch and Gosselink 2015).
Regular Monitoring: Implement a regular monitoring program to identify and address maintenance issues early, ensuring long-term functionality (Vymazal 2011).

5. **Mosquito Breeding**

    Mosquito Control Measures: Introduce natural mosquito predators, such as fish and dragonflies, to the wetland environment. Use biological larvicides as necessary.
    Water Circulation: Design features to promote water circulation and avoid stagnant areas where mosquitoes can breed (Kadlec and Wallace 2009).

6. **Public Perception and Acceptance**

    Aesthetic Design: Incorporate aesthetically pleasing elements such as landscaped edges, walking paths, and educational signage to improve public perception (Mitsch and Gosselink 2015).
    Educational Programs: Conduct community outreach and educational programs to inform the public about the benefits and functions of urban wetlands (Vymazal 2011).

7. **Contaminant Accumulation**

    Periodic Sediment Removal: Schedule regular sediment removal to prevent the buildup of contaminants in the wetland (Kadlec and Wallace 2009).
    Plant Harvesting: Harvest and remove wetland plants periodically to prevent the accumulation of absorbed contaminants and promote new growth (Vymazal 2011).

8. **Climate Change Impacts**

    Resilient Plant Species: Select plant species that are resilient to climate variability, such as those tolerant of both drought and flooding conditions (Mitsch and Gosselink 2015).
    Adaptive Management: Develop adaptive management plans that can be adjusted based on monitoring data and projected climate scenarios (Kadlec and Wallace 2009).

11.6 Conclusion

The integration of wetlands into urban landscapes presents a compelling approach for managing and treating stormwater runoff. These natural systems not only filter pollutants and improve water quality, but also offer additional benefits like flood control, habitat creation, and aesthetic value. Constructed wetland systems specifically designed for urban environments can be tailored to handle the specific challenges of urban runoff, including high volumes and potential contaminant loads.

While wetlands offer a sustainable solution for urban water treatment, challenges remain. Careful planning and design are crucial for ensuring effective treatment capacity and managing potential issues like mosquito breeding. Additionally, integrating wetlands into existing urban infrastructure requires creative solutions to optimize space utilization.

Despite these challenges, the potential benefits of wetland integration are significant. By embracing this approach, cities can move towards a more sustainable water management paradigm, reducing reliance on traditional grey infrastructure and fostering a healthier urban environment. Further research and development focused on optimizing wetland design and function will be crucial for maximizing their effectiveness in treating urban stormwater runoff.

References

D'Arcy BJ, Riddler M, Steel J (2006) The use of wetlands for urban river restoration: case study River Skerne, Darlington, UK River. Res Appl 22(9):987–998

Dale PER, Knight JM (2012) Managing urban wetlands. In: The wetlands handbook. John Wiley & Sons, pp 871–885

Davies G, Day J (2020) Wetland design and engineering: principles and practices. Springer

Day JW, Boesch DF, Clairain EJ, Kemp GP, Laska SB, Mitsch WJ et al (2007) Restoration of the Mississippi Delta: lessons from hurricanes Katrina and Rita. Science 315(5819):1679–1684

Dixon A (2012) Wetland systems in flood control and water management. In: The wetlands handbook. John Wiley & Sons, pp 855–870

Emilsson T, Berndtsson JC, Mellander P-E (2002) The Augustenborg Eco-City project in Malmö, Sweden. In: Proceedings of the third international conference on urban regeneration and sustainability—sustainable urban areas, 17–19 June 2002

Fletcher TD, Shuster W, Hunt WF, Ashley R, Butler D, Arthur S et al (2015) SUDS, LID, BMPs, WSUD and more – the evolution and application of terminology surrounding urban drainage. Urban Water J 12(7):525–542

Gersberg RM, Elkins BV, Goldman CR (1986) Role of aquatic plants in wastewater treatment by artificial wetlands. Water Res 20(3):363–368

Ghosh D (2016) East Kolkata wetlands: a resource recovery system through productive activities. Ecol Eng 36(8):1024–1031

Horne AJ (1995) Nitrogen removal from wastewater treatment pond or activated sludge plant effluents with free-surface wetlands. Water Sci Technol 31(12):341–351

Kadlec RH, Knight RL (1996) Treatment wetlands. CRC Press

Kadlec RH, Wallace SD (2009) Treatment wetlands, 2nd edn. CRC Press, Taylor & Francis Group

Lee SW, Anderson WP (2013) The Cheonggyecheon restoration project: an ecological, cultural, and economic assessment of reclaiming lost space in Seoul, South Korea. J Urban Technol 20(1):15–30

Masi F, Rizzo A, Regelsberger M (2018) The role of constructed wetlands in a new circular economy, resource oriented, and ecosystem services paradigm. J Environ Manag 216:275–284

Mitsch WJ, Gosselink JG (2015) Wetlands. John Wiley & Sons

Mitsch WJ, Gosselink JG (2021) Urban wetlands: a growing imperative in planning and management. Springer

Moomaw WR, Xu Y, Myers C, Antibus R, Sullivan L, McIvor M et al (2018) Wetlands in a changing climate. In: The wetlands handbook. John Wiley & Sons, pp 951–972
PUB (n.d.) Bishan-Ang Mo Kio Park. Retrieved from https://www.pub.gov.sg/bamkp
Rijke J, van Herk S, Zevenbergen C, Ashley R (2012) Room for the river: delivering integrated river basin management in The Netherlands. Int J River Basin Manag 10(4):369–382
Stark TM, Apfelbaum SI, Smith CL, Mayes MA, Gifford MA (2012) Streamflow duration and watershed urbanization influence biogeochemical cycles and microbial community structure in urban stream sediments. Biogeochemistry 111(1–3):121–139
Tanner CC (2001) Vegetation and wetland establishment in stormwater management systems. Water Sci Technol 44(11–12):469–476
Tiner RW (2016) Wetland indicators: a guide to wetland formation, identification, delineation, classification, and mapping. CRC Press
USDA NRCS (2007) The PLANTS website. http://plants.usda.gov. Accessed 070117. National Plant Data Center, Baton Rouge
Vymazal J (2011) Constructed wetlands for wastewater treatment: five decades of experience. Environ Sci Technol 45(1):61–69
Vymazal J (2018) Constructed wetlands for wastewater treatment: a review of the current situation and recent developments. Water 10(8):1007
Wang JH, Lu XG (2007) Urban wetland: its concept, ecological services and protection. Chin J Ecol 26(4):555–560. (in Chinese)
Weiss PT, Gulliver JS, Erickson AJ (2007) Cost and pollutant removal of storm-water treatment practices. J Water Resour Plan Manag 133(3):218–229
Woods-Ballard B, Kellagher R, Martin P, Jefferies C, Bray R, Shaffer P (2007) The SUDS manual. Construction Industry Research and Information Association (CIRIA)

Chapter 12
Applications and Characteristics of Polysaccharide-Based Nanomaterials in Water and Wastewater Remediation

Sesan Abiodun Aransiola, Munachimso Odenakachi Victor-Ekwebelem, Femi Joseph, Mariana Andreina Valle Suárez, and Naga Raju Maddela

Abstract There is an increasing demand for the natural polymers, composites, and engineering nanomaterials (ENMs) in the water and wastewater treatment (WWT). This is mainly due to the fact that the remediation by above listed materials offer undisputable advantages, such as no generation of pollutants, doesn't require extra energy inputs, and no high side costs. In this direction, there is much attention on polysaccharides, which are hydrophilic, non-toxic, relatively cheap polymers and possess an inherent variety of functional groups that allow the better interactions of polysaccharides with pollutants. Up ahead, polysaccharides can be obtained easily in the environment by green processes from agricultural wastes, microbial cells etc. Especially, polysaccharide-based nanomaterials (PNMs) show remarkable potential for the removal of different environmental pollutants. There are different kidneys of polysaccharide materials available such as alginate, pectin, gum, cellulose, starch,

S. A. Aransiola (✉)
Department of Microbiology, Faculty of Science, University of Abuja, Abuja, Nigeria
e-mail: sesan.aransiola@uniabuja.edu.ng

M. O. Victor-Ekwebelem
Department of Microbiology, Alex Ekwueme, Federal University, Ndufu-Alike, Ikwo, Ebonyi State, Nigeria

F. Joseph
Department of Microbiology, Federal University of Technology, Minna, Nigeria

M. A. V. Suárez
Gobierno Autónomo Descentralizado Provincial de Pastaza, Puyo, Pastaza, Ecuador

N. R. Maddela
Departamento de Ciencias Biológicas, Facultad de Ciencias de la Salud, Universidad Técnica de Manabí, Portoviejo, Ecuador

Instituto de Investigación, Universidad Técnica de Manabí, Portoviejo, Ecuador

Grupo de Investigación en Biodiversidad y Ecología de Ecosistemas Acuáticos, Departamento de Acuicultura, Pesca y Recursos Naturales Renovables, Facultad de Ciencias Veterinarias, Universidad Técnica de Manabí, Bahía de Caráquez, Ecuador

B. R. Babaniyi et al. (eds.), *Wetland Ecosystems: Conservation Strategies, Policy Management and Applications*, Wetlands: Ecology, Conservation and Management 12, https://doi.org/10.1007/978-3-031-91982-4_12

chitin, and chitosan. However, cost effectiveness, eco-friendly nature, non-toxicity, biodegradability vary from each other. Therefore, the present chapter emphasize on the following issues for the better implications of polysaccharide-based nanomaterials in the remediation of water and wastewater—(i) Characteristics of different polysaccharide materials, (ii) Mechanism of interaction with the pollutants, (iii) Case studies, and (iv) Future perspectives of polysaccharide-based nanomaterials in the remediation. We are of the opinion that this could be a single source of information for the researchers, environmental engineers, academicians and those who are working in the field of sustainable environmental techniques for the restoration of contaminated media.

Keywords Polysaccharides · Nanomaterials · Remediation · Water and wastewater

12.1 Introduction

Polysaccharides, also called glycan, are formed from several molecules of monosaccharides (minimum of ten) held together by glycosidic bonds to yield a single large molecule. Glycosidic bond, which is an outcome of dehydration reaction, possess an oxygen molecule linking two carbon rings. Polysaccharide are classified into two, Homo-polysaccharide (containing just one type of monomer) and Heteropolysaccharide (containing more than one type of monomer) (Anne 2020). The common participating monomers (residues) in polysaccharide include fructose, glucose, galactose and mannose (Todd 2021). Based on function, polysaccharide could be grouped into two; storage and structural polysaccharide. Storage polysaccharide include starch and glycogen which are used as energy bank by plant and animal respectively. Structural polysaccharides include cellulose and chitin which form structures of plant. During the bonding of the monomers, a linear or branched structure may be formed. Linear polysaccharides are solid in structure e.g. cellulose in wood, while branched structures of polysaccharide are loose, e.g. starch (Anne 2020). Polysaccharide also serve the purpose of sending cellular information (Anne 2020). Polysaccharides are considered as organic compound because of the carbon to carbon bond and carbon to hydrogen bond. The following are the chemical properties of polysaccharide; does not form crystal when desiccated, can be extracted into white powdery form, not sweet, usually insoluble, not osmotically active in cells and has the general chemical formula of $(C_6H_{10}O_5)_n$ (Todd 2021).

Polysaccharides can be used to protect and deliver bioactive molecules such as nanoparticles to targeted sites (Yubia et al. 2021). Polysaccharide in this function presents great advantages which include biodegradability, biocompatibility, facile modification, renewable origin and abundance in nature (Alexander et al. 2021). Considering the concerns for the safety of commonly used materials for nanoparticles, the properties of polysaccharide suggests that it is safe. Polysaccharide are nontoxic and are therefore, healthy for use compared to other synthetic nano materials (Martin et al. 2020). Polysaccharide that can be used as nanoparticles include cellulose, starch, pectin, dextran, chitosan, chitin and alginates. The methods used for forming these particles include spraying, acid hydrolysis, homogenization, ultrasound, ionotropic gelation among others (Erika et al. 2015). Polysaccharide have been used by nature to assemble complex ordered structures and fused materials such as wood. Understanding the principle behind polysaccharide self-assembling presents the possibility of fabricating bio-nanomaterials using polysaccharide as the raw material (Martin et al. 2020). Polysaccharide nanostructures have advantage in the field of controlled release, which influences the concentration, efficacy and location of active pharmaceutical ingredient, nutrients and diagnostics (James et al. 2014).

(a) *Nanoparticle*

Nanoparticles are minute objects that act as a whole unit in properties and in regards to their ability for transportation (Ananya 2013). Their absolute dimension is within the nanoscale (usually bellow 100 nm or 100×10^{-9}). Nanoparticles differs from bulk materials in size, energy absorption, chemical reactivity and biological mobility. The unique size of nano materials makes them crucial in numerous areas of human activities (Salata 2004). They find application in cell and tissue imaging as contrast agents, gene delivery to tumors and drug carriers. Nanoparticles enable therapies and analyses that could ordinarily not be possible. Besides the good nanomaterials serve human, they also pose environmental challenges because of their toxicity (Shashi 2007). Thus, the need for none toxic nanomaterials like polysaccharide. Due to their submicroscopic size, they possess exceptional material features, thus they find application in areas such as engineering, medicine, catalysis, and environmental remediation (Babaniyi et al. 2023a). Nanoparticles are instrumental in curbing the menace of multi-drug resistance pathogens as well as their consequential side effects. The effectiveness of drugs and other agents are improved when trapped, absorbed or linked to nanoparticles. Nanoparticles provide increased surface area for drugs and are also used for synthesizing packaging materials which possess enhanced antimicrobial properties, barriers and mechanical strength (Divya et al. 2016).

(b) *Wastewater*

Availability and access to clean and cheap water is one of the primary humanitarian goals as well a major global twenty-first century challenge. Water is a prized natural resource. Its availability and quality are vital on earth for the survival of living organisms. However, fast industrialization is endlessly downgrading water quality owing to the introduction of huge sum of pollutants into water sources. Thus, water pollution is a threat to life on earth. Thereby, informing the need for water remediation (Aransiola et al. 2021) (removal of pollutant) to recover the polluted water for use (Neeta et al. 2017).

One of the key problems that affects biodiversity in the world is agricultural wastewater. Excess water from irrigation systems is the focal cause of agricultural wastewater which contains residual chemical pesticides and fertilizers. It posts health concern when drained into natural water sources that are used for drinking (Lorenzo et al. 2020; Aransiola et al. 2021).

12.2 Polysaccharide-Based Nanomaterials (PNMs): Characteristics

Polysaccharides are polymer compounds made up of multiple monosaccharide molecules linked together by glycosidic bonds. Homopolysaccharides (made up of single monosaccharides like starch and cellulose) and heteropolysaccharides (made up of two or more monosaccharides) are the two types (composed of different monosaccharides, i.e., mucopolysaccharide). Physical properties of polysaccharides reveal structural differences. Plant material, marine plants, and exogenous metabolites of some bacteria are commonly used to isolate them. For many active compounds, polysaccharides can be used as matrices for encapsulation, immobilization, and controlled release. Food, pharmaceutical, biomedical, and chemical industries all use this matrix. The most abundant biomolecules are polysaccharides, followed by proteins, lipids, and nucleic acids. Chitosan, pectin, alginate, starch, and dextran are among the polysaccharides investigated as encapsulation matrix.

The chemical composition, degree of polymerization, number of branches, and surface charge of polysaccharides suggest distinct structural and physicochemical features, which are advantageous for designing drug carrier systems with adjustable characteristics in terms of drug loading content, release capacity, and biodistribution (Bernd et al. 2004; Jin et al. 2020; Soliman et al. 2020; Babaniyi et al. 2023a).

Polysaccharides are found in a wide range of natural sources, including animals, plants, microbes, and algae, and techniques for isolating and purifying them are essential for developing polysaccharide-based nanomaterials (Mozammil Hasnain et al. 2019). Because proteins and nucleic acids can undergo conformational changes during extraction, polysaccharides are more readily accessible.

Polysaccharide-based nanomaterials outperform synthetic nanocarriers in terms of drug retention, easy access, and ready to use for nanocarrier synthesis without additional preparation. Because of their biocompatibility and drug solubilization, polysaccharide-based nanomaterials have recently been investigated as ideal carriers for enhancing drug bioavailability. Polysaccharide-based nanomaterials can be obtained from different sources such as animal, microbial, algae, plant (phyto-based polysaccharides), and Marine (Marine polysaccharide-based nanomaterials) (De Anda-Flores et al. 2021). Phyto-based polysaccharides can be utilized for drug delivery systems upon graft modification or synergistic interaction *via* non-covalent stabilization with other commercial polysaccharides such as hyaluronic acid (Wang et al. 2019). There is no doubt that leveraging this feature to develop dual-functional polysaccharide-based nanocarriers serving both carrier and therapeutic functions holds great potential from a clinical point of view. This chapter focuses on characteristics of polysaccharide-based nanomaterials and their application in water and wastewater treatment.

12.3 Polysaccharide Nanomaterials (PNMs) Interaction with Pollutants

Wastewater treatment is one of the major issues across the world (Maddela et al. 2019) as this area is critical in the reuse of used water in agriculture and industries. In this direction, several advances have been made in wastewater treatment in recent times (Yang et al. 2017; Roy et al. 2020; Navia-Mendoza et al. 2021; Maddela et al. 2021a, b, c; Chávez-Prado et al. 2022; Mesías et al. 2022). Polysaccharide materials are widely used as adsorbents due to their undisputable advantages over conventional absorbents. By preparing the composites of polysaccharides and nanomaterials, greatly enhance the sorption properties and stability of these novel composites (Asif et al. 2021). Thus, PNMs have a great potential to bind with the pollutants of both inorganic (e.g. heavy metals, nutrients) and organic nature (e.g. dyes, pharmaceuticals, phenols, oils). As these types of pollutants are quite common in the wastewaters/effluents of various chemical industries, PNMs based absorbents are considered as effective tools in the sustainable wastewater treatment facilities. It is also possible to use these materials in different forms according to the requirement. For instance, cellulose-based nanomaterials have been applied in the form of nanofibers, nanoparticles and nanocrystals (Asif et al. 2021).

Polysaccharide based nanoparticles have been widely used in the removal of pollutants. Carboxymethyl cellulose (anionic polysaccharides) stabilized nickel nanoparticles with a bacterial cellulose coating was found to interact efficiently with anionic 2-nitrophenol and cationic methylene blue dye (Kamal et al. 2019). CMC-stabilized magnetite nanoparticles enhanced the uptake of arsenate, and this efficiency was improved by 14% with the magnetite nanoparticles stabilized in starch (Liang et al. 2012). Various polysaccharide-based nanomaterials have been widely applied for the removal of chromium (Orooji et al. 2021). Cellulosic nanomaterial-based adsorbents have been widely used in the removal of chemically diverse pollutants (Nasrollahzadeh et al. 2021). Following are the some of the cellulosic nanomaterials-based adsorbents and the pollutants (*in parenthesis*) removed by them: β-Cyclodextrin modified CNCs@Fe3O4@SiO2 superparamagnetic nanorods (*Procaine, imipramine*); PEG modified CNnCs (*Acetaminophen, sulfamethoxazole, N,N-diethyl-meta-toluamide*); Pristine CNCs (*Chlorpyrifos*); Dialdehyde functionalized CNCs (*Creatinine*); Poly(acrylic acid) modified poly(glycidylmethacrylate) grafted CNCs (*Trypsin*); Poly(methacrylic acid-co-vinyl sulfonic acid)grafted magnetic CNCs (*Hemoglobin, immunoglobulin G*); Aminopropyltriethoxysilane modified, hydroxyl-carbonated apatite modified and

Epoxy modified CNFs (*Hydrogen sulphide*); Carbonated hydroxyapatite modified CNFs (*Phosphate, nitrate*); UiO-66/polydopamine/bacterial cellulose (*Aspirin, tetracycline hydrochloride*); Carboxymethyl cellulose/citric acid aerogel (*Nitrate, nitrite, phosphate*) (Nasrollahzadeh et al. 2021). These insights clearly imply that PNMs have potential in removing the chemically diverse pollutants, and thus PNMs are perfect tools for the sustainable water and wastewater treatment facilities.

Pollutants are removed by PMNs by various mechanisms, such as photoreduction, photocatalytic reduction, reduction, bioreduction and adsorption (Orooji et al. 2021). The adsorption capacities of certain PMNs are extremely high than the traditional absorbents. For example, adsorption capacity of commercial activated carbon and nanoscale zero-valent iron modified carboxymethyl cellulose for Pb(II) are 20.9 and 1237.3 mg g^{-1}, respectively (Wang et al. 2021). Similarly, adsorption efficiency of nano magnetite-chitosan film for Cu(II) is ~17 times higher than that of banana peel biochar. Similar trend has also been reported in the removal of dyes, where higher adsorption capacity has been observed for PMN than either traditional adsorbents or non-PMNs. For instance, adsorption efficiencies of anionic polyacrylamide-modify-chitosan magnetic composite nanoparticles (FS@CS-PAA), biochar produced with Korean cabbage waste, β-cyclodextrin functionalized biogenic zero-valent iron nanoadsorbents, and activated carbon for crystal violet were 2330.3, 1304.0, 454.5, and 271.0 mg g^{-1} respectively (Wang et al. 2021). Similarly, FS@CS-PAA has exhibited ten times higher adsorption efficiency for methylene blue over magnetic graphene oxide modified zeolite. It is also noteworthy that the ionic characters are varied between the different kinds of polysaccharides, which greatly helps in the adsorption of a wide range of pollutants. Starch, cellulose and dextran are neutral types of polysaccharides, and all have –OH functional groups. Carrageenan, alginate and gum tragacanth are anionic type polysaccharides, and have different types of functional groups, such as –OH, OSO_3^-, –OH, and $-COO^-$. Whereas chitosan is a cationic polysaccharide and has –OH and NH_3^+ functional groups (Wang et al. 2021). Thus, polysaccharides have different functional groups (Maddela et al. 2018) which have significant importance in binding with the pollutants in aqueous environment.

Recent advancements in the PMNs further enhanced their pollutant removal efficiencies. Magnetic iron oxide nanoparticles are subjected to surface functionalization by using polysaccharides to enhance adsorption efficiency. This type of surface modified PMNs were found to adsorb antibiotics from the water. For example, hybrid nanoparticles of magnetic quaternary chitosan were efficient in adsorbing diclofenac from water (Soares et al. 2019). Trimethyl chitosan/siloxane-hybrid coated Fe_3O_4 nanoparticles were efficient in the uptake of sulfamethoxazole from water (Soares et al. 2019). Such innovative adsorption studies with PNMs versus certain model pollutants have revealed some interesting insights. Vapor phase deposition of triethyoxyl(octyl)silane on CNFs aerogel has capability to adsorb 139–345 g g^{-1} model organic solvents (Jiang and Hsieh 2014). In fact, addition of hydrophobic silanes to the polysaccharide nanomaterials greatly enhances the pollutant removal efficiencies. Cellulose nanomaterials have been transformed into water-repellent and oleophilic materials in the presence of hydrophobic silanes; this transformation not only resulted in the effective removal of oil either from the top or below the surface of water, but also allowed the reusability of catalyst for six times (Jiang and Hsieh 2014). On the other side, PNMs have proved to be effective in the removal of pollutants from the gaseous phase. Nanocellulose matrix containing hydrogel and graphene oxide have been used to prepare graphene oxide-cellulose nanomaterials composites by H_2 reduction, and these composites were effective in the adsorption of cyclohexane and dimethylformamide (DMF) (Wang et al. 2014). Similarly, microbial surface impregnated with TiO2 nanoparticles effectively degraded (77–88%) the toluene in 40 min in air at ambient temperature under UV irradiation conditions (Rezaee et al. 2016). It should be remembered that volatile organic compounds (VOCs) such as phenol, toluene, xylenes etc. are highly hazardous and have potential human health risk due to their low solubility and volatility. Toluene can also be seen in water (Zeng et al. 2009); therefore, removal of such types of VOCs in water by PNMs has significant importance in avoiding the risk to human health as well as the ecosystem. Nevertheless, nanomaterials based on different polysaccharides such as cellulose, chitin/chitosan, starch, gum, alginate, and pectin are potential in adsorbing different types of inorganic and organic pollutants in water (Nasrollahzadeh et al. 2021). However, in order to make the PNMs based wastewater treatment more sustainable, future research should be emphasized on the following issues—(i) production of PNMs at a relatively low cost; (ii) searching for diverse biowaste nanomaterials for the fabrication of different PNMs; (iii) manipulation of natural polymeric materials of agricultural and animal origin (e.g. bone, bristles, eggshell etc.) for the synthesis of PNMs; (iv) improve the efficiency of catalytic properties of PNMs by producing PMNs under mild conditions; (v) further improvements in the magnetic based PNMs.

12.4 Applications of PNMs in Remediation of Water and WWT: Case Studies

The existence of toxic water contaminants (heavy metals (Aransiola et al. 2019), dyes (Abioye et al. 2015a), phenols, oils (Abioye et al. 2017), pharmaceuticals (Abioye et al. 2015b)) restricts the sustainable supply of clean water globally. Among various economic and accessible wastewater disposal technologies, adsorption is the most extensively practiced approach due to its economic, feasible and sustainable characteristics. In recent years, polysaccharides have been extensively employed for constructing adsorption materials due to their intrinsic properties including high adsorptive ability, low cost, renewability, biodegradability, biocompatibility and ease of modification (Qi et al. 2021; Babaniyi et al. 2023b). Here, a systematic review of polysaccharide-based adsorbents for wastewater treatment is given. This review is aimed at providing guidelines on the rational fabrication and application of polysaccharide-based adsorbents, which will be beneficial to researchers in the field of wastewater remediation. In this section, representative animal, plant and microbial polysaccharides used to fabricate adsorbents for adsorption of heavy metals are discussed and some adsorption results are listed in Table 12.1.

12.4.1 Adsorption of Heavy Metals Using Animal Polysaccharide-Based Adsorbents

Chitosan is the most extensively used animal polysaccharide for designing adsorbents. Chitosan has amino and hydroxyl groups, which are the main active sites for metal adsorption. The adsorption mechanism of chitosan includes chelation, ion-exchange, electrostatic interaction and van der Waals forces (Dinu et al. 2018). However, the native chitosan has poor solubility, hindering its further applications in the highly complicated wastewater environment. The chemical modifications of chitosan can solve this issue. Generally, the modified chitosan-based materials can promote the adsorptive capacity/selectivity/stability and other physicochemical features like mechanical performance and structural stability. For example, Zhang et al. (2020) prepared a recyclable chitosan-based adsorbent for wastewater treatment (Fig. 12.1). To enhance the acid-resistance ability, glutaraldehyde was adopted as a cross-linker to connect the iron oxide (Fe_3O_4) nanoparticles and chitosan. Subsequently, poly (acryloyloxyethyltrimethyl ammonium chloride) (PDAC) was

Table 12.1 Adsorption of heavy metals using polysaccharide-based adsorbents

Polysaccharide used in preparing adsorbents	Heavy metals	Best kinetic models	Best isothermal models	Maximum sorption capacity (mg g^{-1})
Chitin	Cr^{6+}	PSO	Langmuir	387.7
Chitosan	Mg^{2+}	PSO	Freundlich	177.9
Cellulose	Pb^{2+}	PSO	Langmuir	558.7
Pectin	Pb^{2+}	PSO	Langmuir	390.9
Alginate	Pb^{2+}	Elovich	Redlich-Peterson	234.8
Agarose	Co^{2+}	PSO	Langmuir	105.1
κ-Carrageenan	Hg^{2+}	PSO	Freundlich	229.9
Guar gum	Cr^{6+}	PSO	Langmuir	101.0
Konjac glucomannan	Pb^{2+}	PSO	Langmuir	95.2
Starch	Ni^{2+}	PSO and Elovich	Langmuir and sips	83.3
β-Cyclodextrin	Cd^{2+}	PSO	Langmuir	24.7
Pullulan	Cu^{2+}	PSO	Freundlich	100.9
Dextran	Cd^{2+}	PSO	Langmuir	269.0
Salecan	Cd^{2+}	PSO	Langmuir	412.5
Gellan gum	Cr^{3+}	PSO	Freundlich	42.4
Xanthan gum	Pb^{2+}	PSO	Langmuir	1012.2

Retrieved as cited by Liang et al. (2018), Dinari et al. (2020), Wang et al. (2021), Qi et al. (2021)

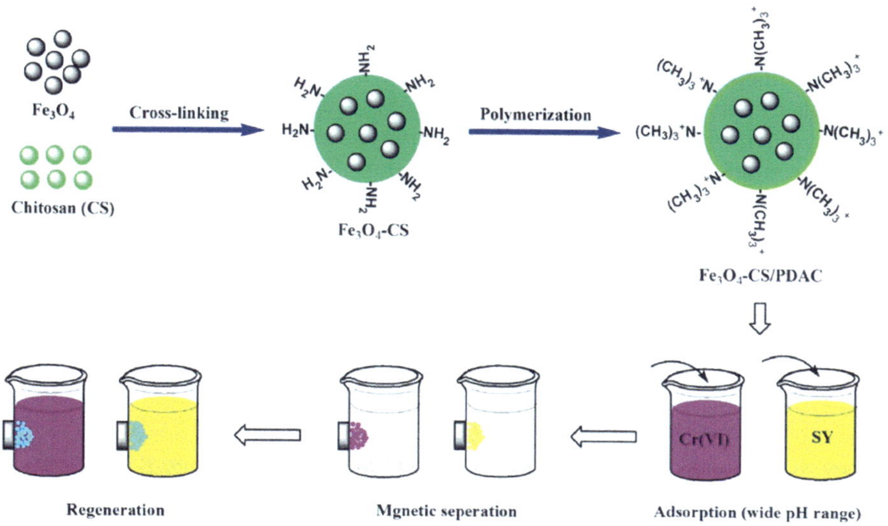

Fig. 12.1 Schematic illustration for the fabrication and adsorption of Fe_3O_4-CS/PDAC. (Reproduced with permission from Elsevier 2020, Zhang et al. (2020))

grafted onto the backbone of chitosan to provide reactive quaternary ammonium sites for adsorption of adsorbates. The obtained adsorption system (denoted as Fe_3O_4-CS/PDAC) was employed for removing Cr^{6+} and organic dye (sunset yellow). The developed Fe_3O_4-CS/PDAC exhibited good adsorption performance (maximum adsorption capacity of 769.2 and 163.9 mg/g for sunset yellow and Cr^{6+}, respectively) and excellent reusability (>90% removal efficiency after five cycles of regeneration). Ahmad et al. (2019) developed a novel diphenyl carbazide (DPC) cross-linked chitosan-based adsorbent for Cu^{2+} ions removal from water. Diphenylcarbazide (DPC) was chosen as a cross-linker to increase the mechanical strength of the adsorbent. The maximum adsorption ability of DPC/chitosan for Cu^{2+} was 185.5 mg/g, which was achieved at solution pH 6.0.

12.4.2 Polymeric Nanoadsorbents

Polymeric adsorbents have been considered as a replacement for conventional adsorbents as a result of their high surface area, adaptable surface chemistry, ideal mechanical rigidity, pore size distribution, and practicable development under some conditions (Zare et al. 2018). Polymeric nanoadsorbents are composite materials that are classified into four major types based on their composition: graphene/polymer nanocomposites, carbon nanotube/polymer nanocomposites, metal and metal oxide/polymer nanocomposites, and dendrimer-based nanocomposite (Bushra 2018). Dendrimers, which has repetitively branched molecules, consider a perfect example for polymeric nanoadsorbents and their impact on removing organic and inorganic contaminates.

Heavy metals can be adsorbed through the external branches, while organic materials can be adsorbed through the internal hydrophobic fractions. Diallo et al. (2005) combined dendrimers with ultrafiltration technique to eliminate copper from the water, which led to removing all copper cations. The shift of pH plays an important role in reactivity of the nanoadsorbent materials. Because of the complex multistage synthesis of dendrimers, only few recent companies produce commercial dendrimers. Some researchers enhanced the efficiency of dendrimers by synthesizing a combined compound like chitosan-dendrimer nanostructure to increasing its capacity in the eliminating of some anionic compounds from wastewater such as dyes (Sadeghi-Kiakhani et al. 2013). They are biodegradable, biocompatible, and non-toxic materials that have great dye removal efficiencies, approximately 99%.

12.5 Characteristics and Applications of Polysaccharide-Based Nanomaterials in Water and Wastewater Remediation

The ecosystem is progressively suffering from pollution due to the activities and increase in human population and urbanization (Danushika et al. 2016). Nanoremediation is the application of nanoparticles for correcting environmental pollution (Ioannis et al. 2020). Due to the ever-growing population on earth, there is need to recover used or waste water to make clean water available for use urbanization. Factors such as population, growth and pollution affect the quality of water negatively. Method such has the use of polysaccharide based nano particle is needed for efficiency and reduced toxicity to the environment is of concern. Amine group and hydroxyl group form the active component of polysaccharides, which could cross link to form insoluble molecules with other groups. This cross-linked polysaccharide is used to treat water as a flocculating or coagulating agent. It could also serve as an absorbent (Arshiya et al. 2022). Polymer nanoparticles have drawn the attention of scientist in purification of water because of enhanced surface area, processability, stability and cost effectiveness. It has the ability to decontaminate faster and its selectivity for removal of divers pollutants is high. It has proved relevant for the removal of dyes, metal ions and microbes from polluted water bodies (Neeta et al. 2017). Heavy metals such as Pb, Cu, Cd, Co, Ni, Hg, and Cr are major pollutants of marine water, industrial effluents and treated wastewaters (Shrivastava et al. 2014). Direct or indirect contact with heavy metals is linked with kidney damage, various cancers, growth retardation and death at exposure to very high concentrations (Manikandan et al. 2017). Environmental pollution by these heavy metals is a prevalent problem rising from growing industrial activities. Intake of polluted water cause various types of diseases such as muscle pain, abdominal pain, weakness, diarrhea. It could also result to cancer and death in many cases. Thus, it is obligatory to eliminate these and other toxic metal ions from water, thereby, the right treatments must be used in polluted waters to accomplish the acceptable environmental standards. Recently, many approaches have been advanced, they include ion exchange, adsorption, nanofiltration, reverse osmosis, membrane permeation, coagulation and oxidation. However, introduction of polysaccharide-based nanoparticles for the removal of inorganic toxicants from drinking water presents the edge of reduced toxicity and improved binding property makes it a better option for remediation (Shrivastava et al. 2014).

Because of the presence of reactive amino groups, considerable attention has been received by CS in water treatment. Chitin and CS have been extensively investigated in the removal of heavy metal ions from wastewater. Fierro et al. studied and discovered that the CS bead immobilized algae system with *Scenedesmus* sp. was better effective in the removal of nitrate and phosphate from water than the conventional free cell system. Manivasagan et al. gave the biosynthesis of silver nanoparticles using polysaccharide-based bioflocculant by *Streptomyces* sp. MBRC-91. Strong antibacterial activity in sewage water was exhibited by the biosynthesized silver nanoparticles and this result can make a new front in wastewater treatment.

12.5.1 Nanoadsorbents for Water/Wastewater Remediation

The nature of an adsorbent is a very important factor in adsorption process where the adsorbent determines the efficiency, versatility and economics of the adsorption method. Although traditional sorbents, such as activated carbons, clay minerals, chelating materials and chitosan/natural zeolites can eliminate heavy metal cations from water/wastewater, however some properties, like the low sorption capacities, can decrease the efficiency in an application. In the last 10 years nanomaterials, substances with a particle size range between 1 and 100 nm have been used to avoid the deficiencies of traditional sorbents.

12.6 Future Perspectives of PNMs in Remediation

The demand for clean water for human around the world necessitates the need for water remediation to recover polluted water for use. Flocculation is the primary industrial method for clarification, removal of suspended particles and wastewater decontamination. However, synthetic flocculants used in wastewater remediation results to grave environmental and health issues, thus, the need for alternative natural and non-toxic material like polysaccharide (Lorenzo et al. 2020). Usually, the remediation methods require additional remediation technique that is environment friendly (Manikandan et al. 2017). The increasing concern for domestic and agricultural wastewater remediation alarm the use of bio-flocculant such as chitosan (a type of polysaccharide) (Auta et al. 2022). Chitosan proved to be highly effective in the removal of pollutant from rivers and waste waters from agriculture. Chitosan effectively reduced the concentration of metals such as Manganese and Iron from waste waters (Lorenzo et al. 2020). There is a greater and unexploited possibilities of polysaccharide nano based particle as an absorbent (Arshiya et al. 2022). The usage of polysaccharide based materials for biosorption, holds high promise for superior deliverables. The benefit of biosorption is that it makes use of industrial wastes and biomass that are abundant and cheap. Biosorbents manipulating the properties of a biosorbent gather the heavy metals from the solution upon desorption in biosorbents regeneration cycle (Manikandan et al. 2017). Use of polysaccharide for biosorption is advantageous to conventional methods. However, biogenic polysaccharides have greater advantages. Thus, the need to further study the properties offered by these natural components of manufactured with ease in nature (Danushika et al. 2016).

The future potentials for biosorption using polysaccharides lies in two important nano-integration domains. First is through nanosizing the polysaccharides, in creating nanosized cellulose, chitosan, alginate nanobeads, nanofibers and pullulan nanofibrils; hence, combining the unique intrinsic properties of polysaccharides with the properties that are associated to nanosized matter. These properties include higher surface area to volume ratio instead of large particles, higher absorption sites and versatility that is concurrent with nanostructures. This results to the evolution of this technique of biosorption to "nanosorption" (Manikandan et al. 2017). Hydroxyapatite (HAp) is a biocompatible compound and has been extensively used in several industrial and biological applications. The formation of porous HAp polymer nanocomposites using carboxymethyl cellulose and chitosan (polysaccharides) were used to synthesis a potential adsorbent nanomaterial with better adsorption ability for textile dyes and heavy metals. The nano material resulting from the synthesis was used to efficiently and specifically remove Pb(II) ions and acid yellow 220, an industrial dye waste (Danushika et al. 2016).

12.7 Conclusions

Nanoparticles differs from bulk materials in size, energy absorption, chemical reactivity and biological mobility. The unique size of nano materials makes them crucial in numerous areas of human activities. They find application in cell and tissue imaging as contrast agents, gene delivery to tumors and drug carriers. Polymer nanoparticles have drawn the attention of scientist in purification of water because of enhanced surface area, processability, stability and cost effectiveness. It has the ability to decontaminate faster and its selectivity for removal of divers pollutants is high. It has proved relevant for the removal of dyes, metal ions and microbes from polluted water bodies.

References

Abioye OP, Aliyu HG, Aransiola SA (2015a) Bio-removal of vat dye from textile effluent by *Candida tropicalis and Candida apis* isolated from soil. Expert Opin Environ Biol 4:3. https://doi.org/10.4172/2325-9655.1000112

Abioye OP, Afolayan EO, Aransiola SA (2015b) Treatment of pharmaceutical effluent by *Saccharomyces cerevisiae* and *Torulaspora delbrueckii* isolated from spoilt water melon. Res J Environ Toxicol 9:188–195. https://doi.org/10.3923/rjet.2015

Abioye OP, Ijah UJJ, Aransiola SA (2017) Chapter 4 – Phytoremediation of soil contaminants by biodiesel plant *Jatropha curcas*. In: Bauddh K et al (eds) Phytoremediation potential of bioenergy plants. Springer, Singapore. https://doi.org/10.1007/978-981-10-3084-0_4

Ahmad M, Zhang B, Wang J, Xu J, Manzoor K, Ahmad S, Ikram S (2019) New method for hydrogel synthesis from diphenylcarbazide chitosan for selective copper removal. Int J Biol Macromol 136:189–198

Alexander P, Zan L, Bernhard VKJS (2021) Polysaccharide nanoparticles: from fabrication to applications. J Mater Chem B:1–33. https://doi.org/10.1039/d1tb00628b

Ananya M (2013) What are nanoparticles? News Medical Life Science

Anne MH (2020) Polysaccharide definition and functions. ThoughtCo

Aransiola SA, Ijah UJJ, Abioye OP, Bala JD (2019) Microbial-aided phytoremediation of heavy metals contaminated soil: a review. Eur J Biol Res 9(2):104–125

Aransiola SA, Ijah UJJ, Abioye OP, Victor-Ekwebelem MO (2021) ANAMMOX in wastewater treatment. In: Maddela NR, García Cruzatty LC, Chakraborty S (eds) Advances in the domain of environmental biotechnology. Environmental and microbial biotechnology. Springer, Singapore. https://doi.org/10.1007/978-981-15-8999-7_15

Arshiya A, Divyanshi M, Saiqa I (2022) Enabling polysaccharide applications for wastewater treatment using carbon nanotubes. In: Innovation in nano-polysaccharides for eco-sustainability from science to industrial applications. Elsevier, Amsterdam, pp 311–324

Asif MB, Zhang Z, Iftekhar S, Lehto VP (2021) Polysaccharide-derived biopolymeric nanomaterials for wastewater treatment. In: Biopolymeric nanomaterials. Elsevier, Amsterdam, pp 447–469

Auta HS, Abioye OP, Aransiola SA, Bala JD, Chukwuemeka VI, Hassan A, Aziz A, Fauziah SH (2022) Enhanced microbial degradation of PET and PS microplastics under natural conditions in mangrove environment. J Environ Manag 304:114273. https://doi.org/10.1016/j.jenvman.2021.114273

Babaniyi BR, Ogundele OD, Thompson SO, Aransiola SA (2023a) Microbial nanomaterial synthesis: types and applications. In: Maddela NR, Rodriguez Diaz JM, Branco da Silva Montenegro MC, Prasad R (eds) Microbial processes for synthesizing nanomaterials. Environmental and microbial biotechnology. Springer, Singapore. https://doi.org/10.1007/978-981-99-2808-8_1

Babaniyi BR, Ogundele OD, Bisi-omotosho A, Babaniyi EE, Aransiola SA (2023b) Remediation approaches in environmental sustainability. In: Maddela NR, Eller LKW, Prasad R (eds) Microbiology for cleaner production and environmental sustainability. CRC Press, Boca Raton. ISBN: 9781032496061. https://www.routledge.com/Microbiology-for-Cleaner-Production-and-Environmental-Sustainability/Maddela-Eller-Prasad/p/book/9781032496061

Bernd AS, Aihara M, Lindsey JD, Weinreb RN (2004) Influence of molecular weight on intracameral dextran movement to the posterior segment of the mouse eye. Investig Ophthalmol Vis Sci 45:480–484. https://doi.org/10.1167/iovs.03-0462

Bushra R (2018) Chapter 11 – Nanoadsorbents-based polymer nanocomposite for environmental remediation. In: New polymer nanocomposites for environmental remediation. Elsevier, Amsterdam, pp 243–260

Chávez-Prado GA, Benavides-García AB, Zambrano-Intriago LA, Maddela NR, Quiroz-Fernández LS, Baquerizo-Crespo RJ, Rodríguez-Díaz JM (2022) Novel application of tagua shell (Phytelephas aequatorialis) as adsorbent material for the removal of Pb (II) ions: kinetics, equilibrium, and thermodynamics of the process. Sustainability 14(3):1309

Danushika CM, Rohini MS, Nalin S, Rivi R (2016) Natural polysaccharides leading to super adsorbent hydroxyapatite nanoparticles for the removal of heavy metals and dyes from aqueous solutions. RSC Adv 6:105618–105630

De Anda-Flores Y, Carvajal-Millan E, Campa-Mada A, Lizardi-Mendoza J, Rascon-Chu A, Tanori-Cordova J, Martínez-López AL (2021) Polysaccharide-based nanoparticles for colon-targeted drug delivery systems. Polysaccharides 2:626–647. https://doi.org/10.3390/polysaccharides203003

Diallo MS, Christie S, Swaminathan P, Johnson JH, Goddard WA (2005) Dendrimer enhanced ultrafiltration. 1. Recovery of Cu(II) from aqueous solutions using PAMAM dendrimers with ethylenediamine core and terminal NH_2 groups. Environ Sci Technol 39:S1366–S1377

Dinari M, Shirani MA, Maleki MH, Tabatabaeian R (2020) Green cross-linked bionanocomposite of magnetic layered double hydroxide/guar gum polymer as an efficient adsorbent of Cr(VI) from aqueous solution. Carbohydr Polym 236:116070

Dinu MV, Dinu IA, Lazar MM, Dragan ES (2018) Chitosan-based ion-imprinted cryo-composites with excellent selectivity for copper ions. Carbohydr Polym 186:140–149

Divya A, Nisha S, Vishal S, Vidushi A, Ravi S, Sundeep J (2016) An update on polysaccharide-based nanomaterials for antimicrobial applications. Appl Microbiol Biotechnol 100:2603–2615

Erika AL, Aurora HG, Maribel CM, Hernández S (2015) Polysaccharide-based nanoparticles. In: Food nanoscience and nanotechnology. Springer, Cham, pp 59–68

Ioannis M, Elisavet S, Agathangelos S, Eugenia B (2020) Environmental and health impacts of air pollution: a review. Front Public Health 8:14

James MM, Venkat KV, Sitaraman K (2014) Self-assembled polysaccharide nanostructures for controlled-release applications. Nanotechnol Rev 3(4):319–346

Jiang F, Hsieh YL (2014) Amphiphilic superabsorbent cellulose nanofibril aerogels. J Mater Chem A 2(18):6337–6342

Jin Y, Zhou Q, Li Z, Yang Z, Fan HJS (2020) Calcium-cross linked polysaccharide microcapsules for controlled release and antimicrobial polysaccharide-based nanocarriers for ocular applications. Colloids Surf A Physicochem Eng Asp 600:125025. https://doi.org/10.1016/j.colsurfa.2020.125025

Kamal T, Ahmad I, Khan SB, Asiri AM (2019) Anionic polysaccharide stabilized nickel nanoparticles-coated bacterial cellulose as a highly efficient dip-catalyst for pollutants reduction. React Funct Polym 145:104395

Liang Q, Zhao D, Qian T, Freeland K, Feng Y (2012) Effects of stabilizers and water chemistry on arsenate sorption by polysaccharide-stabilized magnetite nanoparticles. Ind Eng Chem Res 51(5):2407–2418

Liang X, Fan X, Li R, Li S, Shen S, Hu D (2018) Efficient removal of Cr(VI) from water by quaternized chitin/branched polyethylenimine biosorbent with hierarchical pore structure. Bioresour Technol 250:178–184

Lorenzo APC, Juan ISS, Jose PRL, Grégorio C, Eduardo HO, Ariel GMJ (2020) Environment-friendly approach toward the treatment of raw agricultural wastewater and river water via flocculation using chitosan and bean straw flour as bioflocculants. ACS Omega 5(8):3943–3951

Maddela NR, Zhou Z, Yu Z, Zhao S, Meng F (2018) Functional determinants of extracellular polymeric substances in membrane biofouling: experimental evidence from pure-cultured sludge bacteria. Appl Environ Microbiol 84(15):e00756-18

Maddela NR, Sheng B, Yuan S, Zhou Z, Villamar-Torres R, Meng F (2019) Roles of quorum sensing in biological wastewater treatment: a critical review. Chemosphere 221:616–629

Maddela NR, Cruzatty LCG, Chakraborty S (eds) (2021a) Advances in the domain of environmental biotechnology: microbiological developments in industries, wastewater treatment and agriculture. Springer, Singapore

Maddela NR, Gan Z, Meng Y, Fan F, Meng F (2021b) Occurrence and roles of comammox bacteria in water and wastewater treatment systems: a critical review. Engineering 17:196–206

Maddela NR, Scalvenzi L, Radice M (2021c) Novel insights of microbial exopolysaccharides as bio-adsorbents for the removal of heavy metals from soil and wastewater. In: Microbial exopolysaccharides as novel and significant biomaterials. Springer, Cham, pp 265–283
Manikandan M, Hui-Fen W, Judy G, Iyyakkannu S, Sechul C (2017) Exploiting microbial polysaccharides for biosorption of trace elements in aqueous environments—scope for expansion via nanomaterial intervention. Polymers (Basel) 9(12):721
Martin G, Peter S, Thomas H (2020) Nanoparticles based on hydrophobic polysaccharide derivatives—formation principles, characterization techniques, and biomedical applications. Macromol Biosci 20(4):e1900415
Mesías JAA, Macías LLD, Delgado EAR, Maddela NR (2022) Adsorción de metales pesados utilizando sustancias poliméricas extracelulares inmovilizadas en un polisacárido aniónico. Colón Ciencias, Tecnologia y Negocios 9(1):114–133
Mozammil Hasnain SM, Hasnain MS, Nayak AK (2019) Chapter 1 – Natural polysaccharides: sources and extraction methodologies. In: Hasnain MS, Nayak AK (eds) Natural polysaccharides in drug delivery and biomedical applications. Academic Press, Cambridge, MA, pp 1–14
Nasrollahzadeh M, Sajjadi M, Iravani S, Varma RS (2021) Starch, cellulose, pectin, gum, alginate, chitin and chitosan derived (nano) materials for sustainable water treatment: a review. Carbohydr Polym 251:116986
Navia-Mendoza JM, Zambrano-Intriago LA, Maddela NR, Duarte MMMB, Quiroz-Fernández LS, Baquerizo-Crespo RJ, Rodríguez-Díaz JM (2021) Advances in the application of nanocatalysts in photocatalytic processes for the treatment of food dyes: a review. Sustainability 13(21):11676
Neeta P, Shukla SK, Singh NB (2017) Water purification by polymer nanocomposites: an overview. Nanocomposites 3(2):47–66
Orooji Y, Nezafat Z, Nasrollahzadeh M, Kamali TA (2021) Polysaccharide-based (nano) materials for Cr (VI) removal. Int J Biol Macromol 188:950–973
Qi X, Tong X, Pan W, Zeng Q, You S, Shen J (2021) Recent advances in polysaccharide-based adsorbents for wastewater treatment. J Clean Prod 315(2021):1–27
Rezaee A, Pourtagi G, Hossini H, Loloi M (2016) Microbial cellulose as a support for photocatalytic oxidation of toluene using TiO2 nanoparticles. J Appl Polym Sci 133(8):1–5
Roy K, Mukherjee A, Maddela NR, Chakraborty S, Shen B, Li M, Du D, Peng Y, Lu F, Cruzatty LCG (2020) Outlook on the bottleneck of carbon nanotube in desalination and membrane-based water treatment—a review. J Environ Chem Eng 8(1):103572
Sadeghi-Kiakhani M, Mokhtar AM, Gharanjig K (2013) Dye removal from coloredtextile wastewater using chitosan-PPI dendrimer hybrid as a biopolymer: optimization, kinetic, and isotherm studies. J Appl Polym Sci 127:2607–2619
Salata OV (2004) Applications of nanoparticles in biology and medicine. J Nanobiotechnol 2:3
Shashi KM (2007) Nanoparticles in modern medicine: state of the art and future challenges. Int J Nanomedicine 2(2):129–141
Shrivastava RB, Singh P, Bajpai J (2014) Polysaccharide-based nanosorbents in water remediation. In: Application of nanotechnology in water research. Wiley, Hoboken. https://doi.org/10.1002/9781118939314.ch5
Soares SF, Fernandes T, Trindade T, Daniel-da-Silva AL (2019) Trimethyl chitosan/siloxane-hybrid coated Fe3O4 nanoparticles for the uptake of sulfamethoxazole from water. Molecules 24(10):1958
Soliman OY, Alameh MG, De Cresenzo G, Buschmann MD, Lavertu M (2020) Efficiency of chitosan/hyaluronan-based mRNA delivery systems in vitro: influence of composition and structure. J Pharm Sci 109:1581–1593. https://doi.org/10.1016/j.xphs.2019.12.020
Todd S (2021) Polysaccharide. BiologyOnline
Wang Y, Yadav S, Heinlein T, Konjik V, Breitzke H, Buntkowsky G, Schneider JJ, Zhang K (2014) Ultra-light nanocomposite aerogels of bacterial cellulose and reduced graphene oxide for specific absorption and separation of organic liquids. RSC Adv 4(41):21553–21558

Wang Y, Bai F, Luo Q, Wu M, Song G, Zhang H et al (2019) Lycium barbarum polysaccharides grafted with doxorubicin: an efficient pH-responsive anticancer drug delivery system. Int J Biol Macromol 121:964–970. https://doi.org/10.1016/j.ijbiomac.2018.10.103

Wang XD, Li Y, Dai TT, He XM, Chen MS, Liu CM, Liang RH, Chen J (2021) Preparation of pectin/poly(m-phenylenediamine) microsphere and its application for Pb2+ removal. Carbohydr Polym 260:117811

Yang X, Zhou Z, Raju MN, Cai X, Meng F (2017) Selective elimination of chromophoric and fluorescent dissolved organic matter in a full-scale municipal wastewater treatment plant. J Environ Sci 57:150–161

Yubia DAF, Elizabeth CM, Alma CM, Jaime LM, Agustin RC, Judith TC, Ana LML (2021) Polysaccharide-based nanoparticles for colon-targeted drug delivery systems. Polysaccharide 2(3):626–647

Zare EN, Motaharib A, Sillanpää M (2018) Nanoadsorbents based on conducting polymer nanocomposites with main focus on polyaniline and its derivatives for removal of heavy metal ions/dyes: a review. Environ Res 162:173–195

Zeng X, Wu J, Zhang D, Li G, Zhang S, Zhao H, An T, Wang X, Fu J, Sheng G (2009) Degradation of toluene gas at the surface of ZnO/SnO2 photocatalysts in a baffled bed reactor. Res Chem Intermed 35(6):827–838

Zhang M, Zhang Z, Peng Y, Feng L, Li X, Zhao C, Sarfaraz K (2020) Novel cationic polymer modified magnetic chitosan beads for efficient adsorption of heavy metals and dyes over a wide pH range. Int J Biol Macromol 156:289–301

Chapter 13
Multivariate Analysis of Ecological Resilience and Vulnerability in "La Segua" Wetland: Implications for Conservation and Sustainable Natural Resource Management

Gerardo J. Cuenca-Nevárez, Maritza Viviana Talledo-Solórzano, Danny L. Cuenca-Nevárez, Juan Carlos Vélez Chica, Rodolfo Patricio Panta Vélez, Anderson Javier Pazmiño Castro, Juan José Bernal Zambrano, Víctor Hugo Nevárez-Barberán, and Eduardo Antonio Caicedo Coello

Abstract The tropical coastal wetland of "La Segua", located in the province of Manabí, Ecuador, is an area that exhibits a range of ecological stressors arising from climate change and anthropogenic activities. The primary objective of this research was to explore the condition of the ecosystem by employing a multivariate study that included Spearman correlations, biplots, and hierarchical dendrograms. A total of 35 sampling units were studied, where, during a multi-focal workshop with the community, dominant variables were selected, such as the wetland's sensitivity level, the intensity of environmental impacts produced by different categories of activities, and protection practices. According to the assessment, 45.27% of the wetland is highly sensitive, 30.62% is moderately sensitive, and 24.11% of the wetland's ecological state is in optimal condition. The involvement of the local community, especially women, becomes a socio-environmental strategy for protect-

G. J. Cuenca-Nevárez (✉) · M. V. Talledo-Solórzano · J. C. V. Chica · R. P. P. Vélez · J. J. B. Zambrano
Departamento de Recursos Naturales Renovables, Facultad de Acuicultura y Ciencias del Mar, Universidad Técnica de Manabí, Sucre, Ecuador
e-mail: gerardo.cuenca@utm.edu.ec

D. L. Cuenca-Nevárez
GenteOil Ecuador, Quito, Ecuador

A. J. P. Castro
Departamento de Procesos Agroindustriales, Facultad de AgroCiencias, Universidad Técnica de Manabí, Universidad Técnica de Manabí, Chone, Ecuador

V. H. Nevárez-Barberán · E. A. C. Coello
Universidad Laica Eloy Alfaro de Manabí, Extensión Sucre, Sucre, Ecuador

B. R. Babaniyi et al. (eds.), *Wetland Ecosystems: Conservation Strategies, Policy Management and Applications*, Wetlands: Ecology, Conservation and Management 12, https://doi.org/10.1007/978-3-031-91982-4_13

ing the wetland. The findings lead to the recommendation of implementing adaptive management practices that would reduce pressures on sensitive areas and expand resilient sites as a protection approach to ensure the sustainability and biodiversity of the La Segua wetland under a climate change scenario.

Keywords Conservation · Dendogram · Resilience · Vulnerability · Wetland

13.1 Introduction

There are few ecosystems in the world that can play a key role in mitigating global climate change due to their ability to capture and store atmospheric carbon, known as "blue carbon" (Hilmi et al. 2021). These coastal wetlands, including mangroves and salt marshes, perform this task and maintain resilience against global warming (Cahoon et al. 2021).

Unlike other ecosystems, coastal wetlands are specialized and efficient at sequestering carbon. This efficiency is due to the anaerobic conditions present in the waterlogged soils of these ecosystems, which slow down the decomposition of organic matter, allowing carbon to settle in the soil for centuries or even millennia (Hopkinson et al. 2023). Studies show that coastal wetlands can sequester three to five times more carbon per unit area than tropical forests, underscoring their importance in climate mitigation strategies (Anderegg et al. 2020; Gopalakrishna et al. 2022).

Coastal wetlands have a high capacity to act as carbon sinks, directly influencing the regulation of the carbon cycle (Kirwan et al. 2023). By sequestering large amounts of carbon dioxide (CO_2) from the atmosphere, these ecosystems also function as reducers of greenhouse gas (GHG) levels in the ecosystem, thereby mitigating global warming in two ways (Adiaha et al. 2020; Li et al. 2020). Additionally, coastal wetlands also regulate methane (CH_4) and nitrous oxide (N_2O). All these gases are directly and indirectly involved in agricultural activities carried out in the areas surrounding the wetland's waterbody, bringing about a direct negative impact on aquatic ecosystems and their biodiversity (Cantonati et al. 2020).

In the case of N_2O, it is a gas primarily produced by two microbial processes in the soil: nitrification, where soil bacteria aerobically oxidize ammonium (NH_4^+) to nitrate (NO_3^-), and denitrification, which, in contrast to the first process, is anaerobic, with nitrate-reducing bacteria converting nitrates into gaseous nitrogen using N_2O as an intermediary (Bahram et al. 2022). It is worth mentioning that these two essential processes in the nitrogen cycle are influenced by factors such as soil moisture, temperature, pH, excessive irrigation or the presence of poorly drained soils in these areas can create anoxic conditions (Tang et al. 2020; Yousaf et al. 2021; Mo et al. 2024).

Nitrogen leached from agricultural soils enters wetlands through runoff and groundwater, triggering the biological process known as eutrophication. In this process, the concentration of nutrients in water bodies increases, leading to algal blooms and a reduction in dissolved oxygen in the water (hypoxia). This results in the suffocation of all organisms living in the water body, such as crustaceans, mollusks, and fish, all of which are secondary members of the trophic net (Walton et al. 2020).

According to Wang et al. (2021); despite the capacity of wetlands to mitigate climate change, these ecosystems are highly vulnerable, causing their role as carbon sinks to potentially shift to becoming net sources of greenhouse gases (GHGs). Additionally, it is important to highlight that intensive agriculture, urbanization, aquaculture, pollution, and rising sea levels are some of the main threats these ecosystems face (Taillardat et al. 2020). The significance of the ecological role played by coastal wetlands has led to their inclusion in national and international climate policies (Brown et al. 2021). However, the effective implementation of these policies requires an integrative approach that combines conservation with the sustainable development of coastal communities (Dada et al. 2021).

13.2 Wetland Biodiversity: Fauna and Flora

Ecuadorian coastal wetlands are ecosystems considered to have high biological productivity, as they harbor a wide variety of flora and fauna species that have evolved to adapt to the specific conditions of these environments (Rivera et al. 2022). Therefore, the development of effective conservation and sustainable management strategies for these species would be carried out through the precise characterization of the flora and fauna that interact dynamically within this ecosystem (Stephenson et al. 2022).

Coastal wetlands exhibit vegetation adapted to the periodic flooding conditions typical of these areas, as well as to salinity gradients (Anderson et al. 2022). Species such as *Typha domingensis* (totora), *Eichhornia crassipes* (water hyacinth), *Lemna spp.* (duckweed) and *Phragmites australis* (common reed) are two of the emblematic and dominant species of "La Segua" wetland. Meanwhile, in the transition zone, that is, between freshwater and brackish water, mangroves dominate the landscape. Species such as *Rhizophora mangle* (red mangrove) and *Avicennia germinans* (black mangrove) form plant associations that not only stabilize coastlines but also create complex habitats for fauna species (Cahoon et al. 2021).

Coastal wetlands are areas where a great variety and abundance of birds, fish, and reptiles thrive, due to the high productivity and rich resources available for feeding and breeding (Balwan and Kour 2021). Birds such as *Butorides striata* (striated heron) and *Dendrocygna autumnalis* (black-bellied whistling duck) are prominent examples.

Regarding fish, *Centropomus undecimalis* (snook) and *Dormitator latifrons* (pacific fat sleeper) are very important species, as they play a significant ecological role and also provide economic sustenance for many families living in these areas (Gonzalez-Martinez et al. 2021). Reptiles such as *Caiman crocodilus* (caiman) are a faithful indication that "La Segua" wetland still retains well-preserved remnants of habitat necessary for this species, which requires extensive and unoccupied areas to survive (Catenazzi and von May 2021).

On the other hand, mammals like *Lontra longicaudis* (neotropical river otter) are semi-aquatic species that interact with other aquatic species (Haro-Carrión et al. 2021).

13.3 Ecological Management and Conservation Strategies for the Wetland

The ecological benefits of wetlands include maintaining environmental health and mitigating climate change. Wetlands act as natural regulator during periods of intense rainfall, as they channel excess water and release it slowly during dry periods. This characteristic helps prevent the extreme effects of floods and droughts.

Additionally, they provide freshwater to the human communities living in these areas. Furthermore, wetlands offer unique conditions for the formation of ecological niches that allow the coexistence of aquatic, terrestrial, and aerial species, making these areas vital wildlife refuges. They also form part of the essential migratory routes for birds (Balwan and Kour 2021).

"La Segua" wetland has proven to be capable of adapting to natural environmental changes, but it is currently facing unprecedented challenges due to climate change (Rivera et al. 2022). Unpredictable variations in rainfall and rising temperatures are affecting the water cycle in "La Segua", endangering the benefits this ecosystem provides. In recent years, the wetland has experienced droughts that threaten the biological diversity both within and around its area, as well as its function as a natural flood regulator. Despite this, "La Segua's" resilience has not been completely diminished (Rodríguez 2022).

Through various management and preservation measures that have had the active collaboration of academia, local authorities and the communities that depend on the wetland, the resilience of "La Segua" wetland has been strengthened in recent times.

Chen et al. (2024) determine that reforestation of trees and aquatic vegetation in degraded areas and margins of the wetland has reduced soil erosion, improved water quality and provided refuge for native species. Thus, strengthening the resilience of the wetland. Furthermore, environmental education and awareness campaigns have played a fundamental role in protecting the wetland, fostering a feeling of responsibility and commitment among local inhabitants; These elements are essential to implement resistance and conservation measures effectively.

Climate change poses significant challenges that require continued adaptation and long-term planning. It is important to keep in mind that the previously described attempts will only be effective if the watershed that feeds this wetland is managed (Hurlimann et al. 2021).

"La Segua" wetland is an ecosystem that has known how to adapt to ecological changes. However, the importance of balanced human intervention that integrates conservation with the development needs of local communities is recognized. Preserving and strengthening the resilience of "La Segua" wetland is essential for protecting the environment, thus ensuring the sustainability of this ecosystem increasingly affected by the impacts of climate change.

13.4 Description of the Study Area and Methodology Used

"La Segua" wetland is located on the Ecuadorian coast, specifically in the province of Manabí, in the canton of Chone. According to the WGS84 reference system, with UTM projection in zone 17S, the approximate coordinates of this wetland are 0°42.5′–0°44.3′ S and 80°09′–80°12.2′ W with total area of 18.36 km^2. The climate of "La Segua" wetland is humid tropical, with a marked seasonality between the dry season and the rainy season.

The wetland's proximity to the coast and its position within the Chone river basin significantly shape its microclimate, creating conditions that promote the diversity of flora and fauna (Zambrano Mero et al. 2022) (Fig. 13.1).

The temperature in the "La Segua" wetland region is relatively constant throughout the year, ranging between 24 and 28 °C on average. Temperatures can rise a little more during the dry season, reaching peaks close to 30 °C, while in the rainy

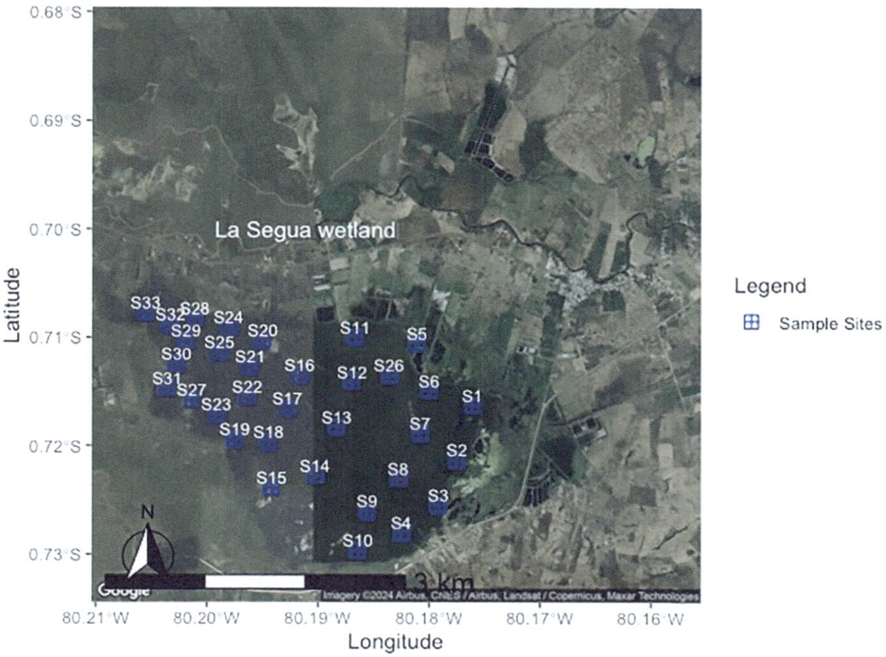

Fig. 13.1 Flood-prone area of the tropical coastal "La Segua" wetland. Manabí, Ecuador

season they tend to be more moderate. While the precipitation, varies significantly between the seasons. The wetland receives an average annual rainfall that ranges between 1000 and 1500 mm. Most of this precipitation occurs during the rainy season, which runs from approximately December to May, with the months of February and March being the wettest (Montilla et al. 2017).

For this study, a data frame was designed based on detailed surveys applied to 35 sampling units distributed throughout most of "La Segua" wetland. These surveys captured information on a series of categorical ecological variables, specifically selected to assess farm resilience in the context of climate change and other ecological stressors (Celis et al. 2023).

In evaluating the resilience of an ecosystem as complex as a "La Segua" coastal wetland, the selection of ecological variables is essential. These variables must be able to effectively capture changes in ecological and management conditions that affect the wetland's ability to resist and recover from disturbances, particularly those associated with climate change (Table 13.1).

The variables used in the study were obtained from multifocal meetings with community members in the area (Table 13.2). These meetings were key to identifying the main concerns and characteristics of the stations and the wetland, allowing a participatory approach in the selection of the most relevant variables for the analysis. The methodology included an inclusive approach, where the active participation of women was emphasized, recognizing their fundamental role in the management and conservation of the natural resources of the wetland (Dimitrovski et al. 2021).

Table 13.1 Variables included in the analysis of different critical aspects of the ecological state and management of "La Segua" wetland

Categorical variable	Responses			
	a	b	c	d
Wetland condition	Dry	Flooded	Degraded	Stable
Type of environmental impact	Water pollution	Biodiversity loss	Soil erosion	GHG emissions
Protection measures	Reforestation	Environmental education	Monitoring and surveillance	Creation of protected areas
Introduced species	Identified invasive species	Unidentified invasive species	Exotic plant species	Exotic animal species
Management responsibility	National governments	Local governments	Private sector	NGOs
Source of pollution	Agrochemicals	Urban liquid and solid waste	Oil spills	Sediments
Level of involvement	High	Medium	Low	None
Fire frequency	Annual	Every 5 years	Every decade	Occasional
Conservation programs	Yes	No	In development	Suspended
Species condition	Critically endangered	Vulnerable	Recovering	Stable

Table 13.2 Coded variables in the analysis of different critical aspects of the ecological status and management of "La Segua" wetland

Categorical variable	Code	Responses			
		a	b	c	d
Wetland condition	LS1	LS1a	LS1b	LS1c	LS1d
Type of environmental impact	LS2	LS2a	LS2b	LS2c	LS2d
Protection measures	LS3	LS3a	LS3b	LS3c	LS3d
Introduced species	LS4	LS4a	LS4b	LS4c	LS4d
Management responsibility	LS5	LS5a	LS5b	LS5c	LS5d
Source of pollution	LS6	LS6a	LS6b	LS6c	LS6d
Level of involvement	LS7	LS7a	LS7b	LS7c	LS7d
Fire frequency	LS8	LS8a	LS8b	LS8c	LS8d
Conservation programs	LS9	LS9a	LS9b	LS9c	LS9d
Species condition	LS10	LS10a	LS10b	LS10c	LS10d

The multivariate statistical analysis applied in the characterization of "La Segua" wetland began with a Spearman correlation analysis to evaluate the relationships between the categorical ecological variables. The analysis of the response variables was deepened using Multiple Correspondence Analysis (MCA), a multivariate technique that facilitated the exploration of complex relationships between the categories of the variables. This analysis was complemented with a hierarchical analysis of the response variables using Ward's method, based on the height and similarity of variances. The resulting dendrogram was a key tool to understand the differences and similarities in ecological parameters and sampling unit resilience, and develop specific management strategies adapted to the needs of each identified group (Demir Yetis and Akyuz 2021; Dar et al. 2021; Yang et al. 2021).

13.5 Multivariate Analysis of Ecological Resilience and Vulnerability

13.5.1 Spearman Correlation of Ecological Variables

After carrying out the respective multivariate analyses, we can determine how the deterioration of this precious ecosystem is occurring and what measures are being taken to strengthen its resilience in the face of vulnerability to climate change; Thus, first of all we are going to visualize a correlation analysis of the response variables, as detailed below (Fig. 13.2).

Figure 13.2, shows the network of interactions that sustain or threaten the delicate resilience of this ecosystem in the face of climate change. We must remember that the "La Segua" wetland, like any other ecosystem, is a living and dynamic entity, where changes in one part can resonate throughout the network. This is how each pair of variables that intersect here influence each other in their struggle to

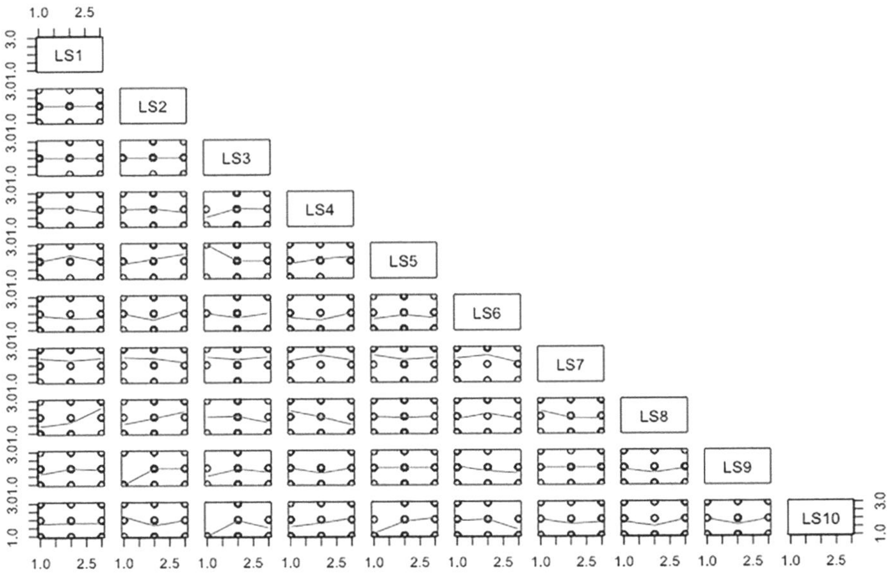

Fig. 13.2 Spearman correlation matrix on the response variables analyzed in the tropical coastal "La Segua" wetland

maintain balance in an environment marked by vulnerability to climate change (Kasperson et al. 2022). It is essential to understand the relationship between these variables to establish connections within the ecosystem and how different environmental and management factors can affect the overall health of the wetland (Xu et al. 2020).

Thus, the correlation between the "State of the Wetland" (LS1) and the "Source of Pollution" (LS6), shows that when the wetland is in a degraded state, sources of pollution such as agrochemicals or solid waste and urban liquids tend to have an even more destructive impact, exacerbating ecosystem vulnerability (Kolawole and Iyiola 2023).

Analyzing the interaction between "Fire Frequency" (LS8) and other variables such as "Species Condition" (LS10); We see that this is a determining factor in the degradation of the ecosystem. The recurrence of fires, although occasional, can have devastating effects if species do not have time to recover between one event and another. This underscores the need for fire management strategies that not only focus on prevention, but also post-fire ecological restoration (Schumann III et al. 2020).

Likewise, we note that the correlation between the "Level of participation" (LS7) and the "State of the wetlands" (LS1) could be indicating that a low level of community participation is correlated with more precarious conditions of the wetlands. This highlights the crucial importance of local participation and management policies that not only include, but empower, the communities that directly depend on the wetland (Hurlimann et al. 2021).

Therefore, the correlation between LS7 (degree of participation) and LS3 (protective measures) might indicate that in regions with higher community involvement, the need for protective measures could be reduced or implemented with less intensity. This could be a sign of more effective local management or lower levels of human impact (Dawson et al. 2021). In a similar context, when examining the correlation between LS2 and LS5, it becomes evident that environmental impacts are closely linked to responsible management, implying that more conscientious management practices might lead to fewer negative effects on the wetland (Kasperson et al. 2022).

Ultimately, the correlation between "Protection Measures" (LS3) and "Status of the Species" (LS10) suggests that successfully implementing actions like reforestation or establishing protected areas is associated with the recovery of species that were once near extinction. This demonstrates that when human intervention is both appropriate and timely, it has the potential to undo the harm caused by years of neglect and exploitation, thereby enhancing the wetland's resilience to climate change (Wang et al. 2021).

13.5.2 Multiple Correspondence Analysis (MCA) at the Stations and Ecological Variable Level

However, to obtain a more holistic understanding of how these relationships manifest spatially and how they influence the overall dynamics of the ecosystem, we will turn to multiple correspondence analysis, which is key to understanding the resilience and vulnerability of wetlands to climate change.

The biplot (Fig. 13.3), represents the sampling sites and it allows us to visualize the associations between them within the La Segua wetland depending on the variables considered. Each blue point on the graph represents a specific sites, and its location within this two-dimensional space gives us clues about its behavior and characteristics (Dar et al. 2021).

The farms that are grouped near the origin, where the coordinates are closer to the intersection of the axes, indicate that they share average characteristics in relation to the variables analyzed. These sampling sites can be considered representative of the "norm" within the wetland, without large deviations or extreme behavior in any direction. However, as we move further away from the origin, we find more distant sites that stand out, revealing unusual or extreme behaviors or conditions in the context of the wetland.

For example, sites located in the furthest quadrants may be associated with more extreme conditions, whether in terms of degradation, vulnerability, or perhaps their ability to withstand climate impacts (Van Hensbergen and Cedergren 2021). The distribution of these points allows us to identify "critical zones" within the wetland, areas that could require urgent intervention or specialized management to improve their ecological resilience. In summary, this biplot offers us a clear vision of how the

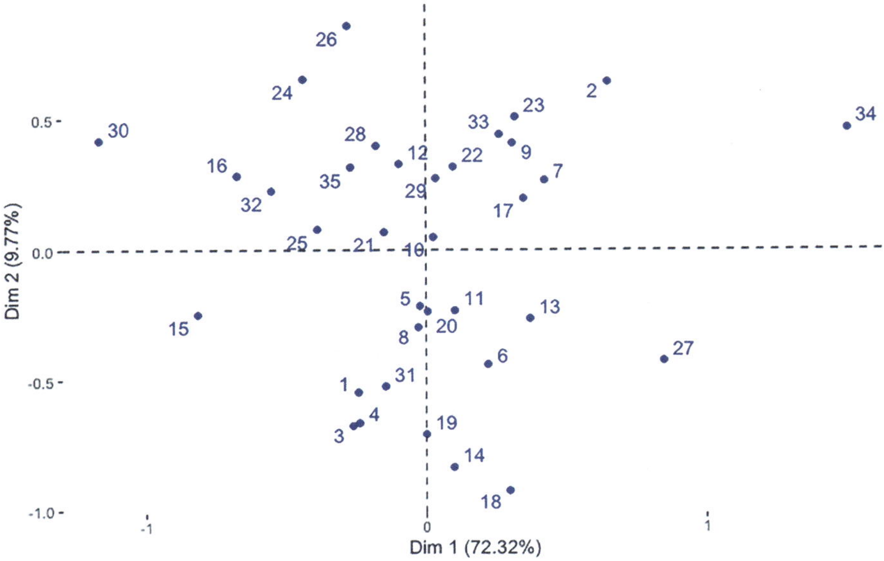

Fig. 13.3 Biplot of dimensions of the sampling units and analyzed in the tropical coastal "La Segua" wetland

farms distributed in "La Segua" wetland area present risk patterns and management opportunities.

The second biplot (Fig. 13.4), on the other hand, focuses on ecological variables and shows us a complex cartography. Here, each red dot represents a specific category within the variables such as "Wetland State", "Type of Environmental Impact", "Protection Measures", among others.

By examining the dispersion of these points, what emerges is a landscape of associations (Dar et al. 2021). Variables that are grouped in proximity suggest that they tend to coexist on the same sampling point or that they are strongly correlated. For example, if "Wetland status: Degraded" is close to "Pollution source: Agrochemicals", this could tell us that farms in a state of degradation are particularly affected by chemical contamination, revealing a possible causality that deserves attention in the management strategies.

Ecological variables in more extreme positions on the graph indicate responses that may be less common but when they occur, they do so significantly and could have a disproportionate impact on the ecosystem. These variables may be associated with rare but severe events, such as fires or floods, that could dramatically alter wetland structure and function if not managed appropriately (Schumann III et al. 2020).

Therefore, the biplot of response variables offers us a powerful visual tool to understand not only the most prevalent ecological relationships, but also to identify those that, although less frequent, could have critical effects on the resilience of the wetland. This analysis is essential for long-term planning, as it allows ecosystem

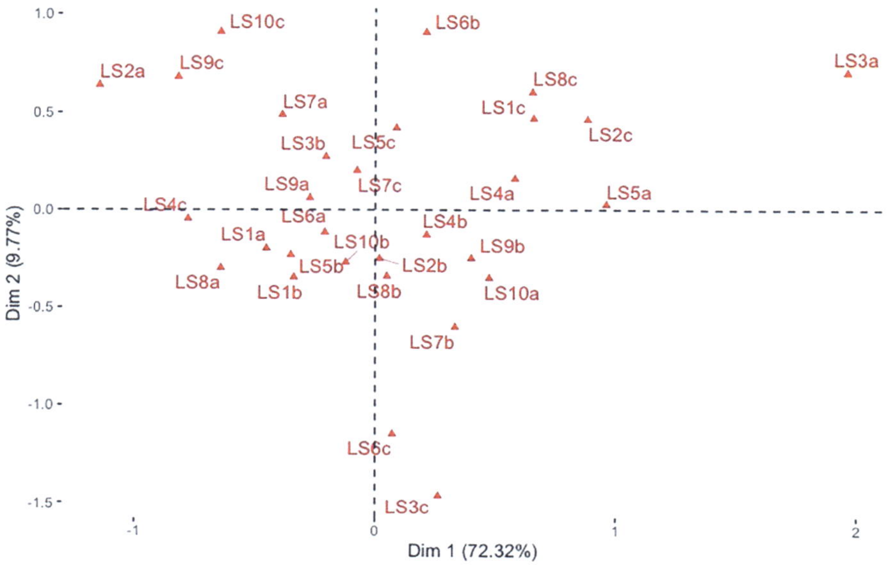

Fig. 13.4 Biplot of dimensions of the response variables analyzed in the tropical coastal "La Segua" wetland

managers to prioritize interventions in the areas that present the greatest threats and greatest opportunities for the conservation and recovery of the "La Segua" wetland (King et al. 2021).

While biplots offer us a two-dimensional view of the connections between variables and farms, dendrograms go a step further, decomposing these relationships into hierarchical levels that allow us to see how variables are grouped based on their similarities Yang et al. (2021), as demonstrated below continuation.

13.5.3 Hierarchical Analysis at the Sampling Sites and Ecological Variable Level

The dendrogram before us is a visual representation that breaks down the sampling sites of the "La Segua" wetland into groups or clusters based on their ecological similarities, which demonstrates how various areas of the wetland are connected by their shared characteristics and how these connections can influence biodiversity and the protection of renewable natural resources (Fig. 13.5).

The groups that were formed from this hierarchical analysis are:

- Red Group (Sites 30, 28, 6, 11, 15, 8, 13)

This group may represent sites facing extreme or similar environmental conditions, perhaps in terms of soil degradation, exposure to pollution sources or low

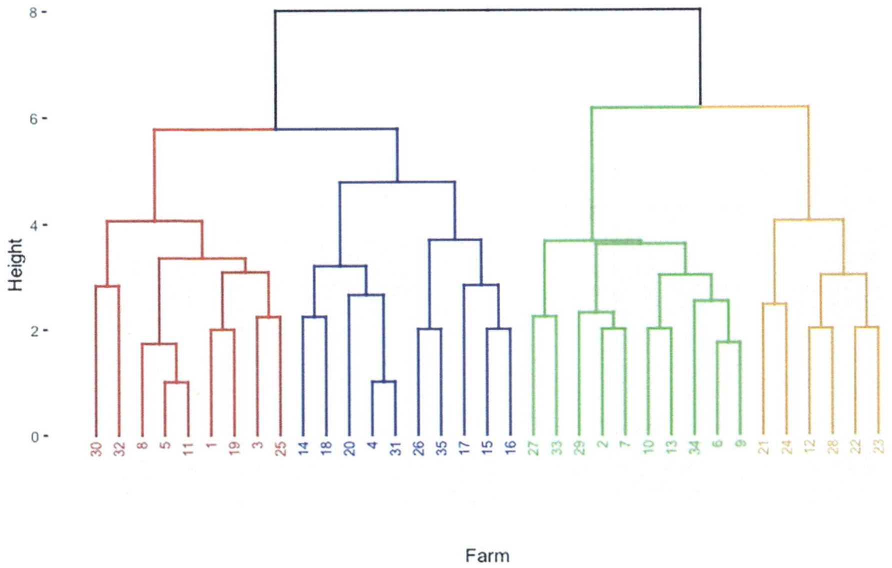

Fig. 13.5 Dendrogram of hierarchical clusters (Ward method) of the farms selected for this study in the tropical coastal "La Segua" wetland

levels of biodiversity. The grouping of these farms suggests that they share a common ecology that could be related to a particular vulnerability to climate change or intensive human activities.

- Blue Group (Sites 14, 18, 35, 16, 26, 32, 34, 19)

Sampling sites in this group could be located in areas where human intervention has had a significant impact, but where some resistance or resilience still persists. The proximity of these points in the dendrogram suggests that they share characteristics that could be related to similar conservation or management efforts, which could have implications for local biodiversity.

- Green Group (Sites 17, 7, 10, 4, 29, 27, 33)

This group may indicate sites that are in a better ecological state or that have greater biodiversity. If these sampling sites are less degraded and host a greater diversity of species, they could represent key areas for conservation, which should be prioritized in efforts to protect renewable natural resources.

- Orange Group (Sites 24, 21, 3, 9, 31, 2, 23)

Sampling sites grouped here could share vulnerable characteristics or could be exposed to a particular risk, such as the invasion of exotic species or soil erosion. These points, being grouped in the same group, suggest that interventions to improve protection and biodiversity in these areas should be addressed together.

This dendrogram is crucial to understanding how protection efforts should be prioritized within the wetland. Sites that are in separate groups may need different management strategies. For example, sites in the red group may require more urgent and aggressive interventions to prevent further degradation and restore lost biodiversity. Instead, sites in the green group could benefit more from monitoring and conservation programs designed to maintain and improve their current state of ecological health (Rodríguez 2022).

For its part, the dendrogram of hierarchical clusters of coded responses decomposes the responses associated with the different categorical variables that we have been analyzing (Fig. 13.6); revealing how these responses are grouped according to the similarity of their variances (Yang et al. 2021).

The dendrogram we observed breaks down the ecological responses of the "La Segua" wetland into groups that reflect their similarity, offering a visual representation of how these responses are grouped and what implications they have for the ecology of the wetland. Each set of coded responses reveals crucial patterns that allow us to better understand underlying ecological dynamics and guide conservation and management efforts.

- Red group

This group includes responses such as "Wetland status: dry" (LS1a), "Pollution source: agrochemicals" (LS6a) and "Species status: critically endangered" (LS10a). The clustering of these responses suggests high vulnerability in wetland areas where these conditions are observed. Sampling site in these areas may be severely

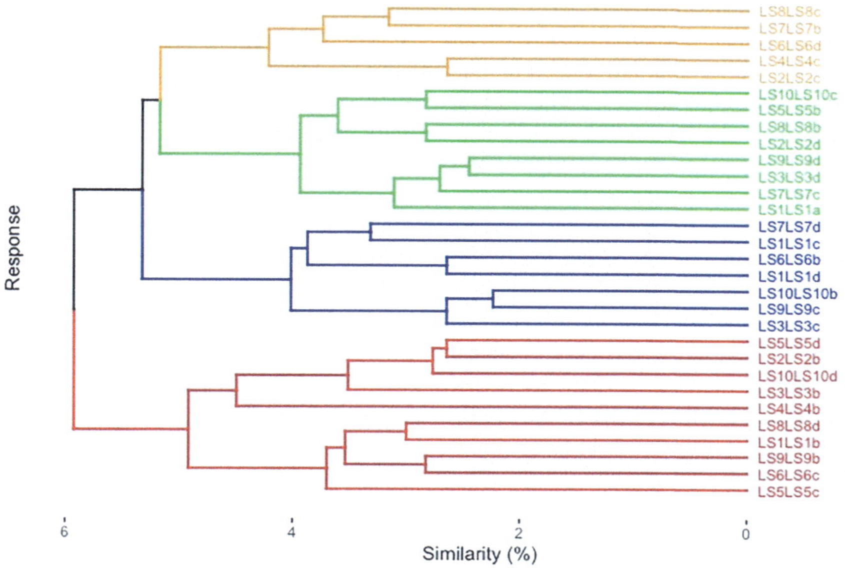

Fig. 13.6 Dendrogram of hierarchical clusters (Ward method) of the categorical variables selected for this study in the tropical coastal La Segua wetland

degraded, with high levels of agrochemical contamination and biodiversity critically threatened. This group points to the urgent need for interventions to mitigate damage and restore the ecological health of these areas.

- Blue group

Responses such as "Fire frequency: Annual" (LS8a), "Introduced species: Identified invasive species" (LS4a) and "Type of environmental impact: Loss of biodiversity" (LS2b) are grouped here. This group seems to represent a set of responses that describe an ecological environment under significant stress, with recurrent presence of fires, identified invasive species and a clear loss of biodiversity. This group could be indicating areas within the wetland that are more likely to suffer permanent ecological damage if conservation and restoration measures are not quickly implemented.

- Green group

Responses in this group, such as "Status of the species: Recovery" (LS10c), "Protection measures: Reforestation" (LS3a) and "Management responsibility: NGO" (LS5d), suggest a more resilient context. This group appears to represent areas of the wetland where conservation interventions are having a positive impact. Reforestation, active management by NGOs and species recovery indicate that these areas may be in the process of restoration, which is an encouraging sign for the biodiversity and ecological stability of the wetland.

- Orange group

This group includes responses such as "Fire frequency: Occasional" (LS8d), "Level of impact: Medium" (LS7b) and "Source of contamination: Sediment" (LS6d). The farms that are related to these responses seem to be in an intermediate situation. Although not as severely affected as those in the red or blue groups, these areas still face challenges that could intensify if not properly addressed. Interventions here can focus on prevention, improving land management and monitoring levels of community participation to ensure these areas do not deteriorate.

This dendrogram highlights the diversity of conditions that exist within the "La Segua" wetland and how these conditions are interrelated. The formation of these groups provides us with a clear roadmap for wetland management: identify the most critical areas (red and blue groups) that require urgent intervention to avoid irreversible loss of biodiversity, while continuing to support and expand practices. Successful in the most resilient areas (green group).

The hierarchical analysis employed here also underlines the importance of adaptive management that is able to respond not only to immediate threats but also to emerging trends that can be identified through continuous monitoring. By grouping responses that capture both threats and opportunities, this graph serves as a strategic tool to prioritize actions, allocate resources efficiently, and maximize the impact of conservation policies (Wiedenfeld et al. 2021).

From the results we can estimate that approximately 45.27% of the wetland is in a state of high vulnerability. This estimate is based on the strong grouping of variables related to soil degradation, the high frequency of fires, contamination by agrochemicals, and the presence of critically endangered species. These areas, which make up almost half of the ecosystem, represent areas that require immediate intervention to avoid irreversible degradation (Almar et al., 2021).

On the other hand, 30.62% of the wetland shows a state of moderate resilience. These areas are characterized by a combination of challenges and conservation efforts that, while not completely solving ecological problems, are managing to maintain a certain level of stability. The recovery of species on some farms, along with the implementation of protection measures such as reforestation and environmental education, reflect a recovery potential that can be strengthened with appropriate interventions, as highlighted by Mackey et al. (2020) regarding the role of tropical ecosystems in climate change mitigation.

Finally, 24.11% of the wetland is in a state of relatively good ecological health. These areas, mainly associated with the green group in the dendrogram, are showing clear signs of recovery and stability thanks to the active participation of NGOs and effective conservation programs. These areas represent cores of biodiversity that can serve as sustainable management models for the rest of the wetland.

13.6 Final Considerations

This multivariate analysis of multiple correspondences allows us to see that the active participation of the community settled in the wetland area is crucial to improve and maintain the health status of "La Segua" wetland. The empowerment of women, in particular, must be a central axis in preservation strategies. Women, who are often on the front lines of natural resource management in rural communities, have invaluable knowledge and significant influence on sustainable resource use practices. This can be achieved through environmental education programs focused on women's empowerment, creating support networks among women conservation leaders, and promoting community initiatives led by women.

The "La Segua" wetland is located at a critical ecological crossroads. The results of our analysis reveal an ecosystem divided between extreme vulnerability and emerging resilience. The most degraded areas, besieged by pollution and loss of biodiversity, are warning signs that cannot be ignored if the ecological integrity of the wetland is to be preserved. Only through a concerted and adaptive effort, which recognizes both the threats and potential of "La Segua" wetland, can we ensure that this wetland not only survives, but thrives in a future dominated by the challenges of climate change. "La Segua" wetland, with its ecological wealth and its crucial role in the environmental balance of the region, should not only be seen as a resource to protect, but as a legacy that we must restore and preserve for future generations.

References

Adiaha MS, Buba AH, Tangban EE, Okpoho AN (2020) Mitigating global greenhouse gas emission: the role of trees as a clean mechanism for CO_2 sequestration. J Agric Sci (Sri Lanka) 15(1):101

Almar R, Morand P, Menard F (2021) Towards West African coastal social-ecosystems sustainability: interdisciplinary approaches. Ocean Coast Manag 211:105746

Anderegg WR, Trugman AT, Badgley G, Anderson CM, Bartuska A, Ciais P et al (2020) Climate-driven risks to the climate mitigation potential of forests. Science 368(6497):eaaz7005

Anderson SM, Ury EA, Taillie PJ, Ungberg EA, Moorman CE, Poulter B et al (2022) Salinity thresholds for understory plants in coastal wetlands. Plant Ecol 223:1–15

Bahram M, Espenberg M, Pärn J, Lehtovirta-Morley L, Anslan S, Kasak K et al (2022) Structure and function of the soil microbiome underlying N_2O emissions from global wetlands. Nat Commun 13(1):1430

Balwan WK, Kour S (2021) Wetland-an ecological boon for the environment. East Afr Scholars J Agric Life Sci 4(3):38–48

Brown CJ, Adame MF, Buelow CA, Frassl MA, Lee SY, Mackey B et al (2021) Opportunities for improving recognition of coastal wetlands in global ecosystem assessment frameworks. Ecol Indic 126:107694

Cahoon DR, McKee KL, Morris JT (2021) How plants influence resilience of salt marsh and mangrove wetlands to sea-level rise. Estuar Coasts 44(4):883–898

Cantonati M, Poikane S, Pringle CM, Stevens LE, Turak E, Heino J et al (2020) Characteristics, main impacts, and stewardship of natural and artificial freshwater environments: consequences for biodiversity conservation. Water 12(1):260

Catenazzi A, von May R (2021) Systematics and conservation of neotropical amphibians and reptiles. Diversity 13(2):45

Celis N, Casallas A, Lopez-Barrera EA, Felician M, De Marchi M, Pappalardo SE (2023) Climate change, forest fires, and territorial dynamics in the Amazon rainforest: an integrated analysis for mitigation strategies. ISPRS Int J Geo Inf 12(10):436

Chen G, Gu X, Mo Y, Cui B (2024) Monospecific mangrove reforestation changes relationship between benthic mollusc diversity and biomass: implication for coastal wetland management. J Environ Manag 353:120140

Dada O, Almar R, Morand P, Menard F (2021) Towards West African coastal social-ecosystems sustainability: interdisciplinary approaches. Ocean Coast Manag 211:105822

Dar SA, Hamid A, Rashid I, Bhat SU (2021) Identification of anthropogenic contribution to wetland degradation: insights from the environmetric techniques. Stoch Env Res Risk A 36:1–15

Dawson NM, Coolsaet B, Sterling EJ, Loveridge R, Gross-Camp ND, Wongbusarakum S et al (2021) The role of Indigenous peoples and local communities in effective and equitable conservation. Ecol Soc 26:19

Demir Yetis A, Akyuz F (2021) Water quality evaluation by using multivariate statistical techniques and pressure-impact analysis in wetlands: Ahlat Marshes, Turkey. Environ Dev Sustain 23:969–988

Dimitrovski D, Lemmetyinen A, Nieminen L, Pohjola T (2021) Understanding coastal and marine tourism sustainability – a multi-stakeholder analysis. J Destin Mark Manag 19:100554

Gonzalez-Martinez A, De-Pablos-Heredero C, González M, Rodriguez J, Barba C, García A (2021) Usefulness of discriminant analysis in the morphometric differentiation of six native freshwater species from Ecuador. Animals 11(1):111

Gopalakrishna T, Lomax G, Aguirre-Gutiérrez J, Bauman D, Roy PS, Joshi PK, Malhi Y (2022) Existing land uses constrain climate change mitigation potential of forest restoration in India. Conserv Lett 15(2):e12867

Haro-Carrión X, Johnston J, Bedoya-Durán MJ (2021) Landscape structure and seasonality: effects on wildlife species richness and occupancy in a fragmented dry forest in coastal Ecuador. Remote Sens 13(18):3762

Hilmi N, Chami R, Sutherland MD, Hall-Spencer JM, Lebleu L, Benitez MB, Levin LA (2021) The role of blue carbon in climate change mitigation and carbon stock conservation. Front Clim 3:710546
Hopkinson C, Weston N, Cai WJ (2023) Global change and estuarine carbon dynamics. In: Kennish MJ, Paerl HW (eds) Climate change and estuaries, CRC Press, pp 183–202
Hurlimann A, Moosavi S, Browne GR (2021) Urban planning policy must do more to integrate climate change adaptation and mitigation actions. Land Use Policy 101:105188
Kasperson JX, Kasperson RE, Turner BL, Hsieh W, Schiller A (2022) Vulnerability to global environmental change. In: Social contours of risk. Routledge, pp 245–285
King SL, Laubhan MK, Tashjian P, Vradenburg J, Fredrickson L (2021) Wetland conservation: challenges related to water law and farm policy. Wetlands 41(5):54
Kirwan ML, Megonigal JP, Noyce GL, Smith AJ (2023) Geomorphic and ecological constraints on the coastal carbon sink. Nat Rev Earth Environ 4(6):393–406
Kolawole AS, Iyiola AO (2023) Environmental pollution: threats, impact on biodiversity, and protection strategies. In: Sustainable utilization and conservation of Africa's biological resources and environment. Springer, Singapore, pp 377–409
Li M, Cui Y, Fu Y, Li N, Tang X, Liu X, Run Y (2020) Simulating the potential sequestration of three major greenhouse gases in China's natural ecosystems. Forests 11(2):128
Mackey B, Kormos CF, Keith H, Moomaw WR, Houghton RA, Mittermeier RA et al (2020) Understanding the importance of primary tropical forest protection as a mitigation strategy. Mitig Adapt Strateg Glob Chang 25(5):763–787
Mo F, Li C, Zhou Q (2024) The pivotal role of phosphorus level gradient in regulating nitrogen cycle in wetland ecosystems. Sci Total Environ 943:173646
Montilla A, Zambrano M, Reyna C (2017) Análisis de las condiciones geográficas y ecológicas del humedal La Segua, provincia de Manabí, Ecuador. La Técnica: Revista de Las Agrociencias 18:70. ISSN 2477-8982
Rivera G, Gonzales S, Aponte H (2022) Wetlands of the South American pacific coast: a bibliometric analysis. Wetl Ecol Manag 30(4):869–877
Rodríguez FVL (2022) Mangroves of Ecuador. In: Mangroves: biodiversity, livelihoods and conservation. Springer, Singapore, pp 489–519
Schumann RL III, Mockrin M, Syphard AD, Whittaker J, Price O, Gaither CJ et al (2020) Wildfire recovery as a "hot moment" for creating fire-adapted communities. Int J Disaster Risk Reduct 42:101354
Stephenson PJ, Londoño-Murcia MC, Borges PA, Claassens L, Frisch-Nwakanma H, Ling N et al (2022) Measuring the impact of conservation: the growing importance of monitoring fauna, flora and funga. Diversity 14(10):824
Taillardat P, Thompson BS, Garneau M, Trottier K, Friess DA (2020) Climate change mitigation potential of wetlands and the cost-effectiveness of their restoration. Interface Focus 10(5):20190129
Tang S, Liao Y, Xu Y, Dang Z, Zhu X, Ji G (2020) Microbial coupling mechanisms of nitrogen removal in constructed wetlands: a review. Bioresour Technol 314:123759
Van Hensbergen H, Cedergren J (2021) Forest-related disasters – three case studies and lessons for management of extreme events, vol 17. Food & Agriculture Organization, Rome
Walton CR, Zak D, Audet J, Petersen RJ, Lange J, Oehmke C et al (2020) Wetland buffer zones for nitrogen and phosphorus retention: impacts of soil type, hydrology and vegetation. Sci Total Environ 727:138709
Wang F, Sanders CJ, Santos IR, Tang J, Schuerch M, Kirwan ML et al (2021) Global blue carbon accumulation in tidal wetlands increases with climate change. Natl Sci Rev 8(9):nwaa296
Wiedenfeld DA, Alberts AC, Angulo A, Bennett EL, Byers O, Contreras-MacBeath T et al (2021) Conservation resource allocation, small population resiliency, and the fallacy of conservation triage. Conserv Biol 35(5):1388–1395
Xu X, Chen M, Yang G, Jiang B, Zhang J (2020) Wetland ecosystem services research: a critical review. Glob Ecol Conserv 22:e01027

Yang X, Liu S, Jia C, Liu Y, Yu C (2021) Vulnerability assessment and management planning for the ecological environment in urban wetlands. J Environ Manag 298:113540

Yousaf A, Khalid N, Aqeel M, Noman A, Naeem N, Sarfraz W et al (2021) Nitrogen dynamics in wetland systems and its impact on biodiversity. Nitrogen 2(2):196–217

Zambrano Mero JD, Delgado Párraga AG, Zambrano Mero ET, Peñaherrera Villafuerte SL (2022) Contaminantes biológicos presentes en fuentes de agua del centro-sur de la provincia de Manabí, Ecuador. Siembra 9(2):e4011

Chapter 14
Constructed Wetlands for Wastewater Treatment and Role of Wetlands in Water Quality Improvement and Filtration

Orimisan Emmanuel Abata, Kemi Catherine Adesorioye, Ebunoluwa Elizabeth Babaniyi, and Joshua Ibukun Adebomi

Abstract Natural wetlands offer a remarkable natural service. They act as natural filters, removing pollutants and sediments from water through various mechanisms. The dense vegetation in wetlands slows down water flow, allowing heavy sediment particles to settle out. This reduces turbidity and protects aquatic life. With the decline of natural wetlands, constructed wetlands have emerged as a sustainable and efficient alternative for wastewater treatment. These engineered systems mimic the natural processes of natural wetlands, utilizing a combination of plants, soil, and microbes to remove pollutants from wastewater sources such as sewage, stormwater runoff, and industrial effluents. Unlike traditional wastewater treatment plants, constructed wetlands operate with lower energy demands, smaller environmental footprints, and higher biodiversity. They also serve as valuable ecological resources, providing habitat for wildlife and performing other ecosystem services such as sediment reduction, nutrient management, and pollution removal. The ability of constructed wetlands to remove emerging contaminants like pharmaceuticals and personal care products (PPCPs) is still under investigation. More research is needed to optimize treatment for these pollutants. However, as our understanding of these systems grows, we can further optimize their design and operation to address a wider range of wastewater challenges.

Keywords Marshes · Biodiversity hotspots · Vegetation · Buffer zone

---

O. E. Abata · K. C. Adesorioye (✉)
Department of Chemistry, Federal University of Technology, Akure, Nigeria

E. E. Babaniyi
Department of Biology Education, Adeyemi College of Education Ondo (Obafemi Awolo University Ile Ife), Ondo, Nigeria

J. I. Adebomi
Department of Biological and Chemical Engineering, University of Saskatchewan, Saskatoon, SK, Canada

B. R. Babaniyi et al. (eds.), *Wetland Ecosystems: Conservation Strategies, Policy Management and Applications*, Wetlands: Ecology, Conservation and Management 12, https://doi.org/10.1007/978-3-031-91982-4_14

14.1 Introduction

A wetland is a distinct ecosystem that is characterized by its water-saturated soil and low-growing plant species. Wetlands are areas where water is the primary factor controlling the environment and the associated plant and animal life. They can be found all over the world and are often located in transitional zones between land and water, such as the edges of rivers, lakes, and coasts (Cherry 2011). There are several types of wetlands, including swamps, marshes, bogs, and fens. These environments support a diverse range of plant and animal species, including many that are specially adapted to survive in such conditions (EPA 2000). Wetlands play an important role in maintaining water quality, preventing soil erosion, providing flood control, and serving as habitats for various organisms. Some examples of wetlands include the Florida Everglades in the United States, the Okavango Delta in Botswana, and the Sundarbans in Bangladesh and India. These areas are renowned for their unique ecosystems and rich biodiversity, which support numerous species, from waterfowl and amphibians to larger mammals like hippopotamuses and crocodiles.

14.1.1 What Is Wetland

Wetlands are lands transitional between terrestrial and aquatic systems where the water table is usually at or near the surface or the land is covered by shallow water.

For purposes of this classification wetlands must have one or more of the following three attributes:

1. at least periodically, the land supports predominantly hydrophytes;
2. the substrate is predominantly undrained hydric soil;
3. the substrate is saturated with water or covered by shallow water at some time during the growing season of each year (Reed 2005).

14.1.1.1 Ramsar Definition of Wetland

The Ramsar Convention, formally known as the Convention on Wetlands of International Importance, especially as Waterfowl Habitat, provides a broad definition of wetlands to encompass the diverse range of ecosystems that fall under this category. According to the Ramsar Convention, wetlands are defined as:

"areas of marsh, fen, peatland or water, whether natural or artificial, permanent or temporary, with water that is static or flowing, fresh, brackish or salt, including areas of marine water the depth of which at low tide does not exceed six meters." (Paudel 2023).

This definition includes various wetland types, such as swamps, marshes, bogs, fens, river deltas, mangroves, tidal flats, and coral reefs. The Ramsar Convention recognizes the importance of wetlands for their ecological, scientific, cultural, and economic values and aims to promote their conservation and wise use through international cooperation and designation of Ramsar sites.

14.1.2 Types of Wetlands

There are several types of wetlands, each with its unique characteristics, vegetation, and water conditions. Some of the major types of wetlands include:

Marshes: These wetlands are dominated by soft-stemmed vegetation, such as grasses, reeds, and sedges. Marshes can be found in both freshwater and saltwater environments, with water levels that may fluctuate seasonally or daily due to tides. Examples include the Florida Everglades and the Camargue marshes in France (Mitsch and Gosselink 2015).

Swamps: Swamps are dominated by woody plants, such as trees and shrubs, and are typically found in low-lying areas near rivers, lakes, or streams. They can be freshwater or brackish and often feature slow-moving or stagnant water. Examples include the Okefenokee Swamp in the United States and the Great Dismal Swamp in Virginia and North Carolina (Keddy 2010).

Bogs: Bogs are characterized by spongy, acidic, nutrient-poor soils that support peat accumulation. These wetlands receive most of their water from precipitation and are often dominated by mosses, such as sphagnum moss, and shrubs like Labrador tea. An example is the Great Kemeri Bog in Latvia (Rydin and Jeglum 2013).

Fens: Fens are peat-forming wetlands that receive water from both precipitation and groundwater. They have more alkaline soils than bogs and often support a diverse range of plant species, including grasses, sedges, and wildflowers. The Norfolk Broads in England are an example of a fen (Wheeler and Proctor 2000).

Mangrove swamps: These coastal wetlands are dominated by mangrove trees, which are adapted to thrive in saltwater environments. Mangrove swamps are found in tropical and subtropical regions, such as the Sundarbans in Bangladesh and India and the Everglades in Florida (Alongi 2020).

Tidal marshes: Tidal marshes are found along coastlines and estuaries, where they experience daily tidal fluctuations. They are typically dominated by salt-tolerant vegetation, such as cordgrass, and provide important habitats for fish, birds, and other wildlife. Examples include the Chesapeake Bay tidal marshes in the United States (Bertness et al. 2001).

Floodplain wetlands: These wetlands are located along rivers and streams and are periodically flooded during high water events. Floodplain wetlands provide important habitats for fish, waterfowl, and other wildlife, and they also help to mitigate flood impacts. Examples include the floodplains of the Amazon River (Junk et al. 1989).

Peatlands: Peatlands are wetlands that accumulate partially decayed organic matter, known as peat. They include both bogs and fens and are significant carbon sinks, playing an essential role in the global carbon cycle. Examples include the Great Kemeri Bog in Latvia and the West Siberian Lowland in Russia (Gorham 1991).

Wet meadows: Wet meadows are wetlands dominated by grasses and wildflowers that can tolerate periodic flooding and saturated soils. They provide habitat for various species of birds, mammals, and insects. An example is the Sierra Nevada meadows in California, USA (Kattelmann 1996).

14.1.3 Functions of Wetland

Wetlands perform a variety of crucial functions that contribute to the health of the environment and provide numerous benefits to both humans and wildlife. Some of the primary functions of wetlands include:

Biodiversity hotspots: Wetlands support a diverse range of plant and animal species, including many that are endangered or endemic to these habitats. They provide crucial habitats for migratory birds, amphibians, reptiles, fish, and mammals. The high productivity and nutrient cycling in wetlands contribute to their rich biodiversity (Barbier et al. 1997).

Carbon sequestration: Wetlands play a significant role in mitigating climate change by sequestering and storing carbon. Peatlands, a type of wetland, store approximately one-third of the world's terrestrial carbon, even though they cover only about 3% of the Earth's land surface. Preserving wetlands is essential for maintaining their carbon storage function and mitigating greenhouse gas emissions (Gorham 1991).

Water purification: Wetlands act as natural filters, trapping sediment and pollutants in the water that flows through them. They can remove excess nutrients, heavy metals, and other contaminants, improving water quality and reducing the risk of harmful algal blooms in downstream ecosystems (Kadlec and Knight 1996).

Flood control: By absorbing and retaining water during heavy rainfall, wetlands can help reduce the impact of floods on surrounding areas. This natural flood control function can minimize property damage, protect lives, and reduce the need for costly engineered solutions (Bullock and Acreman 2003).

Recreation and tourism: Wetlands offer opportunities for outdoor recreation, such as birdwatching, fishing, and boating. These activities can generate revenue for local communities through tourism, contributing to the economic value of wetlands (Barbier et al. 1997).

Erosion control: The vegetation in wetlands helps stabilize shorelines and reduce soil erosion by slowing down water flow and trapping sediments (Mitsch and Gosselink 2015).

Food production: Wetlands support the growth of various edible plants and provide habitats for fish, shellfish, and other food sources (Zhang et al. 2007).

Groundwater recharge: Some wetlands can recharge groundwater aquifers by absorbing and filtering surface water, which helps maintain water availability for human consumption and irrigation (Winter et al. 1998).

Education and research: Wetlands serve as natural laboratories for scientists, students, and educators to study ecological processes, species interactions, and the impacts of human activities on the environment (Beltman et al. 2014).

14.1.4 Threats and Conservation

Wetlands face numerous threats, including land conversion for agriculture, urban development, pollution, and climate change. Efforts to conserve wetlands involve a combination of protection, restoration, and wise use strategies, as well as international cooperation and policy measures, such as the Ramsar Convention on Wetlands.

The Ramsar Convention is an international treaty focused on the conservation and wise use of wetlands. Designated Ramsar sites are wetlands of international importance, and member countries commit to promoting their conservation and sustainable use (Table 14.1).

Table 14.1 Wetlands in Africa

Wetland name	Location	Wetland type	Key features
Okavango Delta	Botswana	Inland delta	Unique ecosystem in Kalahari Desert; abundant wildlife
Lake Chad	Chad, Cameroon, Niger, Nigeria	Shallow lake	Vital resources for local communities
Nile Delta	Egypt	River delta	Fertile, triangular-shaped area; important agricultural region
Zambezi Delta	Mozambique	Inland delta	Diverse wetland system; numerous bird, mammal, and fish species
Sudd	South Sudan	Swamp	Vast swamp intricate network of channels, islands, and papyrus vegetation
Lake Victoria	Kenya, Tanzania, Uganda	Large lake	Africa's largest lake; essential freshwater resource; diverse aquatic species
Banc d'Arguin National Park	Mauritania	Coastal wetland	Critical habitat for migratory birds, Mediterranean monk seal, and fish
Inner Niger Delta	Mali	Inland delta	Vast floodplain wetland; resources for irrigation, fisheries, and pasturelands
Lake Turkana	Kenya, Ethiopia	Desert lake	World's largest permanent desert lake; endemic fish species
Lake Tanganyika	Tanzania, DR Congo, Burundi, Zambia	Deep lake	World's second-deepest lake; exceptional diversity of endemic fish species

14.2 Constructed Wetland for Wastewater Treatment

Constructed wetland represent an ecological engineering approach for wastewater treatment, utilizing natural geochemical and biological mechanisms within a wetland environment to remediate contaminants. The complex interaction among water, vegetation, soil/gravel substrate, microorganisms, and solar energy plays a crucial role in driving the overall wastewater treatment process (Vymazal 2011a, 2011b).

Semi-natural treatment wetlands (SNTWs), are wetland systems that have undergone alterations to cater to the treatment of wastewater. The adjustments implemented in these systems typically involve enhancing water storage capacity through the installation of dams and the creation of channels to direct the influent and effluent. These engineered systems are present in both freshwater and coastal wetland environments. The operational principles of SNTWs bear resemblance to those observed in surface flow constructed wetlands (Ji et al. 2016).

14.2.1 Components of Constructed Wetlands

Constructed wetlands for wastewater treatment typically consist of several key components, each playing a critical role in the overall performance and efficiency of the system. These components include:

Vegetation: Plants play a crucial role in constructed wetlands, contributing to pollutant removal through nutrient uptake and providing a substrate for biofilm growth. Commonly used plants in constructed wetlands include cattails, bulrushes, reeds, and water hyacinths (Kadlec and Wallace 2009).

Substrate: The substrate serves as a medium for plant growth and supports microbial communities involved in pollutant degradation. It also aids in filtration and sedimentation. Gravel, sand, and soil are commonly used substrate materials (Kadlec and Wallace 2009).

Inlet and outlet structures: These structures control the flow of wastewater into and out of the wetland system. The inlet design should ensure uniform distribution of wastewater, while the outlet should prevent the release of floating debris or particulate matter.

Water level control structures: These components enable the adjustment of water levels within the wetland system to optimize treatment performance and accommodate variations in wastewater flow. Examples include water level control weirs, adjustable gates, or pumps (EPA 2000).

Internal baffles or berms: In larger constructed wetlands, internal baffles or berms may be used to create multiple treatment cells or to improve flow distribution, enhancing treatment efficiency.

Liner: A liner, typically made of geomembrane materials, is often used to prevent leakage of wastewater into the surrounding environment and maintain hydraulic control within the wetland system (Vymazal 2011a, 2011b).

Buffer zone: A buffer zone around the constructed wetland can help prevent erosion, minimize the impact on surrounding ecosystems, and act as a filter for surface runoff.

Monitoring and maintenance infrastructure: To ensure proper functioning and performance, constructed wetlands may include infrastructure for monitoring water levels, flow rates, and water quality. Access paths or platforms for maintenance activities may also be provided (EPA 2000).

14.2.2 Types of Constructed Wetlands

Constructed wetlands are specifically engineered to capitalize on a multitude of the same biological and ecological processes that naturally take place in wetlands; however, they do so in a meticulously regulated and controlled setting. In the realm of wastewater treatment, Constructed wetlands can be categorized based on the dominant life form of the macrophytes present, which encompass systems featuring free-floating, floating leaved, rooted emergent, and submerged macrophytes. Moreover, a more detailed classification can be established by considering the hydrology of the wetland, distinguishing between free water surface and subsurface systems, while subsurface flow Constructed wetlands could be further segmented based on the direction of flow, either horizontal or vertical (Vymazal 2018).

14.2.3 Classification and Configuration of Constructed Wetlands for Wastewater Treatment

14.2.3.1 Constructed Wetlands with Surface Flow

Constructed wetlands employing surface flow, also known as free water surface CWs, encompass zones characterized by open water and various types of plants, including floating, submerged, and emergent species (Wu et al. 2023). The shallow depth of the water, slow flow velocity, and the presence of plant stalks and organic debris all contribute to regulating water flow. Especially in elongated channels, these factors ensure enhanced water purification efficiency. The primary applications of surface flow CWs are typically for the tertiary treatment of municipal wastewater, as well as the management of stormwater runoff and mine drainage (Kadlec and Wallace 2009).

14.2.3.2 Constructed Wetlands with Subsurface Flow

In horizontal subsurface flow constructed wetlands, wastewater is introduced through an inlet and gradually permeates through the porous medium beneath the vegetation-covered bed surface towards the outlet. At the outlet, the treated water is collected before being discharged through a water level control structure. Throughout this process, the wastewater encounters various zones, including aerobic, anoxic, and anaerobic areas (Vymazal 2005). The majority of the bed remains anoxic or anaerobic due to continuous saturation, while aerobic zones are present around roots and rhizomes that facilitate oxygen transfer into the substrate. These systems are often lined to prevent seepage and control outflow. While primarily used for the secondary treatment of municipal wastewater, horizontal subsurface flow constructed wetlands have shown versatility in other applications as documented in literature. Despite insufficient oxygen transport for aerobic decomposition, anaerobic processes are crucial in these systems, and some are capable of functioning in colder conditions due to their insulating properties (Vymazal 2018) (Fig. 14.1).

14.2.3.3 Vertical Subsurface Flow Constructed Wetlands

Vertical subsurface flow constructed wetlands (VSSF CWs) are characterized by a flatbed composed of graded gravel and sand hosting macrophytes. These systems receive intermittent large volumes of wastewater that inundate the bed surface, percolate through the media, and are collected at the bottom through a drainage network. The complete drainage of the bed allows for air replenishment, enabling greater oxygen transfer and yielding a nitrified effluent rich in NO_3^-. However, denitrification to gaseous nitrogen forms, which escape into the atmosphere, is limited in VSSF Constructed wetlands. Tidal flow systems, a recent innovation in this field, ensure improved contact between wastewater and the microorganisms on the media, enhancing purification processes significantly (Vymazal 2018).

14.2.3.4 Hybrid Constructed Wetlands

To achieve higher removal efficiencies, especially for nitrogen, different types of constructed wetlands can be combined in hybrid systems. This design typically comprises two stages: multiple parallel vertical flow (VF) beds followed by a series of 2 or 3 horizontal flow (HF) beds (VSSF-HSSF system). The vertical subsurface flow wetland targets organic and suspended solids removal and promotes nitrification, while the horizontal subsurface flow wetland focuses on denitrification and further elimination of organics and suspended solids (Vymazal 2018).

Another arrangement involves a HSSF-VSSF system where a sizable HSSF bed is initially installed to eliminate organics and suspended solids while encouraging denitrification. Following this, a periodically loaded small VF bed is utilized to further eliminate organics and suspended solids, as well as to convert ammonia into

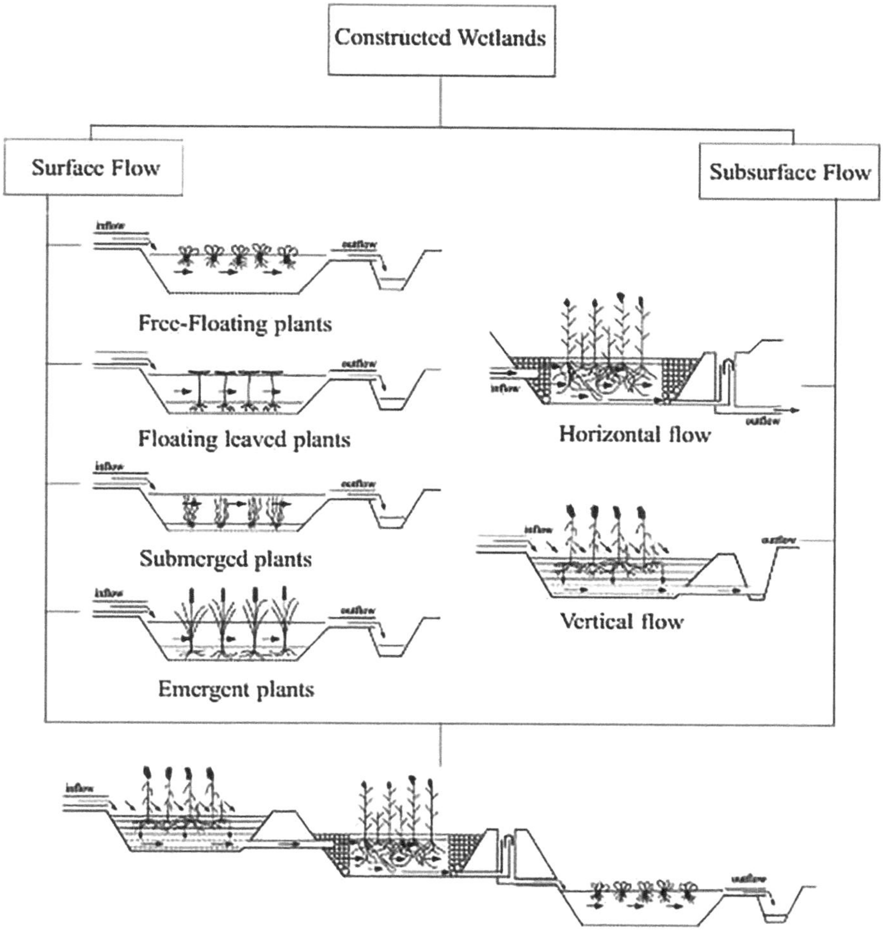

Fig. 14.1 Image showing classification and configuration or constructed wetlands for wastewater treatment. (Adapted from Vymazal 2007, Vymazal 2011a, 2011b)

nitrate through nitrification. Nevertheless, in order to enhance the elimination of total N, the nitrified effluent from the VF bed needs to be returned to a sedimentation tank. The VSSF-HSSF and HSSF-VSSF constructed wetlands are the most prevalent hybrid systems; however, in general, various types of constructed wetlands could be amalgamated to achieve a heightened treatment effectiveness (Vymazal 2018).

14.2.4 *Treatment Performance*

The mechanisms available for enhancing water quality in a constructed wetland system are numerous and often interconnected. These mechanisms comprise various processes such as suspended particulate matter settling, filtration, chemical precipitation, and interactions with substrate, litter, plants, sediment, and microorganisms. Additionally, pollutants are broken down and transformed through microbial and plant activities, while nutrients undergo transformation, and pathogens are naturally eliminated (Kadlec and Wallace 2009).

Constructed wetland treatment technology integrates key elements of wetland ecosystems to facilitate contaminant degradation and control through plant processes, microbial degradation, and physical processes like sorption, filtration, and precipitation (Hassan et al. 2021). The choice of technology depends on the treatment requirements and involves selecting appropriate designs like surface or subsurface flow, single or multiple cells, and parallel or series flow. Constructed wetland systems are sometimes incorporated into treatment sequences that include settling ponds, oil/water separators, and physical/chemical treatments (Wu et al. 2015). Removal mechanisms within these wetlands can target contaminants individually, sequentially, or simultaneously. For example, volatile organic compounds in groundwater are mainly removed through diffusion-volatilization. Other mechanisms like adsorption, photochemical oxidation, and biological degradation also contribute to contaminant removal in constructed wetlands. Physical removal processes in these systems include settling, sedimentation, and volatilization, with gravitational settling being the primary method for removing suspended solids. Effective treatment wetlands are those that facilitate these removal mechanisms (EPA 2000).

14.2.4.1 Organics and Suspended Solids

Removal of organic matter is highly efficient in all categories of constructed wetlands. In free water surface (FWS) and vertical flow (VF) constructed wetlands, the predominant microbial degradation processes are mostly aerobic, whereas in horizontal flow (HF) constructed wetlands, anoxic and anaerobic processes are more prevalent. Treatment efficacy is comparable between FWS and HF constructed wetlands, with VF constructed wetlands exhibiting higher efficiency percentages due to elevated influent concentrations. Typically, VF constructed wetlands are utilized for primary or secondary treatment, while FWS are commonly employed for tertiary treatment and HF wetlands are specifically utilized for treating wastewater mixed with stormwater runoff. Nevertheless, outflow concentrations in secondary treatment systems are similar across all types of constructed wetlands. The elimination of suspended solids is remarkably efficient in all constructed wetlands. Phosphorus retention levels are generally low in constructed wetlands, and these wetlands are rarely designed with phosphorus removal as the primary objective. Research indicates that the accumulation of soil/peat is the primary long-term phosphorus sink in wetlands (Vymazal 2018). Among different types of constructed wetlands, soil build-up is observed only in FWS wetlands due to the lack of vegetation harvesting and direct contact of wastewater with the topsoil layer. Despite this, phosphorus retention levels are relatively modest compared to typical wastewater loads. In sub-surface flow wetlands, the primary mechanisms for phosphorus removal are adsorption and precipitation. However, the commonly used materials in sub-surface flow wetlands, such as washed gravel or crushed rock, have limited sorption and precipitation capacities. Recent studies have explored alternative filtration materials like LECA (lightweight clay aggregates) and industrial by-products such as furnace steel slags in constructed wetlands. These substrates exhibit high phosphorus removal rates, although it is crucial to note that sorption and precipitation have saturation limits, leading to reduced sorption capabilities over time.

Within free water surface (FWS) constructed wetlands, nitrogen elimination primarily occurs through nitrification in the aerobic water column and subsequent denitrification in the anoxic litter layer on the bed surface. Volatilization can also play a significant role in nitrogen removal in constructed wetlands with open water surfaces, where algal communities can elevate pH levels during daylight hours through photosynthesis. Vertical-flow (VF) constructed wetlands are characterized by robust nitrification processes; however, denitrification does not occur due to the entirely aerobic conditions within the vertical bed. To achieve comprehensive nitrogen removal, a combination of VF and horizontal-flow (HF) constructed wetlands may be necessary, as HF wetlands do not nitrify but provide conducive environments for nitrate reduction formed during nitrification in VF beds. Plant uptake in all types of constructed wetlands is effective when plants are harvested, yet the amount sequestered in aboveground biomass is typically minimal and usually does not surpass 10% of the incoming nutrient load (Vymazal 2018).

14.3 Challenges of Constructed Wetland Systems

14.3.1 Balancing Nature and Engineering

While constructed wetlands offer a sustainable and efficient approach to wastewater treatment, they are not without their challenges. These engineered ecosystems require careful design, operation, and maintenance to function optimally.

14.3.1.1 Climate

The extreme hydraulic events, upwelling, and heavy precipitation adversely impact constructed wetlands performance. The extremes in ultraviolet radiation (UV), precipitation, and temperature are threatening challenges to constructed wetlands operations. An increase in the rate of precipitation and run-off significantly induce high pathogens and microbial discharge into constructed wetland-ecosystem. A 96–115.5% increase in the population of the coliform bacteria (CB) through runoff from soil to water induced by a 53.3% increase in precipitation while using a modified SWAT model to evaluate the influence of climate change on the fate and transport of coliform bacteria. The significant rise in temperature due to the direct UV radiation, particularly in FWS-CW has been linked to a logarithmic increase in the population of pathogens in constructed wetlands. Similarly, in wetlands under a Mediterranean climate, Sanchez-Ramos et al. (2019) studied the effects of meteorological variability and flooding regime on pollutant depuration efficiency and reported that reducing flood risk area significantly results in decreased contaminant concentrations. The future effectiveness of constructed wetlands essentially depends on the effective forecasting of climate change variables including droughts, flooding, and deforestations (Sanchez-Ramos et al. 2019).

The general water scarcity is also another constructed wetland challenge that increases the propensity of treated water reuse with a significant increase in constructed wetlands water demand. The water stress due to scarcity induces substrate clogging in constructed wetlands. Al-Layla et al. (2016) evaluated the environmental and health risks of wastewater reuse based on a comprehensive pollution index (CPI) approach and reported high seasonal variations in both salinity and ionic concentrations (e.g., Na^{+}), with significantly high values during spring and winter seasons. The elevated level of total coliform (TC) beyond the mandatory limit remarkably affects the public acceptance of constructed wetland effluent for commercial irrigation agriculture (Al-Layla et al. 2016).

14.3.2 Design and Implementation

Scaling-up: Translating successful small-scale projects into large-scale systems can be difficult. Factors like maintaining consistent flow rates and ensuring adequate oxygen supply become more complex with increasing size. (Garcia-Ochoa and Gomez 2009).

Climate Dependence: Constructed wetlands can be sensitive to fluctuations in temperature and precipitation. Cold climates can slow down treatment processes, while heavy rainfall events may overwhelm the system. (Vymazal 2011a, 2011b).

Site Selection: The location for a constructed wetland must have suitable land, appropriate topography, and adequate space to accommodate the system. Site selection also needs to consider potential impacts on surrounding ecosystems, aesthetic concerns, and accessibility for maintenance.

Hydraulic design: Proper hydraulic design is crucial for effective wastewater treatment in constructed wetlands. This includes sizing the wetland to ensure adequate retention time, designing the flow pattern (e.g., horizontal or vertical flow), and determining the appropriate water depth and hydraulic loading rate.

Vegetation selection: Choosing the right plants is essential for nutrient uptake and overall treatment performance. Plants must be suitable for local climate conditions, have a high tolerance to wastewater constituents, and provide habitat for beneficial microorganisms. (U.S. Environmental Protection Agency 1993).

Substrate selection and sizing: The substrate type and size affect pollutant removal, plant growth, and hydraulic performance. Selecting the appropriate substrate material, depth, and gradation is essential for optimal treatment efficiency and long-term performance. (Vymazal 2011a, b).

Water level control: Constructed wetlands must be designed with water level control structures that enable operators to maintain appropriate water levels for plant growth, pollutant removal, and overall system performance. (U.S. Environmental Protection Agency 1993).

Construction and materials: The construction of a constructed wetland must ensure proper containment of wastewater and prevent leakage. This may require the use of impermeable liners, which can add to construction costs and complexity. (U.S. Environmental Protection Agency 1993).

Local regulations and guidelines: Design and implementation must comply with local regulations and guidelines for wastewater treatment systems, which may vary between jurisdictions. Obtaining the necessary permits and approvals can be a complex and time-consuming process. (U.S. Environmental Protection Agency 1993).

Stakeholder engagement and public acceptance: Involving stakeholders, such as local residents, environmental organizations, and regulatory agencies, in the design and implementation process is crucial for addressing concerns, gaining support, and ensuring successful integration of the constructed wetland into the local community. (Everard et al. 2021).

Operation and maintenance planning: Designing a constructed wetland must consider long-term operation and maintenance requirements, such as access for maintenance activities, monitoring systems, and operational protocols for managing water levels and vegetation (Brown and Seager 2018 & Corral et al. 2017).

14.3.3 Operational Challenges

Clogging: The substrate material used in the wetland can become clogged over time due to organic matter build-up. This reduces water flow and treatment efficiency. Regular maintenance and proper substrate selection are crucial. (Wu et al. 2014).

Nutrient Removal Optimization: While constructed wetlands are effective in removing some pollutants, nitrogen removal can be less efficient. Additional treatment steps might be necessary depending on the discharge requirements. (Vymazal 2007).

Water level management: Maintaining appropriate water levels is crucial for sustaining vegetation health and ensuring efficient pollutant removal. Operators must balance water inputs and outputs, as well as account for precipitation and evapotranspiration to maintain optimal conditions. (U.S. Environmental Protection Agency 1993).

14.3.3.1 Plant Selection and Management

Sediment and debris removal: Over time, sediment and debris can accumulate within the wetland, reducing treatment capacity and potentially leading to clogging. Operators must periodically Choosing the right plant species for the specific type of wastewater and ensuring the health and productivity of wetland vegetation requires regular monitoring, harvesting, and replanting as needed. Operators must also control invasive species and address pest or disease outbreaks that could harm wetland plants.remove accumulated sediment and debris to maintain system performance.

Monitoring and performance assessment: Regular monitoring of water quality parameters and overall system performance is essential for assessing treatment efficiency and identifying potential issues. This can require significant resources and expertise to collect and analyze data.

Climate variability and resilience: Constructed wetlands must be resilient to fluctuations in climate and weather conditions, such as droughts, floods, or extreme temperatures. Operators must adapt water level management and maintenance practices to ensure system performance under variable conditions.

Safety and access: Constructed wetlands can pose safety hazards for operators and visitors, such as drowning risks or exposure to wastewater contaminants. Adequate safety measures must be implemented, and access to the wetland must be controlled.

Odor and vector control: Constructed wetlands can emit odors or harbor disease-carrying insects, such as mosquitoes. Operators must manage water levels, vegetation, and organic matter accumulation to minimize these nuisances.

Training and knowledge requirements: Operating and maintaining a constructed wetland requires specialized knowledge and skills related to wastewater treatment, ecology, and engineering. Operators must be properly trained and have access to ongoing professional development opportunities.

Regulatory Compliance: Constructed wetlands must comply with local regulations and discharge permits for wastewater treatment systems. Operators must maintain accurate records, monitor water quality, and report to regulatory agencies as required. (Brown and Seager 2018 & Deng et al. 2020).

14.3.4 Monitoring and Maintenance

Long-term Performance Evaluation: Assessing the long-term effectiveness of constructed wetlands can be challenging due to the complex biological processes involved. Consistent monitoring of water quality and system health is necessary.

Expertise and Training: Operating and maintaining a constructed wetland system requires specialized knowledge of wetland ecology, wastewater treatment processes, and plant management. Training personnel and having access to expert support is crucial.

Public Perception: Constructed wetlands may not be aesthetically pleasing to everyone, and concerns about odor or mosquito breeding can arise. Public education and proper design can help address these concerns. (Dubourguier et al. 2019).

14.3.5 Emerging Contaminants

Limited Data: The ability of constructed wetlands to remove emerging contaminants like pharmaceuticals and personal care products (PPCPs) is still under investigation. More research is needed to optimize treatment for these pollutants (Matamoros and Bayona 2006).

Variable removal rates: Removal rates of PPCPs can vary widely depending on factors such as the type of compound, wetland design, and environmental conditions. Some PPCPs may be effectively removed through processes like sorption, biodegradation, and plant uptake, while others may be more persistent and resistant to removal (Hijosa-Valsero et al. 2010).

Optimization challenges: Optimizing constructed wetlands for enhanced removal of PPCPs may require modifications to design parameters, vegetation selection, or operational strategies. This could involve balancing multiple objectives and constraints to achieve effective treatment for both conventional pollutants and emerging contaminants (Zhang et al. 2023).

Monitoring and assessment: Detecting and assessing the presence of PPCPs in constructed wetlands can be challenging due to their low concentrations and the complex mixture of pollutants in wastewater. Advanced analytical methods and monitoring strategies may be needed to track the fate of PPCPs in these systems and evaluate their potential impacts on ecosystem health and water quality.

14.4 Role of Wetlands in Water Quality Improvement and Filtration

Wetlands are often referred to as the "kidneys of the landscape" for a reason. These vital ecosystems, where water meets land, play an important role in maintaining healthy freshwater resources by acting as natural filters. This can be achieved by several process.

14.4.1 Physical Filtration

This process involves the capture and retention of suspended particles, such as sediment, organic matter, and associated pollutants, within the wetland system. This filtration process not only clarifies the water but also promotes sedimentation and nutrient cycling within the wetland ecosystem.

14.4.1.1 The Main Factors Contributing to Physical Filtration in Wetlands Are

Vegetation: The dense growth of vegetation in wetlands, such as plants, grasses, and trees, creates a physical barrier that traps suspended particles as water flows through the system. The roots and stems of wetland plants reduce water velocity, causing particles to settle out of the water column and promoting filtration. Additionally, the above-ground biomass of vegetation contributes to filtration by providing a surface area for particle capture.

Substrate: The substrate material in wetlands, such as gravel, sand, or soil, plays a significant role in physical filtration. As water flows through the substrate, particles become trapped within the porous spaces, effectively removing them from the water. Substrate characteristics, such as particle size and porosity, influence filtration efficiency.

Flow Velocity: The slow flow of water through wetlands allows more time for particles to settle out of the water column and be filtered by vegetation and substrate materials. The meandering flow paths and complex topography of wetlands further contribute to reduced flow velocity and enhanced filtration. (Britz and Sridharan 2007; Florsheim et al. 2010).

14.4.2 Biological Filtration

It involves the removal, transformation, and degradation of pollutants through various biological processes mediated by the diverse microbial communities and plant species found in wetland ecosystems.

The main factors contributing to biological filtration in wetlands are:

Microbial degradation: Wetland systems support a wide range of microorganisms, such as bacteria, fungi, and algae, which play a crucial role in breaking down organic matter and transforming pollutants. These microorganisms utilize pollutants as a source of energy and nutrients, effectively removing them from the water. For example, denitrifying bacteria convert nitrate into harmless nitrogen gas, while other bacteria can degrade various organic pollutants.

Plant uptake: Wetland plants absorb nutrients, such as nitrogen and phosphorus, from the water for their growth and metabolism. This nutrient uptake reduces the nutrient load in the water, preventing eutrophication and associated water quality issues. Furthermore, plants can also take up heavy metals and other pollutants, sequestering them within their tissues and removing them from the water.

Biofilm formation: Microorganisms in wetlands often form biofilms on the surfaces of plants, substrate materials, and other structures. These biofilms create a network of microorganisms that can effectively filter pollutants from the water through processes like adsorption, absorption, and biodegradation. Biofilms also provide a habitat for other beneficial microorganisms, further enhancing biological filtration (Jia et al., 2021).

Predation and food web interactions: The complex food webs in wetlands contribute to biological filtration by promoting nutrient cycling and the consumption of pollutant-degrading organisms. For example, zooplankton and small invertebrates graze on algae and bacteria, helping to maintain water clarity and nutrient balance.

Natural die-off and decomposition: In wetlands, the natural die-off and decomposition of plants, algae, and microorganisms release nutrients back into the ecosystem. This process helps maintain nutrient cycling within the wetland and prevents excess nutrients from entering downstream water bodies (Carvalho et al. 2010).

14.4.3 Chemical Transformation

Chemical transformation processes in wetlands contribute to pollutant removal and water quality improvement by changing the form, mobility, or toxicity of contaminants. These processes work in concert with physical and biological filtration mechanisms to create a comprehensive water treatment system that mitigates pollution in aquatic environments.

14.4.3.1 The Primary Chemical Transformation Processes in Wetlands Include

Oxidation-reduction reactions: Wetland systems often have fluctuating redox conditions due to the presence of aerobic and anaerobic zones. These varying conditions facilitate various oxidation-reduction reactions that transform pollutants. For example, the reduction of oxidized forms of heavy metals (e.g., chromium, arsenic) makes them less mobile and toxic, while the oxidation of ammonia to nitrate enables its subsequent removal through denitrification (Kadlec and Wallace 2009).

Precipitation and coagulation: In wetlands, dissolved pollutants can react with other elements or compounds to form insoluble precipitates or coagulate into larger particles, which can then settle out of the water column. For example, the precipitation of phosphorus with calcium or aluminum compounds helps remove this nutrient from the water, reducing eutrophication potential (Vymazal 2011a, b).

Adsorption and ion exchange: Wetland substrate materials and biofilms provide surfaces for adsorption and ion exchange processes, which help remove pollutants from the water. These processes occur when pollutants adhere to the surfaces of particles or exchange ions with them, effectively trapping contaminants and preventing their movement in the environment (Kadlec and Wallace 2009).

Photolysis and photochemical reactions: Exposure to sunlight can drive photochemical reactions in wetlands, transforming pollutants into less harmful or more easily removable forms. For example, sunlight can break down certain organic pollutants through a process called photolysis, while photochemical reactions can transform heavy metals, such as mercury, into less toxic forms.

Plant-mediated chemical transformations: Wetland plants can release various enzymes, organic acids, and other compounds that contribute to chemical transformations in the environment. These plant-mediated processes can enhance the degradation of organic pollutants, influence redox conditions, and impact nutrient cycling within the wetland system (Reddy et al. 2019).

Biogeochemical cycling: The cycling of nutrients and other elements in wetlands involves various chemical transformations that influence water quality. For example, the nitrogen cycle includes processes like nitrification, denitrification, and plant uptake, which help remove excess nitrogen from the water.

Pollutant sequestration: Some pollutants, such as heavy metals and certain organic compounds, can accumulate in wetland substrates or be taken up by plants. This sequestration process effectively removes contaminants from the water, improving overall water quality (Asano 2009).

14.4.4 Buffer Zone and Erosion Control

Buffer Zone:Wetlands act as natural buffer zones between land and water bodies, such as rivers, lakes, and estuaries. They help intercept surface runoff and reduce the direct input of pollutants and sediments into the water bodies by:

Slowing down runoff: The dense vegetation and complex topography of wetlands slow down the flow of surface runoff, providing more time for pollutants to settle and be taken up by plants or degraded by microorganisms.

Filtering pollutants: The physical, biological, and chemical filtration processes within wetlands help remove contaminants from the runoff, such as nutrients, heavy metals, organic compounds, and pathogens. This filtration helps prevent these pollutants from reaching and degrading water quality in adjacent water bodies (Mitsch and Gosselink 2015).

Erosion Control: Wetlands also play a sibnificant role in controlling erosion, which helps prevent sedimentation and the associated pollutants from entering water bodies:

Stabilizing soil: The extensive root systems of wetland plants help stabilize soil and prevent erosion by anchoring the sediment in place (Mitsch and Gosselink 2015).

Reducing runoff velocity: By slowing down runoff, wetlands decrease its erosive force, preventing the dislodging and transport of sediment and associated pollutants.

Trapping sediment: The vegetation and topography of wetlands act as natural sediment traps, capturing and retaining eroded soil particles. This process helps prevent sediment from entering water bodies, where it can cause water turbidity, reduce sunlight penetration, and harm aquatic habitats (Brix 1993).

14.4.5 Carbon Sequestration

Carbon sequestration in wetlands plays a vital role in not only mitigating climate change but also indirectly contributing to water quality improvement. Wetlands are highly efficient at sequestering carbon due to their unique combination of hydrology, vegetation, and soils. The primary mechanisms of carbon sequestration in wetlands are:

Photosynthesis: Wetland plants, such as trees, grasses, and algae, capture atmospheric carbon dioxide (CO2) through the process of photosynthesis and convert it into organic matter.

Organic matter accumulation: The organic matter produced by wetland plants accumulates in the form of plant biomass and is eventually deposited in the soil. Due to the anaerobic conditions in wetland soils, decomposition is slowed down, leading to long-term carbon storage.

Burial and sedimentation: Over time, the accumulation of organic matter in wetland soils leads to a build-up of carbon-rich sediment. This process effectively removes carbon from the atmosphere and stores it in the wetland ecosystem (Mitsch and Gosselink 2015).

14.4.5.1 Impacts of Carbon Sequestration on Water Quality

Carbon sequestration in wetlands indirectly improves water quality through various mechanisms:

Reduced greenhouse gas emissions: By storing carbon in their biomass and soils, wetlands help reduce the concentration of CO_2 in the atmosphere. Lower atmospheric CO_2 levels can help mitigate climate change, which, in turn, can reduce the potential impacts of extreme weather events on water quality, such as increased runoff and erosion (Mitsch and Gosselink 2015).

Nutrient removal: Wetland plants take up nutrients such as nitrogen and phosphorus during photosynthesis. By removing excess nutrients from the water, wetlands help prevent eutrophication, a process that can lead to harmful algal blooms and oxygen depletion in water bodies (Kayranli et al. 2010).

Enhanced biodiversity: Carbon sequestration contributes to the overall health and resilience of wetland ecosystems, which support a diverse range of plant and animal species. A healthy and biodiverse wetland is more capable of maintaining water quality through processes like filtration, nutrient cycling, and pollutant degradation (Mitsch et al. 2013).

14.4.6 Groundwater Recharge

Wetlands helps maintain water quantity and also contributes to water quality improvement by facilitating groundwater recharge through several mechanisms:

Infiltration: The presence of porous soils and dense vegetation in wetlands promotes water infiltration into the ground. As water percolates through the soil, it can replenish groundwater aquifers, providing a valuable source of water for local ecosystems and human communities (Mitsch and Gosselink 2015).

Enhanced recharge capacity: The unique hydrological and geomorphological characteristics of wetlands, such as shallow water depths, slow water movement, and small-scale topographic variations, enhance their groundwater recharge capacity compared to other ecosystems (Acreman and Holden 2013).

14.4.6.1 Impacts of Groundwater Recharge on Water Quality

Groundwater recharge in wetlands indirectly improves water quality through various mechanisms:

Dilution: By recharging groundwater aquifers, wetlands help maintain water levels and dilute pollutants in the groundwater system. This process can reduce the concentration of contaminants in the groundwater, improving its overall quality (Mitsch and Gosselink 2015).

Filtration: As water infiltrates through the wetland soil, it undergoes physical, chemical, and biological filtration processes. These processes help remove pollutants from the water, ensuring that the recharged groundwater is of higher quality (Acreman and Holden 2013).

Temperature regulation: Groundwater recharge in wetlands can help moderate groundwater temperature, creating more stable habitat conditions for aquatic organisms. Stable water temperatures can also influence the rate of chemical and biological processes, which can further impact water quality (Whitehead et al. 2009).

Reduced vulnerability to saltwater intrusion: In coastal areas, maintaining groundwater levels through recharge can help prevent saltwater intrusion into freshwater aquifers. By preventing saltwater contamination, wetlands contribute to the maintenance of freshwater resources and overall water quality.

14.4.7 Pathogen Removal

Wetlands can effectively remove pathogens such as bacteria, viruses, and parasites from water through various physical, biological, and chemical processes.

Wetlands contribute to pathogen removal by:

14.4.7.1 Physical Removal

Sedimentation and filtration: The slow movement of water through wetlands allows particles, including pathogens attached to suspended solids, to settle out of the water column. This sedimentation process helps remove pathogens from the water, improving its overall quality (Kadlec and Wallace 2009).

Sunlight exposure: Exposure to sunlight, particularly ultraviolet (UV) radiation, can inactivate or destroy pathogens in wetlands. This natural disinfection process can help reduce the concentration of harmful microorganisms in the water.

14.4.7.2 Biological Removal.

Microbial predation: Wetlands are home to diverse communities of microorganisms, including those that can prey on or compete with pathogens. These microbial interactions can help reduce pathogen populations in the water, contributing to improved water quality (Vymazal 2011a, 2011b). Plant uptake and filtration: Wetland plants can absorb and filter pathogens from water as it flows through the system. This plant-mediated removal process helps reduce the concentration of pathogens in the water.

Plant uptake and filtration: Wetland plants can absorb and filter pathogens from water as it flows through the system. This plant-mediated removal process aids in reducing the concentration of pathogens in the water. The physical structure of wetland plants, combined with their natural filtration capabilities, contributes to the overall effectiveness of wetlands in removing pathogens and improving water quality (Mitsch and Gosselink 2015).

14.4.7.3 Chemical Removal

Oxidation-reduction reactions: The varying redox conditions in wetlands can facilitate chemical reactions that inactivate or degrade pathogens. For example, exposure to oxidizing conditions can damage pathogen cell structures, reducing their viability and potential for causing harm (Reddy and Delaune 2008).

Adsorption and ion exchange: Pathogens can be removed from water through adsorption onto wetland substrate materials or ion exchange with other particles. These processes help trap pathogens within the wetland system, preventing their transport to downstream water bodies (Kadlec and Wallace 2009).

14.4.8 Pharmaceuticals and Personal Care Products (PPCPs) Attenuation

Pharmaceuticals and personal care products (PPCPs) are emerging contaminants of concern in aquatic ecosystems due to their potential adverse effects on human health and the environment. Wetlands can play a significant role in attenuating PPCPs, contributing to water quality improvement.

14.4.8.1 Physical Attenuation

Sedimentation: Sedimentation is a process where PPCPs adsorb onto particles suspended in water, such as clays, organic matter, and other minerals. As the water flows through the wetland, these particles settle out of the water column and accumulate in the wetland sediment, effectively reducing the concentration of PPCPs in the water (Brown and Seager 2018).

Filtration: Filtration is another physical process that can contribute to PPCPs removal in wetlands. The dense vegetation and substrate materials in wetlands, such as plants, soil, and sediment, form a network of physical barriers that can trap PPCPs as water flows through the system.

This filtration process can be facilitated by the presence of dense stands of emergent and submerged vegetation. These plants provide a large surface area for PPCPs to adsorb to, which can then be removed as the plant matter decomposes (Brix et al. 2019).

14.4.8.2 Biological Attenuation

Biodegradation: Wetlands serve as natural wastewater treatment systems, containing diverse microbial communities that can metabolize and transform emerging contaminants, including pharmaceutical and personal care products (PPCPs).

Microbial degradation involves the catabolic processes by which microorganisms like bacteria, fungi, and algae transform PPCPs into less hazardous forms or even mineralize them into CO2, H2O, and salts (Dubourguier et al. 2019). Bacteria, for instance, can use PPCPs as a source of carbon, nitrogen, or energy, metabolizing them into simpler compounds.

Plant uptake: Wetland plants can also play a significant role in PPCPs attenuation, through a process known as phytoremediation (Wang et al. 2021). Some plants, such as cattails (Typha species), water hyacinth (Eichhornia crassipes), and reed canary grass (Phalaris arundinacea), have the ability to absorb and accumulate PPCPs in their tissues, effectively removing these contaminants from the water (Deng et al. 2020).

14.4.8.3 Chemical Attenuation

Photodegradation: In addition to microbial degradation, sunlight exposure can contribute to the degradation of PPCPs in wetlands through a process called photodegradation (Li et al. 2022). Photodegradation involves the photochemical reactions that occur when PPCPs are exposed to sunlight, leading to the breakdown of their chemical structures (Sánchez et al. 2022).

Wetlands often receive significant sunlight exposure, providing optimal conditions for photodegradation. Certain functional groups, like aromatic rings, are particularly susceptible to photodegradation, resulting in the formation of reactive oxygen species and/or the cleavage of chemical bonds (Yang et al. 2019).

Adsorption and ion exchange: PPCPs can also interact with the sediment, soil, and suspended particles in wetlands, leading to their adsorption and immobilization (Corral et al. 2017). Adsorption involves the PPCPs being attracted to the surface of mineral particles like clays, silts, and organic matter, forming strong chemical bonds. Ion exchange, another process that occurs in wetlands, involves the PPCPs exchanging ions with other particles in the sediment or soil, which can immobilize the contaminants in place.

Chemical transformations: Chemical transformations refer to the chemical reactions that can change the properties and behavior of PPCPs in wetlands (Kumar et al. 2018). For instance, oxidation-reduction reactions involve the addition or removal of electrons, altering the molecular structure of PPCPs. This can lead to the formation of new compounds, some of which may be more water-soluble, making them easier to remove by other mechanisms (Li and Zhang 2019).

Another chemical transformation, hydrolysis, involves the cleavage of chemical bonds in PPCPs by the reaction with water molecules.

14.5 Conclusion

Wetlands are unique and valuable ecosystems that provide a wide range of ecological, economic, and social benefits. They serve as critical habitats for a diverse array of plant and animal species, many of which are found nowhere else on Earth. Wetlands act as natural filters, improving water quality by removing pollutants, sediments, and nutrients, thereby protecting the health of aquatic ecosystems and the quality of our drinking water supplies.

Wetlands also play a vital role in the global water cycle, storing and releasing water during periods of drought and flooding, which helps to mitigate the impacts of extreme weather events. Additionally, wetlands sequester carbon, helping to mitigate the effects of climate change.

Constructed wetlands have emerged as a promising solution for wastewater treatment, offering an environmentally friendly and cost-effective alternative to traditional treatment methods. These engineered systems harness the natural processes of wetlands to remove contaminants from wastewater, ensuring that it can be safely returned to the environment or reused for various purposes.

Despite the many benefits that wetlands provide, they continue to face significant threats, including habitat loss, pollution, and invasive species. It is crucial that we recognize the value of wetlands and take action to protect and restore these vital ecosystems for the benefit of current and future generations. This can be achieved through increased public awareness, improved regulations and policies, and collaborative efforts between stakeholders, including government agencies, non-governmental organizations, and local communities.

To address these challenges and optimize the performance of constructed wetlands, ongoing research and development are essential. This includes investigating innovative design features, identifying the most effective plant species and substrate materials, and exploring new ways to integrate constructed wetlands into existing urban infrastructure. Furthermore, collaborative efforts between researchers, engineers, policymakers, and stakeholders are crucial for promoting the adoption and successful implementation of constructed wetlands for wastewater treatment.

References

Acreman MC, Holden J (2013) How wetlands affect floods. Wetlands 33(5):773–786. https://doi.org/10.1007/s13157-013-0473-2

Al-Layla MA, Al-Rawi SM, Al-Kubaisi AR (2016) Environmental and health risks associated with reuse of wastewater for irrigation. Environ Monit Assess 188(3):1–12

Alongi DM (2020) The structure and function of mangrove ecosystems. In: Alongi DM (ed) The global ecology of mangroves. CRC Press, pp 13–31. https://doi.org/10.1201/9780429325688-2

Asano T (2009) Photochemical degradation of chlorinated organic compounds in wetlands. Ecol Eng 35(1):74–79

Barbier EB, Acreman M, Knowler D (1997) Economic valuation of wetlands: A guide for policy makers and planners. Ramsar Convention Bureau

Beltman B, Farley J, Grams P, Grosse J, Hanowski J, Sklar F (2014) Understanding, managing and restoring ecosystems of the wetland–upland landscape. Ecol Eng 73:310–318. https://doi.org/10.1016/j.ecoleng.2014.09.002

Bertness MD, Gaines SD, Hay ME (2001) Marine community ecology. Sinauer Associates, Inc. ISBN: 978-0878930908

Britz JR, Sridharan S (2007) Wetland ecosystem services: understanding, quantifying and valuing multiple benefits. Environ Sci Pol 10(6):462–474

Brix H (1993) Macrophyte-mediated oxygen transfer in wetlands: Transport mechanisms and rates. In: Constructed Wetlands for Water Quality Improvement. Lewis Publishers, pp 391–398. https://doi.org/10.1016/B978-1-4832-8333-9.50048-3

Brix JM, Vieira FP, Schmook LP, Rodrigues AG, Kechter C (2019) Treatment of emerging organic contaminants in constructed wetlands: a review. Water Res 153:84–103

Brown L, Seager S (2018) Pharmaceuticals in water: occurrence, behaviour, and removal technologies. Philos Trans R Soc A Math Phys Eng Sci 376(2152):20170452

Bullock A, Acreman M (2003) The role of wetlands in the hydrological cycle. Hydrol Earth Syst Sci 7(3):358–389. https://doi.org/10.5194/hess-7-358-2003

Carvalho CA, Ramos JL, Farley R, Tiedje JM (2010) Bacterial community response and pollutant degradation in constructed wetland mesocosms. Water Res 44(12):3595–3607

Cherry JA (2011) Ecology of wetland ecosystems: water, substrate, and life. Nat Educ Know 3(10):16

Corral LM, Rodríguez MM, Ramos JL, Bañón C (2017) Persistence of pharmaceuticals in soils: An overview. J Hazard Mater 349:645–662

Deng Z, Gao J, Wang Q, Wang J (2020) Phytoremediation of pharmaceuticals and personal care products by wetland plants: a review. Int J Phytoremediation 22(8):822–840

Dubourguier N, Bracken SS, Lim BC, Proia SM (2019) Natural attenuation of pharmaceuticals and personal care products (PPCPs) in terrestrial and aquatic environments

Everard M, Shuker L, Gurnell AM (2021) Integrating artificial urban wetlands into communities: A pathway to socio-ecological resilience. Front Built Environ 7:Article 777383. https://doi.org/10.3389/fbuil.2021.777383

Florsheim MJ, Gubala RA, Tsai KH, Johnson DC (2010) Analysis of wetland stability using morphology and watershed variables: Implications for the Lake Erie Coastal Wetland monitoring network. Water Resour Res 46(1):W01507

Garcia-Ochoa F, Gomez E (2009) Bioreactor scale-up and oxygen transfer rate in microbial processes: an overview. Biotechnol Adv 27(2):153–176. https://doi.org/10.1016/j.biotechadv.2008.10.006

Gorham E (1991) Northern peatlands: Role in the carbon cycle and probable responses to climatic warming. Ecol Appl 1(2):182–195. https://doi.org/10.2307/1941844

Hassan I, Chowdhury SR, Prihartato PK, Razzak SA (2021) Wastewater treatment using constructed wetland: Current trends and future potential. PRO 9(11):1917

Hijosa-Valsero M, Matamoros V, Sidrach-Cardona R, Martín-Villacorta J, Bécares E, Bayona JM (2010) Comprehensive assessment of the design configuration of constructed wetlands for the removal of pharmaceuticals and personal care products from urban wastewaters. Water Res 44(12):3669–3678. https://doi.org/10.1016/j.watres.2010.04.022

Ji GD, Sun T, Yang YH, Zhou Q (2016) A review on the implementation of semi-natural constructed wetlands for wastewater treatment in China. Ecol Eng 87:46–54. https://doi.org/10.1016/j.ecoleng.2015.11.013

Jia W, Yang Y, Yang L, Gao Y (2021) High-efficient nitrogen removal andits microbiological mechanism of a novel carbon self-sufficient constructed wetland. Sci Total Environ 775:145901

Junk WJ, Bayley PB, Sparks RE (1989) The flood pulse concept in river-floodplain systems. In: Dodge DP (ed) Proceedings of the international large river symposium. Canadian Special Publication of Fisheries and Aquatic Sciences, pp 110–127. https://doi.org/10.1139/f89-012

Kadlec RH, Knight RL (1996) Treatment wetlands. CRC Press. ISBN: 9781566700617

Kadlec, R. H., & Wallace, S. D. (2009). Treatment wetlands (2nd ed.). CRC Press. ISBN: 9781420042995

Kattelmann R (1996) Alkaline fen ecosystems: Characteristics, rare species, and restoration. Wetlands 16(3):293–304. https://doi.org/10.1007/BF03160583

Kayranli B, Scholz M, Mustafa A, Hedmark Å (2010) Carbon storage and fluxes within freshwater wetlands: a critical review. Wetlands 30(1):111–124. https://doi.org/10.1007/s13157-009-0003-4

Keddy PA (2010) Wetland ecology: principles and conservation, 2nd edn. Cambridge University Press. https://doi.org/10.1017/CBO9780511778052

Kumar A, Gupta S, Chaudhry R (2018) Phytoremediation: a sustainable approach for pharmaceutical and personal care products. Curr Pharm Biotechnol 19(4):517–534

Li W, Zhang Y (2019) Advanced oxidation processes for the removal of pharmaceuticals and personal care products from wastewater: A review. Sci Total Environ 662:430–441. https://doi.org/10.1016/j.scitotenv.2019.01.118

Li Y, Zhang W, Ma Y, Han W, Liu L (2022) Photodegradation mechanism of bisphenol A (BPA): review on the effects of chemical structure and environmental parameters. J Hazard Mater 412:127800

Matamoros V, Bayona JM (2006) Elimination of pharmaceuticals and personal care products in subsurface flow constructed wetlands. Environ Sci Technol 40(18):5811–5816. https://doi.org/10.1021/es060537q

Mitsch WJ, Gosselink JG (2015) Wetlands, 5th edn. Wiley. https://doi.org/10.1002/9781119159167

Mitsch WJ, Bernal B, Hernandez ME (2013) Ecosystem services of wetlands. Int J Biodivers Sci Ecosystem Serv Manag 9(1):1–16. https://doi.org/10.1080/21513732.2013.768230

Paudel P (2023) Ramsar convention: contribution in global wetland conservation and challenges

Reddy KR, DeLaune RD (2008) Biogeochemistry of wetlands: Science and applications. CRC Press

Reddy NM, Reeder CR, Tufenkji N (2019) Plant-mediated pollutant transformations: A mechanistic review. Environ Sci Technol 53(16):8562–8575

Reed DJ (2005) Wetlands. In: Schwartz ML (ed) Encyclopedia of Coastal Science. Encyclopedia of Earth Science Series. Springer, Dordrecht. https://doi.org/10.1007/1-4020-3880-1_352

Rydin H, Jeglum JK (2013) The biology of peatlands, 2nd edn. Oxford University Press. https://doi.org/10.1093/acprof:oso/9780199608519.001.0001

Sanchez-Ramos I, Gagnon V, Rojas R (2019) Understanding the potential influence of climate change variability on the pollutants removal in water resources and constructed wetlands-a systematic review. Sustain For 11(11):1–31. https://doi.org/10.3390/su11113073

Sánchez M, Ruiz I, Soto M (2022) The potential of constructed wetland systems and photodegradation processes for the removal of emerging contaminants, a review. Environments 9(9):116

U.S. Environmental Protection Agency (1993) A Handbook of Constructed Wetlands. https://www.epa.gov/sites/default/files/2015-10/documents/constructed-wetlands-handbook.pdf

U.S. Environmental Protection Agency (EPA) (2000) Manual: Constructed wetlands treatment of municipal wastewater. EPA/625/R-99/010

Vymazal J (2005) Horizontal sub-surface flow and hybrid constructed wetlands systems for wastewater treatment. Ecol Eng 25(5):478–490

Vymazal J (2007) Removal of nutrients in various types of constructed wetlands. Sci Total Environ 380(1–3):48–65

Vymazal J (2011a) Constructed wetlands for wastewater treatment: Five decades of experience. Environ Sci Technol 45(1):61–69. https://doi.org/10.1021/es1014542

Vymazal J (2011b) Plants used in constructed wetlands with horizontal subsurface flow: A review. Hydrobiologia 674(1):133–156. https://doi.org/10.1007/s10750-011-0738-9

Vymazal J (2018) Types of constructed wetlands for wastewater treatment. In: Naddeo VA, Kadlec DJ, Brix SL, Vymazal J (eds) Wetland Systems to Control Urban Runoff. Elsevier, pp 133–150. https://doi.org/10.1016/B978-0-12-816160-6.00006-5

Wang Y, Liu X, & Zhang J (2021) Phytoremediation of pharmaceuticals and personal care products (PPCPs) in wetlands: A review. Environ Sci Pollut Res, 28(15), 18253–18267

Wheeler BD, Proctor MCF (2000) Ecological gradients, subdivisions and terminology of north-west European mires. J Ecol 88(1):187–203. https://doi.org/10.1046/j.1365-2745.2000.00421

Whitehead PG, Wilby RL, Battarbee RW, Kernan M, Wade AJ (2009) A review of the potential impacts of climate change on surface water quality. Hydrol Sci J 54(1):101–123

Winter TC, Harvey JW, Franke OL, Alley WM (1998) Ground water and surface water: A single resource. USGS Circular 1139. https://pubs.er.usgs.gov/publication/cir1139

Wu S, Wallace S, Brix H, Kuschk P, Kirui WK, Masi F, Dong R (2014) Treatment of industrial effluents in constructed wetlands: Challenges, operational strategies, and overall performance. Environ Pollut 192:245–262. https://doi.org/10.1016/j.envpol.2014.04.010

Wu S, Wallace S, Brix H, Kuschk P, Kirui WK, Masi F, Dong R (2015) Treatment of industrial effluents in constructed wetlands: challenges, operational strategies and overall performance. Environ Pollut 201:107–120

Wu H, Wang R, Yan P, Wu S, Chen Z, Zhao Y et al (2023) Constructed wetlands for pollution control. Nat Rev Earth Environ 4(4):218–234

Yang CW, Yu WJ, Zheng ZL, Jiang H, Chen X (2019) Photodegradation characteristics of emerging water pollutants: Bisphenol A (BPA) and tetracycline (TC). Environ Pollut 242:517–526

Zhang W, Ricketts TH, Kremen C, Carney K, Swinton SM (2007) Ecosystem services and dis-services to agriculture. Ecol Econ 64(2):253–260. https://doi.org/10.1016/j.ecolecon.2007.02.006

Zhang D, Luo J, Wong MH (2023) Insight into pharmaceutical and personal care products removal in constructed wetlands: Mechanisms, influencing factors and optimization strategies. Sci Total Environ 856:159024. https://doi.org/10.1016/j.scitotenv.2022.159024

Part IV
Wetland Protection

Chapter 15
Monitoring and Assessment of Water Treatment in Wetlands

Mustapha Abdulsalam, Musa Ojeba Innocent, Miracle Uwa Livinus, Shehu-Alimi Elelu, Ganiyat Omotayo Ibrahim, Salami Olaitan Lateefat, Auwal Sagir Muhammad, and Abdulhakeem Idris Abdulhakeem

Abstract This chapter offers an in-depth exploration of the monitoring and assessment of water treatment in wetlands, focusing on technological innovations, remote sensing, GIS, and conservation tools. Wetlands are essential for maintaining biodiversity, regulating water quality, and providing ecosystem services through processes such as nutrient cycling and water filtration. The chapter reviews the significant advancements in monitoring technologies, including the use of remote sensing technologies like satellite imagery and unmanned aerial vehicles (UAVs), which provide detailed, real-time data crucial for effective wetland management. Geographic Information Systems (GIS) are highlighted for their role in spatial data analysis, hydrological modeling, and decision-making. The chapter also addresses conservation strategies, emphasizing the importance of strong policy frameworks, community engagement, and adaptive management practices. Recommendations are provided for enhancing wetland conservation efforts, including integrating diverse technologies, fostering stakeholder involvement, and adopting emerging

M. Abdulsalam (✉) · M. O. Innocent
Department of Microbiology, Skyline University Nigeria, Kano, Nigeria

M. U. Livinus
Department of Biochemistry, Skyline University Nigeria, Kano, Nigeria

S.-A. Elelu
Department of Chemistry, Howard University, Washington, DC, USA

G. O. Ibrahim
Department of Chemistry, Nottingham Trent University, Nottingham, UK

S. O. Lateefat
Department of Chemistry and Molecular Biology, University of Gothenburg, Gothenburg, Sweden

A. S. Muhammad
School of Informatics, Xiamen University, Xiamen, China

A. I. Abdulhakeem
Department of Fisheries and Aquaculture, Federal University Dutsin-Ma, Dutsin Wai, Nigeria

B. R. Babaniyi et al. (eds.), *Wetland Ecosystems: Conservation Strategies, Policy Management and Applications*, Wetlands: Ecology, Conservation and Management 12, https://doi.org/10.1007/978-3-031-91982-4_15

technologies such as artificial intelligence and machine learning. The chapter underscores the need for continued innovation and collaborative efforts to tackle environmental challenges and preserve wetland ecosystems for future generations.

Keywords Wetland monitoring · Remote sensing · Geographic Information Systems (GIS) · Conservation strategies · Technological innovations

15.1 Introduction

Wetlands are among the most productive ecosystems on Earth, offering numerous ecological, economic, and social benefits. They act as natural water purifiers, filtering pollutants and sediments from water, which helps maintain water quality in rivers, lakes, and groundwater (Jisha and Puthur 2021). More so, wetlands provide habitat for diverse flora and fauna, including many endangered species, and play a critical role in global carbon cycling by sequestering carbon dioxide and other greenhouse gases. Beyond their ecological functions, wetlands also offer recreational opportunities, support fisheries and agriculture, and protect against flooding by absorbing excess rainwater (Silvius et al. 2000).

Wetlands naturally treat water through a combination of physical, chemical, and biological processes, including sedimentation, filtration, chemical precipitation, adsorption, and biological uptake. Constructed wetlands are designed to mimic these natural processes to treat wastewater from various sources, including municipal, agricultural, and industrial effluents (Rana and Maiti 2020). By utilizing the natural vegetation and microbial communities, constructed wetlands can effectively remove nutrients, pathogens, and heavy metals from contaminated water, making them a sustainable and cost-effective water treatment solution (Yu et al. 2022).

The primary objective of this chapter is to explore and elucidate the latest technological innovations in the monitoring and assessment of water treatment in wetlands. The chapter aims to highlight the significance of wetlands in natural water treatment and ecosystem services. It will provide a comprehensive overview of the current technologies used in wetland monitoring, including in-situ tools, automated systems, and remote sensing technologies. Furthermore, the chapter will discuss the applications of Geographic Information Systems (GIS) in wetland management and conservation. Case studies demonstrating successful implementations of these technologies in wetland restoration and conservation projects will be presented. Finally, the chapter will identify future directions and emerging trends in wetland monitoring and conservation, emphasizing the need for continuous innovation and collaboration among researchers, practitioners, and policymakers (Demarquet et al. 2023). By achieving these objectives, the chapter seeks to equip readers with the knowledge and tools necessary for effective wetland conservation and water treatment monitoring, ultimately contributing to the sustainability of these vital ecosystems.

15.2 Wetland Ecosystems and Their Role in Water Treatment

15.2.1 Types of Wetlands

Wetlands are diverse ecosystems classified based on their hydrology, vegetation, and soil types. The primary types include marshes, swamps, bogs, and fens. Marshes are dominated by herbaceous plants and are frequently or continually inundated with water. Swamps are characterized by woody plants and can be either freshwater or saltwater. Bogs are acidic wetlands with peat soils, typically fed by rainwater, and host unique plant species like sphagnum moss. Fens, similar to bogs, have peat soils but are less acidic and receive water from groundwater sources, making them nutrient-rich and supporting diverse plant and animal life (Lamers et al. 2015). These varying types of wetlands provide different habitats and ecological functions, each contributing uniquely to water treatment processes.

15.2.2 Natural Processes in Wetland Water Treatment

Wetlands utilize a combination of physical, chemical, and biological processes to treat water. Sedimentation allows suspended particles to settle, removing them from the water column. Filtration through plant roots and soil matrices traps additional particles and pollutants. Chemical processes, such as adsorption and precipitation, immobilize heavy metals and other contaminants. Biological uptake by plants and microbial activity further break down and assimilate nutrients and organic pollutants. For instance, plants absorb nitrogen and phosphorus, essential nutrients that, in excess, can cause eutrophication in water bodies. Microbial communities in wetland soils degrade organic matter and transform pollutants through processes like denitrification, which converts nitrates to nitrogen gas, reducing nutrient loads in the water (Witthar 2020).

15.2.3 Benefits and Challenges

Wetlands offer numerous benefits for water treatment and broader environmental health. They enhance water quality by removing pollutants, reducing nutrient loads, and mitigating the impacts of stormwater runoff. Wetlands also provide critical habitats for wildlife, support biodiversity, and contribute to climate regulation by sequestering carbon (Moomaw et al. 2018). Additionally, wetlands offer recreational, educational, and cultural benefits, enhancing human well-being and community resilience.

However, there are challenges associated with wetland water treatment. Wetlands can be sensitive to changes in water quantity and quality, making them vulnerable to pollution, climate change, and human activities such as agriculture and urban development. The effectiveness of wetlands in treating water can be influenced by seasonal variations, hydrological conditions, and the types and concentrations of pollutants (Walaszek et al. 2018). Constructed wetlands, while designed to replicate natural processes, require careful design, monitoring, and maintenance to function effectively. Balancing the ecological integrity of wetlands with their use for water treatment and other human benefits requires integrated management approaches and robust regulatory frameworks.

15.3 Technological Innovations in Wetland Monitoring

15.3.1 Overview of Monitoring Technologies

Monitoring technologies have significantly advanced, providing detailed data on wetland health, water quality, and ecological changes. These technologies range from traditional manual sampling methods to sophisticated automated systems and remote sensing tools. The primary goal of monitoring technologies is to collect accurate, real-time data that can inform conservation efforts, regulatory compliance, and adaptive management strategies. The integration of these technologies enables comprehensive monitoring of physical, chemical, and biological parameters, offering insights into the dynamics of wetland ecosystems (Forio and Goethals 2020).

15.3.2 In-Situ Monitoring Tools

In-situ monitoring involves placing sensors and instruments directly in the wetland to measure various parameters on-site. Common in-situ tools include:

- **Water Quality Sensors**: These sensors measure parameters such as pH, temperature, dissolved oxygen, conductivity, turbidity, and nutrient concentrations (e.g., nitrogen and phosphorus). Advanced sensors can also detect pollutants like heavy metals and organic contaminants (Onojake et al. 2023).
- **Soil and Sediment Samplers**: Tools such as soil corers and sediment traps collect samples for analyzing soil composition, nutrient levels, and contaminant concentrations. These samplers help understand the interactions between water, soil, and vegetation in wetlands (Pierzynski et al. 2020).
- **Biological Monitoring Tools**: Devices like flow meters and cameras monitor the presence and activities of wildlife, providing data on biodiversity and the ecological health of wetlands. These tools are essential for assessing the impacts of water quality on flora and fauna (Miah et al. 2021).

In-situ monitoring tools offer high-resolution, site-specific data, crucial for understanding localized processes and conditions in wetlands. However, they require regular maintenance and calibration to ensure accuracy and reliability.

15.3.3 Automated Monitoring Systems

Automated monitoring systems enhance the efficiency and effectiveness of wetland monitoring by continuously collecting and transmitting data without the need for constant human intervention. These systems include:

- **Data Loggers and Telemetry Systems**: Data loggers record measurements from various sensors over time, while telemetry systems transmit this data to remote servers or monitoring stations. This setup allows for real-time data access and analysis, facilitating timely decision-making and management responses (Djordjević et al. 2020).
- **Unmanned Aerial Vehicles (UAVs)**: UAVs, or drones, equipped with cameras and sensors, provide aerial imagery and data on wetland conditions. They can cover large areas quickly, offering detailed spatial data on vegetation health, water levels, and land use changes (Zuo et al. 2022).
- **Remote Sensing Technologies**: Satellite and airborne remote sensing technologies collect data on wetland extent, water quality, and habitat conditions. These technologies use various spectral bands to detect changes in vegetation, water properties, and land cover, providing comprehensive temporal and spatial data (Lunetta et al. 2022).

Automated monitoring systems reduce the labor and time required for data collection, improve consistency, and enable continuous monitoring. They are particularly valuable for detecting trends, anomalies, and rapid changes in wetland ecosystems, supporting proactive management and conservation efforts.

15.4 Remote Sensing in Wetland Assessment

15.4.1 Introduction to Remote Sensing

Remote sensing is acquiring information about an object or phenomenon without making physical contact with it. This technology employs sensors on satellites, aircraft, or drones to capture data across various electromagnetic spectrum bands. In wetland assessment, remote sensing offers a powerful tool for monitoring and analyzing large and often inaccessible areas. It provides critical insights into wetland extent, health, and changes over time, enhancing our understanding of these complex ecosystems and supporting effective management and conservation efforts (Thamaga et al. 2022).

15.4.2 Types of Remote Sensing Technologies

Remote sensing technologies vary in terms of the platforms they use and the types of sensors they employ:

- **Satellite Remote Sensing**: Satellites orbiting the Earth capture broad-scale imagery using different sensors. Key satellites for wetland monitoring include Landsat, which provides long-term data on land cover and vegetation changes, and Sentinel-2, which offers high-resolution multispectral images suitable for detailed wetland analysis (Chen et al. 2019).
- **Airborne Remote Sensing**: Aircraft and drones equipped with sensors collect high-resolution data from lower altitudes compared to satellites. Airborne systems can use LiDAR (Light Detection and Ranging) to measure vegetation structure and water levels, and hyperspectral sensors to detect subtle changes in wetland vegetation and water quality (Chasmer et al. 2020).
- **Unmanned Aerial Vehicles (UAVs)**: UAVs or drones equipped with cameras and multispectral or thermal sensors provide highly detailed, real-time data over specific areas. They are particularly useful for monitoring small to medium-sized wetlands and capturing high-resolution imagery for vegetation mapping and habitat assessment (Messina and Modica 2020).

15.4.3 Applications of Remote Sensing in Wetland Monitoring

Remote sensing plays a crucial role in various aspects of wetland monitoring:

- **Mapping Wetland Extent and Changes**: Remote sensing enables the creation of detailed maps of wetland boundaries and their changes over time. This is essential for tracking land-use changes, habitat loss, and the impact of human activities on wetlands (Guo et al. 2017).
- **Vegetation Analysis**: By analyzing spectral data from satellites or UAVs, researchers can assess wetland vegetation types, health, and biomass. Remote sensing can detect changes in vegetation cover, identify invasive species, and evaluate the effectiveness of restoration efforts (Arasumani et al. 2021).
- **Water Quality Assessment**: Remote sensing can provide information on surface water quality by detecting parameters such as turbidity, chlorophyll-a concentration, and nutrient levels. Sensors can capture data on the color of the water, which correlates with various water quality indicators (Chebud et al. 2012).
- **Flood Monitoring and Management**: Remote sensing is valuable for monitoring flood events and their impacts on wetlands. It provides data on water levels,

inundation patterns, and the extent of flooding, aiding in flood risk assessment and management (Tahsin et al. 2018).
- **Habitat and Biodiversity Monitoring**: Remote sensing helps in assessing habitat conditions and monitoring biodiversity by mapping habitat types, tracking species distribution, and identifying critical habitats for conservation (Lahoz-Monfort and Magrath 2021).

15.5 Geographic Information Systems (GIS) in Wetland Management

Geographic Information Systems (GIS) are powerful tools for capturing, storing, analyzing, and visualizing spatial and geographic data. GIS integrates hardware, software, and data to manage and analyze spatial information, allowing users to interpret complex relationships and patterns within geographic contexts. In wetland management, GIS is utilized to analyze spatial data related to wetland distribution, land use changes, water quality, and ecological processes. By providing a comprehensive view of spatial data, GIS facilitates informed decision-making and effective management of wetland resources (Sritart and Miyazaki 2022).

15.5.1 GIS Tools and Techniques

GIS tools and techniques encompass a range of functionalities designed to manage and analyze spatial data:

- **Data Layers and Spatial Analysis**: GIS allows for the layering of various types of spatial data, such as land use, vegetation cover, and hydrological features. Spatial analysis techniques, including overlay analysis, buffer analysis, and spatial interpolation, enable the examination of relationships between different data layers. For instance, overlay analysis can be used to assess how changes in land use affect wetland areas, while buffer analysis helps in evaluating the impact of surrounding land uses on wetland health (Gaglio et al. 2017).
- **Remote Sensing Integration**: GIS integrates data from remote sensing technologies, such as satellite imagery and aerial surveys. By combining remote sensing data with GIS, users can enhance wetland mapping, monitor changes over time, and analyze spatial patterns in vegetation and water quality. This integration allows for detailed temporal and spatial analysis, aiding in the assessment of wetland dynamics and health (Waleed et al. 2023).
- **Hydrological Modeling**: GIS tools are used to model hydrological processes and predict the impacts of various factors on wetland hydrology. Techniques such as watershed modeling, flood modeling, and water flow analysis help in understanding water movement, distribution, and availability within wetland

systems. These models can be used to predict the effects of land use changes, climate variability, and water management practices on wetlands (Chun et al. 2020).

- **Decision Support Systems**: GIS supports decision-making through tools that provide spatial analyses and simulations. Decision support systems (DSS) use GIS data to assess different management scenarios, evaluate potential impacts, and deUvelop conservation strategies. These systems facilitate stakeholder engagement and collaborative decision-making by visualizing complex data in an accessible format (John et al. 2020).

15.5.2 Case Studies of GIS in Wetland Conservation

Several case studies demonstrate the effective use of GIS in wetland conservation and management:

- **Everglades Restoration Project, USA**: The Comprehensive Everglades Restoration Plan (CERP) employs GIS to manage and restore the Everglades ecosystem. GIS is used to analyze water flow patterns, monitor vegetation changes, and evaluate the impacts of restoration activities. By integrating spatial data from various sources, CERP aims to restore natural water flows and improve ecosystem health (Dessu et al. 2021).
- **Yangtze River Floodplain, China**: GIS has been utilized to map and assess wetlands in the Yangtze River floodplain. Through integrating remote sensing data and GIS, researchers have monitored changes in wetland extent, vegetation cover, and water quality. The GIS-based analysis has informed conservation strategies and management practices to protect critical wetland habitats (Xiang et al. 2023).
- **Wetland Restoration in the Mississippi Alluvial Valley, USA**: GIS has played a key role in planning and implementing wetland restoration projects in the Mississippi Alluvial Valley. By using GIS to analyze land use changes, hydrological conditions, and habitat quality, managers have developed targeted restoration strategies and evaluated the success of conservation efforts (Murray and Klimas 2013).

15.6 Conservation Tools and Strategies for Wetlands

15.6.1 Policy and Regulatory Frameworks

Policy and regulatory frameworks are essential for effectively conserving and managing wetlands. These frameworks provide the legal and institutional structures necessary to protect wetlands and ensure sustainable use. Key aspects include:

- **International Agreements and Conventions**: International agreements such as the Ramsar Convention on Wetlands are crucial for global wetland conservation. The Ramsar Convention promotes the conservation of wetlands of international importance and encourages member countries to develop national policies for wetland protection and sustainable use (Gell et al. 2023).
- **National Legislation**: Many countries have specific laws and regulations governing wetland conservation. For example, the Clean Water Act in the United States regulates the discharge of pollutants into waters and protects wetlands through permitting processes and water quality standards (Andreen 2024). Similarly, the European Union's Water Framework Directive sets out water quality and wetland conservation objectives across member states (Munné et al. 2015).
- **Local Regulations and Zoning**: At the local level, zoning regulations and land-use planning can influence wetland conservation. Municipalities often implement zoning laws to control development near wetlands and ensure that land use is compatible with conservation goals. Local regulations can include restrictions on construction, requirements for buffer zones, and guidelines for sustainable land use (Petrakovska et al. 2020).

These frameworks help to establish guidelines and enforcement mechanisms that support wetland conservation and sustainable management.

15.6.2 Community Involvement and Stakeholder Engagement

Community involvement and stakeholder engagement are vital for the success of wetland conservation efforts. Effective conservation strategies often include:

- **Public Awareness and Education**: Raising awareness about the importance of wetlands and the benefits they provide is crucial for garnering public support. Educational programs and outreach activities can help inform communities about wetland functions, conservation needs, and ways to participate in protection efforts (Gell et al. 2023).
- **Local Participation in Decision-Making**: Engaging local communities in decision-making processes ensures that conservation strategies are relevant and consider local knowledge and needs. Participatory approaches can involve community consultations, workshops, and collaborative planning, allowing stakeholders to contribute to the development and implementation of conservation plans (Kusters et al. 2018).
- **Partnerships and Collaborations**: Forming partnerships between government agencies, non-governmental organizations (NGOs), businesses, and local communities can enhance conservation efforts. Collaborative initiatives can leverage resources, share expertise, and coordinate actions to address conservation challenges more effectively (Abdulsalam et al. 2024).

Active engagement of stakeholders helps to build consensus, enhance the effectiveness of conservation measures, and promote sustainable management of wetland resources.

15.6.3 Best Practices in Wetland Conservation

Adopting best practices in wetland conservation ensures that conservation efforts are effective, sustainable, and aligned with ecological goals. Key practices include:

- **Restoration and Rehabilitation**: Restoration involves reestablishing the natural functions and structure of degraded wetlands. This can include actions such as removing invasive species, replanting native vegetation, and restoring hydrological regimes. Successful restoration projects require careful planning, monitoring, and adaptive management (Musa et al. 2024).
- **Sustainable Management**: Implementing sustainable management practices involves balancing ecological, economic, and social objectives. This can include sustainable use of wetland resources, such as controlled fishing or ecotourism, and managing land use around wetlands to reduce impacts and support conservation goals (Lawal et al. 2023).
- **Monitoring and Evaluation**: Regular monitoring and evaluation are essential for assessing the effectiveness of conservation measures and making informed decisions. This includes tracking changes in wetland conditions, evaluating the success of restoration efforts, and adapting management practices based on monitoring data (Geneletti, 2004).
- **Adaptive Management**: Adaptive management is an iterative process that incorporates feedback and learning to improve conservation strategies. By continually assessing outcomes and adjusting approaches based on new information, adaptive management helps to address uncertainties and enhance the resilience of wetland ecosystems (Livinus et al. 2024).

15.7 Case Studies and Applications

15.7.1 Successful Wetland Restoration Projects

Wetland restoration projects aim to revive and enhance the ecological functions of degraded wetlands. Successful projects demonstrate effective strategies for restoring wetland health and resilience:

- **Florida Everglades Restoration (USA)**: The Comprehensive Everglades Restoration Plan (CERP) is one of the most extensive wetland restoration projects globally. Initiated in 2000, CERP aims to restore the natural flow of water across the Everglades, address habitat loss, and improve water quality. Key activities include removing barriers to water flow, restoring natural wetlands, and managing water levels to support diverse plant and animal species. CERP has made significant progress in improving water flow and enhancing habitat conditions, though challenges remain in fully achieving restoration goals (Davis and Ogden 1994).

- **The Danube Delta Restoration (Romania/Ukraine)**: The Danube Delta is a UNESCO World Heritage Site and one of Europe's most biodiverse wetlands. Restoration efforts have focused on reconnecting the delta's natural water channels, enhancing floodplain connectivity, and improving water quality. These efforts have led to the recovery of critical habitats for fish and bird species, improved ecological functions, and increased biodiversity. The project highlights the importance of cross-border cooperation and integrated management for successful restoration (Bojinski et al. 2014).
- **The Loess Plateau Watershed Rehabilitation Project (China)**: Although not exclusively a wetland restoration project, this initiative has significantly impacted wetland ecosystems by addressing land degradation and improving watershed health. The project has implemented soil conservation measures, reforestation, and wetland restoration to reduce erosion, enhance water retention, and support local livelihoods. The success of this project demonstrates the potential benefits of integrating wetland restoration with broader landscape management efforts (Shaojun et al. 2004).

15.7.2 Innovative Technologies in Action

Innovative technologies are increasingly being applied to enhance wetland management and restoration:

- **LiDAR Technology**: Light Detection and Ranging (LiDAR) technology is used to create high-resolution digital elevation models and assess vegetation structure in wetlands. LiDAR data helps in mapping wetland topography, monitoring changes in vegetation cover, and evaluating hydrological patterns. For instance, LiDAR has been used in the Mississippi Alluvial Valley to assess floodplain elevations and support restoration planning (van der Most and Hudson 2018).
- **Remote Sensing for Water Quality Monitoring**: Advanced remote sensing technologies, such as hyperspectral and thermal imaging, are used to monitor water quality parameters, including temperature, chlorophyll-a concentration, and turbidity. These technologies provide real-time data on water conditions, enabling timely responses to pollution events and improved management of wetland water quality (Chasmer et al. 2020).
- **Automated Monitoring Systems**: Automated systems equipped with sensors and telemetry units are deployed to continuously monitor wetland conditions. These systems can measure parameters such as water level, flow rate, and soil

moisture in real-time. Automated systems facilitate efficient data collection, reduce the need for manual sampling, and provide continuous insights into wetland dynamics (Lunetta et al. 2022).

15.7.3 Lessons Learned from Various Regions

Case studies from different regions offer valuable lessons for wetland management and restoration:

- **Adaptive Management**: The experience from the Everglades Restoration Project underscores the importance of adaptive management in addressing uncertainties and evolving conditions. Continuous monitoring, evaluation, and flexibility in management strategies are essential for achieving long-term restoration goals (Davis and Ogden 1994).
- **Stakeholder Engagement**: Successful projects, such as the Danube Delta Restoration, highlight the need for active stakeholder engagement and cross-border cooperation. Engaging local communities, governments, and NGOs in planning and implementation ensures that restoration efforts are well-supported and aligned with local needs (Bojinski et al. 2014).
- **Integration with Broader Goals**: The Loess Plateau project demonstrates the benefits of integrating wetland restoration with broader environmental and socio-economic goals. Combining wetland restoration with land management, soil conservation, and livelihood improvements can enhance the overall effectiveness and sustainability of conservation efforts (Jiang et al. 2019).

15.8 Future Directions and Emerging Trends

15.8.1 Advances in Monitoring Technologies

Monitoring technologies are continuously evolving, offering new tools and methods for tracking and managing wetland ecosystems:

- **Satellite and Aerial Remote Sensing**: Advances in satellite and aerial remote sensing technologies are improving the resolution and accuracy of wetland monitoring. Newer satellites, such as the Sentinel-2 mission, provide high-resolution multispectral imagery that allows for detailed vegetation analysis and water quality assessment. More so, the development of hyperspectral sensors enhances the ability to detect subtle changes in wetland ecosystems, such as variations in plant health and nutrient concentrations (Chasmer et al. 2020).
- **Drones and UAVs**: Unmanned Aerial Vehicles (UAVs) or drones are increasingly used for wetland monitoring due to their ability to capture high-resolution

imagery and provide real-time data. Advances in drone technology, including improved flight stability and longer battery life, enable detailed assessments of wetland conditions, vegetation structure, and hydrological changes (Messina and Modica 2020).
- **Smart Sensors and IoT**: The integration of smart sensors and Internet of Things (IoT) technologies is revolutionizing wetland monitoring. These sensors can measure various environmental parameters such as water quality, soil moisture, and atmospheric conditions. IoT networks facilitate the collection of real-time data from remote and inaccessible locations, enabling more responsive management and conservation efforts (Salam 2024).

15.8.2 Integrating New Data Sources

The integration of diverse data sources enhances the understanding and management of wetlands:

- **Big Data and Analytics**: The use of big data and advanced analytics is transforming wetland management. By integrating large volumes of data from remote sensing, in-situ measurements, and historical records, researchers can perform complex analyses to identify trends, predict changes, and assess the impact of various factors on wetland ecosystems (Saheed et al. 2023). Big data analytics can provide insights into patterns of wetland degradation, restoration effectiveness, and ecological interactions.
- **Citizen Science**: Citizen science initiatives are increasingly contributing to wetland monitoring and conservation. Platforms that engage the public in data collection and observation allow for the gathering of large datasets from diverse locations. These contributions can enhance the monitoring of wetland species, track changes in wetland conditions, and increase public awareness and involvement in conservation efforts (Silvertown 2009).
- **Cross-Disciplinary Integration**: Combining data from different disciplines, such as hydrology, ecology, and climatology, provides a more comprehensive understanding of wetland systems. Cross-disciplinary approaches enable the integration of various data types and models, facilitating more accurate predictions and better-informed management strategies (Dandois and Ellis 2010).

15.8.3 Predictions for the Future of Wetland Conservation

The future of wetland conservation will likely be shaped by several emerging trends and predictions:

- **Increased Use of AI and Machine Learning**: Artificial Intelligence (AI) and machine learning are expected to play a significant role in wetland conservation.

These technologies can automate data analysis, improve the accuracy of ecological models, and identify patterns and anomalies in large datasets. AI-driven tools will enhance decision-making processes and facilitate more efficient conservation planning (Raihan 2023).

- **Climate Change Adaptation**: As climate change continues to impact wetland ecosystems, adaptation strategies will become increasingly important. Future conservation efforts will need to focus on enhancing the resilience of wetlands to climate-induced stresses, such as altered hydrological patterns and increased temperatures. Strategies may include developing climate-resilient restoration techniques, managing migration corridors for species, and integrating climate projections into conservation planning (Lahoz-Monfort and Magrath 2021).
- **Emphasis on Ecosystem Services**: There will be a growing emphasis on recognizing and valuing the ecosystem services provided by wetlands. Future conservation efforts will likely incorporate economic valuation of ecosystem services, such as flood regulation, water purification, and carbon sequestration, into decision-making processes. This approach will help to justify investments in wetland conservation and restoration by highlighting the tangible benefits provided by these ecosystems (Goyette et al. 2021).

15.9 Conclusion

This chapter has examined the diverse methods for monitoring and assessing water treatment in wetlands, emphasizing the importance of technological innovations, remote sensing, GIS, and conservation tools. Wetlands play a crucial role in maintaining biodiversity, regulating water quality, and providing essential ecosystem services through natural processes like nutrient cycling and water filtration. Advances in monitoring technologies, including remote sensing, GIS, and in-situ sensors, are transforming how wetlands are observed and managed by offering detailed spatial and temporal data. Remote sensing technologies such as satellite imagery and UAVs provide valuable insights into wetland conditions, while GIS tools support spatial data analysis and hydrological modeling. Effective wetland conservation requires strong policy frameworks, community engagement, and best practices. To enhance conservation efforts, it is recommended that practitioners integrate various technologies and data sources, engage stakeholders, adopt adaptive management practices, and promote supportive policies. Embracing emerging technologies like AI and machine learning will also play a key role. The conservation and management of wetland ecosystems are vital for sustaining their ecological functions and services. Addressing environmental challenges and integrating innovative approaches will be crucial for the future of wetland conservation, with collaborative efforts from practitioners, policymakers, and communities being essential for long-term success.

References

Abdulsalam M, Elelu SA, Innocent MO, Ibrahim GO, Livinus MU, Lateefat SO, Muhammad AS (2024) Women empowerment in environmental conservation. In: Prospects for soil regeneration and its impact on environmental protection. Cham, Springer, pp 351–370

Andreen WL (2024) Water policy in the United States: the clean water act. In: Handbook on the governance and politics of water resources. Edward Elgar Publishing, pp 285–297

Arasumani M, Bunyan M, Robin VV (2021) Opportunities and challenges in using remote sensing for invasive tree species management, and in the identification of restoration sites in tropical montane grasslands. J Environ Manag 280:111759

Bojinski S, Verstraete M, Peterson TC, Richter C, Simmons A, Zemp M (2014) The concept of essential climate variables in support of climate research, applications, and policy. Bull Am Meteorol Soc 95(9):1431–1443

Chasmer L, Mahoney C, Millard K, Nelson K, Peters D, Merchant M et al (2020) Remote sensing of boreal wetlands 2: methods for evaluating boreal wetland ecosystem state and drivers of change. Remote Sens 12(8):1321

Chebud Y, Naja GM, Rivero RG, Melesse AM (2012) Water quality monitoring using remote sensing and an artificial neural network. Water Air Soil Pollut 223:4875–4887

Chen Y, Guerschman JP, Cheng Z, Guo L (2019) Remote sensing for vegetation monitoring in carbon capture storage regions: a review. Appl Energy 240:312–326

Chun X, Qin FY, Zhou HJ, Dan D, Xia YY, Ulambadrakh K (2020) Effects of climate variability and land use/land cover change on the Daihai wetland of Central Inner Mongolia over the past decades. J Mt Sci 17(12):3070–3084

Dandois JP, Ellis EC (2010) Remote sensing of vegetation structure using computer vision. Remote Sens 2(4):1157–1176

Davis S, Ogden JC (1994) Everglades: the ecosystem and its restoration. CRC Press

Demarquet Q, Rapinel S, Dufour S, Hubert-Moy L (2023) Long-term wetland monitoring using the landsat archive: a review. Remote Sens 15(3):820

Dessu SB, Paudel R, Price RM, Davis SE (2021) Using empirical data and modeled scenarios of everglades restoration to understand changes in coastal vulnerability to sea level rise. Clim Chang 168(3):19

Djordjević M, Jovičić B, Marković S, Paunović V, Danković D (2020) A smart data logger system based on sensor and internet of things technology as part of the smart faculty. J Amb Intel Smart En 12(4):359–373

Forio MAE, Goethals PL (2020) An integrated approach of multi-community monitoring and assessment of aquatic ecosystems to support sustainable development. Sustain For 12(14):5603

Gaglio M, Aschonitis VG, Gissi E, Castaldelli G, Fano EA (2017) Land use change effects on ecosystem services of river deltas and coastal wetlands: a case study in Volano–Mesola–Goro in Po river delta (Italy). Wetl Ecol Manag 25:67–86

Gell P, Davidson N, Finlayson M (eds) (2023) Ramsar wetlands: values, assessment, management. Elsevier

Geneletti D (2004) A GIS-based decision support system to identify nature conservation priorities in an alpine valley. Land Use Policy 21(2):149–160

Goyette JO, Cimon-Morin J, Mendes P, Thériault M, Pellerin S, Poulin M (2021) Planning wetland protection and restoration for the safeguard of ecosystem service flows to beneficiaries. Landsc Ecol 36(9):2691–2706

Guo M, Li J, Sheng C, Xu J, Wu L (2017) A review of wetland remote sensing. Sensors 17(4):777

Jiang C, Zhang H, Wang X, Feng Y, Labzovskii L (2019) Challenging the land degradation in China's loess plateau: benefits, limitations, sustainability, and adaptive strategies of soil and water conservation. Ecol Eng 127:135–150

Jisha KC, Puthur JT (2021) Ecological importance of wetland systems. In: Wetlands conservation: current challenges and future strategies. Wiley, pp 40–54

John B, Lang DJ, von Wehrden H, John R, Wiek A (2020) Advancing decision-visualization environments—empirically informed design recommendations. Futures 123:102614
Kusters K, Buck L, de Graaf M, Minang P, van Oosten C, Zagt R (2018) Participatory planning, monitoring, and evaluation of multi-stakeholder platforms in integrated landscape initiatives. Environ Manag 62(1):170–181
Lahoz-Monfort JJ, Magrath MJ (2021) A comprehensive overview of technologies for species and habitat monitoring and conservation. Bioscience 71(10):1038–1062
Lamers LP, Vile MA, Grootjans AP, Acreman MC, van Diggelen R, Evans MG et al (2015) Ecological restoration of rich fens in Europe and North America: from trial and error to an evidence-based approach. Biol Rev 90(1):182–203
Lawal TO, Abdulsalam M, Aliyu M, Sundararajan S (2023) Economic and environmental implications of sustainable agricultural practices in arid regions: a cross-disciplinary analysis of plant science, management, and economics. Int J 10(3):3100–3114
Livinus MU, Bala SZ, Abdulsalam M, Innocent MO, Hassan M, Kini P (2024) Natural occurrences of soil dilapidation. In: Prospects for soil regeneration and its impact on environmental protection. Cham, Springer, pp 205–223
Lunetta RS, Knight JF, Ediriwickrema J, Lyon JG, Worthy LD (2022) Land-cover change detection using multi-temporal MODIS NDVI data. In: Geospatial information handbook for water resources and watershed management, volume II. CRC Press, pp 65–88
Messina G, Modica G (2020) Applications of UAV thermal imagery in precision agriculture: state of the art and future research outlook. Remote Sens 12(9):1491
Miah MR, Sayok AK, Rahman AAMS, Samdany AA, Akhtar F, Azad AK et al (2021) Impact of sensor networks on aquatic biodiversity in wetland: an innovative approach. Geosciences 11(1):10–42
Moomaw WR, Chmura GL, Davies GT, Finlayson CM, Middleton BA, Natali SM et al (2018) Wetlands in a changing climate: science, policy, and management. Wetlands 38(2):183–205
Munné A, Ginebreda A, Prat N (eds) (2015) Experiences from surface water quality monitoring: the EU water framework directive implementation in the Catalan river basin district (part I) (vol. 42). Springer
Murray EO, Klimas CV (2013) A regional guidebook for applying the hydrogeomorphic approach to assessing functions of forested wetlands in the Mississippi Alluvial Valley
Musa IO, Samuel JO, Adams M, Abdulsalam M, Nathaniel V, Maude AM, Tiamiyu AGT (2024) Soil erosion, mineral depletion and regeneration. In: Prospects for soil regeneration and its impact on environmental protection. Cham, Springer, pp 159–172
Onojake MC, Obi C, Mahmoud AED (2023) Water quality monitoring using sensors and models. In: Artificial intelligence and modeling for water sustainability. CRC Press, pp 97–127
Petrakovska O, Trehub M, Trehub Y, Yankin O (2020) Determining and determinable factors influencing the size of the zone of land-use restriction. Min Miner Depos 14:107–111
Pierzynski GM, Hettiarachchi GM, Koelliker JK (2020) Methods for assessing the impacts of soil degradation on water quality. In: Methods for assessment of soil degradation. CRC Press, pp 513–545
Raihan A (2023) Artificial intelligence and machine learning applications in forest management and biodiversity conservation. Nat Resour Conserv Res 6(2):3825
Rana V, Maiti SK (2020) Municipal and industrial wastewater treatment using constructed wetlands. In: Shmaefsky B (ed) Phytoremediation. Concepts and strategies in plant sciences. Springer, Cham, pp 329–367
Saheed YK, Balogun BF, Odunayo BJ, Mustapha A (2023) Microarray gene expression data classification via Wilcoxon sign rank sum and novel Grey Wolf optimized ensemble learning models. IEEE/ACM Trans Comput Biol Bioinform 20:3575–3587
Salam A (2024) Internet of things in water management and treatment. In: The internet of things for sustainable community development: wireless communications, sensing, and systems. Springer, Cham, pp 273–298

Shaojun C, Yue W, Yijie W (2004, May) The loess plateau watershed rehabilitation project. In: Scaling up poverty reduction: a global learning process and conference, Shanghai, pp 25–27
Silvertown J (2009) A new dawn for citizen science. Trends Ecol Evol 24(9):467–471
Silvius MJ, Oneka M, Verhagen A (2000) Wetlands: a lifeline for people at the edge. Phys Chem Earth B 25(7–8):645–652
Sritart H, Miyazaki H (2022) Geographic Information System (GIS) and data visualization. In: Disaster nursing, primary health care and communication in uncertainty. Springer, Cham, pp 297–307
Tahsin S, Medeiros SC, Singh A (2018) Assessing the resilience of coastal wetlands to extreme hydrologic events using vegetation indices: a review. Remote Sens 10(9):1390
Thamaga KH, Dube T, Shoko C (2022) Advances in satellite remote sensing of the wetland ecosystems in sub-Saharan Africa. Geocarto Int 37(20):5891–5913
van der Most M, Hudson PF (2018) The influence of floodplain geomorphology and hydrologic connectivity on alligator gar (Atractosteus spatula) habitat along the embanked floodplain of the lower Mississippi River. Geomorphology 302:62–75
Walaszek M, Bois P, Laurent J, Lenormand E, Wanko A (2018) Urban stormwater treatment by a constructed wetland: seasonality impacts on hydraulic efficiency, physico-chemical behavior and heavy metal occurrence. Sci Total Environ 637:443–454
Waleed M, Sajjad M, Shazil MS, Tariq M, Alam MT (2023) Machine learning-based spatial-temporal assessment and change transition analysis of wetlands: an application of Google earth engine in Sylhet, Bangladesh (1985–2022). Eco Inform 75:102075
Witthar SR (2020) Wetland water treatment systems. In: Constructed wetlands for water quality improvement. CRC Press, pp 147–155
Xiang H, Xi Y, Mao D, Mahdianpari M, Zhang J, Wang M, Wang Z (2023) Mapping potential wetlands by a new framework method using random forest algorithm and big earth data: a case study in China's Yangtze River Basin. Glob Ecol Conserv 42:e02397
Yu G, Wang G, Chi T, Du C, Wang J, Li P, Chen H (2022) Enhanced removal of heavy metals and metalloids by constructed wetlands: a review of approaches and mechanisms. Sci Total Environ 821:153516
Zuo Z, Liu C, Han QL, Song J (2022) Unmanned aerial vehicles: control methods and future challenges. IEEE/CAA J Autom Sinica 9(4):601–614

Chapter 16
Application of Remote Sensing in Wetland Monitoring

Adejoke Blessing Aransiola, Alaaya Abdulrasheed Abdulwaheed, Mamud Baba, Imeime Innocent Uyo, Temitope Oyadokun, and Mohammed Oludare Idrees

Abstract Wetlands are a valuable resource that are natural which offer the ecology several advantages. For this reason, wetlands need to be well defined. So far, a number of review papers have been published for Remote Sensing (RS) based wetland mapping. This review will give an insight to the reader on the usage of remote sensing (RS) in the monitoring of wetlands. Nevertheless, there is a pressing need to classify and analysis wetlands. Several RS techniques for classifying wetlands are examined, along with the benefits and drawbacks of each. Wetlands are amazing ecosystems with a wealth of ecological, commercial, and social advantages. These methods involve the classification of wetlands using aerial, multispectral, synthetic aperture radar (SAR), and various other data sets. This study also investigates object-based and pixel-based wetland classification techniques. The most significant findings from the literature indicate that the optimal optical bands for mapping wetlands are the red edge and near-infrared bands.

Keywords Multi-spectral satellites · Coarse spatial resolution data · Mapping · Photography

---

A. B. Aransiola (✉) · A. A. Abdulwaheed · M. O. Idrees
Department of Surveying and Geoinformatics, Faculty of Environmental Science, University of Abuja, Abuja, Nigeria
e-mail: adejoke.aransiola@uniabuja.edu.ng

M. Baba
Department of Surveying and Geoinformatics, Faculty of Environmental Technology, Federal University of Technology, Minna, Nigeria

I. I. Uyo
Institute of Photogrammetry and Geoinformation, Leibniz University Hannover, Hanover, Germany

T. Oyadokun
African University of Science and Technology, Abuja, Nigeria

B. R. Babaniyi et al. (eds.), *Wetland Ecosystems: Conservation Strategies, Policy Management and Applications*, Wetlands: Ecology, Conservation and Management 12, https://doi.org/10.1007/978-3-031-91982-4_16

16.1 Introduction

Wetland is a terrain that transitions between terrestrial and aquatic systems and has shallow water covering it or a water table that is typically at or near the surface (Mitsch and Gosselink 1993). Wetlands are extremely advantageous to the environment and serve as essential habitats for a number of rare kinds of plants and animals. They provide a variety of advantages, including recreational opportunities, soil and water conservation, protection from natural dangers, and water purification (Ji et al. 2015). Melton et al. (2013) estimate that wetlands occupy at least seven million square kilometers of the planet. However, due to intensive irrigation techniques, exploitation, and other factors, wetlands are vulnerable to an accelerated degradation (Millennium Ecosystem Assessment 2005). Numerous stakeholders have created various wetland inventories for this reason after realizing the importance of wetlands and the great potential of remote sensing (RS) for mapping these priceless natural resources. Wetland inventories are a valuable tool for assessing the efficacy of wetland policies. They are simply maps that show the location and distribution of wetlands across geographic regions. Nevertheless, creating categorization schemes (Fig. 16.1) to specify the kinds of wetland classes that need to be mapped is a prerequisite for creating wetland inventories. As a result, a variety of classification method which may be broadly categorized into field-based and RS methods—were developed for the purpose of implementing categorization schemes. In contrast to field-based techniques, remote sensing (RS) is an affordable instrument that can obtain regular measurements from far locations and deliver data in

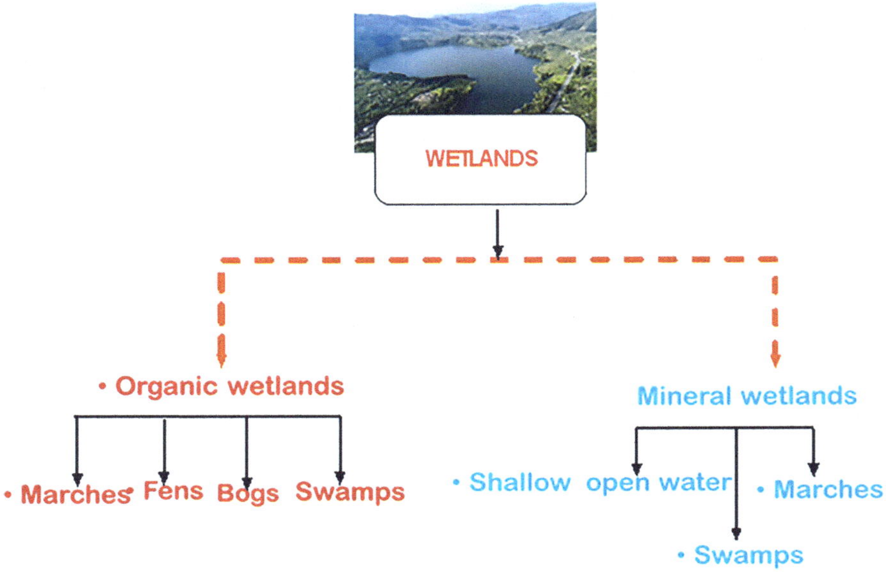

Fig. 16.1 The 2017 Alberta Wetland Policy and the Wetland Classification System (WCS)

real time. Satellite remote sensing (RS) has proven to be the most successful and economical approach for this goal, considering the present requirement for current information and the extensive coverage of wetland maps (Ozesmi and Bauer 2002). Wetland classification remains a difficult task from the perspective of RS, even with numerous advancements in RS technology (Corcoran et al. 2012). The fact that each of the wetland classes has a number of unique traits, but also some ecological parallels with other non-wetland classes and each other, is a major contributing factor to this issue. As a result, spectral and/or backscattering information in RS images is similar across different wetlands (Amani et al. 2017). Furthermore, wetlands differ greatly throughout time and place. Notwithstanding these challenges, RS approaches are preferred, particularly in light of the fact that fieldwork (a tedious, somewhat expensive, and time-consuming task) is the prospective alternative. Consequently, a great deal of research has been done in an effort to create novel and efficient RS techniques for the goal of mapping wetlands with the least amount of in situ data. Many review papers have been published regarding the classification of wetlands and the several issues that come with it. For example, SAR sensors were briefly discussed in Ozesmi and Bauer's (2002) comprehensive assessment of various sensors used for wetland mapping and monitoring. This review paper also examined various approaches to the classification of wetlands, such as aerial photo interpretation and supervised and unsupervised classification techniques. The reader will gain a great deal of understanding from this review of the advantages and restrictions of using SAR sensors to detect wetlands. Additionally, it gives consumers advice on selecting the proper sensor configuration. In a different study, Adam et al. (2010) examined the spectral properties of different types of vegetation found in wetlands and reviewed the classification of wetland using multispectral and hyperspectral RS sensors. Similar to this, Dronova (2015) conducted a comprehensive and helpful analysis of the literature on object-based wetland classification and investigated the benefits and drawbacks of object-oriented mapping of wetlands, as well as the factors that influence its correctness. Lastly, Brisco (2015) looked into surface water and wetland mapping and monitoring using SAR RS techniques. Every review that was previously discussed examined wetland classification from a particular angle. Consequently, a more comprehensive literature study is required, one that includes an overview of the significance of wetlands, the rationale behind the necessity of mapping and monitoring wetlands, and a varied explanation of the numerous approaches to wetlands classification and monitoring through the use of RS. This review gives an insight of the application of RS in wetland monitoring and also the likely future guidelines in wetland ecosystem monitoring.

16.2 Remote sensing in Wetland Monitoring

Land surveys of vegetation, soil and water are the approaches on which the traditional mapping and classification techniques are based on in other to check different wetland characteristics. Though laborious, costly and having some limitation when considering accuracy (for example hydrological regime and wetland connections), a framework for understanding the characteristics of wetland and the management of the resources of wetland can be provided effectively using this approach (Huang et al. 2002).

More severe on a large scale are these limitations for mapping wetland. For some years now, Earth Observation (EO) technologies data has alleviated the problem to a significant ratio by gathering EO time series data for different applications in the environment. Repeatable and consistent observation, wide spatial coverage and high spatial and temporal resolution data ability in particular reduced need for wide field campaigns (Hemati et al. 2023).

16.2.1 Remote Sensing (RS) in Wetland Studies and Their Categories

The spectral, spatial and temporal dynamics of wetland ecosystems essential component which are hydrology, vegetation, and geomorphology, can be best taken with the use of data and technology of remote sensing. According to Schmidt and Skidmore 2003, a major basis of earth surface's cover constitution spatial information is RS. Abundant Earth information are captured by different sensors and can be useful for scientists that have interest in spatial information monitoring in a timely manner (Park et al. 2003).

In many researches involving wetland areas, RS technology has been used during the past five decades for purposes such as: (1) Climate warming in wetland environments and carbon cycle (Holden 2005); (2) wetland regions mapping or land use/cover changes (Giri et al. 2011); (3) peatland fires release of carbon (Rappold et al. 2011); and (4) wetlands hydrology procedures (Slater and Reeve 2002). The various remote sensing categories include.

16.2.1.1 Studying Wetland with Aerial Photographs

This permits the speedy gathering of a huge sum of data and a exceptional impression of an area. The foremost remote sensing technology used in analyzing events on ground surface is the Aerial photography. Satellite remote sensing progress during the 1970s and 1980s has pushed the analyses of aerial-imagery to a new state (Kuenzer et al. 2011).

Wetland plant and land cover identification and classification are of a big challenge due to phenological vegetation variations, the complication of the landscapes of wetland and the impact of biological and physical variables, such as levels of water, density and content of salt (Cline et al. 2011). Aerial photography procedures have been broadly used in studies of wetland because of their outstanding benefits when considering spatial resolution, time and cost, particularly when satellite remote sensing methods were in an initial stage. Since the challenges of data procurement in large zones by aircraft, aerial photography is usually been used for small areas mapping of wetland. After the launching of the satellites, particularly Landsat TM, aerial photography was chiefly used in the valuation of the procedures involved in biomass classification gotten from remote sensing approaches with lower-resolution (Guo et al. 2017).

16.2.1.2 Optical Remote Sensing for Wetland Classification

Multi-spectral (MS) satellites, such as Landsat optical remote sensing data are widely used for the classification of wetland, which result in precise classification maps (Mahdianpari et al. 2018). This is due to the fact that most sensitivity of classes of wetland are in the visible bands and Near-Infrared (NIR) and derived indexes of vegetation such as Normalized Difference Vegetation Index (NDVI) from MS data, which makes them distinct from another. Nevertheless, cloud cover and /night conditions hinder the ability of optical remote sensing sensors. Thus, commonly unusable in geographic areas with high cloud cover are such data (Mahdianpari et al. 2021). As a result, Synthetic Aperture Radar (SAR) has been an attractive substitute, considering its capacity of collecting data in all conditions of weather. Also, the penetration of radar makes it advantageous for studies of vegetation in wetland. The signal of SAR is also sensitive to the surface dielectric and roughness properties thus making it likely to recover data about the size, shape, moisture content and orientation of the target (Mahdianpari et al. 2018).

16.2.1.3 Studying Wetland with LiDAR Data

Light Detection and Ranging (LiDAR) is a remote sensing technology used in monitoring the surface of the Earth. It makes use of a pulsed laser to measure the distance between the target object and the sensor. Three-dimensional data about the ground and its surface can be generated by LiDAR and is therefore useful in deriving the Digital Elevation Model (DEM) (Guo et al. 2017). LiDAR data have been effectively used in elevation mapping of the forest environments using the advantage of acquiring an accurate DEM. The first signal returned is reflected from the canopy while the last signal returned is from the surface of the ground for full waveform LiDAR (Hladik and Alber 2012). Disasters such as flood, storm surges and saltwater intrusion along coastal regions are caused by sea level rise, an important factor for salt marshes stability. Because of the benefits LiDAR data have in deriving accurate elevations of the ground, sea level rise and related research are also key research themes using LiDAR data (Guo et al. 2017).

16.2.1.4 Studying Wetland with Radar (Radio Detection and Ranging) Data

Since the 1970s, Space borne optical data have been extensively used for studying the environment and vegetation. Landsat and Terra/Aqua launching took optical sensors usage to a new level. Though, the principal optical data weakness is the haze and clouds presence. In some regions, because of the cycle of the monsoon, rain and clouds last for a long time; this period is sometimes also significant for the growth of plant. As a result of the thick vegetation cover that leads to signal saturation within wetlands, optical images usually fail to monitor vegetation types (Morandeira et al. 2016). A radar being a sensor that is active transmits, receives and processes the signals of microwave radio. The ability of the radar data to infiltrate clouds is a strong advantage over optical data in the acquisition of ground information. Because of the strength of infiltrating canopy of vegetation and obtaining information of the ground all without the restriction of clouds, radar data, such as RadarSAT, European Remote Sensing (ERS-1), Phased Array L-band Synthetic Aperture Radar (PALSAR), AIRSAR, TerraSAR-X and Japanese Earth Resources Satellite 1 (JERS-1) are uniquely fit to monitor and identify changes in flooding, soil moisture, and above-ground biomass in wetlands (Kasischke et al. 2009).

Multi-sensors were used to by White (2013) in capturing the variation in the community's vegetation of spring wetland and Australian Great Artesian Basin surface expressions. The imagery of airborne hyperspectral was carefully chosen as a reference data to differentiate spring plant communities and surrounding substrate in detail. Rapinel et al. 2015a used the combination of multispectral and multi-seasonal imagery and Light Detection and Ranging (LiDAR) data to map precisely the wetland habitats distribution in northeastern Brittany near the Mont-Saint-Michel Bay, France.

16.3 Resolution of RS for Wetland Monitoring

16.3.1 Studying Wetland with Medium Spatial Resolution Data

Spatial resolution data within the range of 4-30 m (referred to as Medium spatial resolution) in remote sensing are used for studying wetland. They mostly include LandSat TM/Enhanced Thematic Mapper plus (ETM+), Systeme Probatoire D'Observation De La Terre (SPOT 1–4), China & Brazil Earth Resources Satellite (CBERS), Advance Spaceborne Thermal Emission and Reflection Radiometer (ASTER), Advanced Visible and Near Infrared Radiometer type 2(AVNIR-2) and Advanced Land Observation Satellite (ALOS) (Guo et al. 2017).

16.3.2 *Studying Wetland with Coarse Spatial Resolution Data*

Studying Wetland with Coarse spatial resolution data comprises Moderate Resolution Imaging Spectroradiometer (MODIS) and AVHRR data. MODIS instrument offers almost daily coverage of the Earth's surface repeatedly with a swath width of nearly 2330 km and 36 spectral bands. Seven bands are precisely designed for land remote sensing with 250 m (bands 1to 2) and 500 m (bands 3 to 7) spatial resolutions (Pflugmacher et al. 2007).

16.3.3 *Studying Wetland with High Spatial Resolution Optical Data*

Guo et al. 2017 described image of high spatial resolution as images with <4 m spatial resolution, which primarily include SPOT-5, Quickbird, IKONOS, GeoEye and WorldView data. Compared with images of medium-resolution and hyperspectral, high spatial resolution images have more information on texture and geometry on the surface features which can be used to recognize features of the ground easily in wetland areas.

16.3.4 *Studying Wetland with Hyperspectral Data*

Imaging spectrometers are used to obtain Hyperspectral data which provide continuous and complete spectral information with a huge number of narrow bands (less than 10 nm) which fall within 0.38–2.5 μm range. This significantly increases the comprehensive information on vegetation and has been widely used in wetland research as a result of complex composition of the vegetation. Satellite, hyperspectral data and airspace and have been excellently used for mapping wetland, research of plant leaf chemistry, identification of wetland species, wetland analysis of soil property, and other themes (Guo et al. 2017).

16.4 Synthetic Aperture Radar (SAR) for Wetland Monitoring

SAR provides the physical and geometric structure of vegetation, spot the inundation levels of vegetation, track the periodic and annual differences within wetland location while optical data provides the molecular and chemical structure of vegetation. As such, the amalgamation can improve accuracy, increase information content, improve the mapping of the wetland boundaries and enhance temporal coverage

(Hosseiny et al. 2021). The European Space Agency (ESA) launched Copernicus program, brought about the massive advancement for wetland mapping. Precisely, the 12 SAR Sentinel-1 and 10 optical Sentinel-2 (multi-spectral instrument, MSI) sensors availability is remarkable, as it offers an extraordinary opportunity to gather data of high spatial resolution for mapping wetland.

Majority of wetland research used SAR in classifying and mapping wetlands (Bartsch et al. 2009). The algorithms for classification are mostly divided into a group of two: unsupervised and supervised (Melendez-Pastor et al. 2010). K-means ISODATA are instances of unsupervised classification methods, while support vector machine (SVM) (Bourgeau-Chavez et al. 2016), maximum likelihood classification (MLC) (Betbeder et al. 2015), and object-based random forest (RF) (Zhang et al. 2019) are supervised procedures.

Unsupervised classification procedures are frequently in use for classifying types of land cover with the use of optical imagery (Li et al. 2016). While studying wetland, supervised classification procedures are preferred given the high similarity between different classes of wetland. The machine learning techniques availability has allowed large-volume earth observation data analyzes (Tamiminia et al. 2020). Non-restriction of the data input to a distribution that is normal is one of the advantages of machine learning classifiers (Bhat and Huang 2021).

Convolutional neural network (CNN) are one more instance of supervised classification procedure for Deep learning approach. Its classifiers can make use of the spatial spectral information gotten directly from an image as training (Meng et al. 2019). Also, due to rising trend in wetlands lost, applying diverse methods to detect and interpret changes in wetland's dynamic ecosystems is essential (Yuan et al. 2015).

The SAR signal illuminated surface is being altered by changes in wetlands and so therefore change the intensity of the image or mechanism of backscattering. The intensity-based change detection method final result is calculated by differencing before and after phenomena image pixels (Bouvet and Le Toan 2011). This backscattering mechanism change can be considered as a pointer to change in the surface type (Kwoun and Lu 2009).

The estimation of soil moisture is one more way of applying SAR imagery (Fig. 16.2) in monitoring wetlands (Kasischke et al. 2009). SAR is sensitive to dielectric properties signal of the target and so, content of water of a target can be retrieved using SAR data. Three common inversion methods exist for estimation of soil moisture in wetland complexes: empirical, semi-empirical and physically based models (Millard and Richardson 2018). Synthetic aperture radar (SAR) has appeared as a capable tool for fast and precise wetland extent and type monitoring. SAR gives an exceptional potential for wetland monitoring in the acquisition of information on the roughness and moisture content of the surface, (Adeli et al. 2020).

Wetland classification and mapping using SAR data have been one of the most explored application to date (Mohammadimanesh et al. 2019), yet the backscattering similarity of wetland classes presents challenges. Spatiotemporal change detection is another application of SAR wetland monitoring. Of a great benefit is this because the dynamic nature of wetlands needs methods that can monitor changes

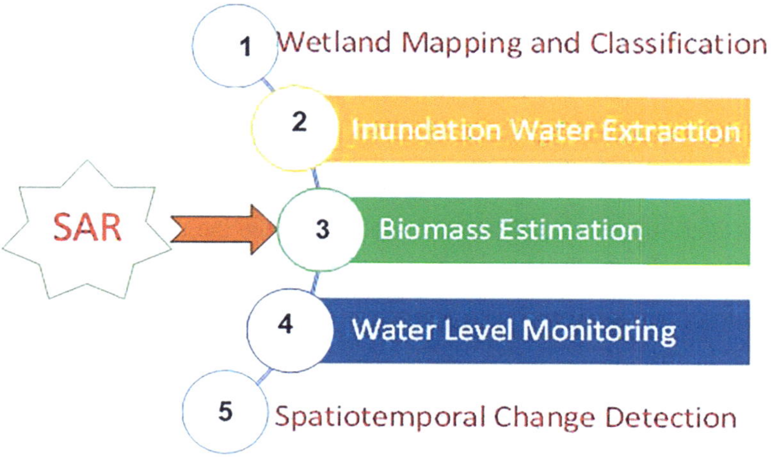

Fig. 16.2 SAR and its application

(Zakharova et al. 2014). Monitoring of water level using Interferometric SAR (InSAR) methods is another, less developed technique of monitoring wetland (Kim et al. 2009). Finally, due to the great influence of wetlands in methane emission and consequently the carbon cycle, estimation of biomass in wetlands using SAR data are also becoming progressively vital (Lee and Fatoyinbo 2015).

16.5 Mapping of Wetland

The anthropogenic activities and climate change have been a threat to the wetlands globally (Aransiola et al. 2024). The changes caused are seen in shrinkage of the wetland area or land cover or vegetation cover. When these changes are understood, it could help in assessing wetland ecosystems and provision of information for department in charge of environmental protection. These trends and sudden change of wetland, dynamic analysis and wetland protection could be understood by long-term change detection studies. On a regional scale, because of the availability of free data and also benefit of spatial and temporal resolution, Landsat is used for research themes in wetland areas mapping, vegetation classification and change detection (Guo et al. 2017). Rapinel et al. (2015b) discovered that with higher accuracy, the images of Landsat 8 OLI can be used in mapping plant communities in coastal marshlands. Also, its data have been used in mapping of flood or building flood models in wetland flood research. Robinove (1978) in Queensland, Australia, initiated the usage of Landsat images in floods mapping. He projected that the areas appearing dark in the Landsat images are wet soil but not flooded regions. Overton (2005), based on Geographic Information System (GIS), Remote sensing, and a hydrological model built an inundation model. The images of Landsat TM were used in monitoring flood inundation extent. It was stated that the model detects flood inundation of a large area and at a lower cost.

16.5.1 *Habitat/Biodiversity*

Wildlife tends to habituate in Wetlands. Depending on the types of species, some endangered birds are constrained to specific wetland habitats. Waterfowls are highly threatened by degradation and habitat loss as a result of anthropogenic activities such as water conservation engineering and reclamation of wetland and climate change such as frequent drought which could cause wetland area shrinkage (Jiang et al. 2014). Landsat data, as a medium-resolution sensor, have been considered to be a good technique in habitat study of wildlife in wetlands at a large area, due to time scales >30 years.

16.6 Wetland Classification

Lush and complex vegetation structure are the major characteristics of wetland and this pose to be a challenge for mapping using traditional optical sensors. Due to the benefit with respect to the band number and continuous reflectance values for the vegetation of the ground, Hyperspectral sensors have been broadly used for classifying wetland. Wetland types, remote sensors used for wetland mapping and their various applications are displayed in Fig. 16.3.

An efficient and effective tool for water body areas detection and inundation of flood extent on a large area is satellite remote sensing. MODIS, due to high temporal resolution and large coverage, has substantial benefit for wetland extent mapping and dynamics at a coarse spatial resolution. In Namibia North-Central, the distribution of surface water was recognized by (Mizuochi et al. 2014) with the use of modified normalized difference water index (MNDWI) of MODIS and normalized difference polarization index (NDPI) of the multi-frequency Advanced Microwave Scanning Radiometer Earth Observing System (AMSR-E). Also, Kaptué et al. (2013), monitored the surface water spatiotemporal variations from 2003 to 2011 in the Soudan-Sahel region of Africa and observed that MODIS could efficiently characterize open water bodies larger than 50 ha. It was concluded that it could also be used in monitoring the surface water seasonal changes of semiarid regions.

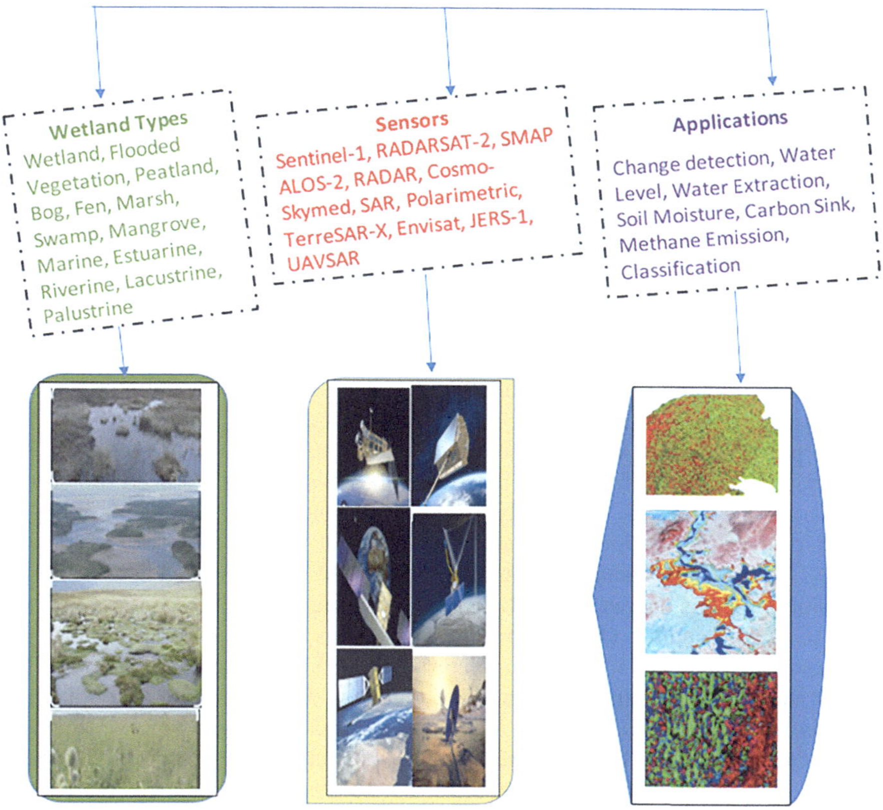

Fig. 16.3 Wetland, sensors and application

16.7 Future Perspective of GIS and Wetland Preservation

Geographic Information Systems (GIS) is an essential tool use for mapping, managing and monitoring environmental resources. GIS tools enhance robust spatial data manipulation, analysis, and management capabilities, which are essential for understanding and reducing the effects of climate change and the negative anthropogenic activity on wetlands. This article seeks to present the future of GIS technique on wetland preservation, with the aim of strengthen wetland conservation through data acquisition, decision support making procedures and data processing approach.

Wetland mapping capabilities have been greatly improved over the past decades with the aid of GIS tools in conjunction with remote sensing. At the early period, wetland mapping were manually based, where aerial photographs in conjunction with ground field survey data are used for mapping and characterization of wetland areas. This technique is time consuming and labor-intensive. Since the first multi-spectral satellite data were made accessible to the general public in the early 1970s, wetland mapping and monitoring in developing nations has advanced tremendously. It has become more promising due to the advancements in GIS and remote sensing technologies, which offer accurate and sufficient information on its ecosystem.

The utilization of active remote sensing technologies is becoming more increasing, LiDAR (Light Detection and Ranging) and satellite altimetry (LaRocque et al. 2020). These technologies enable the comprehensive collection of topographical and subsurface water potential data, particularly in the areas of mapping wetland boundaries and assessing their groundwater quality. Wetland areas that could not be mapped efficiently by traditional mapping approaches, can now be located with high-resolution digital elevation models (DEMs) created using LiDAR that can display seconds fluctuations in the landscape (Rouxinol 2023). Furthermore, changes in vegetation health and soil moisture content can be detected by multispectral and hyperspectral imaging, providing trustworthy data on the biological condition of wetlands (Casamaitjana et al. 2020: Younis and Iqbal 2015).

The preservation of Wetland ecosystem is now more achievable through GIS technique. However, before it can reach its full potential, there's need to address some critical challenges. Especially, in the direction of data characteristics. The quality of dataset used for spatial analyses has great influence on the results outcome. The higher the data resolution, the better the result (Amanda 2023). The accessibility of this data is also a challenge, most open source multispectral image data are of low quality. Overcoming these obstacles will require accessibility of high resolution data and it manipulation process. Also, Training and capacity building for staff is a prerequisite requirements that should be taken into account. Data analysis, and spatial data modeling are professional skills needed to use GIS effectively (Graybill 2024). To guarantee that GIS tools are utilized to their maximum potential, it is imperative to offer resources and training stakeholders involve in the wetland conservation. Collaboration between governmental, non-governmental organization, and academic institutions will go a long way in initiating positive drive, and encourage more innovative ideas in the preservation of wetlands.

An expert in GIS tools has the potential to facilitate well-informed Decision-Support System (DSS) in the protection and preservation of wetland. Through its interface, a tool that enables the simulation and visualization of numerous data sources can help stakeholders comprehend the difficulties the environment faces. An expert in (GIS) can also, quickly evaluate possible conservation difficulties and determine the best course of action. Lin et al. (2006) state that the application of DSS can help to rank the importance of wetland regions in terms of biodiversity, flood danger, and pressure from land development when it comes to their need for protection. A comprehensive and accurate foundation for decision-making is provided by GIS-based DSS, which can make wetland preservation efforts more dependable and efficient.

Wetland preservation has swiftly been made flexible by the constantly improving its analytical capabilities of GIS. Assessing data with deep machine learning algorithms and artificial intelligence (AI) is a newly emerging field. Large dataset processing capacities and trend patterns analysis are capabilities of machine learning algorithms as human analysis may not be able to match. Machine learning enables various applications such as classifying land cover, tracking changes in land surface dynamics, recognizing wetland change dynamics, and forecasting the impact of environmental factors on wetland ecosystems. With the help of deep learning algorithms, wetland dynamics may now be recreated and the findings will be more accurate, GIS-based modeling techniques make this possible (Ahmed et al. 2021). Hydrological models, water management strategies, and aquifer characterization can all be used to forecast subsurface water storage and flow as well as storage in wetlands.

In conclusion, equality and inclusion must be the driving principles behind the use of GIS in wetland preservation in order to guarantee that all parties involved such as organizations, residents, and stakeholders benefit equally from the conservation efforts. This is accomplished by acknowledging the rights of Indigenous peoples and include them in the process of making decisions. GIS can be used to create an inclusive and active approach to wetland protection, leading to more equitable and long-lasting effects.

In summary, GIS can provide better tools for monitoring, understanding, and protecting these vital ecosystems through enhanced data collection, enhanced decision support system and automated analytical techniques processes. As we move forward, the use of inclusive, collaborative methodologies and the simulation of GIS with new spatial technologies will be necessary to realize the full potential of GIS in wetland preservation.

16.8 Conclusion

Wetland is an important ecosystem that harbors diversity of organisms and contribute to stability of ecosystem if maintained. Remote sensing has proven to be a good tool or technology in monitoring the preservation of ecosystem and helps in predicting the outcome of the wetland deterioration.

References

Adam E, Mutanga O, Rugege D (2010) Multispectral and hyperspectral remote sensing for identification and mapping of wetland vegetation: a review. Wetl Ecol Manag 18:281–296. https://doi.org/10.1007/s11273-009-9169-z

Adeli S, Salehi B, Mahdianpari M, Lindi J, Quackenbush I, Brisco B, Tamiminia H, Shaw S (2020) Wetland monitoring using SAR data: a meta-analysis and comprehensive review. Remote Sens 12(14):2190

Ahmed RK, Asker S, Maraandi A, Schuth C (2021) A simple and roburst wetland classification by using optical indice unsupervised and supervised machine learning algorithm. Remote Sens Appl 23:100569

Amanda L (2023) What does data resolution have to do with data accuracy in GIS? Global mapper. Blue marble geographic

Amani M, Salehi B, Mahdavi S, Granger J (2017) Spectral analysis of wetlands in newfoundland using Sentinel 2a and Landsat 8 Imagery. In: Imaging and Geospatial Technology Forum, IGTF 2017-ASPRS Annual conference 2017, Baltimore

Aransiola SA, Babaniyi BR, Aransiola AB, Maddela NR (eds) (2024) Prospects for soil regeneration and its impact on environmental protection, 1st edn. Springer, Cham. https://link.springer.com/book/9783031532696

Bartsch A, Wagner W, Scipal K, Pathe C, Sabel D, Wolski P (2009) Global monitoring of wetlands—the value of ENVISAT ASAR Global mode. J Environ Manag 90:2226–2233

Betbeder J, Rapinel S, Corgne S, Pottier E, Hubert-Moy L (2015) TerraSAR-X dual-pol time-series for mapping of wetland vegetation. ISPRS J Photogramm Remote Sens 107:90–98

Bhat SA, Huang NF (2021) Big data and AI revolution in precision agriculture: survey and challenges. IEEE Access 9:110209–110222

Bourgeau-Chavez LL, Lee YM, Battaglia M, Endres SL, Laubach ZM, Scarbrough K (2016) Identification of woodland vernal pools with seasonal change PALSAR data for habitat conservation. Remote Sens 8:490

Bouvet A, Le Toan T (2011) Use of ENVISAT/ASAR wide-swath data for timely rice fields mapping in the Mekong River Delta. Remote Sens Environ 115:1090–1101

Brisco B (2015) Mapping and monitoring surface water and wetlands with synthetic aperture radar. In: Remote sensing of wetlands: applications and advances. CRC Press, Baco Raton, pp 119–136

Casamaitjana M, Maria C, Madronero T, Bernal-Riobo J, Vanga D (2020) Soil moisture analysis by means of multispectral image according to land use and spatial resolution on andosols in the Colombiam Andes. J Appl Sci 16(10):5540

Cline MD, Feagin RA, Yeager KM, Alstyne JMV (2011) Fault-induced wetland loss at Matagorda, Texas, USA: land cover changes from 1943 to 2008. Geocarto Int 26:633–645

Corcoran J, Knight J, Brisco B, Kaya S, Cull A, Murnaghan K (2012) The integration of optical, topographic, and radar data for wetland mapping in Northern Minnesota. Can J Remote Sens 37:564–582. https://doi.org/10.5589/m11-067

Dronova I (2015) Object-based image analysis in wetland research: a review. Remote Sens 7:6380–6413. https://doi.org/10.3390/rs70506380

Giri C, Ochieng E, Tieszen LL, Zhu Z, Singh A, Loveland T, Masek J, Duke N (2011) Status and distribution of mangrove forests of the world using earth observation satellite data. Glob Ecol Biogeogr 20:154–159

Graybill KJ (2024) Adding map spatial analytical skills. Change the equation and make a graduation memorable

Guo M, Li J, Sheng C, Xu J, Wu L (2017) A review of wetland remote sensing. J Photogramm Remote Sens 130:13–31

Hemati M, Hasanlou M, Mahdianpari M, Mohammadimanesh F (2023) Iranian wetland inventory map at a spatial resolution of 10 m using Sentinel-1 and Sentinel-2 data on the Google Earth engine cloud computing platform. Environ Monit Assess 195(2023):1–21

Hladik C, Alber M (2012) Accuracy assessment and correction of a LIDAR-derived salt marsh digital elevation model. Remote Sens Environ 121:224–235

Holden J (2005) Peatland hydrology and carbon release: why small-scale process matters. Philos Trans R Soc Lond A Math Phys Eng Sci 2005(363):2891–2913

Hosseiny B, Mahdianpari M, Brisco B, Mohammadimanesh F, Salehi B (2021) WetNet: a spatial– temporal ensemble deep learning model for wetland classification using Sentinel-1 and Sentinel-2. IEEE Trans Geosci Remote Sens 60(2021):1–14

Huang C, Davis LS, Townshend JRG (2002) An assessment of support vector machines for land cover classification. Int J Remote Sens 23:725–749

Ji W, Xu X, Murambadoro D (2015) Understanding urban wetland dynamics: cross-scale detection and analysis of remote sensing. Int J Remote Sens 36:1763–1788. https://doi.org/10.1080/01431161.2015.1024895

Jiang F, Qi S, Liao F, Ding M, Wang Y (2014) Vulnerability of Siberian crane habitat to water level in Poyang Lake wetland. China Gisci Remote Sens 51:662–676

Kaptué AT, Hanan NP, Prihodko L (2013) Characterization of the spatial and temporal variability of surface water in the Soudan-Sahel region of Africa. J Geophys Res 118:1–12

Kasischke ES, Bourgeau-Chavez LL, Rober AR, Wyatt KH, Waddington JM, Turetsky MR (2009) Effects of soil moisture and water depth on ERS SAR backscatter measurements from an Alaskan wetland complex. Remote Sens Environ 2009(113):1868–1873

Kim JW, Lu Z, Lee H, Shum CK, Swarzenski CM, Doyle TW, Baek SH (2009) Integrated analysis of PALSAR/Radarsat-1 InSAR and ENVISAT altimeter data for mapping of absolute water level changes in Louisiana wetlands. Remote Sens Environ 113:10–2365

Kuenzer C, Bluemel A, Gebhardt S, Quoc TV, Dech S (2011) Remote sensing of mangrove ecosystems: a review. Remote Sens 3:878–928

Kwoun OI, Lu Z (2009) Multi-temporal RADARSAT-1 and ERS backscattering signatures of coastal wetlands in southeastern Louisiana. Photogramm Eng Remote Sens 75:11–617

Larocque A, Phiri C, Leblon B, Pirolti F, Cannor K, Hanson A (2020) Wetland mapping with landsat 8 OLI, Sentinel-1, Alos-1, Palsar and LIDAR data in Southerm New Brunsswick, Canada. J Remote Space Sci 12(13):2095

Lee SK, Fatoyinbo TE (2015) TanDEM-X Pol-InSAR inversion for mangrove canopy height estimation. IEEE J Sel Top Appl Earth Obs Remote Sens 2015(8):3608–3618

Li H, Zhang S, Ding X, Zhang C, Cropp R (2016) A novel unsupervised bee colony optimization (UBCO) method for remote-sensing image classification: a case study in a heterogeneous marsh area. Int J Remote Sens 37:5726–5748

Lin PJ, Bourne GS, Kleiss AB (2006) Creating a wetland restoration decision support system using GIS tools

Mahdianpari M, Salehi B, Mohammadimanesh F, Brisco B, Mahdavi S, Amani M, Granger JE (2018) Fisher linear discriminant analysis of coherency matrix for wetland classification using PolSAR imagery. Remote Sens Environ 206:300–317

Mahdianpari M, Brisco B, Granger J, Mohammadimanesh F, Salehi B, Homayouni S, Bourgeau-Chavez L (2021) The third generation of pan-Canadian wetland map at 10 m resolution using multisource Earth observation data on cloud computing platform. IEEE J Sel Top Appl Earth Obs Remote Sens 14(2021):8789–8803. https://doi.org/10.1109/JSTARS.3105645

Melendez-Pastor I, Navarro-Pedreno J, Koch M, Gomez I (2010) Multi-resolution and temporal characterization of land-use classes in a Mediterranean wetland with land-cover fractions. Int J Remote Sens 31:5365–5389

Melton J, Wania R, Hodson E, Poulter B, Ringeval B, Spahni R, Bohn T, Avis C, Beerling D, Chen G (2013) Present state of global wetland extent and wetland methane modelling: conclusions from a model inter comparison project (WETCHIMP). Biogeosciences 10:753–788. https://doi.org/10.5194/bg-10-753-2013

Meng X, Zhang S, Zang S (2019) Lake wetland classification based on an SVM-CNN composite classifier and high-resolution images using wudalianchi as an example. J Coast Res 93:153–162

Millard K, Richardson M (2018) Quantifying the relative contributions of vegetation and soil moisture conditions to polarimetric C-Band SAR response in a temperate peatland. Remote Sens Environ 2018(206):123–138

Millennium Ecosystem Assessment (2005) Ecosystems and human well-being: wetlands and water. World Resources Institute, Washington, DC

Mitsch W, Gosselink JG (1993) Wetlands, 2nd edn. Van Nostrand Reinhold, New York, NY

Mizuochi H, Hiyama T, Ohta T, Nasahara K (2014) Evaluation of the surface water distribution in North-Central Namibia based on MODIS and AMSR series. Remote Sens 2014(6):7660–7682

Mohammadimanesh F, Salehi B, Mahdianpari M, Brisco B, Gill E (2019) Full and simulated compact polarimetry sar responses to canadian wetlands: separability analysis and classification. Remote Sens 11:516

Morandeira NS, Grings F, Facchinetti C, Kandus P (2016) Mapping plant functional types in floodplain wetlands: an analysis of C-band polarimetric SAR data from RADARSAT-2. Remote Sens 8:1–17
Overton IC (2005) Modelling floodplain inundation on a regulated river: integrating GIS, Remote Sensing and hydrological models. River Res Appl 21:991–1001
Ozesmi SL, Bauer ME (2002) Satellite remote sensing of wetlands. Wetl Ecol Manag 10:381–402. https://doi.org/10.1023/A:1020908432489
Park NW, Chi KH, Kwon BD (2003) Geostatistical integration of spectral and spatial information for land-cover mapping using remote sensing data. Geosci J 7:335–341
Pflugmacher D, Krankina ON, Cohen WB (2007) Satellite-based peatland mapping: potential of the MODIS sensor. Glob Planet Chang 56:248–257
Rapinel S, Bouzillé JB, Oszwald J, Bonis A (2015a) Use of bi-seasonal Landsat-8 imagery for mapping marshland plant community combinations at the regional scale. Wetlands 35:1–12
Rapinel S, Hubert-Moy L, Clément B (2015b) Combined use of LiDAR data and multispectral earth observation imagery for wetland habitat mapping. Int J Appl Earth Obs Geoinf 37:56–64
Rappold AG, Stone SL, Cascio WE, Neas LM, Kilaru VJ, Carraway MS, Szykman JJ, Ising A, Cleve WE, Meredith JT et al (2011) Peat bog wildfire smoke exposure in rural north Carolina is associated with cardiopulmonary emergency department visits assessed through syndromic surveillance. Environ Health Perspect 119:1415–1420
Robinove CJ (1978) Interpretation of a Landsat image of an unusual flood phenomenon in Australia. Remote Sens Environ 7:219–225
Rouxinol A (2023) LIDAR vs. Traditional surveying: clesh of technology. Aerial Precision Blogg
Schmidt K, Skidmore A (2003) Spectral discrimination of vegetation types in a coastal wetland. Remote Sens Environ 85:92–108
Slater LD, Reeve A (2002) Investigating peatland stratigraphy and hydrogeology using integrated electrical geophysics. Geophysics 67:365–378
Tamiminia H, Salehi B, Mahdianpari M, Quackenbush L, Adeli S, Brisco B (2020) Google Earth Engine for geo-big data applications: a meta-analysis and systematic review. ISPRS J Photogramm Remote Sens 164:152–170
White D (2013) Defining ecosystem processes of the Australian Great Artesian Basin springs from multi-sensor synergies. In Proceedings of the 20th international congress on modelling and simulation, Adelaide, Australia, 1–6 December pp 1938–1944
Younis ZMS, Iqbal J (2015) Estimation of soil moisture using multispectral and FTIR technique. J Remote Sens Space Sci 2(18):116–161
Yuan T, Lee H, Jung HC (2015) Toward estimating wetland water level changes based on hydrological sensitivity analysis of PALSAR backscattering coefficients over different vegetation fields. Remote Sens 7:3153–3183
Zakharova EA, Kouraev AV, Rémy F, Zemtsov VA, Kirpotin SN (2014) Seasonal variability of the Western Siberia wetlands from satellite radar altimetry. J Hydrol 512:366–378
Zhang M, Chen F, Tian B, Liang D (2019) Multi-temporal SAR image classification of coastal plain wetlands using a new feature selection method and random forests. Remote Sens Lett 10:312–321

Chapter 17
Legal Frameworks and Policies for Wetland Protection: International Conventions and Regulations

Daniel Oraeloka Udorah, Chinedu Christian Anyene, and Chukwuka O. Igwegbe

Abstract This study examines the influence of international conventions, such as the Ramsar Convention, on national policies for wetland protection. It investigates how different countries integrate international guidelines into their domestic legal frameworks and the challenges they face in doing so. The study highlights the successes and shortcomings of these conventions in promoting sustainable wetland management. By analyzing case studies from various countries, the research assesses the effectiveness of international agreements in shaping and enhancing national legal frameworks. Key findings suggest that while international conventions provide a strong foundation, their impact is often limited by inadequate implementation at the national level. It concludes with recommendations for improving global wetland conservation efforts, emphasizing the need for tailored approaches that consider the unique socio-economic and environmental contexts of each country. The research calls for stronger enforcement mechanisms and increased international cooperation to ensure the effective protection of wetlands globally.

Keywords Wetlands · Conventions and policies · Legal frameworks · Global biodiversity and ecosystems · Wetland protection and management

---

D. O. Udorah (✉)
Department of Civil and Environmental Engineering, Federal University of Technology, Akure, Ondo State, Nigeria

Department of Agricultural and Bio-environmental Engineering, Federal Polytechnic Oko, Oko, Anambra State, Nigeria
e-mail: danieludorah@gmail.com; udorah.daniel@federalpolyoko.edu.ng

C. C. Anyene
Department of Agricultural and Bio-environmental Engineering, Federal Polytechnic Oko, Oko, Anambra State, Nigeria

C. O. Igwegbe
University of Hull, Hull, UK

B. R. Babaniyi et al. (eds.), *Wetland Ecosystems: Conservation Strategies, Policy Management and Applications*, Wetlands: Ecology, Conservation and Management 12, https://doi.org/10.1007/978-3-031-91982-4_17

17.1 Introduction

Wetlands are portions of the earth that are made up of the lands saturated and submerged with artificially or naturally formed water, and are of either temporary or permanent life span. Wetlands are among the world's most productive ecosystems, with water saturating the soil permanently or periodically. This particular hydrological state supports a wide range of flora adapted to such conditions. It encompasses swamp land inundated with fresh, brackish or marine water of depth not exceeding six metres (Ramsar Convention on Wetlands 2016; Ahmad et al. 2019; Vasumathi et al. 2022). Wetlands are complex and dynamic systems with highly unique hydrologic and biological characteristics. The naturally existing wetlands are categorized into estuarine (areas where freshwater meets saltwater, often dominated by mangroves and salt marshes), marine, lacustrine (inland wetlands associated with lakes), riverine and palustrine wetlands (inland wetlands associated with shallow, and non-tidal waters); while the artificially constituted wetlands include the farm ponds, waste-water treatment ponds, fish ponds, irrigated agricultural lands and water-holding and storage structures like dams, reservoirs and swamp reservoirs. Ecologically, they are symbolized by the totality of the physical, chemical and biological components of the submerged ecosystem, and their relative interactions that sustains the characteristics, functions and products of such ecosystem (Millennium Ecosystem Assessment 2005).

The study aims to investigate the legal structures and policies that control wetland protection at both the international and national levels. It examines the efficiency of various international conventions, treaties, and regulations in protecting wetland ecosystems. By understanding the strengths and weaknesses of these legal instruments, the research seeks to provide recommendations for improving wetland conservation efforts globally.

17.2 The Need for Wetland Protection

17.2.1 Importance of Wetland

The multifarious roles and values of wetland ecosystems to the universe have been of serious research concerns in the present era (Rwanda Environment Management Authority 2020). Wetlands are critical to our planet's health, giving several ecological, economic, and social advantages (Zacharias et al. 2005). The United Nation's position on global Sustainable Development Goals (SDGs) also indicated that wetlands owe critical roles in the effective actualization of the ecosystem, water resources, marine resources, and climate related SDGs (IPBES 2017; Jaramillo et al. 2019; Ransar Convention on Wetlands 2018; Convention on wetlands 2021). The wetland ecosystem serve as critical habitats for a diverse range of species by providing huge support for different species of aerial, terrestrial and aquatic animals, and equally serves as an essential pool for plant genetic resources (Dahle 2023). Unsustainable practices towards the flow of water in ecosystems culminate to unsafe decline in the supposed benefits of wetlands to humans, other animals and the environment (Ekeruche 2024). Wetlands play an important role in water purification because they operate as natural filters, capturing sediments, nutrients, and contaminants and thereby improving water quality.

Wetlands are productive ecosystems that are essential to human livelihoods. They support fisheries, which are critical to the local and national economies. Wetlands also help agriculture by supplying water for irrigation and grazing areas for cattle. Additionally, they provide raw materials like reeds, peat, and medicinal plants (Kingsford et al. 2016; Balqis et al. 2023; Ramsar Convention on Wetlands 2018). As depicted in Fig. 17.1, wetlands constitute about 2983 million ha of the earth surface (Xu et al. 2019; Ballut-Dajud et al. 2022) with respective 980 million ha, 795 million ha, 565 million ha and 56 million ha coverage distribution across Asia, South American, North American and African (Boone and Bhomia 2017; Ballut-Dajud et al. 2022). The ecosystem serves as the fountain of biodiversity that sustains the survival of countless plants and animals species. Wetland ecosystems are potent in the provision of prominent services to global biodiversity and ecosystems (Kingsford et al. 2016; Dahle 2023), and creates the opportunities for recreational and tourism activities (Agbasi and Odiaka 2016; US-EPA 2016; FAO 2021; Sima et al. 2022; Balqis et al. 2023). They present several ecological services to human beings including the supply of water and fish resources, regulation of the climate variables, purification of water resource, regulation of flood, protection of coastal land (Sánchez-Espinosa and Schröder 2019; Ballut-Dajud et al. 2022). Wetlands also feature in the respective purification and filtration of water resource and agricultural pollutants (Sica et al. 2016. Furthermore, wetlands also serve as substantial carbon sinks. They sequester significant amounts of CO_2 by accumulating plant debris and peat, and helps in the prevention of climate change consequences through atmospheric CO_2 sequestration (Were et al. 2019).

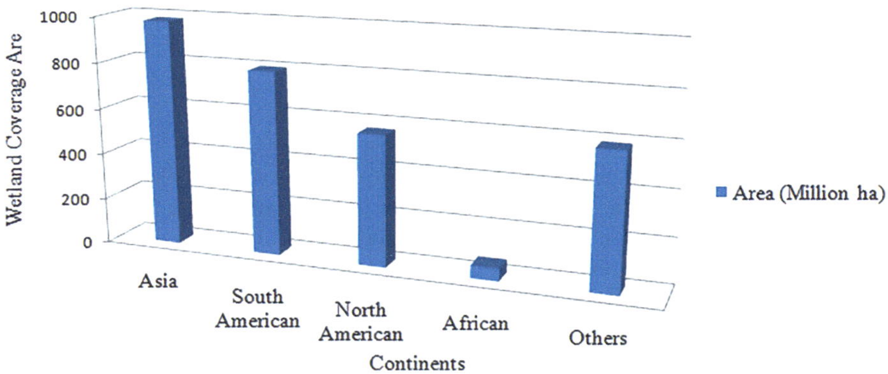

Fig. 17.1 Wetland coverage area of different continents. (Ballut-Dajud et al. 2022, adapted)

17.2.2 Threats to Wetlands

Despite their ecological, economic, and social importance, wetlands face a number of threats to their health and survival. These challenges, largely manmade but also natural, cause the degradation and loss of wetland ecosystems. Some of the specific threats bedeviling wetland ecosystems are pollution, urbanization, climate change, and overexploitation of wetland resources (Zacharias et al. 2005). Pollution is a major threat to wetlands around the world. Wetlands are especially susceptible to contamination because they frequently act as natural filters for toxins. However, overwhelming pollution might overcome their capacity to cleanse water, resulting in severe environmental damage. The primary sources of pollution in wetlands are the unreasonable channeling of effluents from agricultural runoff, industrial activities, and urban runoff and wastewater. They inject surplus nutrients (nitrogen and phosphorus) from fertilizers, pesticides, herbicides through the release of heavy metals, toxic compounds, and other contaminants into wetland habitats, amounts to serious damages to wetlands. (Paul and Meyer 2001; Schuster et al. 2002; Rabalais et al. 2002).

Furthermore, climate change offers a multidimensional threat to wetlands, affecting their hydrology, ecology, and ability to deliver ecosystem services. Coastal wetlands, such mangroves and salt marshes, are especially vulnerable to sea-level rise. As sea levels rise, these wetlands may be inundated or destroyed, causing habitat loss for numerous species (Ghosh et al. 2015). Changes in precipitation patterns, such as increasing frequency and intensity of droughts and/or more intense rainfall, can also contribute to changes in wetland hydrology, including reduced water availability in wetlands, flooding, and increased pollutant runoff.

The overexploitation of wetland resources for commercial, recreational, and subsistence interests has the potential to deplete these critical ecosystems. The indiscriminate use of the water resources of wetlands, disposal of municipal wastes in wetlands, unsustainable fishing and hunting, and draining of wetlands for the construction of power substations, gas plants and oil pipelines, peat mining and timber harvesting can degrade and destroy habitats (Page et al. 2002; FAO 2007).

17.2.3 Wetland Degradation

Wetlands and humans are interdependent due to their economic, recreational and scientific significance to the human race (Akinbode and Jean-Claude 2022). Several of them exist as 'international systems' that are located along the boundaries regions or river basins in the regions. The safeness of wetlands is related to the amount and quality of the trans-boundary water that is being supplied from the water bodies (rivers, springs, streams and lakes) and saturated underground water bearing masses (aquifers). Despite the official international precedence of wetland protection in 169 nations under Ransar Convention ratification (Ramsar 2018), many wetlands are still under the threats of land drainage and reclamation for agriculture, industrialization and urbanization (Davidson 2014; Li et al. 2017; Materu et al. 2018; Amadu 2021).

The wetland ecosystem serves as a habitat to both terrestrial and non-terrestrial living creatures and also serves a great deal in ecological balance, purification and conservation. Population growth and the resultant increasing anthropogenic activities are the major degradation drivers of natural and artificial wetlands. Ballut-Dajud et al. (2022) reported that such anthropogenic activities as agriculture, urbanization, aquaculture and industrialization affect natural wetlands to the respective tune of 25%, 16.8%, 10.7% and 7.6%. Poorkhaghan et al. (2022) outlined the specific wetland degradation causative factors to include poor management of water resources, unsustainable land use techniques, pollution incidences, indiscriminate hunting and fishing practices, land development and conversion, climate change, disturbances from tourism and outdoor activities establishments, and pollution resulting from spillages and improperly managed sewer and effluents.

Unsustainable forestry and agricultural practices amplify soil erosion and poor water retention capacity problems. Change in climate variables (temperature and precipitation), upland and lowland agricultural activities, and high livestock grazing practices pose adverse effects on wetlands water quality, periodic and seasonal flows, loss of biodiversity and natural utility. Pollution from solid, pesticides, nitrates, polycyclic aromatic hydrocarbons (PAHs), polychlorinated biphenyls (PCBs), heavy metals and phosphates causes serious impact to wetlands (Poorkhaghan et al. 2022).

Unguarded and indiscriminate exploitation of the water resources from the wetlands pose serious adverse effects on both humans and their surroundings (Wetland international 2010). The non-commensurate demands and supply of water in the present era of population explosion across the globe causes obvious instability in food production, safe water availability, economic progressions, human health sustenance and geopolitical cohesions (Debaere and Amanda 2017; Emran et al. 2021; UN 2023). These impasses have awakened the advocacy towards the development and optimization of water allocation, supply and utilization management legislations and legal frameworks (Agbasi and Odiaka 2016). Government, non-governmental organizations, corporate bodies as well as concerned individuals have continued to expend capital resources to conserve the existing wetlands and/or revive the wetlands with impaired biological and hydrological features (Ransar Convention on Wetland 2016). With the continual increase in global population projections, the possible adaptation of wetlands to the unpredictable changing conditions is critical for the sustenance of the human and wildlife communities (Raburu et al. 2012).

17.2.4 Wetland Conservation: Halting the Loss of Wetland Biodiversity

According to Finlayson et al. (2019), swift action is needed to address the precarious state of wetlands and the species that inhabit them worldwide, and they include:

(i) putting a high priority on creating interconnected, well-understood, and well-managed networks of protected areas for a sizable portion of the world's wetland habitats;
(ii) keeping wetlands intact and reestablishing them while stopping their conversion to other land uses;
(iii) Large-scale restoration of wetland plant communities;
(iv) rewilding wetlands with natural species—including top predators—in order to bring back the dynamics and processes of the ecosystem;
(v) creating and implementing appropriate policy tools to stop the exploitation and trade of vulnerable species, as well as the overfishing and poaching that are causing the loss of wetland creatures;
(vi) minimizing the amount of food that is wasted by improving infrastructure and education.
(vii) encouraging dietary changes to lessen the amount of cattle overgrazing on wetlands;
(viii) expanding adult and child education about wetlands outdoors; and involving the general public, particularly local and indigenous communities, in wetlands management

17.3 Legal Frameworks and Policies for Wetland Protection

There is a dire need for directing legal frameworks and policies towards the ever-increasing infringement on and loss of wetlands. Legal frameworks are set of rules governing the rights and responsibilities of individuals, organizations and government (Riegner 2014; NRGI 2015; African Union 2018; Ger 2021). They vary from region to region and comprise of regulations, legislation, policies and contracts. For every exploration of natural resources that has to do with land, there are set legal framework. The persons that are involved in constituting the legally instituted regulations and the complementary documents differ from region to region. In modern dispensations, the constitutions of most regions in the world saddle the government with the responsibility of environmental protection with reference to 'right to a healthy environment' (Poorkhaghan et al. 2022). Different nation have assembles for an international treaty in order to adopt the Ransar Conventions on the establishment of regulations for sustaining wetland protection and reliable use (Dupuy and Vinuales 2018).

17.3.1 International Legal frameworks for Wetland Protection

Legal frameworks and policies measures must be instituted effectively to conserve and enforce the wise utilization of wetland resources (Aminzadeh et al. 2017). Presently, policy bodies and conventions are developed regionally and globally as a medium to ventilate information on negligence towards the harvestable roles, resources and services of wetlands and the numerous factors threatening their extinction (Rwanda Environment Management Authority 2020). The individual efforts and approaches of different governments toward wetlands sustainability are liable to fail for lack of conventions or frameworks for inter-regional interactions, harmonization and synergy towards mutual recompense. Governments of different nations across the globe are now teaming up with known wetlands related conventions and policy making body in the fight to overturn the negative narratives on wetland degradation and eventual loss (Ramsar Convention on Wetlands 2018). The formidability of such efforts requires mutual inter-governmental engagements.

17.3.1.1 Ramsar Convention on Wetlands

The Ramsar Convention on Wetlands has been extensively reported to be at the forefront in providing such frameworks for local, national and international wetland conservation. It is formally known as the Convention on Wetlands of International Importance especially as Waterfowl Habitat, and it is the premier international treaty focused on the conservation and sustainable use of wetlands. It was adopted in 1971 in the city of Ramsar, Iran, and it is one of the oldest modern global environmental agreements, reflecting early recognition of the critical importance of wetlands (Ramsar Convention on Wetlands 2016; Davidson 2023). It was established in response to growing awareness of the deteriorating state of wetlands and its adverse effects on wetland habitats and the associated fauna, since its inception in 1971, more than 2000 wetlands, spanning over 2000,000 km^2, have been designated as internationally important by 172 countries (Gell et al. 2023). The Ramsar Convention serves as a fundamental pillar of international wetland conservation, providing a comprehensive framework designed to prevent global wetland degradation. The primary objective of the Ramsar Convention is to halt the worldwide loss of wetlands and to ensure their conservation through the principles of wise use and management of wetland, designation of wetland of international importance, and international cooperation (Ramsar 2018; Votteler and Muir 1996; Agbasi and Odiaka 2016). The implementation of the Ramsar Convention has led to significant global efforts for wetland conservation in terms of Ramsar sites, national wetland policies, and capacity building, education and public awareness (Kaller et al. 2013). However, addressing the challenges that are related to compliance, climate integration, funding, and monitoring is needful for the continued effectiveness and enhancement of the convention in the face of evolving environmental pressures. These advances empower the Convention to support and enable signatory nations to achieve social and economic objectives while continuing to advance the conservation of natural assets within the global network of significant wetlands (Gell et al. 2023).

17.3.1.2 The Convention on Biological Diversity (CBD)

The Convention on Biological Diversity (CBD) is a pivotal international legal framework designed to address the widespread decline in global biodiversity, including the protection and sustainable management of wetlands. It was established during the Earth Summit in Rio de Janeiro in 1992 (Ekardt et al. 2023). The CBD has three primary objectives: the conservation of biological diversity, the sustainable use of its components, and the fair and equitable sharing of benefits arising from genetic, especially with indigenous communities and local stakeholders involved in wetland conservation (UNE 2020; UK-NEA 2014; Wetland international 2022). The conservation of biodiversity is a shared concern among humans since it encompasses all domains that are directly or indirectly related to biodiversity, including ecosystems, species, and genetic resources. For example, the Democratic Republic of the Congo has the highest amount of biological diversity in Africa. Despite the fact that this rich natural wealth is of local and worldwide significance, the IUCN Red List of Threatened Species now classifies 190 species as critically endangered, endangered, or vulnerable. (UNE 2020). Hence the need for commitment to CBD's objectives. However, ensuring compliance with CBD commitments as well as financial constraints, faced by many countries that hinders their ability to protect and manage wetlands effectively, remains a significant challenge. This is because the CBD relies on voluntary national commitments, leading to variable implementation and enforcement across countries (Ekardt et al. 2023). Strengthening international cooperation and creating mechanisms for accountability are therefore essential to enhance compliance. Many countries face financial constraints that hinder their ability to protect and manage wetlands effectively.

17.3.1.3 United Nations Framework Convention on Climate Change (UNFCCC)

The United Nations Framework Convention on Climate Change (UNFCCC) is also an international treaty (mechanism) established to combat climate change and its adverse effects on global ecosystems, including wetlands (Thokchom 2020). It was adopted at the United Nations Conference on environment and development in 1992. It provides a framework for negotiating specific international agreements (protocols) that may set binding limits on greenhouse gases by addressing the impacts of climate change on wetlands (Stephenson 2018; Pulselli 2008). It birthed both the Kyoto Protocol and the Paris Agreement. Its main goal is to stabilize greenhouse gas concentrations at a level that would prevent dangerous human interference with the climate system (Thokchom 2020; Minx et al. 2021). It states that climate change is expected to be contained in a time frame that: permits ecosystems to take natural adjustment to climate change; doesn't entrammel sustainable development; and asserts agricultural productivity. By promoting mitigation, adaptation, and resilience building, the convention supports strategies to protect these vital ecosystems through common but differentiated responsibilities especially to developed and developing nations; data gathering and reporting processes especially about greenhouse gases (GHG) emission; and administrative institutions, structures and process (Hannah 2011; Kuh 2018; Sage 2003; Noordwijk 2014).

17.3.2 Regional Frameworks and Policies

Developed legal frameworks and policies are set in force to compel the member nations to apply tactfulness and care in the exploitations of wetlands. Enshrined in the international conventions are the regional and national frameworks which help to drive home the key objectives of the different conventions. While some nations may have constituted frameworks and pass them through parliamentary processes to be established as specific policies, other nations might have them enshrined in the laws of the environment. The legal frameworks and policies pertaining to wetlands are continually reviewed, modified and reconstituted through the proceedings of international wetlands conventions and regulations, in order to ensure that there are no antagonistic laws besetting the wetlands and environmental legal frameworks and policies (Ramsar Convention on Wetlands 2016). Some of them include:

17.3.2.1 European Union Directives and Regulations

The European Union (EU) has established a comprehensive legal framework to protect wetlands and ensure sustainable water management. This framework includes various directives and regulations designed to preserve biodiversity, enhance natural capital, and achieve environmental sustainability (Farantouris 2009; EP 2024). Key among these are the Water Framework Directive (WFD) and the Habitats Directive, both of which play crucial roles in wetland protection (WECOOP 2023). These directives work in conjunction with other EU policies to form a comprehensive strategy for wetland conservation.

(a) **Water Framework Directives (WFD):** The WFD, adopted in 2000, aims to achieve "good status" for all European waters, including rivers, lakes, coastal waters, and groundwater, by 2015 (Mcinnes 2016). Although the initial deadline has passed, ongoing efforts continue to meet this goal. This directive takes a holistic approach to water management, emphasizing the integration of water protection into other policy areas and promoting sustainable water use. The WFD mandates public consultation and involvement in water management decisions, ensuring that stakeholders, including local communities, have a say in the protection of water resources (Ignar and Grygoruk 2015). They must also establish monitoring programs to assess water status and report on progress toward the objectives of the WFD. The WFD has led to significant improvements in water quality and ecosystem health across Europe. It has facilitated cross-border cooperation on water issues and promoted the restoration of degraded wetlands, contributing to biodiversity conservation and climate resilience (Mcinnes 2016).

(b) **The Habitats Directive**: It was adopted in 1992 and it aims to promote the maintenance of biodiversity by requiring member states to take measures to maintain or restore natural habitats and wild species at a favorable conservation status. This includes many types of wetlands (EP 2024). The directive established the Natura 2000 network, a coherent European ecological network of protected areas, including many wetland sites. Member states must designate Special Areas of Conservation (SACs) and implement management plans to protect these areas (Ignar and Grygoruk 2015). Apart from the Natura 2000 Network, the directive also lists species that require strict protection, including many that depend on wetland habitats. Member states must establish monitoring and reporting mechanisms to ensure the conservation of these species. The Habitats Directive has been instrumental in safeguarding Europe's most valuable and threatened species and habitats. The creation and management of Natura 2000 sites have enhanced habitat connectivity and ecological integrity, benefiting wetland ecosystems and the species they support (WECOOP 2023).

17.3.2.2 African Union Policies and Initiatives

The African Union (AU), comprising 55 member states, is a continental body committed to fostering unity, cooperation, and development across Africa. The AU has developed several comprehensive policies and initiatives aimed at addressing environmental challenges, promoting sustainable development, and ensuring regional integration. These frameworks play a crucial role in the protection of natural resources, including wetlands, and in improving the livelihoods of Africa's population. Some of the frameworks include:

(a) **Comprehensive Africa Agriculture Development Programme (CAADP):** CAADP is an initiative launched in 2003 under the New Partnership for Africa's Development (NEPAD) framework. It aims to eliminate hunger, reduce poverty, and stimulate economic growth through agriculture-led development (AU 2023). It encourages member states to allocate at least 10% of their national budgets to agriculture and achieve a minimum of 6% annual growth in agricultural productivity. Also, it emphasizes the sustainable use of natural resources, including wetlands, to enhance agricultural productivity and resilience (EU 2024). CAADP has led to substantial policy reforms, increased investments, and improved agricultural productivity. These efforts have contributed to enhanced food security, economic growth, and sustainable land and water management practices across the continent (AU 2023).

(b) **African Land Policy Centre (ALPC):** The ALPC, a joint initiative by the AU Commission, the African Development Bank, and the United Nations Economic Commission for Africa, aims to promote effective land governance, secure land rights, and sustainable land use (ALPC 2023). It supports the formulation and implementation of inclusive land policies that recognize the importance of wetlands and other critical ecosystems. It also provides technical assistance and capacity-building programs to enhance land governance and management at both national and regional levels (AU 2023). The ALPC has facilitated the development of inclusive and sustainable land policies, improved land tenure security, and promoted the sustainable use of land resources. These efforts have had a positive impact on wetland conservation and biodiversity protection (ALPC 2023; Gardner et al. 2009).

17.3.2.3 Association of Southeast Asian Nations (ASEAN)

The Association of Southeast Asian Nations (ASEAN) has developed several frameworks and policies aimed at protecting wetlands, which are crucial for biodiversity and ecosystem services in the region (ASEAN 2024). These initiatives are integral to sustainable development and environmental conservation efforts across ASEAN member states (Prusty et al. 2017; Lim 2023). The frameworks for wetland protection include:

(a) **ASEAN Centre for Biodiversity (ACB):** ACB plays a vital role in coordinating regional efforts to conserve biodiversity and ecosystems, including wetlands, through research, capacity-building, and policy advocacy. It implements flagship programs that focus on securing the health and wellbeing of communities dependent on wetlands, emphasizing ecosystem-based approaches (ASEAN 2024; Lim 2023).

(b) **ASEAN Environmental Legal Framework:** ASEAN has established legal frameworks that address environmental conservation and management, indirectly impacting wetland protection. These include agreements on biodiversity conservation and sustainable development (Lipman et al. 2015).

Despite the challenge of implementing cohesive wetland conservation strategies across ASEAN member, efforts such as the ACB's programs and ASEAN agreements have led to improved regional cooperation, increased awareness, and enhanced capacity to manage and protect wetlands (ASEAN 2024; Irini et al. 2012).

17.3.3 National Legal Framework

National legal frameworks play a crucial role in the protection and management of wetlands worldwide. These frameworks typically include legislation, regulations, and policies aimed at conserving wetland ecosystems and their associated biodiversity. In some African countries like Nigeria, legal frameworks and policies for wetland protection must follow the trends of the existing laws in relations to such other areas that are likely to affect wetlands as oil spills, fishing, pollution, plant and animal husbandry, etc. (Ignar and Grygoruk 2015). Developing countries that is not sufficient in the manpower and technological requirements for effectively managing and regulating wetland resources should tactically complement international frameworks and policies with their domestic frameworks for better resource control and protection (Snethlage 2015). In order to ascertain optimal results, environmental extension services must be adequately empowered and provided with the necessary resources for efficient awareness creation to rural communities who are direct custodians of the wetlands on improved conservation and protection techniques. Also, the Land Use Act has been applied in combination with national policies in some areas to check illegal and unsustainable exploitation and exploration of wetlands (Agbasi and Odiaka 2016).

Below are some case studies demonstrating effective national policies for wetland protection.

1. **The Ramsar Convention and National Wetland Policies:** Many countries have adopted comprehensive wetland policies as part of their commitment to the Ramsar Convention on Wetlands. These policies typically include measures for wetland conservation, restoration, and sustainable use. For instance, South Africa's National Wetland Policy integrates wetland protection into broader environmental and land-use planning processes. This policy promotes the conservation of wetland ecosystems by incorporating wetland considerations into development planning and ensuring that activities that may impact wetlands are subject to stringent environmental assessments (Sinthumule 2024).
2. **United States: The Clean Water Act:** The United States has one of the most robust legal frameworks for wetland protection, primarily through the Clean Water Act (CWA). The CWA regulates the discharge of pollutants into waters, including wetlands, and requires permits for any activities that might result in the discharge of dredged or fill material into wetlands. The Act has been instrumental in reducing wetland loss and degradation. Additionally, the U.S. Fish and Wildlife Service and the Environmental Protection Agency (EPA) work collaboratively to enforce wetland protection laws and restore degraded wetlands (BCN 2024).
3. **Australia: The Environmental Protection and Biodiversity Conservation Act:** Australia's Environmental Protection and Biodiversity Conservation (EPBC) Act provides a framework for the protection of wetlands of international importance. The EPBC Act requires environmental assessments and approvals for actions that may have a significant impact on listed wetlands. The Act also facilitates the development of recovery plans and conservation advice for threatened species and ecological communities associated with wetlands. This legislative framework has contributed to the protection and management of key wetland sites across Australia (Peters 2020).
4. **Nigeria: National Wetland Policy:** Nigeria's National Wetland Policy focuses on the sustainable management and conservation of wetlands to enhance their ecological and socio-economic benefits. The policy emphasizes the importance of wetlands for biodiversity conservation, flood control, and livelihood support for local communities. Implementation of this policy involves the participation

of multiple stakeholders, including government agencies, local communities, and non-governmental organizations. The policy framework aims to integrate wetland management into national development plans and promotes the restoration of degraded wetland areas (Obona 2016; Agbasi and Odiaka 2016; Ewebuke and Albrecht 2021).

5. **European Union: Water Framework Directive:** The European Union's Water Framework Directive (WFD) is a comprehensive policy that aims to protect and enhance the status of aquatic ecosystems, including wetlands. The WFD requires member states to achieve good ecological status for all water bodies and to prevent further deterioration. It promotes integrated water resources management and encourages the restoration of degraded water bodies. The Directive also supports the development of river basin management plans, which include specific measures for wetland protection and restoration (Van 2013).

17.3.4 Appraisal of Wetland Policy Mechanisms

The Ramsar Convention is a cornerstone of international wetland conservation, with its comprehensive framework aimed at halting global wetland degradation. It is now widely acknowledged that wetlands undergo constant changes across various time scales, with direct human activities being just one of many influencing factors. Ongoing monitoring of these changes enables the Convention and its signatory nations to adjust the framework in response to evolving insights into wetland dynamics (Gell et al. 2013). The governance of the Ramsar Convention involves multiple stakeholders. The Conference of the Parties (COP), composed of the Contracting Parties, meets approximately every 3 years and serves as the primary decision-making body. At each COP, Resolutions are adopted to provide guidance and instructions for implementing the Convention. Between COP sessions, the Standing Committee, comprising Contracting Parties with regional representation, assumes decision-making responsibilities. Administrative support is provided by the Convention's Secretariat, based in Switzerland, while scientific guidance is offered by the Scientific and Technical Review Panel. Additionally, the Convention is supported by International Organisation Partners, which include both intergovernmental and non-governmental organizations. Ramsar operates primarily through a collaborative and consensus-driven approach (Gardner et al. 2023).

The wetland policy mechanisms at both national and international levels are crucial for the conservation and sustainable management of Ramsar Wetlands (Gell et al. 2013). As Gardner et al. (2016) critically appraise, the effectiveness of the convention lies in its emphasis on policy mechanisms that intertwine national responsibilities with robust international cooperation. Their analysis underscores the necessity for cohesive strategies that align global objectives with the nuanced challenges faced by individual nations. National obligations ensure that countries take responsibility for their wetlands, while international cooperation fosters a collaborative approach to addressing the challenges these ecosystems face. The integration of historical data and ongoing monitoring are essential components of effective wetland management, enabling nations to respond to changes and work towards the restoration of ecological character. This emphasizes on interdisciplinary collaboration underscores the need for a comprehensive approach to wetland conservation that transcends borders and disciplines (Gell et al. 2013).

The appraisal of wetland policy mechanisms, particularly in the context of national obligations and international cooperation, reveals a multifaceted approach to the conservation and sustainable use of wetland ecosystems. Concentrating solely on the wetland site risks overlooking the broader context. Similarly, when examining wetland policies, it is valuable to look beyond domestic frameworks—such as national and local laws and policies—and take global wetland policies into account (Gardner and Davidson 2011). According to Gardner et al. (2016), the Ramsar Convention on Wetlands serves as a foundational framework for these efforts, emphasizing the importance of both national and international actions. This dual approach is essential for addressing the complex challenges facing wetland ecosystems today, ultimately contributing to biodiversity conservation and the well-being of communities that depend on these vital resources. The dual approach include:

(a) National Obligations:

   (i) Legal frameworks: Countries that are Contracting Parties to the Ramsar Convention are encouraged to review and strengthen their national legislation related to wetland conservation. This includes ensuring that laws at various levels (national, provincial, and local) do not conflict and effectively promote the wise use of wetlands.
   (ii) National inventories: The Convention recommends that countries maintain national inventories of wetlands, which serve as critical tools for assessing the status and trends of wetland ecosystems. This inventory process is essential for informed decision-making and prioritizing conservation efforts.
   (iii) Management planning: National obligations also extend to the development of management plans for designated Ramsar Sites. These plans should address conservation status, ecological character, and the sustainable use of resources, ensuring that local communities benefit while maintaining ecological integrity.
   (iv) Reporting and Accountability: Contracting Parties are required to submit national reports on their implementation of the Convention, which fosters accountability and allows for the sharing of best practices among nations.

(b) International Cooperation:

   (i) Shared knowledge and resources: The Ramsar Convention facilitates international cooperation by providing a platform for sharing knowledge, experiences, and resources among member countries. This collaboration is vital for addressing transboundary wetland issues and promoting best practices in wetland management.

(ii) Capacity building: The Convention supports capacity-building initiatives that enhance the ability of countries to implement effective wetland policies. This includes training programs, workshops, and the development of guidelines that assist nations in fulfilling their obligations.

(iii) Global frameworks and resolutions: The Convention's resolutions, such as those addressing risk assessment frameworks and strategic guidelines for wetland management, provide a cohesive international framework that guides national policies and actions. These resolutions encourage countries to adopt a unified approach to wetland conservation.

(iv) Partnerships with other organizations: The Ramsar Convention collaborates with various international organizations, including the IUCN and UNEP, to promote integrated approaches to wetland conservation. These partnerships enhance the effectiveness of conservation efforts and leverage additional resources for implementation.

17.3.5 Challenges in Implementation and Enforcement of Legal Frameworks

The implementation and enforcement of legal frameworks for wetland protection face numerous and multifaceted challenges. These challenges significantly impact the effectiveness of policies designed to conserve wetlands and their ecological functions, which include:

1. **Inadequate Resources and Funding:** One of the primary challenges in implementing and enforcing wetland protection laws is the lack of adequate resources and funding. Effective enforcement requires sufficient financial resources to support monitoring, compliance, and enforcement activities. Many countries, especially those with limited economic resources, struggle to allocate the necessary funds to adequately protect wetlands (Prusty et al. 2017).

   For example, developing nations often face significant financial constraints that hinder their ability to enforce environmental regulations. This lack of funding can lead to insufficient staffing, inadequate training for enforcement personnel, and limited capacity to conduct regular inspections and monitoring. These financial limitations make it difficult for authorities to maintain an active and effective presence in wetland areas, leading to lapses in enforcement and increased vulnerability of wetlands to illegal activities such as unregulated development and pollution (Ger 2021).
2. **Weak Institutional Capacity**: Institutional capacity is critical for the effective implementation of wetland protection laws. Weak institutional frameworks, characterized by fragmented responsibilities and lack of coordination among agencies, can undermine enforcement efforts. In many countries, multiple agencies may have overlapping mandates for wetland management, leading to confusion and inefficiency (Kaller et al. 2013).

For instance, in Nigeria, the implementation of public policies often faces challenges due to weak institutional frameworks and lack of coordination among government agencies. This fragmentation can result in inconsistent enforcement and gaps in policy implementation, ultimately weakening the overall effectiveness of wetland protection measures. Streamlining institutional responsibilities and enhancing inter-agency cooperation are essential steps towards improving enforcement capacity and ensuring cohesive management of wetlands (IPBES 2017).

3. **Lack of Public Awareness and Participation:** Public awareness and participation are essential components of successful wetland conservation efforts. However, in many regions, there is a lack of awareness among local communities and stakeholders about the importance of wetlands and the legal frameworks designed to protect them. This lack of awareness can lead to non-compliance and resistance to conservation measures (Thokchom 2020).

   For example, in some countries, local communities may engage in activities that degrade wetlands, such as illegal logging, agriculture, or infrastructure development, due to a lack of understanding of the environmental impacts. Increasing public awareness and involving communities in wetland management can enhance compliance and support for conservation efforts. Education campaigns, community engagement initiatives, and participatory management approaches are effective strategies for fostering public support and active participation in wetland conservation (Junk et al. 2013).

4. **Enforcement Challenges:** Enforcement of wetland protection laws often encounters significant challenges, including corruption, lack of political will, and insufficient penalties for non-compliance. Corruption within enforcement agencies can undermine the effectiveness of legal frameworks, allowing illegal activities to continue unchecked. Bribery, favoritism, and other forms of corruption can create an environment where violators evade punishment and regulations are selectively enforced (Riegner 2014).

   Additionally, weak legal penalties and lack of political will to enforce regulations can discourage compliance. In some cases, powerful economic interests may exert influence over enforcement agencies, leading to selective enforcement and impunity for violators. Strengthening anti-corruption measures, ensuring transparency and accountability in enforcement processes, and establishing stringent penalties for non-compliance are crucial for overcoming these challenges and enhancing the effectiveness of wetland protection laws (Kuh 2018).

5. **Climate Change and Environmental Pressures:** Climate change poses a significant challenge to wetland conservation, exacerbating existing environmental pressures. Rising temperatures, changing precipitation patterns, and increased frequency of extreme weather events can alter wetland ecosystems, making it more difficult to implement and enforce protection measures (Hannah 2011).

   For instance, climate change can lead to the degradation of wetland habitats, reducing their resilience and capacity to provide essential ecosystem services. This makes it imperative for legal frameworks to adapt and incorporate climate resilience measures to ensure long-term protection. Developing adaptive management strategies, integrating climate change considerations into wetland protection policies, and promoting the use of nature-based solutions are essential for enhancing the resilience of wetlands to climate change impacts (Noordwijk 2014).

17.3.6 Role of Indigenous and Local Communities in International Conventions and Regulations

According to Gardner et al. (2023), during the Convention's first 50 years, significant effort was dedicated to establishing institutional frameworks and creating technical guidelines. However, as highlighted in the Global Wetland Outlook, the Contracting Parties have yet to successfully prevent wetland loss and degradation. This therefore boils down to the indigenous and local communities. Indigenous and local communities hold an integral position in addressing the multifaceted challenges and issues intersecting wetland conservation and sustainable development. Their traditional knowledge, holistic approaches to land management, and intimate connection to their environments offer invaluable insights (Tommasoli 2012). However, integrating their perspectives and effectively overcoming the challenges they face require addressing several key issues.

1. **Traditional Knowledge and Sustainable Practices:** Indigenous and local communities possess rich traditional knowledge systems and sustainable practices that have been honed over centuries. This knowledge includes deep understandings of local ecosystems, biodiversity, and hydrological cycles, often surpassing conventional scientific knowledge in specificity and contextual relevance (Vasumathi et al. 2022)

   For example, indigenous practices such as controlled burning, rotational farming, and water management reflect a sophisticated understanding of ecological balance. These practices are crucial for maintaining the health of wetland ecosystems. In many regions, such as the Amazon Basin, indigenous communities use agroforestry systems that integrate crop cultivation with forest conservation, promoting biodiversity and soil health (Ignar and Grygoruk 2015).
2. **Empowerment and Participation:** Empowering indigenous and local communities to actively participate in wetland management is essential for achieving sustainable outcomes. This empowerment involves recognizing their rights, ensuring their voices are heard in decision-making processes, and providing them with the necessary resources and support. Community-based management approaches can foster a sense of ownership and responsibility, leading to more effective and sustainable conservation efforts (Were et al. 2019)

   For instance, participatory approaches that engage communities in monitoring and managing wetlands can lead to more context-specific and culturally appropriate solutions. By involving local communities in the development and implementation of management plans, it is possible to create strategies that are both effective and sustainable. This participatory approach not only enhances conservation efforts but also strengthens community resilience and cohesion (Frieden 2014).

3. **Legal and Institutional Recognition:** A significant challenge faced by indigenous and local communities is the lack of legal and institutional recognition of their rights and contributions. In many regions, these communities lack formal recognition of their land and resource rights, making it difficult for them to protect their territories and the wetlands within them. Strengthening legal frameworks to recognize and protect the rights of indigenous and local communities is crucial for empowering them to participate in wetland conservation (Peters 2020).

   For example, the establishment of legal mechanisms that secure land tenure for indigenous peoples can greatly enhance their ability to manage and conserve wetlands. In countries like Brazil and Australia, legal recognition of indigenous land rights has led to improved conservation outcomes and greater protection of wetland ecosystems (Tommasoli 2012).

4. **Climate Change and Environmental Pressures:** Climate change poses a significant threat to wetlands and the communities that depend on them. Indigenous and local communities are often on the front lines of climate impacts, experiencing changes in weather patterns, rising sea levels, and increased frequency of extreme weather events. These changes can exacerbate existing environmental pressures and pose additional challenges to wetland conservation (Thokchom 2020).

   Traditional livelihoods such as fishing, farming, and hunting can be severely affected by climate change, leading to increased vulnerability and reduced resilience. Integrating indigenous and local knowledge into climate adaptation strategies is essential for developing effective responses that enhance the resilience of both communities and wetlands. For instance, combining traditional knowledge with scientific research can lead to innovative solutions for managing water resources and preserving biodiversity in the face of climate change (NAP 2007).

5. **Education and Capacity Building**: Building the capacity of indigenous and local communities through education and training is vital for enabling them to effectively participate in wetland conservation. This includes providing access to scientific knowledge, technical skills, and resources that can complement traditional practices. Education and capacity-building initiatives can empower communities to engage in monitoring, management, and advocacy activities (Tommasoli 2012).

   For example, programs that facilitate the exchange of knowledge between indigenous communities and scientists can lead to the co-creation of innovative conservation solutions. Capacity-building efforts that focus on leadership, governance, and sustainable livelihoods can further enhance the ability of communities to manage wetlands sustainably. In this regard, educational initiatives that integrate traditional knowledge with modern scientific understanding can create a more comprehensive approach to conservation (Bansal et al. 2024; Biagini and Sagar 2004).

17.4 Recommendations for Strengthening Legal Frameworks

Strengthening legal frameworks for wetland protection is crucial for ensuring sustainable management and conservation of these vital ecosystems. The following recommendations aim to enhance the effectiveness of existing policies and introduce new measures where necessary:

1. **Develop Stand-Alone Wetland Policies:** One of the major gaps in current legal frameworks is the absence of dedicated wetland policies. Developing comprehensive, stand-alone wetland policies would provide clear guidelines and objectives specific to wetland conservation. These policies should integrate principles of sustainable management, conservation priorities, and community engagement to ensure a holistic approach (Sengupta 2016).
2. **Enhance Policy Integration and Coordination:** Wetland protection often involves multiple sectors, including agriculture, water management, and urban development. Strengthening the integration and coordination of policies across these sectors can prevent conflicts and ensure that wetland conservation is considered in all relevant policy decisions. Creating inter-agency task forces or committees can facilitate better coordination and information sharing (Bell-James et al. 2023).
3. **Strengthen Enforcement Mechanisms**: Effective enforcement of wetland protection laws is essential for their success. This requires adequate funding and resources for regulatory agencies, as well as training for enforcement officers. Strengthening penalties for non-compliance and ensuring that enforcement actions are timely and consistent can also improve adherence to wetland protection laws (Obasuyi and Sabbo 2022).
4. **Incorporate Traditional and Indigenous Knowledge:** Incorporating the knowledge and practices of indigenous and local communities can enhance wetland management strategies. These communities often possess valuable insights into sustainable practices that have been refined over generations. Legal frameworks should recognize and protect the rights of these communities to manage their traditional lands and waters, and involve them in decision-making processes (Sinthumule 2024).
5. **Promote Public Awareness and Participation**: Public awareness and participation are critical for the success of wetland conservation efforts. Legal frameworks should include provisions for public education campaigns to inform citizens about the importance of wetlands and the need for their protection. Additionally, mechanisms for public participation in decision-making processes can ensure that policies reflect the concerns and priorities of local communities (Sengupta 2016).
6. **Establish Monitoring and Evaluation Systems:** Robust monitoring and evaluation systems are necessary to track the effectiveness of wetland protection measures. Legal frameworks should mandate regular assessments of wetland health

and the impact of conservation strategies. This data can inform adaptive management practices, ensuring that policies remain effective in the face of changing environmental conditions (Bell-James et al. 2023; UK-NEA 2014).

17.5 Conclusion

Wetland protection is an essential aspect of environmental conservation that requires a multifaceted approach. This research has highlighted the importance of wetlands for biodiversity, climate regulation, and human well-being, demonstrating the critical need for robust legal and policy frameworks to ensure their conservation and sustainable management. The analysis of international frameworks (also at regional and national levels) reveal significant progress in wetland protection, yet gaps remain. The current policies and regulations, while comprehensive in certain areas, often lack the coordination and enforcement mechanisms needed to be fully effective. To address these challenges, several recommendations have been proposed, including the development of standalone wetland policies, enhanced inter-sectoral coordination, improved enforcement mechanisms, incorporation of indigenous and local knowledge, public awareness campaigns, and the establishment of robust monitoring and evaluation systems. Implementing these recommendations requires concerted efforts from policymakers, stakeholders, and communities at all levels.

In conclusion, strengthening legal frameworks for wetland protection is imperative for achieving sustainable development goals and safeguarding ecological integrity. By adopting a holistic and integrated approach, we can ensure the long-term conservation of wetlands, thereby preserving their invaluable contributions to biodiversity and human well-being.

References

African Union (2018) Providing a legal framework for a national public health institute (NPHI). https://ianphi.org/_includes/documents/sections/tools-resources/all-frameworks/legalframeworkenglish_version8.pdf

Agbasi MN, Odiaka NO (2016) The legal framework for the protection of wetlands in Nigeria. J Law Policy Glob 54:40–49

Ahmad Z, Hussain A, Shakeel A (2019) Economics importance of wetlands, their benefits and values case of Pakistan. MPRS – Munich personal RePEc archive, paper no. 95260, p 13. https://mpra.ub.uni-muenchen.de/95260/1/MPRA_paper_95260.pdf

Akinbode O, Jean-Claude S (2022) The effectiveness of the law and environmental policies towards protecting upper Niger lake wetlands in West Africa. Afr J Emerg Issues 4(12):10–20

ALPC-African Land Policy Center (2023) African land policy. Retrieved from https://www.uneca.org/african-land-policy-centre

Amadu I (2021) The roles of laws and policies influencing anthropogenic activities on wetlands: a study of the Iture-Abakam wetlands in Cape Coast. A master's thesis submitted to the University of Cape Coast. p 121

Aminzadeh E, Shareq ZS, Rezazadeh H (2017) Responsibility in conservation of wetlands by studying of Hoor-al Azim/Hoor-al-Howzeh. J Res Energy Law Stud 3(2):193–217. https://doi.org/10.22059/jrels.2018.243251.149

ASEAN Centre for Biodiversity (2024) World Wetlands Day 2024: securing the health and wellbeing of ASEAN communities. Retrieved from https://asean.chm-cbd.net/news/world-wetlands-day-2024-securing-health-and-wellbeing-asean-communities

AU- African Union. Continental Frameworks (2023). Agenda 2063. Retrieved from https://au.int/en/agenda2063/continental-frameworks

Ballut-Dajud GA, Sandoval Herazo LC, Fernández-Lambert G, Marín-Muñiz JL, López Méndez MC, Betanzo-Torres EA (2022) Factors affecting wetland loss: a review. Land 11:434. https://doi.org/10.3390/land11030434

Balqis A, May LS, Sumarni I, Siti SH (2023) Sustainable resilient design challenges for ecotourism education in Setiu wetland. Int J Acad Res Bus Soc Sci 13(10):312–325

Bansal S, Sarker T, Yadav A, Garg I, Gupta M, Sarvaiya H (2024) Indigenous communities and sustainable development: a review and research agenda. Glob Bus Organ Excell 43(4):65–87

BCN- BetterCare Network (2024) Legal and policy framework. Retrieved from https://bettercarenetwork.org/practitioner-library/legal-and-policy-framework/national-policies

Bell-James J, Foster R, Lovelock CE (2023) Identifying priorities for reform to integrate coastal wetland ecosystem services into law and policy. Environ Sci Policy 142:164–172. Retrieved from https://www.sciencedirect.com/science/article/pii/S1462901123000461

Biagini B, Sagar A (2004) Nongovermental organizations (NGOs) and energy. Encyclopedia of Energy

Boone JK, Bhomia RK (2017) Ecosystem carbon stocks of mangroves across broad environmental gradients inWest-Central Africa: global and regional comparisons. PLoS One 12:e0187749

Convention on Wetlands (2021) Global wetland outlook: special edition 2021. Secretariat of the Convention on Wetlands; Gland

Dahle ML (2023) Developing a mitigation hierarchy framework to conserve wetland biodiversity under pressure from development. Master's thesis submitted to The Arctic University of Norway. p 72

Davidson NC (2014) How much wetland has the world lost? Long-term and recent trends in global wetland area. Mar Freshw Res 65(10):934–941

Davidson NC (2023) An introduction to the Ramsar convention on wetlands. Ramsar Wetland: values, assessments, management, pp 1–36. https://doi.org/10.1016/B978-0-12-817803-4.00018-8

Debaere P, Amanda K (2017) Decomposing US water withdrawal since 1950. J Assoc Environ Resour Econ 4(1):155–196

Dupuy PM, Vinuales JE (2018) International environmental law, 2nd edn. Canbridge University Press

Ekardt F, Günther P, Hagemann K, Garske B, Heyl K, Weyland R (2023) Legally binding and ambitious biodiversity protection under the CBD, the global biodiversity framework, and human rights law. Environ Sci Eur 35(1):80. https://doi.org/10.1186/s12302-023-00786-5

Ekeruche F (2024) NCF cautions against indiscriminate use of wetlands. Nigerian Conservation Foundation (NCF) article. https://www.environewsnigeria.com/ncf-cautions-against-indiscriminate-use-of-wetlands/

Emran GI, Barma R, Khan AH, Roy M (2021) Reasons behind the water crisis and its potential health outcomes. Eur J Dev Stud. arXiv preprint arXiv:2403.07019

EP- Environment Policy (2024) Environment policy: general principles and basic framework. Retrieved from https://www.europarl.europa.eu/factsheets/en/sheet/71/environment-policy-general-principles-and-basic-framework

EU-European Union (2024) Sub-Saharan African Multi-annual indications. Retrieved from https://international-partnerships.ec.europa.eu/document/download/44e52b4e-a332-4c31-bb41-c119ba73f295_en

Ewebuke E, Albrecht E (2021) Threats to Niger-Delta wetlands: a case study of Apoi Creek Forest. Open J Ecol 11(2):145–160

FAO (2007) The World's Mangroves 1980-2005. FAO Forestry PaperNo. 153. Rome, Forest Resources Division, FAO, pp. 77

FAO (2021) FAOSTAT: land cover. Food and Agricultural Organization, Rome. http://www.fao.org/faostat/en/#data/LC

Farantouris NE (2009) The international and EU legal framework for the protection of wetlands with particular reference to the Mediterranean basin. Macquarie J Int Comp Envir Law 6:51–50

Finlayson CM, Davies GT, Moomaw WR, Chmura GL, Natali SM, Perry JE et al (2019) The second warning to humanity–providing a context for wetland management and policy. Wetlands 39:1–5

Frieden TR (2014) Six component necessary for effective public health program implementation. Am J Public Health 104(1):17–22

Gardner RC, Davidson NC (2011) The Ramsar convention. In: LePage B (ed) Wetlands. Springer, Dordrecht. https://doi.org/10.1007/978-94-007-0551-7_11

Gardner RC, Connolly KD, Bamba A (2009) African wetlands of international importance: assessment of benefits associated with designations under the Ramsar convention. Georget Int Environ Law Rev 21:257–294

Gardner RC, Davidson NC, Gell PA, Bouquerel L, Grobicki A et al (2016) Ramsar handbooks: International cooperation on wetlands: an introduction to the Ramsar convention on wetlands, 7th edn (previously The Ramsar Convention Manual), Secretariat, Gland

Gardner RC, Okuno E, Pritchard D (2023) Ramsar convention governance and processes at the international level. In: Ramsar Wetlands. Elsevier, pp 37–67

Gell P, Burge O, Flower R (2013) Ramsar wetlands: understanding change in ecological character. PAGES Workshop Rep 22(2):107

Gell PA, Davidson NC, Finlayson CM, Herb AM, McInnes RJ, Pittock J, Pritchard D (2023) 20-Wetlands and future change—Implications and opportunities with the Ramsar Convention. In: Gell PA, Davidson NC, Finlayson CX (eds) Ramsar Wetlands. Elsevier, pp 555–561. ISBN 9780128178034. https://doi.org/10.1016/B978-0-12-817803-4.00017-6

Ger S (2021) Legal framework analysis: national report: Netherlands. Cooperative Europ: ICA-EU Partnership. p 16. https://coops4dev.coop/sites/default/files/2021-08/Netherlands%20Legal%20Framework%20Analysis%20Report.pdf

Ghosh T, Roy C, Hazra S (2015) Application of the DPSIR framework for coastal vulnerability assessment in the lower part of the Indian Sundarban delta. Environ Urban Asia 6(1):1–18. https://doi.org/10.1177/0975425314563952

Hannah L (2011) Climate change, connectivity, and conservation success. Conserv Biol 25(6):1139–1142

Ignar S, Grygoruk M (2015) Wetlands and water framework directive, GeoPlanet: Earth and Planetary Sciences. Springer. https://doi.org/10.1007/978-3-319-13764-31

IPBES (2017) Media release: wetlands are critical to the success of the SDGs. Secretariat of the intergovernmental science-policy platform on biodiversity and ecosystem services (IPBES). https://www.ipbes.net/sites/default/files/downloads/pdf/20170202_media_rele ase_world_wetlands_day.pdf

Irini I, Noralina AA, Norha AH (2012) The laws of wetness: the legislative framework in Malaysia regarding wetlands conservation. Procedia Soc Behav Sci 50:574–581

Jaramillo F, Desormeaux A, Hedlund J, Jawitz JW, Clerici N, Piemontese L, Rodríguez-Rodriguez JA, Anaya JA, Blanco-Libreros JF, Borja S, Celi J, Chalov S et al (2019) Priorities and interactions of sustainable development goals (SDGs) with focus on wetlands. Water 11(3):619. https://doi.org/10.3390/w11030619

Junk WJ, An S, Finlayson CM, Gopal B, Květ J, Mitchell SA, Mitsch WJ, Robarts RD (2013) Current state of knowledge regarding the world's wetlands and their future under global climate change: A synthesis. Aquat Sci 75:151–167

Kaller MD, Kelso WE, Trexler JC (2013) Wetland fish monitoring and assessment. In Wetland techniques: volume 2: organisms, 197–263. https://doi.org/10.1007/978-94-007-6931-1_6
Kingsford RT, Basset A, Jackson L (2016) Wetlands: conservation's poor cousins. Aquat Conserv Mar Freshwat Ecosyst 26(5):892–916
Kuh KF (2018) Climate change. In United Nations framework convention on climate change. Encyclopaedia of the Anthropocene
Li J, Yang L, Pu R, Liu Y (2017) A review on anthropogenic geomorphology. J Geogr Sci 27(1):109–128
Lim TM (2023) ASEAN: inspiring action for wetlands restoration. ASEAN centre for biodiversity. Retrieved from https://asean.chm-cbd.net/news/asean-inspiring-action-wetlands-restoration
Lipman Z, Alam S, Nurhidayah L (2015) The influence of international law upon ASEAN approaches in addressing transbouindary haze pollution in Southeast Asia. Contemp Southeast Asia 37(2):183–210
Materu SF, Urban B, Heise S (2018) A critical review of policies and legislation protecting Tanzanian wetlands. Ecosyst Health Sustain 4(12):310–320. https://doi.org/10.1080/2096-4129.2018.1549510
Mcinnes R (2016) European Union water framework directive and wetlands. In: The Wetland Book. https://doi.org/10.1007/978-94-007-6172-8_129-1. Retrieved from https://www.researchgate.net/publication/312333548_European_Union_Water_Framework_Directive_and_Wetlands
Millennium Ecosystem Assessment (2005) Ecosystems and human Well-being: wetlands and water synthesis. World Resources Institute, Washington, DC
Minx JC, Lamb WF, Andrew RM, Canadell JG, Crippa M, Döbbeling N, ... Ziegler M (2021) A comprehensive dataset of greenhouse gas emissions by sector 1970–2018. Sci Data 8(1):1–15
NAP- National Academies Press (2007) Engaging privacy and information technology in a digital Age. Retrieved from https://nap.nationalacademies.org/read/11896/chapter/11
Noordwijk M (2014) Climate change: agricultural mitigation. encyclopaedia of agriculture and food systems, pp 220–231
NRGI – Natural Resource Governance Institute (2015) Legal framework navigating the web of laws and contracts governing extractive industries. NRGI Reader, p 4. https://resourcegovernance.org/sites/default/files/nrgi_Legal-Framework.pdf
Obasuyi A, Sabbo J (2022) The effectiveness of the law and environmental policies towards protecting upper Niger lake wetlands in West Africa. Afr J Emerg Issues 4(12):10–20
Obona EE (2016) Effective implementation of public policies in Nigeria: issues and challenges. Afr J Pol Adm Stud 9(1):122–141
Page SE, Siegert F, Rieley JO, Boehm HV, Jaya A, Limin S (2002) The amount of carbon released from peat and forest fires in Indonesia during 1997. Nature 420(6911):61–65
Paul MJ, Meyer JL (2001) Streams in the urban landscape. Annu Rev Ecol Syst 32(1):333–365
Peters D (2020) Policies and legal framework. SSWM University Course. Retrieved from https://sswm.info/sswm-university-course/module-3-ecological-sanitation-and-natural-systems-wastewater-treatment-2/policies-and-legal-framework
Poorkhaghan Z, Raise L, Hedary H (2022) Environmental legal study of wetland protection with emphasis on Shadegan International Wetland. Int Stud J (ISJ) 19(3):127–144. https://doi.org/10.22034/isj.2023.285147.1500
Prusty BA, Chandra R, Azeez PA (2017) Wetland science: conservation of coastal wetlands: an appraisal of the policy and legal framework in south Asian Nations. Springer. https://doi.org/10.1007/978-81-322-3715-0_27
Pulselli FM (2008) City out of chaos: Urban self-organization and sustainability. WIT Press. ISBN: 9781845641358
Rabalais NN, Turner RE, Wiseman WJ (2002) Gulf of Mexico hypoxia, aka "The dead zone". Annu Rev Ecol Syst 33(1):235–263
Raburu PO, Kwena F, Nyandiga CO (2012) Challenges and lessons learnt on community based wetland management. In: Raburu PO, Okeyo-Owuor JB, Kwena F (eds) Community based

approach to the management of Nyando Wetland, Lake Victoria Basin, Kenya. Kenya disaster concern – VIRED – UNDP, pp 122–131

Ramsar (2018) The list of wetlands of international importance. Accessed 23 Apr 2024. https://www.ramsar.org/sites/default/files/documents/library/sitelist.pdf

Ramsar Convention on Wetlands (2016) An introduction to the Ramsar convention on wetlands, 7th edn. Ramsar Convention Secretariat, Gland

Ramsar Convention on Wetlands (2018) Global wetland outlook: state of the world's wetlands and their services to people. Ramsar Convention Secretariat, Gland

Ransar Convention on Wetlands (2018) Scaling up wetland conservation, wise use and restoration to achieve the sustainable development goals. Ramsar convention on Wetlands https://www.ramsar.org/sites/default/files/documents/library/wetlands_sdgs_e.pdf

Riegner M (2014) Legal frameworks and general principles for indicators in sovereign debt restructuring. Paper prepared for the Third Session of the UNCTAD Working Group on a Debt Workout Mechanism. p 37. https://unctad.org/system/files/non-official-document/gds_sd_2014-03-19_bp_legal_en.pdf

Rwanda Environment Management Authority (2020) National wetlands management framework for Rwanda. In: Centre of Excellence in biodiversity and natural resource management. University of Rwanda, Rwanda

Sage C (2003) Social embeddedness and relations of regard: alternatives to commodity exchange. In: Hinchliffe S, et al. (eds.), Geographies of rural culturesand societies. Ashgate

Sánchez-Espinosa A, Schröder J (2019) Land use and land cover mapping in wetlands one step closer to the ground: sentinel-2 versus landsat 8. J Environ Manag 247:484–498

Schuster PF, Krabbenhoft DP, Naftz DL, Cecil LD, Olson ML, Dewild JF, Aiken GR (2002) Atmospheric mercury deposition during the last 270 years: a glacial ice core record of natural and anthropogenic sources. Environ Sci Technol 36(11):2303–2310

Sengupta S (2016) Wetland protection policy needs to be strengthened: Javadekar. DownToEsrth: water. Retrieved from https://www.downtoearth.org.in/water/framework-to-protect-wetlands-needs-to-be-strengthened-javadekar-52674

Sica YV, Quintana RD, Radeloff VC, Gavier-Pizarro GI (2016) Wetland loss due to land use change in the lower Paraná River Delta, Argentina. Sci Total Environ 568:967–978

Sima S, Hosein K, Akbar R (2022) Complexity and the ecological impacts of tourism on wetlands ecosystem and rural sustainability: a case study of Zrebar wetlands. Am J Agric Biol Sci 18:22–38

Sinthumule NI (2024) An analysis of policy and legal framework on wetland conservation and management in South Africa. Watershed Ecol Environ 4:63–69

Snethlage M (2015) Short topic assessment: integrated approaches for wetland conservation

Stephenson M (2018) The coming industrial revolution? Fossil fuels and developing countries. Energy and Climate management ETC/BD report to the EEA, p 25. Elsevier

Thokchom B (2020) Water-related problem with special reference to global climate change in India. In: Water conservation and wastewater treatment in BRICS nations. Elsevier, pp 37–60

Tommasoli M (2012) Rule of law and democracy: addressing the gap between policies and practices. UN Chronicles. Retrieved from https://www.un.org/en/chronicle/article/rule-law-and-democracy-addressing-gap-between-policies-and-practices

UK National Ecosystem Assessment (UK- NEA) (2014) The UK national ecosystem assessment. UNEP-WCMC, LWEC. http://uknea.unep-wcmc.org/Default.aspx

UN (2023) Global issues: water. Retrieved May 4, 2024, from United Nations Web site: https://www.un.org/en/global-issues/water

UNE- United Nations Environment (2020) International Day for biological diversity (22 May): convention on biological diversity, key international instrument for sustainable development. Retrieved from https://www.un.org/en/observances/biological-diversity-day/convention

US-EPA – United State Environmental Protection Agency (2016) Wetland functions and values. Watershed academy web. p 23. Retrieved from https://www.epa.gov/sites/default/files/2016-02/documents/wetlandfunctionsvalues.pdf

Van H (2013) Developing legal frameworks to promote fiscal responsibilities: design matters. IMF eLibrary. Retrieved from https://www.elibrary.imf.org/view/book/9781475531091/ch002.xml

Vasumathi V, Kalpana R, Pazhanivelan S, Priya MV, Pon Arasan A (2022) Classification, distribution, features, biodiversity and functions of wetlands: a review. Agric Rev. https://doi.org/10.18805/ag.R-2559

Votteler TH, Muir TA (1996) Wetland protection legislation. Wetland management and research: USGS professional paper, pp 1–2. Retrieved from https://www.researchgate.net/publication/374114264

WECOOP- EU-Central Asia Cooperation on Water-Environment-Climate Change (2023) EU environment policy: the 8th environment action programme till 2030. Retrieved from https://wecoop.eu/regional-knowledge-centre/eu-policies-regulations/

Were D, Kansiime F, Fetahi T, Cooper A, Jjuuko C (2019) Carbon sequestration by wetlands: a critical review of enhancement measures for climate change mitigation. Earth Syst Environ 3:327–340

Wetland International (2010) Wetlands and water, sanitation and hygiene (WASH) – understanding the linkages. Wetlands International, Ede. https://www.ircwash.org/sites/default/files/WetlandsInt-2010-Wetlands.pdf

Wetland International (2022) Post-2020 global biodiversity framework: ensuring protection and restoration of inland waters. Retrieved from www.wetlands.org/publication/post-2020-global-biodiversity-framework-ensuring-protection-and-restoration-of-inland-waters/

Xu S, Liu X, Li X, Tian C (2019) Soil organic carbon changes following wetland restoration: a global meta-analysis. Geoderma 353:89–96

Zacharias I, Dimitriou E, Koussouris T (2005) Integrated water management scenarios for wetland protection: application in Trichonis Lake. Environ Model Softw 20:177–185

Chapter 18
Future Challenges and Opportunities for Wetland Preservation

Olanrewaju Isola Eresanya, Babafemi Raphael Babaniyi, Oluwafemi Temitope Ojo, and Kehinde Oluwasiji Olorunfemi

Abstract Wetlands are essential ecosystems renowned for their ecological, economic, cultural, and spiritual significance. They facilitate vital natural activities such as carbon sequestration, biodiversity support, water regulation, and flood control. Despite their importance, wetlands face increasing threats from climate change, pollution, habitat destruction, and over-extraction of resources. Addressing these challenges requires a multifaceted approach that includes conservation, restoration, and sustainable management practices. Effective strategies for preserving wetlands involve restoring degraded areas, establishing protective buffer zones, implementing constructed wetlands projects, and enforcing pollution control measures. Additionally, integrating wetlands into land use planning, engaging local communities, and recognizing their cultural and spiritual significance can enhance conservation efforts. Economic benefits such as sustainable livelihoods and recreational opportunities further underscore the need for adequate funding and international synergy, exemplified by frameworks like the Ramsar Convention. Ongoing research and monitoring are crucial for adapting management strategies to emerging challenges. This comprehensive approach ensures that wetlands continue to sustain their invaluable activities and maintain their resilience against environmental threats and anthropogenic activities.

Keywords Wetlands · Land use · Environmental management and conservation

O. I. Eresanya · B. R. Babaniyi · K. O. Olorunfemi (✉)
Bioresources Development Centre, National Biotechnology Development Agency (NABDA), Abuja, Nigeria
e-mail: olorunfemiko@futa.edu.ng

O. T. Ojo
Bioresources Development center, National Biotechnology Development Agency, Abuja, Nigeria

B. R. Babaniyi et al. (eds.), *Wetland Ecosystems: Conservation Strategies, Policy Management and Applications*, Wetlands: Ecology, Conservation and Management 12, https://doi.org/10.1007/978-3-031-91982-4_18

18.1 Introduction

Wetland is one of the major ecosystems on the planet earth and of inestimable value. As a naturally occurring multifaceted system, formed as a result of land and water interaction. Wetlands can be defined by Ramsar convention, as an area of marsh, fen, peatland or water, whether natural or artificial, short-termed or long-termed, with static or flowing water, fresh, brackish or salt, including areas of marine water, the depth at which at low tide does not exceed 6 metres (Ye et al. 2022). The Convention on Wetlands is an intergovernmental treaty adopted on 2 February 1971 in the Iranian city of Ramsar, on the southern shore of the Caspian Sea. Thus, though the name of the Convention is written “Convention on Wetlands (Ramsar, Iran, 1971)”, it has come to be known popularly as the “Ramsar Convention”. The very vast benefits of wetlands, requires that it is preserved and sustained. Globally, almost 50% of the nation’s original wetlands area has been lost, by the conscious strategy of preservation, the rate of wetland losses will be minimized. Lack of wetland preservation have shown to negatively affect biodiversity, as well as endanger species which are wetland-dependent. Wetland preservation can be defined as the embodiment of steps and action taken to protect, conserve, and restore wetlands ecosystem, to maintain their ecological integrity, functions and biodiversity populations. The motive behind wetland preservation is to safeguard these valuable natural environments. From the Ramsar Convention perspective, wise use or sustainable use of wetlands, through effective land-use planning, water-resource management planning, or development planning, will support wetlands preservation. However, wetland preservation is increasingly challenging due to evolving environmental, social, and economic conditions. These ecosystems, vital for biodiversity, water purification, and flood control, are under constant threat from various factors (Xu et al. 2019). Wetlands plays a crucial role in mitigating climate change, and conversely, climate change affects wetlands. Also, how wetlands preservation, management and sustainability is affected by global sea level, urbanization, and water pollution will be delved into.

18.1.1 Climate Change

Wetlands play a critical role in regulating atmospheric greenhouse gas concentrations, including methane, carbon dioxide, and nitrous oxide, which contribute approximately 60%, 20%, and 6% to global warming potential, respectively (IPCC 2007; Kingsford et al. 2016). Various biotic and abiotic factors influence wetland functionality, but climate change stands out as a significant threat. It alters wetland ecosystems by raising temperatures and modifying hydrological patterns, which can, in turn, affect the biogeochemistry of these environments (Erwin 2009). Wetlands are among the most productive ecosystems, actively sequestering carbon through photosynthesis and storing it as plant biomass or organic matter in soil. Due to their waterlogged and anoxic conditions, decomposition rates are lower than production rates, leading to substantial carbon accumulation. This makes wetlands a crucial carbon sink (Laiho 2006).

Climate change can affect wetlands by direct and indirect effects of rising temperature, changes in rainfall intensity and frequency, extreme climatic events such as drought, flooding and the frequency of storms. Altered hydrology and rising temperature can change the biogeochemistry and function of the wetland to the degree that some important services might be turned into disservices (Reid et al. 2019). This means that they will no longer provide a water purification service and may decompose and release nutrients to the surface water causing such problems as eutrophication, acidification and brownification in the water bodies (Stets and Cotner 2008; Corman et al. 2018).

18.1.2 Global Sea Level Rise

Global sea level refers to the increase in the average height of the world's oceans. Sea level is measured in two ways: relative sea level (RSL), which is the height of the sea relative to the solid Earth's surface, and geocentric sea level, measured with respect to a geocentric reference like the reference ellipsoid (Mengesha 2017). RSL estimates have been derived from tide gauges and geological records over the past few centuries, while geocentric sea level has been measured using satellite altimetry over the last two decades. When averaged globally, these measurements provide the global mean sea level (GMSL).

The physical processes driving GMSL rise and regional variations in RSL are interconnected but not identical. The primary contributors to the current rise in GMSL include the thermal expansion of seawater due to warming, land ice melt from glaciers and ice sheets, and the exchange of freshwater between oceans, land water reservoirs, and groundwater storage changes (Levitus et al. 2009; Ishii and Kimoto 2009; Domingues et al. 2008; Church et al. 2011; IPCC AR4; Cazenave and Remy 2011). These recent trends are largely attributed to climate change driven by anthropogenic greenhouse gas emissions. The ongoing global sea level rise poses significant risks, including increased coastal flooding, shoreline erosion, saltwater intrusion into freshwater resources, and adverse impacts on coastal and marine ecosystems (IPCC AR4). Addressing these challenges requires a comprehensive understanding of the drivers of sea level change and effective adaptation and mitigation strategies.

18.1.3 Urbanization

Urban expansion frequently leads to the drainage, filling, and degradation of wetlands to accommodate development needs. By 2050, the global population is projected to reach approximately 9.8 billion, with the majority residing in urban areas. Without careful planning and management, urbanization poses a significant threat to urban wetlands, resulting in their destruction, contamination, and conversion for housing, agriculture, and industrial purposes (Moore et al. 1998; Zedler and Kercher 2005). Urbanization presents several critical challenges to wetlands, compromising their ecological, hydrological, and biological integrity. Key impacts include:

(i) **Habitat Loss and Fragmentation:** Urban development often leads to the direct conversion of wetland areas into residential, commercial, and industrial zones, fragmenting habitats and disrupting ecological connectivity. This fragmentation reduces habitat quality, impedes species movement, and leads to declines in biodiversity (Chen et al. 2019).

(ii) **Water Quality Degradation:** Urban runoff is a major source of pollution, introducing contaminants such as heavy metals, hydrocarbons, pesticides, and excessive nutrients into wetland systems. These pollutants can lead to eutrophication—where nutrient over-enrichment causes algal blooms, depleting oxygen levels and harming aquatic life (Findlay and Houlahan 1997).

(iii) **Hydrological Alterations:** The proliferation of impervious surfaces, such as roads and buildings, increases surface runoff while reducing groundwater recharge. These changes alter the natural hydrological dynamics of wetlands, impacting their capacity to regulate water flow, support groundwater recharge, and maintain essential ecological functions (Calhoun et al. 2017).

(iv) **Invasive Species:** Urbanization facilitates the introduction and spread of invasive species, which can outcompete native flora and fauna, disrupt ecosystem processes, and alter wetland habitats. These invasions often result in significant biodiversity losses and shifts in ecological balance (Davidson 2014).

(v) **Climate Change:** Urban areas contribute to climate change through elevated greenhouse gas emissions. Climate change further impacts wetlands by altering precipitation patterns, increasing temperatures, and raising sea levels. These changes can lead to habitat loss, shifts in species composition, and altered hydrological cycles, compounding the stress on already vulnerable wetland ecosystems (US EPA 1993; Kantrud et al. 1989). To safeguard urban wetlands, it is crucial to implement integrated planning and conservation strategies that address these threats, preserving the vital ecological services wetlands provide to urban communities.

18.1.4 Water Pollution

Wetlands are critical ecosystems that offer invaluable ecological services, such as water filtration, flood control, carbon storage, and providing habitat for diverse species, including many that are endangered. Despite their importance, wetlands are highly vulnerable to water pollution, which threatens their health, functionality, and ability to sustain biodiversity.

18.1.4.1 Major Sources of Water Pollution Affecting Wetlands

(a) **Agricultural Runoff:** Agricultural runoff is one of the leading sources of wetland pollution, significantly affecting the health and functionality of these ecosystems. Fertilizers, pesticides, and herbicides commonly used in modern agriculture are frequently washed away from fields by rainwater or irrigation, making their way into wetland areas through surface runoff (Alikhani et al. 2021). This process introduces high concentrations of nutrients, primarily nitrogen and phosphorus, along with a variety of toxic chemicals into wetland environments, causing multiple detrimental effects. One of the most severe impacts of agricultural runoff is nutrient overload, which leads to eutrophication—a process where nutrient enrichment stimulates excessive growth of algae and aquatic plants. In wetlands, this rapid algal growth forms dense mats on the water surface, blocking sunlight and disrupting photosynthesis for submerged plants. As the algae and plants die and decompose, they consume large amounts of oxygen, creating hypoxic or anoxic conditions where aquatic life cannot survive. This oxygen depletion results in "dead zones" that collapse local food webs and cause a significant decline in biodiversity (Sarkar et al. 2016).

The toxic chemicals from pesticides and herbicides carried into wetlands through runoff are highly damaging to aquatic life, even in low concentrations. These substances can poison fish, amphibians, and invertebrates, causing mortality, behavioral changes, and reproductive failures. Herbicides also harm native wetland vegetation, which plays a critical role in providing habitat, stabilizing soil, and maintaining water quality. The impact of these chemicals can persist over time as they accumulate in sediments and bioaccumulate in the tissues of aquatic organisms, magnifying their harmful effects throughout the food chain. Agricultural runoff also disrupts the natural nutrient cycles of wetlands, causing imbalances that alter ecosystem functions. The influx of excess nitrogen and phosphorus can change the composition of wetland plant communities, often favoring invasive or fast-growing species that outcompete native vegetation (Bănăduc et al. 2022). This shift in plant dominance alters the structure and function of the wetland, reducing its ability to support diverse wildlife and perform essential services such as water filtration and carbon storage. In addition, nutrient pollution and chemical contaminants degrade water quality, making it harmful for both wildlife and human use. Elevated levels of nitrates and phosphates can contaminate drinking water sources, posing health risks such as methemoglobinemia, or "blue baby syndrome," in infants. Pesticides in water are also hazardous to human health, with potential acute and chronic effects, including endocrine disruption and cancer.

Agricultural practices often contribute to increased soil erosion, with sediments carried by runoff into wetlands. This sedimentation reduces water clarity, smothers aquatic habitats, and disrupts the breeding grounds of fish and amphibians (Boyd and Boyd 2020). The accumulation of sediment also alters wetland hydrology, affecting water flow patterns and diminishing the wetland's capacity to store and filter water. The cumulative effects of agricultural runoff weaken wetland resilience, making these ecosystems more vulnerable to additional stressors, such as climate change and invasive species. By degrading the structural and functional integrity of wetlands, agricultural runoff diminishes their capacity to provide critical ecosystem services, including flood protection, water purification, and habitat support. Mitigating the impact of agricultural runoff requires integrated land and water management practices, such as establishing buffer zones, adopting sustainable farming techniques, and reducing the use of chemical inputs. Protecting wetlands from agricultural pollution is essential for preserving their ecological health and the invaluable services they offer to both nature and human communities (Newton et al. 2020).

(b) **Industrial Discharges:** Factories and industrial plants are major contributors to wetland pollution, as they frequently discharge pollutants directly into nearby water bodies, including wetlands. These discharges often contain a mix of harmful substances, such as heavy metals, toxic chemicals, and other hazardous materials like solvents and hydrocarbons. When released into wetland environments, these pollutants do not simply dissipate; instead, they tend to accumulate in sediments and water, posing long-term and severe risks to both wildlife and human health (Bulonza 2015).

Heavy metals, including mercury, lead, cadmium, and arsenic, are common industrial pollutants that pose significant threats to wetland ecosystems. These metals are highly toxic even at low concentrations and can persist in the environment for extended periods due to their non-biodegradable nature. Once deposited in wetland sediments, heavy metals can be taken up by aquatic plants and animals, leading to bioaccumulation and biomagnification throughout the food web (Kumar et al. 2021). This process not only affects individual species, causing reproductive failures, developmental abnormalities, and mortality, but also disrupts entire ecosystems, leading to declines in biodiversity and altering the structure of wetland communities (Davis et al. 2016). Toxic chemicals, such as polychlorinated biphenyls (PCBs), dioxins, and phenols, are also prevalent in industrial discharges and pose serious ecological and health concerns. These substances are known for their persistence in the environment and their ability to cause a wide range of harmful effects, including endocrine disruption, carcinogenicity, and immune system damage in both wildlife and humans. In wetlands, these chemicals can bind to sediments, making them available for uptake by benthic organisms, which are often at the base of the food chain. As these contaminants move up the food web, they can have compounding effects on predators, including birds, fish, and mammals, ultimately threatening the health of entire ecosystems.

Hydrocarbons, including oil and its byproducts, are another class of pollutants commonly found in industrial discharges. When released into wetlands, hydrocarbons can create slicks on the water surface that block sunlight and reduce oxygen exchange, which is critical for the survival of aquatic plants and animals. These compounds can coat the feathers of waterfowl, impairing their ability to regulate body temperature and fly, and can smother aquatic habitats, affecting fish and invertebrates. Over time, the breakdown of hydrocarbons releases additional toxic compounds into the water and sediment, exacerbating the damage to wetland ecosystems. The cumulative impact of industrial pollution on wetlands extends beyond ecological damage, posing direct risks to human health as well. Wetlands serve as natural filters, trapping pollutants before they can reach larger bodies of water, including those used for drinking water. However, when wetlands are overwhelmed with industrial pollutants, their capacity to filter and detoxify contaminants diminishes, allowing harmful substances to enter groundwater and surface water supplies. This can lead to a range of health issues in nearby communities, including increased risks of cancer, neurological disorders, and other chronic health conditions linked to exposure to toxic industrial chemicals. Industrial pollution also disrupts the natural functions of wetlands, reducing their ability to provide essential services such as water filtration, flood control, and habitat provision. Contaminated wetlands are less effective at supporting the diverse plant and animal life that relies on these ecosystems, leading to a decline in ecosystem resilience and functionality. Addressing the impact of industrial discharges on wetlands requires stringent regulations, effective waste management practices, and the implementation of advanced treatment technologies to prevent pollutants from entering these vulnerable ecosystems in the first place. Protecting wetlands from industrial pollution is not only critical for maintaining biodiversity but also for safeguarding human health and the broader environmental benefits that wetlands provide.

(c) **Urban Runoff:** Urban areas are significant contributors to wetland pollution, primarily through stormwater runoff that carries a wide array of pollutants into wetland environments. During rainfall or snowmelt, water flows over impervious surfaces such as roads, parking lots, and rooftops, picking up contaminants like oil, grease, heavy metals, road salts, and various forms of debris. This runoff often makes its way directly into nearby wetlands without treatment, introducing pollutants that degrade water quality and disrupt the delicate balance of these ecosystems.

Oil and grease from vehicles, as well as other urban sources, create slicks on the water surface, impeding gas exchange and reducing oxygen levels in wetland habitats. This can suffocate aquatic plants and animals, affecting the entire food web (Anaero-Nweke 2013). Heavy metals such as lead, zinc, copper, and cadmium, commonly found in runoff from roads and industrial areas, accumulate in wetland sediments where they pose long-term toxic threats to aquatic life. These metals can impair the growth, reproduction, and survival of fish, amphibians, and invertebrates, and can bioaccumulate in the tissues of organisms, leading to further ecological harm as they move up the food chain. Road salts used for de-icing in urban areas are another common pollutant carried into wetlands by stormwater runoff. High concentrations of salts can alter the salinity of wetland water, making it inhospitable for many freshwater species. This salinity shift can stress or kill sensitive plants and animals, disrupt breeding cycles, and reduce overall biodiversity. Additionally, road salts can change the soil chemistry of wetlands, affecting nutrient availability and altering plant communities. Debris and litter, including plastics and microplastics, are also carried by urban runoff into wetlands, where they pose physical and chemical threats to wildlife (Arias and Botte 2020). Larger debris can obstruct water flow, trap animals, and degrade the aesthetic and recreational value of these environments. Microplastics, which break down into smaller particles, can be ingested by aquatic organisms, leading to physical harm, chemical contamination, and disruption of feeding behavior. The cumulative impact of urban runoff pollutants compromises the ability of wetlands to perform essential functions such as water purification, flood control, and habitat provision. Polluted wetlands are less effective at filtering contaminants, which can then flow downstream, affecting larger water bodies and drinking water sources (Akpor et al. 2014). To mitigate these impacts, urban areas need to adopt sustainable stormwater management practices, such as green infrastructure, permeable surfaces, and improved waste disposal systems, to reduce the flow of pollutants into wetland ecosystems. Protecting wetlands from urban runoff is crucial for maintaining their ecological health, supporting biodiversity, and ensuring the continued delivery of the valuable services they provide to human communities.

(d) **Wastewater Treatment Plants:** Even after undergoing treatment, wastewater can still contain various pollutants, including pharmaceuticals, hormones, and personal care products, which are often resistant to conventional treatment methods. These substances, known as emerging contaminants, are not fully removed during typical wastewater treatment processes and frequently find their way into wetlands through effluent discharge. Pharmaceuticals such as antibiotics, painkillers, and antidepressants, along with hormones from birth control and hormone replacement therapies, pose significant risks to wetland ecosystems due to their bioactive nature. Once introduced into wetland environments, these contaminants can disrupt the endocrine systems of aquatic species, leading to altered reproductive and developmental processes. For example, exposure to even low concentrations of synthetic hormones can cause feminization of male fish, reduced fertility, and skewed sex ratios, threatening population stability. Other pharmaceuticals can interfere with growth, immune function, and behavior, making aquatic organisms more susceptible to disease and predation. Personal care products, including antimicrobial agents, fragrances, and UV filters, add another layer of complexity to wetland pollution. These substances can accumulate in sediments and bioaccumulate in the tissues of aquatic organisms, leading to chronic toxicity and long-term ecological impacts. The presence of these pollutants can alter the composition of microbial communities in wetlands, disrupting nutrient cycling and other critical ecosystem functions. The persistence and bioactive nature of these contaminants make them particularly problematic, as they can exert effects at very low concentrations, often undetected by traditional monitoring methods. Their impact on wetlands extends beyond individual species, potentially disrupting entire food webs and altering ecosystem dynamics. Mitigating the entry of these pollutants into wetlands requires advanced treatment technologies, improved pharmaceutical disposal practices, and increased public awareness of the ecological consequences of everyday chemical use. Protecting wetlands from these emerging contaminants is essential to preserve their ecological integrity and the vital services they provide to both wildlife and human communities (Mitsch and Gosselink 2015; NRC 2001; U.S. EPA 2020; Vymazal 2010; Carpenter et al. 1998).

(e) **Plastic and Microplastic Pollution:** Plastics and microplastics, originating from sources such as packaging, personal care products, and synthetic textiles, are increasingly found accumulating in wetland environments, where they pose significant physical and chemical threats to wildlife (Bamigboye et al. 2024). These pollutants, ranging in size from large plastic debris to microscopic fragments, can cause a variety of harmful effects on the organisms that inhabit these ecosystems. Large plastic debris, such as bags, bottles, and discarded fishing gear, can physically obstruct and entangle wildlife, leading to injury, impaired movement, or death. Birds, fish, and other animals often mistake plastic pieces for food, leading to ingestion that can cause internal blockages, malnutrition, and starvation, as the ingested plastics occupy space in the digestive system but provide no nutritional value. Additionally, plastics can cause physical abrasions or damage to sensitive tissues in the digestive tracts of organisms, exacerbating their health impacts. Microplastics, which are tiny plastic particles less than 5 millimeters in size, pose unique threats due to their small size and widespread presence. These particles can be ingested by a wide range of organisms, from tiny plankton to larger fish and birds, often entering the food web at its very base. The ingestion of microplastics can lead to sub-lethal effects such as reduced feeding efficiency, altered behavior, and compromised growth and reproduction, which can have cascading impacts on wetland biodiversity and ecosystem stability (Chaukura et al. 2021).

Beyond their physical effects, plastics and microplastics are also carriers of toxic chemical additives and pollutants. Many plastics contain additives such as phthalates, bisphenol A (BPA), and flame retardants, which can leach out into the environment, contaminating water and sediments. These chemicals are known endocrine disruptors and can interfere with the hormonal systems of wildlife, leading to reproductive, developmental, and immune system disorders. Additionally, microplastics can adsorb persistent organic pollutants (POPs) from the surrounding environment, such as pesticides and industrial chemicals, effectively acting as vectors for these toxic substances. When ingested, these contaminated particles can transfer harmful chemicals through food webs, magnifying their impacts on higher trophic levels, including predators and, ultimately, humans. The accumulation of plastics and microplastics in wetlands disrupts not only individual organisms but also the broader ecological balance, altering nutrient cycles, habitat quality, and the natural processes that wetlands provide. Addressing this growing problem requires comprehensive waste management strategies, pollution prevention efforts, and greater public awareness of the environmental impacts of plastic use. Reducing plastic pollution in wetlands is essential for preserving these critical habitats and protecting the wildlife and ecosystem services they support (Sreelakshmi and Chitra 2021).

18.1.5 Mining and Extraction Activities

Activities such as oil drilling, natural gas extraction, and mining have profound impacts on wetland ecosystems, primarily through habitat destruction and pollution. These industrial operations often require the removal of extensive vegetation, leading to significant habitat loss and fragmentation. The clearing of vegetation not only destroys the physical habitat but also disrupts the ecological processes and connections that wetlands provide, severely affecting the flora and fauna that depend on these environments. Mining operations, in particular, are known for their extensive environmental footprint (Ali et al. 2024). The removal of large areas of land for mining often results in the direct destruction of wetland habitats, which are crucial for supporting diverse plant and animal species. This destruction can lead to the fragmentation of wetland ecosystems, making it challenging for species to migrate, find food, and reproduce. The loss of these habitats diminishes biodiversity and impairs the ability of wetlands to perform essential ecological functions.

In addition to physical habitat destruction, these activities frequently release a variety of harmful pollutants into the environment. For instance, oil drilling and natural gas extraction can result in spills and leaks that introduce hydrocarbons into wetland areas. These pollutants can contaminate water sources, leading to toxic conditions for aquatic life and impacting plant growth. The release of heavy metals, such as mercury, lead, and arsenic, from mining activities further exacerbates the problem, as these metals can accumulate in sediments and pose long-term health risks to wildlife.

Mining operations can also produce acidic runoff, known as acid mine drainage, which occurs when sulfide minerals exposed during mining react with oxygen and water, creating sulfuric acid. This acid runoff lowers the pH of water bodies, leading to a condition known as acidification. Acidification can severely disrupt aquatic ecosystems by harming organisms that are sensitive to changes in pH, such as fish, insects, and plants, and by degrading the water quality essential for the health of wetland environments (Gupta and Pandey 2024). The combined effects of habitat destruction and pollution from these industrial activities significantly compromise the integrity of wetland ecosystems. To mitigate these impacts, it is crucial to implement effective environmental management practices, including habitat restoration, pollution control measures, and regulatory frameworks that ensure sustainable resource extraction. Protecting wetlands from the adverse effects of industrial activities is essential for maintaining their ecological health and the valuable services they provide to both wildlife and human communities.

18.1.6 Over-extraction of water

Excessive groundwater extraction for agricultural, industrial, or domestic purposes can significantly impact wetlands by lowering water tables and depriving these ecosystems of essential water inputs. When groundwater levels drop, wetlands can experience reduced water availability, leading to the drying out of these habitats. This reduction in water levels can cause the loss of plant and animal species that rely on consistent moisture, as well as deteriorate water quality by concentrating pollutants and reducing dilution. In addition to groundwater depletion, excessive extraction from rivers and streams can diminish surface water flow into wetlands. This disruption in hydrology can alter the natural water cycle, affecting the ecological balance and overall health of wetland environments. Without adequate surface water input, wetlands may struggle to maintain their critical functions, such as water purification, flood control, and habitat support. Coastal wetlands face unique challenges due to freshwater over-extraction. In these areas, excessive withdrawal of freshwater can lead to increased salinization through seawater intrusion. When the natural balance between freshwater and seawater is disrupted, higher salinity levels can penetrate wetlands, impacting species that are adapted to specific salinity ranges. This change in salinity can significantly alter the structure and function of coastal wetland ecosystems, affecting both plant and animal species and reducing biodiversity (Zedler and Kercher 2005; Williams 2001).

Beyond the issues of water management, wetlands face additional challenges. Regulatory changes can impact wetland protection and management policies, potentially leading to inadequate conservation measures. The lack of public awareness about the importance of wetlands can hinder efforts to protect and restore these ecosystems. Economic pressures often drive development and land use changes that encroach upon wetland areas, while limited funding can restrict conservation and restoration efforts. Natural disasters, such as floods and storms, can cause immediate and severe damage to wetlands, altering their structure and function. Legal and property rights issues can complicate efforts to manage and protect wetland areas, especially when ownership and jurisdictional disputes arise. Biodiversity loss is a significant concern, as the decline of species within wetlands can disrupt ecological balance and reduce the resilience of these ecosystems. Furthermore, the mismanagement of carbon sequestration in wetlands can undermine their role as carbon sinks. Wetlands are critical for storing carbon and mitigating climate change, but improper management practices can lead to the release of stored carbon and negate their climate benefits. Addressing these challenges requires a comprehensive approach that includes effective regulation, public education, adequate funding, and adaptive management strategies to ensure the continued health and functionality of wetland ecosystems (Acreman and Ferguson 2010).

18.2 Opportunities for Wetland Preservation

To protect and sustain wetland ecosystems effectively, various strategies and approaches can be employed. These strategies are designed to address the multifaceted environmental, economic, and social importance of wetlands while mitigating the impacts of human activities that threaten their health. Key opportunities for wetland preservation include:

(i) **Establishing Protected Areas:** Designating wetlands as protected areas or reserves is a crucial strategy for preserving these vital ecosystems. By formally recognizing wetlands as protected, it becomes possible to prevent development and other activities that could degrade their ecological integrity. This designation serves as a safeguard against immediate threats such as urban expansion, industrial activities, and agricultural encroachment, which might otherwise lead to habitat loss and environmental degradation. Protected status ensures that wetlands are managed according to conservation objectives, focusing on maintaining their ecological functions and values. This can involve implementing regulations that restrict land use and prohibit harmful activities, such as draining, filling, or polluting the wetland areas. It also provides a framework for monitoring and enforcement, ensuring that any activities within or near the protected areas adhere to conservation guidelines. Furthermore, establishing protected areas offers a foundation for long-term conservation efforts by creating a stable environment where wetlands can recover from past disturbances and continue to function as critical habitats for diverse species. It facilitates habitat restoration projects, supports scientific research, and promotes public awareness and education about the importance of wetlands

(ii) **Implementing Sustainable Management Practices:** Adopting sustainable land use and water management practices is essential for minimizing the negative impacts of agriculture, industry, and urbanization on wetlands. These practices aim to balance development needs with the preservation of wetland ecosystems and their vital functions.

Controlled groundwater extraction involves carefully managing the amount and rate at which groundwater is withdrawn for agricultural, industrial, or domestic purposes. By implementing measures such as monitoring groundwater levels, setting extraction limits, and promoting water-efficient technologies, it is possible to reduce the depletion of aquifers and maintain adequate water supplies for wetlands.

Sustainable agriculture practices are designed to minimize the impact of farming on wetland ecosystems. This includes techniques such as precision farming, which optimizes the use of fertilizers and pesticides to reduce runoff into wetlands, and conservation tillage, which preserves soil structure and reduces erosion. Additionally, practices like crop rotation and agroforestry can enhance soil health and reduce the need for chemical inputs, further protecting wetland water quality.

Green infrastructure refers to the use of natural systems and processes to manage stormwater runoff and reduce its impact on wetlands. Examples include creating wetlands or rain gardens to capture and filter runoff, implementing permeable paving to allow water infiltration, and restoring riparian buffers along waterways to intercept pollutants. These solutions help manage excess stormwater, prevent erosion, and filter pollutants before they reach wetland areas. By integrating these sustainable practices into land use and water management strategies, it is possible to mitigate the adverse effects of human activities on wetlands. This approach helps preserve wetland ecosystems, maintain their ecological functions, and support their resilience in the face of ongoing environmental changes.

(iii) **Restoring Degraded Wetlands:** Restoration projects are vital for rehabilitating wetlands that have been damaged or lost due to human activities. These projects aim to restore the ecological health and functionality of wetlands, enabling them to recover their essential functions and support biodiversity. The restoration process typically involves several key activities:

1. **Replanting Native Vegetation:** One of the primary components of wetland restoration is replanting native plant species that are suited to the local environment. Native vegetation helps to stabilize soil, provide habitat for wildlife, and support the natural nutrient cycles of the wetland. By reintroducing native plants, restoration projects can enhance the ecological integrity of the wetland and promote the return of associated animal species.
2. **Restoring Natural Hydrology:** Wetlands depend on specific hydrological conditions to maintain their ecological functions. Restoration efforts often focus on reestablishing natural water flow patterns by removing barriers such as drainage ditches, levees, or culverts that have altered the hydrology of the wetland. This may also involve reintroducing natural water sources or modifying land use practices to ensure that the wetland receives the appropriate amount of water.
3. **Removing Contaminants:** Contaminants from previous land use, such as heavy metals, pesticides, or excess nutrients, can degrade wetland health and function. Restoration projects may include measures to remove or reduce these contaminants, such as implementing soil and water treatment technologies, or removing contaminated sediments. By addressing pollution, these projects help to improve water quality and create a healthier environment for plants and animals.
4. **Enhancing Habitat Features:** In addition to replanting and hydrology restoration, restoration projects may involve creating or enhancing habitat features such as ponds, mudflats, and marshes. These features can provide

critical resources and refuge for wildlife, support diverse plant communities, and improve the overall ecological functioning of the wetland.

5. **Monitoring and Adaptive Management:** Restoration projects require ongoing monitoring to assess their effectiveness and make necessary adjustments. This involves tracking changes in vegetation, water quality, and wildlife populations to ensure that the restoration goals are being met. Adaptive management allows for the modification of restoration strategies based on monitoring results and changing conditions.

   Through these restoration activities, degraded wetlands can be rehabilitated to their natural state, improving their ecological health and enhancing their ability to provide important ecosystem services such as water filtration, flood control, and habitat for wildlife.

(iv) **Enhancing Regulatory Frameworks:** Strengthening and enforcing environmental regulations and policies that protect wetlands is crucial for addressing and mitigating issues such as pollution, habitat destruction, and over-extraction of resources. Effective regulatory frameworks are essential for ensuring the long-term conservation of wetlands and the preservation of their ecological functions.

Strengthening Regulations: Developing and updating comprehensive environmental regulations and policies specific to wetland protection can establish clear guidelines and standards for the management and conservation of these ecosystems. The Ramsar Convention, established in 1971, is the key international treaty dedicated to wetland conservation. It sets a framework for protecting and managing wetlands through national and international efforts. Member countries commit to designating at least one wetland as a Wetland of International Importance, known as a Ramsar Site, and ensuring its effective management. As of 2024, over 2400 Ramsar Sites cover more than 250 million hectares worldwide, reflecting a global commitment to safeguarding these critical ecosystems (Ramsar Convention on Wetlands 1971). This includes setting limits on pollutant discharges, controlling land use activities, and establishing protected zones to prevent encroachment and degradation. Regulations should address both direct and indirect impacts on wetlands, such as those from agriculture, industry, and urban development.

Enforcing Policies: Enforcement is key to ensuring that environmental regulations are implemented effectively. This involves monitoring compliance with regulations, conducting inspections, and addressing violations promptly. Enforcement actions may include issuing fines, mandating corrective measures, or prosecuting legal cases against violators. Strong enforcement mechanisms help deter illegal activities and ensure that stakeholders adhere to conservation requirements (Zedler and Kercher 2005).

Monitoring: Rigorous monitoring programs are essential for tracking the health and condition of wetland ecosystems. Monitoring efforts should include regular assessments of water quality, habitat conditions, and biodiversity. This data provides valuable information on the effectiveness of regulations and helps identify emerging threats or areas of concern. Continuous monitoring also supports adaptive management by informing necessary adjustments to regulations and management practices.

Stakeholder Engagement: Engaging stakeholders, including local communities, businesses, and environmental organizations, is important for the successful implementation and enforcement of regulations. Collaborative approaches can help build support for conservation efforts, promote compliance, and address conflicts. Public awareness and education about the importance of wetlands and regulatory requirements can further enhance the effectiveness of protection measures (Mitsch and Gosselink 2015; Ramsar 2016; Owen and Pickering 2006).

Integrating Science and Policy: Effective regulations should be informed by scientific research and data on wetland ecosystems. Integrating scientific knowledge into policy development ensures that regulations are based on the best available evidence and are capable of addressing the specific needs and challenges of wetlands. This approach helps to align regulatory measures with ecological realities and improves their effectiveness. Through strengthening and rigorously enforcing environmental regulations and policies, it is possible to address the various threats facing wetlands and promote their conservation and sustainability. These measures play a critical role in safeguarding wetlands' ecological integrity, supporting biodiversity, and maintaining the essential services they provide to both the environment and human communities.

(v) **Promoting Public Awareness and Education:** Raising awareness about the value and importance of wetlands is a crucial step in fostering community support and encouraging responsible behavior towards these vital ecosystems. By increasing public understanding of wetlands' ecological, economic, and social significance, communities are more likely to engage in conservation efforts and adopt practices that protect these areas. Educational programs and outreach initiatives play a key role in this process. These programs can be tailored to various audiences, including schools, local communities, and policymakers, to highlight the diverse benefits wetlands provide, such as water purification, flood control, and habitat for wildlife. Interactive workshops, seminars, and educational materials can help convey complex information in an accessible and engaging manner.

Outreach efforts can also involve community events, such as clean-up drives, nature walks, and citizen science projects, which offer hands-on experiences and promote active participation in wetland conservation. Public campaigns, media coverage, and social media platforms can further amplify messages about the importance of wetlands and encourage individuals to take actions like reducing pollution, supporting conservation policies, and participating in local restoration projects.

(vi) **Encouraging Community Involvement:** Engaging local communities in wetland conservation efforts is essential for developing effective and sustainable management practices. When communities are actively involved in conservation, they bring valuable local knowledge, resources, and a vested interest in protecting their natural environments. This collaborative approach fosters a sense of ownership and responsibility, leading to more impactful and enduring conservation outcomes. Community-based programs play a pivotal role in this engagement. These programs can include activities such as volunteer clean-up events, habitat restoration projects, and educational workshops tailored to local needs and interests. By involving community members in hands-on conservation work, these programs help build skills, raise awareness, and strengthen the connection between people and their local wetlands.

Partnerships between communities, government agencies, non-profit organizations, and businesses can further enhance conservation efforts. Such collaborations leverage diverse expertise and resources, facilitating comprehensive approaches to wetland management. For instance, partnerships can support the development of management plans that reflect both scientific knowledge and local insights, ensuring that conservation strategies are both practical and effective. Promoting stewardship and advocacy within communities is also crucial. Community leaders and local organizations can champion wetland protection, advocate for sustainable practices, and mobilize public support. Efforts to engage and educate local stakeholders about the importance of wetlands and the benefits of conservation can lead to grassroots movements that drive policy changes and support conservation initiatives. Harnessing local knowledge and fostering a culture of stewardship, community engagement not only enhances the effectiveness of wetland management but also ensures that conservation practices are culturally relevant and supported by those who are most directly impacted by them. This approach leads to more resilient and sustainable wetland ecosystems, benefiting both the environment and the communities that depend on them.

(vii) **Integrating Wetlands into Land Use Planning:** Incorporating wetlands into land use and urban planning is crucial for ensuring that these important ecosystems are considered in development decisions. Integrating wetlands into planning efforts helps balance economic development with environmental protection, preserving the critical functions and values that wetlands provide.

Effective land use and urban planning should include:

1. **Wetland Mapping and Assessment:** Identifying and mapping wetlands within planning areas is the first step in ensuring their protection. Comprehensive assessments of wetland functions, values, and ecological health can inform planning decisions and highlight areas that require special attention.
2. **Integrating Wetlands into Zoning and Land Use Policies:** Zoning regulations and land use policies should incorporate provisions to protect wetlands. This includes designating wetland areas as conservation zones, establishing buffer zones around wetlands, and restricting development

activities that could harm these ecosystems. Planning policies can also encourage the use of green infrastructure and sustainable practices that reduce impacts on wetlands.

3. **Environmental Impact Assessments (EIAs):** Requiring EIAs for proposed developments can help evaluate the potential effects on wetlands and other natural resources. These assessments should include measures to avoid, minimize, or mitigate adverse impacts on wetlands, ensuring that development projects do not compromise their ecological integrity.
4. **Promoting Sustainable Development Practices:** Planning efforts should encourage sustainable development practices that minimize harm to wetlands. This includes promoting low-impact development techniques, such as permeable surfaces and stormwater management systems, which reduce runoff and pollution entering wetland areas.
5. **Public Participation and Stakeholder Engagement:** Engaging the public and stakeholders in the planning process is essential for ensuring that wetlands are effectively integrated into land use decisions. Public consultations and participatory planning approaches can help gather input, address concerns, and build support for conservation measures.
6. **Adaptive Management and Monitoring:** Implementing adaptive management practices allows for the ongoing evaluation and adjustment of planning strategies based on new information and changing conditions. Regular monitoring of wetland health and development impacts helps ensure that planning efforts remain effective and responsive to emerging challenges.

By incorporating wetlands into land use and urban planning, communities can achieve a balanced approach to development that safeguards these vital ecosystems while accommodating growth and economic activities. This integrated approach helps preserve wetland functions such as water filtration, flood control, and habitat provision, ensuring their long-term sustainability and benefits to both people and nature.

(viii) **Supporting Scientific Research and Monitoring:** Ongoing research and monitoring are critical for gaining a comprehensive understanding of wetland ecosystems and their responses to environmental changes. This continuous process provides essential insights into how wetlands function, how they are affected by various stressors, and how they adapt to shifts in climate and land use. Supporting scientific studies helps generate valuable data that informs effective management and conservation strategies. Research can reveal trends in wetland health, changes in biodiversity, and variations in water quality, enabling the development of targeted interventions. Additionally, monitoring programs can track the success of conservation efforts, identify emerging threats, and adjust management practices as needed. By investing in research and monitoring, we can enhance our knowledge of wetland dynamics and improve our ability to protect and sustain these vital ecosystems. This approach ensures that conservation strategies are based on up-to-date, evidence-based information, leading to more effective and resilient management of wetland resources.

(ix) **Promoting Climate Change Mitigation and Adaptation:** Recognizing the role of wetlands in climate regulation and carbon sequestration is crucial for guiding efforts to address climate change impacts effectively. Wetlands play a significant role in mitigating climate change by acting as carbon sinks, where they store large amounts of carbon dioxide (CO_2) in their soil and vegetation. This carbon sequestration helps reduce the concentration of greenhouse gases in the atmosphere, thereby moderating global warming.

To harness the full potential of wetlands in climate regulation, several strategies can be employed:

1. **Preserving Existing Wetlands:** Protecting and maintaining existing wetlands is essential for preserving their carbon storage capacity. Conservation efforts should focus on preventing habitat destruction, mitigating pollution, and controlling invasive species that threaten wetland health. By safeguarding these ecosystems, we ensure that they continue to function effectively as carbon sinks.
2. **Restoring Degraded Wetlands:** Restoration projects can rehabilitate wetlands that have been damaged or degraded, enhancing their ability to sequester carbon. Restoration activities may include reestablishing natural hydrology, replanting native vegetation, and removing contaminants. Restored wetlands can regain their carbon storage capacity and contribute to climate change mitigation.
3. **Implementing Adaptive Management:** As climate conditions change, wetlands may face new challenges such as altered hydrological patterns and increased temperatures. Adaptive management involves adjusting conservation and management practices based on ongoing research and monitoring to address these emerging challenges. This approach ensures that wetlands remain resilient and continue to provide their ecological functions in the face of climate change.
4. **Incorporating Wetlands into Climate Strategies:** Integrating wetlands into broader climate action plans and policies can enhance their role in climate regulation. This includes recognizing their value in carbon accounting and incorporating wetland conservation and restoration into climate mitigation and adaptation strategies.
5. **Promoting Research and Monitoring:** Supporting research on wetland carbon dynamics and climate interactions can provide valuable insights into how wetlands respond to changing conditions. Monitoring programs can track changes in carbon sequestration rates and inform management decisions to optimize the climate benefits of wetlands.

(x) **Securing Funding and Resources:** Adequate funding is essential for the success of wetland conservation and restoration projects. Securing financial resources ensures that effective conservation strategies can be implemented and maintained, ultimately supporting the long-term sustainability of wetland ecosystems. Funding is necessary for a variety of aspects in wetland management, including:

1. **Project Implementation:** Financial resources are required to carry out on-the-ground conservation and restoration activities, such as habitat restoration, pollution control, and invasive species management. This includes costs related to materials, labor, and equipment necessary for these projects.
2. **Research and Monitoring:** Ongoing research and monitoring programs depend on adequate funding to gather data, assess wetland health, and evaluate the effectiveness of conservation efforts. These programs provide the scientific basis for informed decision-making and adaptive management.
3. **Capacity Building:** Funding supports the development of technical expertise, training, and capacity building for personnel involved in wetland management. This includes educating local communities, stakeholders, and decision-makers about wetland conservation and best practices.
4. **Public Engagement and Education:** Financial resources are needed to conduct outreach and educational campaigns that raise awareness about the importance of wetlands and encourage community involvement in conservation efforts. This can include workshops, seminars, and informational materials.
5. **Policy and Advocacy:** Funding can support advocacy efforts aimed at influencing policy and securing regulatory protections for wetlands. This includes lobbying for stronger conservation policies and ensuring that wetland issues are addressed in land use planning and development decisions.
6. **Long-Term Maintenance:** Ensuring the long-term sustainability of wetland projects requires ongoing financial support for maintenance and management. This includes monitoring the health of restored wetlands, addressing emerging threats, and making necessary adjustments to conservation strategies.

Securing funding can be achieved through various means, including applying for grants from governmental and non-governmental organizations, forming partnerships with private sector entities, and seeking investments from conservation-focused foundations. Engaging with a diverse range of funding sources can help build a robust financial base for wetland conservation efforts.

By securing adequate funding, we can effectively implement and sustain conservation and restoration projects, ensuring that wetlands continue to provide their invaluable ecological, economic, and social benefits for future generations.

18.3 Climate Change Mitigation

Wetlands are among the most productive ecosystems on Earth, providing essential services such as carbon sequestration, biodiversity support, water regulation, and climate stabilization. However, climate change—manifested through rising temperatures, sea-level rise, and altered precipitation patterns—poses significant risks to these vital ecosystems. To mitigate these impacts, it is crucial to implement strategies that minimize the effects of climate change on wetlands.

Climate change mitigation focuses on reducing or preventing greenhouse gas emissions to limit the extent and pace of climate change. Wetlands, including swamps, marshes, bogs, and fens, are key players in global carbon sequestration but are also highly susceptible to climate-related stressors. Therefore, effective mitigation is essential to preserve these ecosystems and maintain their ecological functions. Several strategies can be employed to protect wetlands from the effects of climate change (Lane et al. 2018). These include conserving existing wetlands, restoring degraded areas, promoting reforestation and afforestation, encouraging sustainable agriculture, and implementing effective water management systems. Additionally, initiatives like wetland carbon credits and blue carbon projects can support these efforts by recognizing and rewarding the carbon storage capabilities of wetlands. These combined approaches are critical for safeguarding wetlands and ensuring their continued contribution to environmental health and stability.

18.4 Water Quality Improvement

Wetland preservation is critically important for maintaining the health of these ecosystems, which serve as natural water filtration systems. By enhancing water quality, wetlands can continue to support diverse flora and fauna, sustain ecological balance, and provide essential services to human communities (Craft 2022). Wetlands act as natural purifiers by trapping pollutants, sediments, and excess nutrients that might otherwise flow into rivers, lakes, and oceans, causing water quality degradation. They achieve this through a combination of physical (sedimentation and filtration), chemical (nutrient uptake), and biological (decomposition of organic matter) processes.

To improve water quality in wetlands, several strategies can be employed:

Restoring degraded wetlands is vital for enhancing water quality. This involves reestablishing natural hydrology, replanting native vegetation, and removing invasive species to restore the wetland's filtration capacity (Spieles 2022).

Buffer zones of vegetation around wetlands can help reduce pollutant inflow. These zones act as initial filters for runoff from agricultural and urban areas, trapping sediments and absorbing nutrients before they enter the wetland. Constructed wetlands are engineered systems designed to replicate the functions of natural wetlands. They treat wastewater, stormwater, and agricultural runoff, proving particularly useful in areas where natural wetlands have been lost or are overwhelmed by pollutant loads (Rodrigo 2021).

Pollution control measures, such as reducing fertilizer and pesticide use, controlling industrial discharges, and managing stormwater, can significantly improve water quality before it reaches wetlands. This reduces the burden on wetlands and enhances their ability to process any remaining pollutants. The importance of preserving wetlands through water quality improvement cannot be overstated. Wetlands are among the most productive ecosystems on Earth, providing habitat for a diverse range of species, supporting biodiversity, and offering essential services like water purification, flood control, and carbon sequestration. Healthy wetlands act as buffers against pollution, protect downstream ecosystems, and ensure the availability of clean water for human use (Zedler and Kercher 2005; Mitsch et al. 2005; Verhoeven and Meuleman 1999).

18.5 Flood Control

Wetlands function as natural sponges, absorbing and storing excess rainfall, which helps mitigate flood risks. Effective flood control through wetland preservation involves managing water flow to prevent damage to wetlands and leveraging their natural flood-reducing capabilities to protect surrounding communities.

Wetlands contribute to natural flood control in several crucial ways:

Wetlands serve as reservoirs, absorbing substantial amounts of water during heavy rainfall or snowmelt (Roulet and Woo 1986). By slowing the movement of water, they help reduce the severity of downstream flooding, which is particularly important in areas prone to flash floods. They regulate water flow, stabilizing water levels in rivers and streams. This buffering effect is essential for maintaining consistent water supplies and mitigating the extremes of both droughts and floods. Wetlands trap sediments and nutrients carried by floodwaters, improving water quality and reducing sediment loads downstream. This helps prevent riverbeds from rising and channels from losing capacity, which could otherwise exacerbate flooding.

To preserve wetlands and enhance flood control, several strategies can be employed:

Structural approaches include building levees and dikes to protect wetlands and shield human settlements from flooding, using floodwater diversion methods such as channels and spillways to redirect water away from sensitive wetland areas, and constructing retention basins to temporarily hold floodwaters. Nonstructural approaches involve restoring degraded wetlands, conserving existing wetlands, implementing proper land use planning to restrict development in flood-prone areas, and adopting sustainable agriculture and forestry practices to minimize runoff.

The benefits of preserving wetlands for flood control extend beyond flood mitigation:

Wetland preservation supports biodiversity conservation by protecting habitats for a diverse range of species, thereby safeguarding entire ecosystems. It improves water quality, as wetlands filter sediments and nutrients, leading to cleaner water in rivers, lakes, and oceans. This is vital for drinking water supplies and the health of aquatic ecosystems. Wetlands act as carbon sinks, capturing carbon dioxide from the atmosphere and storing it in plant biomass and soil, which helps mitigate climate change.

Economically, healthy wetlands offer valuable resources such as fish, timber, and recreational opportunities. They also reduce costs associated with flood damage to infrastructure, agriculture, and communities. Economic benefits of wetland preservation also include supporting sustainable livelihoods and subsistence for communities, particularly in developing regions where wetlands provide essential resources like fish, timber, and medicinal plants. Additionally, wetlands offer recreational and aesthetic value, attracting tourists for activities such as birdwatching and boating, which benefits local businesses and enhances property values. Overall, wetland preservation plays a crucial role in flood control and provides significant economic and ecological benefits, contributing to the health of both natural and human communities (Costanza et al. 1997; Brander et al. 2006).

18.6 Cultural and Spiritual Significance

The cultural and spiritual significance of wetlands profoundly influences their preservation and underscores the deep connections communities maintain with these ecosystems. Beyond their vital ecological functions, wetlands are imbued with intrinsic value that reflects cultural heritage and spiritual beliefs.

Wetlands often hold deep cultural importance for local and indigenous communities, manifesting in various aspects of their daily lives and traditions. Traditional practices and knowledge frequently intertwine with wetland environments. Many indigenous peoples rely on wetlands for subsistence activities such as fishing, hunting, and gathering. For example, the Cree people in Canada utilize wetlands for trapping beaver, a practice that is not only essential for their livelihood but also integral to their cultural heritage and identity (Houle 2012). This reliance on wetlands extends beyond mere sustenance; it forms a core part of their cultural practices and traditions. Wetlands also contribute significantly to the cultural landscapes that shape community identities. These ecosystems are often regarded as sacred landscapes that embody cultural values and narratives. In Japan, the Matsushima Bay wetlands are celebrated as a cultural landscape that has inspired countless works of art and poetry over centuries, reflecting the deep reverence and aesthetic appreciation for these natural areas (Kurokawa and Ueno 2018). Such cultural landscapes play a critical role in shaping the collective identity and cultural expression of communities.

Furthermore, wetlands are often integral to various rituals and ceremonies. Many cultures have historically incorporated wetlands into their religious and ceremonial practices. For instance, Mesoamerican civilizations have used wetland environments in agricultural rituals and ceremonies dedicated to water deities, illustrating how these ecosystems are intertwined with spiritual practices and beliefs (McClung de Tapia 2010). These rituals not only reflect the spiritual significance of wetlands but also reinforce their role in cultural continuity. The spiritual significance of wetlands further enhances their value, often making them crucial to religious and spiritual practices. Wetlands are frequently regarded as sacred spaces with deep spiritual meaning. For many cultures, these ecosystems are seen as integral to their cosmology and spiritual beliefs. In the Amazon, wetlands are central to the spiritual and cosmological frameworks of local indigenous peoples, who view these areas as vital to their spiritual practices and worldview. Similarly, the Osun-Osogbo Sacred Grove in Nigeria, dedicated to the Yoruba goddess Osun, is protected through traditional religious practices that prohibit deforestation and other harmful activities. This reverence ensures the grove's preservation and highlights its spiritual significance (Chernela 2012). Spiritual beliefs often attribute divine or ancestral significance to wetlands, viewing them as inhabited by spirits or deities. Indigenous Australian groups, such as the Yolngu, have spiritual narratives that depict wetlands as places of ancestral significance, reflecting the deep spiritual connection between these ecosystems and their cultural heritage (Gammage 2011). These spiritual beliefs contribute to the preservation of wetlands by fostering a sense of sacredness and respect.

Moreover, wetlands frequently symbolize concepts of fertility, renewal, and abundance. In Hinduism, wetlands are associated with deities such as Varuna, the god of water, and are seen as symbols of fertility and renewal. This symbolism underscores the role of wetlands in spiritual and cultural contexts, reinforcing their importance beyond their ecological functions (Cohen 2007). Integrating the cultural and spiritual significance of wetlands into conservation strategies can enhance preservation efforts and foster greater community engagement. By acknowledging and valuing the deep cultural and spiritual connections people have with wetlands, conservation initiatives can be more effective and resonate more deeply with local communities, ultimately contributing to the long-term protection and sustainability of these vital ecosystems.

18.7 Conclusion

Wetlands are among the most crucial and diverse ecosystems on Earth, offering a wealth of ecological, economic, cultural, and spiritual benefits. Their roles in carbon sequestration, biodiversity support, water regulation, and flood control underscore their importance in maintaining ecological balance and human well-being. However, these vital ecosystems face significant threats from climate change, pollution, habitat destruction, and over-extraction of resources. To ensure the preservation and sustainability of wetlands, it is imperative to adopt a multifaceted approach that integrates conservation, restoration, and sustainable management practices. Strategies such as restoring degraded wetlands, establishing buffer zones, implementing constructed wetlands, and enforcing pollution control measures are essential for improving water quality and maintaining the ecological functions of these areas. Additionally, incorporating wetlands into land use planning and engaging local communities in conservation efforts can enhance the effectiveness of preservation initiatives.

Recognizing the cultural and spiritual significance of wetlands further strengthens conservation efforts. Many communities have deep-rooted cultural and spiritual connections to wetlands, viewing them as sacred spaces or integral to their traditional practices. These connections can be leveraged to foster community support and advocacy for wetland preservation, ensuring that conservation efforts are not only ecologically sound but also culturally sensitive. The economic benefits of wetlands, including their contributions to sustainable livelihoods, recreational opportunities, and climate regulation, highlight the importance of integrating economic considerations into conservation strategies. Adequate funding, supported by international frameworks like the Ramsar Convention, is crucial for implementing effective conservation and restoration projects. Ongoing research and monitoring are vital for understanding the responses of wetland ecosystems to environmental changes and for developing data-driven management strategies. By continuously supporting scientific studies and adapting to emerging challenges, we can enhance our ability to protect and sustain these critical ecosystems. Safeguarding wetlands requires a comprehensive approach that balances ecological, economic, cultural, and spiritual dimensions. Through concerted efforts involving conservation, community engagement, and policy implementation, we can ensure that wetlands continue to provide their invaluable services and remain resilient in the face of environmental challenges.

References

Acreman MC, Ferguson AJD (2010) Environmental flows and the European Water Framework Directive. Freshw Biol 55(1):32–48

Akpor OB, Okolomike UF, Olaolu DT, Aderiye BI (2014) Remediation of polluted wastewater effluents: hydrocarbon removal. Trends Appl Sci Res 9(4):160–173

Ali N, Khan MH, Ali M, Ahmad S, Khan A, Nabi G, Ali F, Bououdina M, Kyzas GZ (2024) Insight into microplastics in the aquatic ecosystem: properties, sources, threats and mitigation strategies. Sci Total Environ 169489:169489

Alikhani S, Nummi P, Ojala A (2021) Urban wetlands: a review on ecological and cultural values. Water 13(22):3301

Anaero-Nweke GN (2013) Impact of oil refinery effluent on the water quality: case study of Ekerikana creek in Nigeria (Doctoral dissertation)

Arias AH, Botte SE (eds) (2020) Coastal and deep ocean pollution. CRC Press, London

Bamigboye O, Alfred MO, Bayode AA, Unuabonah EI, Omorogie MO (2024) The growing threats and mitigation of environmental microplastics. Environ Chem Ecotoxicol
Bănăduc D, Simić V, Cianfaglione K, Barinova S, Afanasyev S, Öktener A, Curtean-Bănăduc A (2022) Freshwater as a sustainable resource and generator of secondary resources in the 21st century: stressors, threats, risks, management and protection strategies, and conservation approaches. Int J Environ Res Public Health 19(24):16570
Boyd CE, Boyd CE (2020) Water quality protection. In: Water quality: an introduction, pp 379–409
Brander LM, Florax RJ, Vermaat JE (2006) The empirics of wetland valuation: a comprehensive summary and a meta-analysis of the literature. Environ Resour Econ 33(2):223–250
Bulonza N (2015) Impacts of human activities on wetland resources management at Nakivubo Wetland, Kampala-Uganda
Calhoun AJ, Mushet DM, Bell KP, Boix D, Fitzsimons JA, Isselin-Nondedeu F (2017) Temporary wetlands: challenges and solutions to conserving a 'disappearing 'ecosystem. Biol Conserv 211:3–11
Carpenter SR, Caraco NF, Correll DL, Howarth RW, Sharpley AN, Smith VH (1998) Nonpoint pollution of surface waters with phosphorus and nitrogen. Ecol Appl 8(3):559–568. https://doi.org/10.1890/1051-0761
Cazenave A, Remy F (2011) Sea level and climate: measurements and causes of changes. WIREs Clim Change 2:647–662
Chaukura N, Kefeni KK, Chikurunhe I, Nyambiya I, Gwenzi W, Moyo W, Abulude FO (2021) Microplastics in the aquatic environment—the occurrence, sources, ecological impacts, fate, and remediation challenges. Pollutants 1(2):95–118
Chen W, He B, Nover D, Lu H, Liu J, Sun W, Chen W (2019) Farm ponds in southern China: challenges and solutions for conserving a neglected wetland ecosystem. Sci Total Environ 659:1322–1334
Chernela J (2012) Sacred Wetlands: spiritual and ecological importance. University of Arizona Press
Church JA, Gregory JM, White NJ, Platten SM, Mitrovica JX (2011) Understanding and projecting sea-level change. Oceanography 24:130–143. https://doi.org/10.5670/oceanog.2011.33
Cohen R (2007) Hinduism and the sacred geography of India. Routledge
Corman JR, Bertolet BL, Casson NJ, Sebestyen SD, Kolka RK, Stanley EH (2018) Nitrogen and phosphorus loads to temperate seepage lakes associated with allochthonous dissolved organic carbon loads. Geophys Res Lett 45:5481–5490. https://doi.org/10.1029/2018GL077219
Costanza R, d'Arge R, de Groot R et al (1997) The value of the world's ecosystem services and natural capital. Nature 387(6630):253–260
Craft C (2022) Creating and restoring wetlands: from theory to practice. Elsevier
Davidson NC (2014) How much wetland has the world lost? Long-term and recent trends in global wetland area. Mar Freshw Res 65(10):934–941
Davis AM, Pearson RG, Brodie JE, Butler B (2016) Review and conceptual models of agricultural impacts and water quality in waterways of the Great Barrier Reef catchment area. Mar Freshw Res 68(1):1–19
Domingues CM, Church JA, White NJ, Gleckler PJ, Wijffels SE, Barker PM, Dunn JR (2008) Improved estimates of upper-ocean warming and multi-decadal sea-level rise. Nature 453(7198):1090–1093. https://doi.org/10.1038/nature07080
Erwin KL (2009) Wetlands and global climate change: the role of wetland restoration in a changing world. Wetl Ecol Manag 17:71–84. https://doi.org/10.1007/s11273-008-9119-1
Findlay S, Houlahan J (1997) Anthropogenic impacts on biodiversity, ecosystem function, and water quality in urban wetlands. Environ Manag 21(5):687–695
Gammage B (2011) The biggest estate on earth: how Aborigines Made Australia. Allen and Unwin
Gupta AP, Pandey P (2024) Microplastics in wetland ecosystem: a complex nexus and way forward. In: Microplastics in African and Asian environments: the influencers, challenges, and solutions. Springer Nature Switzerland, Cham, pp 317–330
Houle C (2012) Indigenous rights and wetland preservation in Canada. University of Alberta Press

Intergovernmental Panel on Climate Change (2007) Technical summary. In: Solomon S et al (eds) Climate change 2007: the physical science basis. Contribution of working group I to the fourth assessment report of the intergovernmental panel on climate change. Cambridge University Press, Cambridge, pp 19–91

Ishii M, Kimoto M (2009) Re-evaluation of historical ocean heat content variations with an XBT depth bias correction. J Oceanogr 65:287–299. https://doi.org/10.1007/s10872-009-0027-7

Kantrud HA, Millar JB, Van-der-Valk AG (1989) Vegetation of wetlands of the prairie pothole region. In: Northern Prairie Wetlands, pp 132–187

Kingsford RT, Basset A, Jackson L (2016) Wetlands: conservation's poor cousins. Aquat Conserv Mar Freshwat Ecosyst 26(5):892–916

Kumar N, Kumar A, Marwein BM, Verma DK, Jayabalan I, Kumar A, Kumar N (2021) Agricultural activities causing water pollution and its mitigation–a review. Int J Mod Agric 10(1):590–609

Kurokawa K, Ueno K (2018) Wetlands and Cultural Heritage in Japan. Springer

Laiho R (2006) Decomposition in peatlands: reconciling seemingly contrasting results on the impacts of lowered water levels. Soil Biol Biochem 38:2011–2024. https://doi.org/10.1016/j.soilbio.2006.02.017

Lane CR, Leibowitz SG, Autrey BC, LeDuc SD, Alexander LC (2018) Hydrological, physical, and chemical functions and connectivity of non-floodplain wetlands to downstream waters: a review. J Am Water Resour Assoc 54(2):346–371

Levitus S, Antonov JI, Boyer TP, Garcia HE, Locarnini RA, Mishonov AV, Garcia HE (2009) Global ocean heat content 1955–2008 in light of recently revealed instrumentation problems. Geophys Res Lett 36:L07608. https://doi.org/10.1029/2008GL037155

McClung de Tapia E (2010) Mesoamerican Wetland Landscapes and Rituals. University of Texas Press

Mengesha TA (2017) Review on the natural conditions and anthropogenic threats of Wetlands in Ethiopian. Glob J Ecol 2(1):006–014

Mitsch WJ, Gosselink JG (2015) Wetlands. Wiley, New York, p 736

Mitsch WJ, Zhang L, Anderson CJ (2005) Creating riverine wetlands: ecological succession, nutrient retention, and pulsing effects in floodplains. Ecol Eng 25(5):510–527

Moore TR, Roulet NT, Waddington JM (1998) Uncertainty in predicting the effect of climatic change on the carbon cycling of Canadian peatlands. Clim Chang 40:229–245. https://doi.org/10.1023/A:1005408719297

National Research Council (2001) Compensating for wetland losses under the clean water act. National Academies Press

Newton A, Icely J, Cristina S, Perillo GM, Turner RE, Ashan D, Kuenzer C (2020) Anthropogenic, direct pressures on coastal wetlands. Front Ecol Evol 8:144

Owen J, Pickering C (2006) Ecotourism as a tool for promoting conservation of threatened native plants and habitats: a case study from the Australian Alps. J Sustain Tour 14(4):323–337

Ramsar Convention Secretariat (1971) An Introduction to the Ramsar Convention on Wetlands, 6th edn. Ramsar Convention provides guidelines on involving communities in wetland conservation and management

Ramsar Convention Secretariat (2016) An Introduction to the Ramsar Convention on Wetlands, 7th ed. (previously The Ramsar Convention Manual). Ramsar Convention Secretariat, Gland, Switzerland. Ramsar Convention provides guidelines on involving communities in wetland conservation and management

Reid AJ, Carlson AK, Creed IF, Eliason EJ, Gell PA, Johnson PT, Cooke SJ (2019) Emerging threats and persistent conservation challenges for freshwater biodiversity. Biol Rev 94(3):849–873

Rodrigo MA (2021) Wetland restoration with hydrophytes: a review. Plan Theory 10(6):1035

Roulet NT, Woo MK (1986) Hydrology of a wetland in the continuous permafrost region. J Hydrol 89(1–2):73–91

Sarkar UK, Nag SK, Das MK, Karnatak G, Sudheesan D (2016) Conserving wetlands–An effective climate change adaptation in India, Bulletin No. NICRA/CIFRI/2015-16/2. ICAR-Central Inland Fisheries Research Institute, Barrackpore

Spieles DJ (2022) Wetland construction, restoration, and integration: a comparative review. Land 11(4):554

Sreelakshmi T, Chitra K (2021) Microplastics contamination in the environment: an ecotoxicological concern. Intern J Zool Invest 7(1):230–258

Stets EG, Cotner JB (2008) The influence of dissolved organic carbon on bacterial phosphorus uptake and bacteria-phytoplankton dynamics in two Minnesota lakes. Limnol Oceanogr 53:137–147. https://doi.org/10.4319/lo.2008.53.1.0137

U.S. Environmental Protection Agency Office (1993) Natural wetlands and urban stormwater: potential impacts and management; O. a. WWD Office of Wetlands, Ed.; U.S. Environmental Protection Agency Office: Washington, DC, USA

U.S. Environmental Protection Agency (2020) Wetlands and Water Quality. Retrieved from EPA

Verhoeven JT, Meuleman AF (1999) Wetlands for wastewater treatment: opportunities and limitations. Ecol Eng 12(1–2):5–12

Vymazal J (2010) Constructed wetlands for wastewater treatment. Water 2(3):530–549. https://doi.org/10.3390/w2030530

Williams WD (2001) Anthropogenic salinization of inland waters. Hydrobiologia 466(1):329–337

Xu T, Weng B, Yan D, Wang K, Li X, Bi W et al (2019) Wetlands of international importance: status, threats, and future protection. Int J Environ Res Public Health 16(10):1818

Ye S, Pei L, He L, Xie L, Zhao G, Yuan H, Ding X, Pei S, Yang S, Li X, Laws EA (2022) Wetlands in China: evolution, carbon sequestrations and services, threats and preservation/restoration. Water 14(7):1152. https://doi.org/10.3390/w14071152

Zedler JB, Kercher S (2005) Wetland resources: status, trends, ecosystem services, and restorability. Annu Rev Environ Resour 30:39–74

Index

B. R. Babaniyi et al. (eds.), *Wetland Ecosystems: Conservation Strategies, Policy Management and Applications*, Wetlands: Ecology, Conservation and Management 12, https://doi.org/10.1007/978-3-031-91982-4

The manufacturer's authorised representative in the EU is Springer
Nature Customer Service Centre GmbH, Europaplatz 3, 69115 Heidelberg,
Germany. If you have any concerns regarding our products, please
contact ProductSafety@springernature.com

Printed and bound by CPI Group (UK) Ltd, Croydon, CR0 4YY
05/10/2026
02228130-0018